Molecular and Diagnostic Procedures in Mycoplasmology

Volume I
MOLECULAR CHARACTERIZATION

Edited by

SHMUEL RAZIN

Department of Membrane and Ultrastructure Research
The Hebrew University–Hadassah Medical School
Jerusalem, Israel

JOSEPH G. TULLY

Mycoplasma Section
Laboratory of Molecular Microbiology
National Institute of Allergy and Infectious Diseases
Frederick Cancer Research and Development Center
Frederick, Maryland

ACADEMIC PRESS

San Diego   New York   Boston   London   Sydney   Tokyo   Toronto

Front cover photograph: Vero cells infected with *M. hyorhinis*, stained by a double stain method using DNAF and fluoresceinated anti-*M. hyorhinis* antibodies, and viewed with the filter set for DNAF. Courtesy of Dr. Gerald K. Masover and Frances A. Becker, Genentech, Inc., South San Francisco, CA.

This book is printed on acid-free paper.

Copyright © 1995 by ACADEMIC PRESS, INC.

All Rights Reserved.
No part of this publication may be reproduced or transmitted in any form or by any means, electronic or mechanical, including photocopy, recording, or any information storage and retrieval system, without permission in writing from the publisher.

Academic Press, Inc.
A Division of Harcourt Brace & Company
525 B Street, Suite 1900, San Diego, California 92101-4495

United Kingdom Edition published by
Academic Press Limited
24-28 Oval Road, London NW1 7DX

Library of Congress Cataloging-in-Publication Data

Molecular and diagnostic procedures in mycoplasmology / edited by
  Shmuel Razin, Joseph G. Tully.
      p.   cm.
    Includes indexes.
    Contents: v. 1. Molecular characterization -- v. 2. Diagnostic procedures.
    ISBN 0-12-583805-0 (v. 1: alk. paper)   ISBN 0-12-583806-9 (v. 2: alk. paper)
    1. Mycoplasma diseases--Diagnosis. 2. Mycoplasma diseases--Molecular aspects. I. Razin, Shmuel. II. Tully, Joseph G.
    [DNLM: 1. Mycoplasma--physiology. 2. Mycoplasma--pathogenicity.
  3. Molecular Biology--methods. 4. Mycoplasma Infections--diagnosis.
  QW 143 M718 1995]
  QR201.M97M63 1995
  589.9--dc20
  DNLM/DLC
  for Library of Congress   95-4586
                                                                                CIP

PRINTED IN THE UNITED STATES OF AMERICA
95 96 97 98 99 00 BC 9 8 7 6 5 4 3 2 1

Contents

Contributors   xi
Preface   xv
Contents of Volume II   xvii

Molecular Properties of Mollicutes: A Synopsis   1
Shmuel Razin

SECTION A   Cultivation and Morphology

A1   Introductory Remarks   29
Joseph G. Tully

A2   Culture Medium Formulation for Primary Isolation and Maintenance of Mollicutes   33
Joseph G. Tully

A3   Cultivation of Spiroplasmas in Undefined and Defined Media   41
Kevin J. Hackett and Robert F. Whitcomb

A4   Insect Cell Culture Approaches in Cultivating Spiroplasmas   55
Kevin J. Hackett and Dwight E. Lynn

A5   Measurement of Mollicute Growth by ATP-Dependent Luminometry   65
Janet A. Robertson and Gerald W. Stemke

A6   Intracellular Location of Mycoplasmas   73
David Taylor-Robinson

A7	Localization of Mycoplasmas in Tissues	81
	Douglas J. Wear and Shyh-Ching Lo	
A8	Localization of Antigens on Mycoplasma Cell Surface and Tip Structures	89
	Duncan C. Krause and Marla K. Stevens	

SECTION B Genome Characterization and Genetics

B1	Introductory Remarks	101
	Shmuel Razin	
B2	Isolation of Mycoplasma-like Organism DNA from Plant and Insect Hosts	105
	Bruce C. Kirkpatrick, Nigel A. Harrison, Ing-Ming Lee, Harold Neimark, and Barbara B. Sears	
B3	Mollicute Chromosome Size Determination and Characterization of Chromosomes from Uncultured Mollicutes	119
	Harold Neimark and Patricia Carle	
B4	Physical and Genetic Mapping	133
	Thomas Proft and Richard Herrmann	
B5	Characterization of Virus Genomes and Extrachromosomal Elements	159
	Kevin Dybvig	
B6	Plasmid and Viral Vectors for Gene Cloning and Expression in *Spiroplasma citri*	167
	J. Renaudin and J. M. Bové	
B7	Artificial Transformation of Mollicutes via Polyethylene Glycol- and Electroporation-Mediated Methods	179
	Kevin Dybvig, Gail E. Gasparich, and Kendall W. King	

B8	DNA Methylation Analysis	185
	Aharon Razin and Paul Renbaum	
B9	Identification and Characterization of Genome Rearrangements	195
	Bindu Bhugra and Kevin Dybvig	
B10	Expression of Mycoplasmal Genes in *Escherichia coli*	201
	Paul Renbaum and Aharon Razin	

SECTION C Membrane Characterization

C1	Introductory Remarks	215
	Shmuel Razin	
C2	Posttranslational Modification of Membrane Proteins	217
	Åke Wieslander, Susanne Nyström, and Anders Dahlqvist	
C3	Variant Membrane Proteins	227
	Kim S. Wise, Mary F. Kim, and Robyn Watson-McKown	
C4	Membrane Fusion	243
	Shlomo Rottem and Mark Tarshis	
C5	Mycoplasma Membrane Potentials	251
	Ulrich Schummer and Hans Gerd Schiefer	
C6	Ion Flow and Cell Volume	265
	Shlomo Rottem	

SECTION D Cell Metabolism

D1	Introductory Remarks	275
	Shmuel Razin	

D2	Methods for Testing Metabolic Activities in Mollicutes	277
	J. Dennis Pollack	
D3	Rapid Microcalorimetric and Electroanalytical Measurements of Metabolic Activities	287
	R. J. Miles	
D4	Characterization of Heat Shock Proteins	297
	Christopher C. Dascher and Jack Maniloff	
D5	Nucleolytic Activities of Mycoplasmas	305
	F. Chris Minion and Karalee J. Jarvill-Taylor	
D6	Proteolytic Activities	315
	Tsuguo Watanabe and Ken-ichiro Shibata	
D7	Phospholipase Activity in Mycoplasmas	325
	Shlomo Rottem and Michael Salman	

SECTION E Taxonomy and Phylogeny

E1	Introductory Remarks	335
	Shmuel Razin	
E2	Minimal Standards for Description of New Species of the Class Mollicutes	339
	Joseph G. Tully and Robert F. Whitcomb	
E3	Ribosomal RNA Sequencing and Construction of Mycoplasma Phylogenies	349
	William G. Weisburg	
E4	Restriction Endonuclease Analysis	355
	Shmuel Razin and David Yogev	

Contents ix

E5 Southern Blot Analysis and Ribotyping   361
    David Yogev and Shmuel Razin

E6 Phylogenetic Classification of Plant Pathogenic
    Mycoplasma-like Organisms or Phytoplasmas   369
    *Bernd Schneider, Erich Seemueller, Christine D. Smart,
    and Bruce C. Kirkpatrick*

E7 Determination of Cholesterol and Polyoxyethylene
    Sorbitan Growth Requirements of Mollicutes   381
    Joseph G. Tully

SECTION F   Pathogenicity

F1 Introductory Remarks   393
    Shmuel Razin

F2 Mycoplasma Adherence to Host Cells: Methods
    of Quantifying Adherence   397
    Itzhak Kahane and Enno Jacobs

F3 Mycoplasma Adherence to Host Cells:
    Epitope Mapping of Adhesins   407
    Enno Jacobs

F4 Oxidative Damage Induced by Mycoplasmas   415
    Itzhak Kahane

F5 Activation of Macrophages and Monocytes by Mycoplasmas   421
    *Ruth Gallily, Ann Avron, Gerlinde Jahns-Streubel, and
    Peter F. Mühlradt*

F6 Identification, Characterization, and Purification
    of Mycoplasmal Superantigens   439
    Barry C. Cole and Curtis L. Atkin

Contents

F7	Mycoplasmal B-Cell Mitogens	451
	Yehudith Naot	
F8	Modulation of Expression of Major Histocompatibility Complex Molecules by Mycoplasmas	461
	P. Michael Stuart and Jerold G. Woodward	
F9	Interaction of Mycoplasmas with Natural Killer Cells	469
	Wayne C. Lai and Michael Bennett	

Index 481

Contributors

Numbers in parentheses indicate the pages on which the authors' contributions begin.

Curtis L. Atkin (439), Department of Internal Medicine, Division of Rheumatology, University of Utah School of Medicine, Salt Lake City, Utah 84132

Ann Avron (421), The Lautenberg Center for General and Tumor Immunology, The Hebrew University–Hadassah Medical School, Jerusalem 91120, Israel

Michael Bennett (469), Department of Pathology, Division of Comparative Medicine, University of Texas Southwestern Medical Center, Dallas, Texas 75235

Bindu Bhugra (195), Department of Molecular Genetics, Microbiology, and Biochemistry, University of Cincinnati, Cincinnati, Ohio 45267

J. M. Bové (167), Laboratoire de Biologie Cellulaire et Moléculaire, Centre de Recherches de Bordeaux, Institut National de la Recherche Agronomique, 33883 Villenave d'Ornon Cedex, France

Patricia Carle (119), Laboratoire de Biologie Cellulaire et Moléculaire, Centre de Recherches de Bordeaux, Institut National de la Recherche Agronomique, 33883 Villenave d'Ornon Cedex, France

Barry C. Cole (439), Department of Internal Medicine, Division of Rheumatology, University of Utah School of Medicine, Salt Lake City, Utah 84132

Anders Dahlqvist (217), Department of Biochemistry, Umeå University, S-901 87 Umeå, Sweden

Christopher C. Dascher (297), Department of Rheumatology and Immunology, Brigham and Women's Hospital, Harvard Medical School, Boston, Massachusetts 02115

Kevin Dybvig (159, 179, 195), Department of Comparative Medicine, University of Alabama at Birmingham, Birmingham, Alabama 35294

Ruth Gallily (421), The Lautenberg Center for General and Tumor Immunology, The Hebrew University–Hadassah Medical School, Jerusalem 91120, Israel

Gail E. Gasparich (179), Insect Biocontrol Laboratory, United States Department of Agriculture, Agricultural Research Service, Beltsville, Maryland 20705

Kevin J. Hackett (41, 55), Insect Biocontrol Laboratory, United States Department of Agriculture, Agricultural Research Service, Beltsville, Maryland 20705

Nigel A. Harrison (105), Fort Lauderdale Research and Education Center, University of Florida, Institute of Food and Agricultural Sciences, Fort Lauderdale, Florida 33314

Richard Herrmann (133), Zentrum für Molekulare Biologie, Universität Heidelberg, 69120 Heidelberg, Germany

Enno Jacobs (397, 407), Institute for Medical Microbiology and Hygiene, University of Freiburg, D-79104 Freiburg, Germany

Gerlinde Jahns-Streubel (421), Department of Hematology and Oncology, Medical Center, University of Göttingen, 37075 Göttingen, Germany

Karalee J. Jarvill-Taylor (305), Veterinary Medical Research Institute, College of Veterinary Medicine, Iowa State University, Ames, Iowa 50011

Itzhak Kahane (397, 415), Department of Membrane and Ultrastructure Research, The Hebrew University–Hadassah Medical School, Jerusalem 91120, Israel

Contributors

Mary F. Kim (227), Department of Molecular Microbiology and Immunology, School of Medicine, University of Missouri–Columbia, Columbia, Missouri 65212

Kendall W. King (179), Animal Health Discovery Research, The Upjohn Company, Kalamazoo, Michigan 49001

Bruce C. Kirkpatrick (105, 369), Department of Plant Pathology, University of California at Davis, Davis, California 95616

Duncan C. Krause (89), Department of Microbiology, University of Georgia, Athens, Georgia 30602

Wayne C. Lai (469), Department of Pathology, Division of Comparative Medicine, University of Texas Southwestern Medical Center, Dallas, Texas 75235

Ing-Ming Lee (105), Molecular Plant Pathology Laboratory, United States Department of Agriculture, Agricultural Research Service, Beltsville, Maryland 20705

Shyh-Ching Lo (81), Department of Infectious and Parasitic Disease Pathology, American Registry of Pathology, Division of Molecular Pathobiology, Armed Forces Institute of Pathology, Washington, District of Columbia 20306

Dwight E. Lynn (55), Insect Biocontrol Laboratory, United States Department of Agriculture, Agricultural Research Service, Beltsville, Maryland 20705

Jack Maniloff (297), Department of Microbiology and Immunology, University of Rochester Medical Center, Rochester, New York 14642

R. J. Miles (287), Division of Life Sciences, King's College, London W8 7AH, United Kingdom

F. Chris Minion (305), Veterinary Medical Research Institute, College of Veterinary Medicine, Iowa State University, Ames, Iowa 50011

Peter F. Mühlradt (421), Gesellschaft für Biotechnologische Forschung, GBF, D-38124 Braunschweig, Germany

Yehudith Naot (451), Department of Immunology, The Bruce Rappaport Faculty of Medicine, Technion, Haifa 31096, Israel

Harold Neimark (105, 119), Department of Microbiology and Immunology, State University of New York Health Science Center at Brooklyn, Brooklyn, New York 11203

Susanne Nyström (217), Astra Hässle, S-901 24 Umeå, Sweden

J. Dennis Pollack (277), Department of Medical Microbiology and Immunology, Ohio State University, Columbus, Ohio 43210

Thomas Proft (133), Department of Molecular Medicine, School of Medicine, University of Auckland, Auckland, New Zealand

Aharon Razin (185, 201), Department of Cellular Biochemistry, The Hebrew University–Hadassah Medical School, Jerusalem 91120, Israel

Shmuel Razin (1, 101, 215, 275, 335, 355, 361, 393), Department of Membrane and Ultrastructure Research, The Hebrew University–Hadassah Medical School, Jerusalem 91120, Israel

J. Renaudin (167), Laboratoire de Biologie Cellulaire et Moléculaire, Centre de Recherches de Bordeaux, Institut National de la Recherche Agronomique, 33883 Villenave d' Ornon Cedex, France

Paul Renbaum (185, 201), Department of Cellular Biochemistry, The Hebrew University–Hadassah Medical School, Jerusalem 91120, Israel

Contributors

Janet A. Robertson (65), Department of Medical Microbiology and Infectious Diseases, Faculty of Medicine, University of Alberta, Edmonton, Alberta, Canada T6G 2H7

Shlomo Rottem (243, 265, 325), Department of Membrane and Ultrastructure Research, The Hebrew University–Hadassah Medical School, Jerusalem 91120, Israel

Michael Salman (325), Department of Membrane and Ultrastructure Research, The Hebrew University–Hadassah Medical School, Jerusalem 91120, Israel

Hans Gerd Schiefer (251), Institut für Medizinische Mikrobiologie, Klinikum der Justus Liebig Universität, D-35392 Giessen, Germany

Bernd Schneider (369), Faculty of Science, Northern Territory University, Darwin, NT 0909, Australia

Ulrich Schummer (251), Institut für Medizinische Mikrobiologie, Klinikum der Justus Liebig Universität, D-35392 Giessen, Germany

Barbara B. Sears (105), Department of Botany and Plant Pathology, Michigan State University, East Lansing, Michigan 48824

Erich Seemueller (369), Biologische Bundesanstalt für Land und Forstwirtschaft, Institut für Pflanzenschutz im Obstbau, D-69216 Dossenheim, Germany

Ken-ichiro Shibata (315), Department of Oral Bacteriology, Hokkaido University School of Dentistry, Sapporo, Hokkaido 060, Japan

Christine D. Smart (369), Department of Plant Pathology, University of California at Davis, Davis, California 95616

Gerald W. Stemke (65), Department of Biological Sciences, Faculty of Science, University of Alberta, Edmonton, Alberta, Canada T6G 2H7

Marla K. Stevens (89), Department of Microbiology, University of Texas Southwestern Medical Center, Dallas, Texas 75235

P. Michael Stuart (461), Department of Ophthalmology and Visual Sciences, Washington University Medical Center, St. Louis, Missouri 63110

Mark Tarshis (243), Department of Membrane and Ultrastructure Research, The Hebrew University–Hadassah Medical School, Jerusalem 91120, Israel

David Taylor-Robinson (73), Department of Genitourinary Microbiology and Medicine, The Jefferiss Wing, St. Mary's Hospital, Paddington, London W2 1NY, United Kingdom

Joseph G. Tully (29, 33, 339, 381), Mycoplasma Section, Laboratory of Molecular Microbiology, National Institute of Allergy and Infectious Diseases, Frederick Cancer Research and Development Center, Frederick, Maryland 21702

Tsuguo Watanabe (315), Department of Oral Bacteriology, Hokkaido University School of Dentistry, Sapporo, Hokkaido 060, Japan

Robyn Watson-McKown (227), Department of Molecular Microbiology and Immunology, School of Medicine, University of Missouri–Columbia, Columbia, Missouri 65212

Douglas J. Wear (81), Department of Infectious and Parasitic Disease Pathology, American Registry of Pathology, Division of Molecular Pathobiology, Armed Forces Institute of Pathology, Washington, District of Columbia 20306

William G. Weisburg (349), GENE-TRAK Systems, Framingham, Massachusetts 01701

Robert F. Whitcomb (41, 339), Insect Biocontrol Laboratory, United States Department of Agriculture, Agricultural Research Service, Beltsville, Maryland 20705

Åke Wieslander (217), Department of Biochemistry, Umeå University, S-901 87 Umeå, Sweden

Kim S. Wise (227), Department of Molecular Microbiology and Immunology, School of Medicine, University of Missouri–Columbia, Columbia, Missouri 65212

Jerold G. Woodward (461), Department of Microbiology and Immunology, Albert B. Chandler Medical Center, University of Kentucky, Lexington, Kentucky 40536

David Yogev (355, 361), Department of Membrane and Ultrastructure Research, The Hebrew University–Hadassah Medical School, Jerusalem 91120, Israel

Preface

The two volumes of *Methods in Mycoplasmology* published by Academic Press in 1983 have gained wide recognition as the most comprehensive and authoritative treatise on mycoplasma methodology, and are highly cited in the mycoplasma literature. These volumes have provided researchers and laboratory workers with well-tried and standardized procedures for the recovery, identification, and characterization of mycoplasmas.

The developments in mycoplasmology which have taken place since the publication of these volumes have been outstanding due mainly to the application of molecular genetic methodology to mycoplasmas. Introduction of this methodology has had a significant impact on our understanding of the cell structure, genetics, metabolism, taxonomy, and phylogeny of mycoplasmas, as well as of the mechanisms of pathogenicity and the interaction of mycoplasmas with the immune system. These advances have found expression in the development of new diagnostic procedures, including those based on DNA probes and DNA amplification. As could be expected, significant developments have also taken place in the more "classical" procedures, those dealing with the cultivation, serological characterization, and pathogenicity testing of mycoplasmas.

The two volumes of *Molecular and Diagnostic Procedures in Mycoplasmology* focus on the new procedures developed during the past decade, particularly those based on the new molecular methodology. This volume outlines the approaches, techniques, and procedures applied to cell and molecular biology studies of mycoplasmas. The second volume deals with the new genetic and immunological tools applied to the diagnosis of mycoplasma infections of humans, animals, plants, insects, and cell cultures. We are well aware that techniques outlined for rapidly moving subdisciplines may soon become dated. Yet experience gained through the use of the *Methods in Mycoplasmology* volumes confirms that the majority of methods detailed in the new volumes will continue to be useful for years to come.

We thank our colleagues who were most helpful at the initial stages of selecting the topics and procedures to be covered and in developing the volume outlines. Considering the large number of chapters and contributors, keeping to the deadlines set by the publisher can be considered an outstanding achievement. Obviously, this could not have been accomplished without the cooperation of the many contributors. We express our gratitude and appreciation for their friendly cooperation in this endeavor.

Shmuel Razin
Joseph G. Tully

Contents of Volume II

Diagnostic Procedures

Mollicute–Host Interrelationships: Current Concepts
and Diagnostic Implications
Joseph G. Tully

SECTION A DIAGNOSTIC GENETIC PROBES

A1 Introductory Remarks
Shmuel Razin
A2 Oligonucleotide Probes Complementary to 16S rRNA
Karl-Erik Johansson
A3 Cloned Genomic DNA Fragments as Probes
David Yogev and Shmuel Razin
A4 PCR: Selection of Target Sequences
Rémi Kovacic, Odile Grau, and Alain Blanchard
A5 PCR: Preparation of DNA from Clinical Specimens
*Bertille de Barbeyrac, Christiane Bébéar,
and David Taylor-Robinson*
A6 PCR: Amplification and Identification of Products
Bertille de Barbeyrac and Christiane Bébéar
A7 PCR: Application of Nested PCR to Detection
of Mycoplasmas
Ryô Harasawa
A8 PCR: Random Amplified Polymorphic DNA Fingerprinting
Steven J. Geary and Mark H. Forsyth

SECTION B IMMUNOLOGICAL TOOLS

B1 Introductory Remarks
Joseph G. Tully
B2 ELISA in Small Animal Hosts, Rodents, and Birds
M. B. Brown, J. M. Bradbury, and J. K. Davis
B3 ELISA in Large Animals
J. Nicolet and J. L. Martel
B4 ELISA in Human Urogenital Infections and AIDS
Richard Yuan-Hu Wang and Shyh-Ching Lo
B5 ELISA in Respiratory Infections of Humans
Gail H. Cassell, Ginger Gambill, and Lynn Duffy
B6 Monoclonal Antibodies as Diagnostic Tools
*Chester B. Thomas, Monique Garnier, and
John T. Boothby*

B7 Microimmunofluorescence
   David Taylor-Robinson
B8 Immunoblots and Immunobinding
   David Thirkell and Bernard L. Precious
B9 Differentiation of *Mycoplasma genitalium* from *Mycoplasma pneumoniae* by Immunofluorescence
   Joseph G. Tully

SECTION C ANTIBIOTIC SENSITIVITY TESTING

C1 Introductory Remarks
   Christiane Bébéar
C2 Problems and Opportunities in Susceptibility Testing of Mollicutes
   George E. Kenny
C3 Determination of Minimal Inhibitory Concentration
   Christiane Bébéar and Janet A. Robertson
C4 Cidal Activity Testing
   David Taylor-Robinson

SECTION D DIAGNOSIS OF SPECIFIC DISEASES

D1 Introductory Remarks
   Joseph G. Tully
D2 Laboratory Diagnosis of *Mycoplasma pneumoniae* Infection
   R. J. Harris, J. Williamson, C. Hahn, and B. P. Marmion
D3 Diagnosis of Sexually Transmitted Diseases
   David Taylor-Robinson
D4 Diagnosis of Neonatal Infections
   Ken B. Waites and Gail H. Cassell
D5 Mycoplasmas in AIDS Patients
   Shyh-Ching Lo
D6 Mycoplasma Infections of Cattle
   Ed A. ter Laak and H. Louise Ruhnke
D7 Mycoplasma Infections of Goats and Sheep
   A. J. DaMassa
D8 Mycoplasma Infections of Swine
   Richard F. Ross and Gerald W. Stemke
D9 Mycoplasma Infections of Poultry
   Stanley H. Kleven and Sharon Levisohn
D10 Diagnosis of Spiroplasma Infections in Plants and Insects
    C. Saillard, C. Barthe, J. M. Bové, and R. F. Whitcomb
D11 Detection of Phytoplasma Infections in Plants
    E. Seemueller and B. C. Kirkpatrick
D12 Identification of Mollicutes from Insects
    Robert F. Whitcomb and Kevin J. Hackett

SECTION E EXPERIMENTAL INFECTIONS

- E1 Introductory Remarks
  Joseph G. Tully
- E2 Experimental Mycoplasmal Respiratory Infections in Rodents
  Gail H. Cassell and A. Yancey
- E3 Urogenital Infections in Rodents
  Patricia M. Furr and David Taylor-Robinson
- E4 Experimental Models of Arthritis
  Leigh Rice Washburn
- E5 Experimental Infections in Poultry
  Janet M. Bradbury and Sharon Levisohn
- E6 Experimental Infections of Swine
  Marylène Kobisch and Richard F. Ross
- E7 Experimental Infections in Cattle
  Ricardo F. Rosenbusch and H. Louise Ruhnke
- E8 Experimental Infections of Plants by Spiroplasmas
  X. Foissac, J. L. Danet, C. Saillard, R. F. Whitcomb, and J. M. Bové
- E9 Experimental Phytoplasma Infections in Plants and Insects
  Alexander H. Purcell
- E10 Mycoplasmas and *in Vitro* Infections of Cell Cultures with HIV
  Shyh-Ching Lo and Alain Blanchard

SECTION F DIAGNOSIS OF MYCOPLASMA INFECTIONS OF CELL CULTURES

- F1 Introductory Remarks
  Joseph G. Tully
- F2 Isolation of Mycoplasmas from Cell Cultures by Axenic Cultivation Techniques
  Richard A. Del Giudice and Joseph G. Tully
- F3 Detection of Mycoplasmas by DNA Staining and Fluorescent Antibody Methodology
  Gerald K. Masover and Frances A. Becker
- F4 Detection of Mycoplasma Infection by PCR
  Connie Veilleux, Shmuel Razin, and Laurie H. May
- F5 Antibiotic Treatment of Mycoplasma-Infected Cell Cultures
  Richard A. Del Giudice and Roberta S. Gardella
- F6 Prevention and Control of Mycoplasma Infection of Cell Cultures
  Ann Smith and Jon Mowles

Appendix: Tables I–V
Index

MOLECULAR PROPERTIES OF MOLLICUTES: A SYNOPSIS
Shmuel Razin

Introduction

Mycoplasmas are distinguished phenotypically from other bacteria by their minute size and total lack of a cell wall. Taxonomically, the lack of cell walls is used to separate mycoplasmas from other bacteria in a class named Mollicutes (*mollis,* soft; *cutis,* skin, in Latin). The current classification of Mollicutes and the properties distinguishing the currently established taxa are presented in Chapter E2 (this volume). While the trivial terms mycoplasmas or mollicutes have been used interchangeably to denote any species included in Mollicutes, the trivial names ureaplasmas, entomoplasmas, mesoplasmas, spiroplasmas, acholeplasmas, asteroleplasmas and anaeroplasmas are routinely used for members of the corresponding genus. The molecular characterization of the uncultured plant and insect mycoplasma-like organisms has provided strong experimental support for their inclusion in the class Mollicutes. Although formal taxonomy of these mollicutes has not been established as yet, the trivial term phytoplasmas has been proposed to replace the cumbersome and rather vague name mycoplasma-like organisms (Chapter E6, this volume).

The purpose of this introductory chapter is to provide the reader with an outline of the issues brought up in the various sections of this volume. Because of space limitations, this synopsis is sketchy, far from being comprehensive, and the selection of references is largely arbitrary, giving preference to the most recent ones. The interested reader is referred to a number of books on various aspects of mycoplasmology published since the early 1980s (Razin and Barile, 1985; Whitcomb and Tully, 1989; Maniloff *et al.*, 1992; Rottem and Kahane,

1993; Kahane and Adoni, 1993). A wealth of information can be found in the Proceedings of the biannual meetings of the International Organization for Mycoplasmology (the last three issues were named IOM Letters, Vols. 1–3) and in a special issue of *Journal of Clinical Infectious Diseases* devoted to the role of mycoplasmas in respiratory disease and AIDS (acquired immunodeficiency syndrome) (Vol. 17, Suppl. 1, 1993). Reviews covering different aspects of the molecular biology and genetics of mycoplasmas, as well as their general properties and taxonomy, are also available (Dybvig, 1990; Razin, 1978, 1985, 1991, 1992a,b; Tully and Whitcomb, 1991; Bove, 1993).

Morphology

The mollicutes vary in shape from spherical or pear-shaped structures (0.3–0.8 μm in diameter) to branched or helical filaments. Genome replication precedes, but is not necessarily synchronized with, cell division. Thus, budding forms, filaments, and chains of beads can be observed in mycoplasma cultures. The spiroplasmas are shaped as thin helical filaments exhibiting rotatory motility with flexional and twitching movements (Razin, 1978). Although most mycoplasmas are nonmotile and have no flagella, some *Mycoplasma* species exhibit gliding motility on liquid-covered surfaces. These gliding mycoplasmas usually possess a specialized tip structure or organelle that plays a role in the adhesion of mycoplasmas to host cells (Razin and Jacobs, 1992) and apparently in the penetration into host cells of the newly discovered human mycoplasma *M. penetrans* (Lo *et al.*, 1992). The tip organelles exhibit an ultrastructure different from that of the rest of the cell. Following dissolution of the cell membrane of *M. pneumoniae* by a mild detergent, such as Triton X-100, the tip structure reveals a striated rod with associated microfilaments, possibly representing a primitive cytoskeleton (Triton shell).

Habitat

Mollicutes are widespread in nature as parasites of humans, mammals, reptiles, fish, arthropods, and plants (Razin, 1992b). The range of hosts known to harbor mycoplasmas is continuously increasing as does the number of established mollicute species, exceeding 150 at the time this chapter was written (Table I in Chapter E2, this volume, and Appendix, Vol. II). A general agreement exists among mycoplasmologists that the mollicutes that have already been characterized and taxonomically defined constitute only a part, apparently a minor one, of the mollicutes living in nature. The fact that only recently has a new human mycoplasma, *M. penetrans,* been isolated from AIDS patients (Lo *et al.*, 1992) illustrates this point well. It may be speculated that weakening of the immune system in immunocompromised hosts may provide better grounds for the proliferation of mollicutes that are usually suppressed or limited to a very specific niche in the host.

The primary habitats of human and animal mycoplasmas are the mucous surfaces of the respiratory and urogenital tracts, the eyes, alimentary canal, mammary glands, and the joints. The obligatory anaerobic anaeroplasmas have so far been found only in the bovine and ovine rumen. Spiroplasmas and phytoplasmas are widespread in the gut, hemocoel, and salivary glands of arthropods; through sap-sucking insects, the spiroplasmas and phytoplasmas may be introduced into the phloem tissues of plants, causing disease (see the introductory chapter of Vol. II). As stated earlier, most human and animal mycoplasmas adhere tenaciously to the epithelial linings of the respiratory and urogenital tracts, and may thus be considered as typical surface parasites (Razin and Jacobs, 1992). Although there is no question as to the ability of spiroplasmas and phytoplasmas to become intracellular during some phases of infection, the intracellular location of human and animal mycoplasmas was disputable for a long time. The recent demonstration of the intracellular location of some human mycoplasmas, including *M. fermentans* in tissues of AIDS patients (Chapters A6 and A7, this volume) and penetration of *M. penetrans* into cells, indicates that a variable percentage of the population of some mycoplasmas may occupy an intracellular location. This location presumably offers better protection of the mycoplasmas against host immune system components as well as against antimycoplasmal agents, in line with chronicity of mycoplasmal infections in animals and their persistent infection of cell cultures (Section F, Vol. II).

In Vitro Culture

There can be little doubt that the number of taxonomically defined mollicute species will continue to rise with the improvement in culture media and with the increasing weight given to molecular tools in bacterial taxonomy and phylogeny. Thus, as can be seen in Chapter E6 (this volume), the uncultured phytoplasmas have already been placed in 12 distinct subclades, mostly on the basis of their 16S rRNA gene sequences. This preliminary phylogenetic classification will apparently be replaced in the near future by proper binomial names (Chapter E2, this volume). The possibility that uncultured mollicutes are also present in humans and animals appears most plausible. Thus, the uncultured grey lung agent of mice shows the morphologic and genomic characteristics of a mollicute, including genome size and a characteristic 16S rRNA gene sequence (Neimark *et al.*, 1994).

The failure of current mycoplasma culture media to support good or even minimal growth can be illustrated by the long incubation time and the frequent failure to cultivate *M. pneumoniae* from clinical specimens (Chapter D2, Vol. II), by the extreme difficulty to cultivate *M. genitalium* (Jensen *et al.*, 1994), and by the low titers of ureaplasmas cultivated on the best mycoplasma media available (Chapters A1 and A2, this volume). A novel approach to improve the chances of *in vitro* cultivation of extremely fastidious mollicutes is based on coculture with eukaryotic cell lines. In this way, unculturable spiroplasmas, such as the sex ratio organism of *Drosophila,* have been successfully cocultured on insect cell lines (Chapter A4, this volume). Application of the cell-assisted culture approach has resulted in the cultivation of several new *M. genitalium* clinical isolates (Jensen *et al.,* 1994), a significant feat in light of numerous, mostly fruitless, efforts to cultivate this human mycoplasma from clinical material.

Molecular Biology and Genetics

Genome Size

Major molecular properties distinguishing the mollicutes from eubacteria are summarized in Table I. The application of pulse-field gel electrophoresis (PFGE) to mollicute genome size determinations (Chapter B3, this volume) has provided a much more accurate and labor-saving procedure than the previously used renaturation kinetics method, resulting in a wealth of genome size data. Data show a continuum of genome sizes among mollicutes, ranging from less than 600 to over 2200 kbp, with overlapping values between mollicute genera.

TABLE I

PROPERTIES DISTINGUISHING MOLLICUTES FROM OTHER EUBACTERIA[a]

Property	Mollicutes	Other eubacteria
Cell wall	Absent	Present
Plasma membrane	Cholesterol present in most species	Cholesterol absent
Genome size	580–2220 kbp	1450–>6000 kbp
G + C content of genome	23–41 mol%	25–75 mol%
No. of rRNA operons	1–2[b]	1–10
5S rRNA length	104–113 nucleotides	>114 nucleotides
No. of tRNA genes	30 (*M. capricolum*)	51 (*B. subtilis*)
	33 (*M. pneumoniae*)	78 (*E. coli*)
UGA codon usage	Tryptophan codon in *Mycoplasma, Ureaplasma, Spiroplasma,* and *Mesoplasma*	Stop codon
RNA polymerase	Resistant to rifampicin	Rifampicin sensitive

[a] Adapted from Razin (1991).
[b] Three rRNA operons in *Mesoplasma lactucae* (Bove, 1993).

Thus, genome sizes of *Mycoplasma* species range from 580 kbp for *M. genitalium* to 1380 kbp for *M. mycoides* subsp. *mycoides* LC, whereas the genome sizes of the helical *Spiroplasma* species range from 950 kbp for *S. monobiae* to 2220 kbp for *S. ixodetis* (Bove, 1993; Carle *et al.,* 1995; Table I in Chapter E2, this volume). Clearly, genome size cannot be used anymore as the definitive taxonomic criterion used previously to distinguish higher taxa in Mollicutes (Razin, 1991). Yet, as a general rule, *Acholeplasma* and *Spiroplasma* species, considered phylogenetically as "early" mollicutes, have larger genome sizes than *Mycoplasma* and *Ureaplasma* species, considered as phylogenetically more recent mollicutes. This is an agreement with the notion that Mollicutes have evolved by degenerative, reductive, or regressive evolution, accompanied by significant losses of genomic sequences (Woese, 1987; Chapter E3, this volume). Genome sizes are variable not only within the same genus, but even among strains of the same species. One reason for this variability is the frequent occurrence of repetitive elements differing in size and number in the genome of many mollicutes, such as in *M. pneumoniae,* and the presence of variable numbers of viral sequences integrated into the mollicute genome, as found in *S. citri,* where these sequences may account for up to 150 kbp, that is one-twelfth of the entire genome (Bove, 1993).

Despite the finding of mollicutes with relatively large genomes, the conclusion that the smallest genomes of self-replicating, free-living organisms are found among the mollicutes is still valid. Thus, *M. genitalium* with a genome size of less than 600 kbp and *M. pneumoniae* with a genome size of about 800 kbp can be considered closest to the concept of a "minimum cell." Assuming that 20–30% of the genome consists of intergenic spacers, transcription signals, etc., the estimated number of genes in these two mycoplasmas cannot be much higher than 600–700 genes, promoting the use of these mycoplasmas as adequate models for the molecular definition of an entire functioning machinery of a living cell (Morowitz, 1984; Razin, 1992a). How could organisms with such an extremely limited amount of genetic information perform all the essential functions required for self-replication? Economization in genetic information is the answer. It takes shape in a variety of ways. Thus, the number of gene copies is kept to a minimum in mollicutes, even with genes that are very important for cell growth. Thus, the number of rRNA gene copies in mollicutes is one or a maximum of two (Table I) compared to 7 copies in *Escherichia coli* and 10 in *Bacillus subtilis.* Gene saving, in this case, comes at the expense of the ability of rapid synthesis of ribosomes and, consequently, the failure of mollicutes to grow rapidly, particularly when transferred to a new medium. The dependence of the parasitic mollicutes on host-derived nutrients has led to the saving of a considerable amount of genetic information. The marked dependence of mollicutes on exogenous nutrients finds its expression in the fastidious nature of these organisms, reflected by the great difficulties encountered in their cultivation *in vitro.* Significant economization of genetic information was also achieved through the loss of the cell wall during mollicute evolution. In this way, a significant number of the genes involved in the synthesis of cell wall polymers could be saved. Again, the loss of a protective cell wall had its price, as expressed by the fragility and marked osmotic sensitivity of mollicutes.

Genome Mapping

Application of rare cutting restriction endonucleases and PFGE has facilitated the establishment of physical (restriction) maps of increasing numbers of mollicutes. The use of cloned conserved prokaryotic genes as hybridization probes has helped in locating the corresponding genes on the physical maps and converting them into genetic maps. The ongoing human genome project has provided the stimulus, and some funding, for the complete sequencing and genetic mapping of prokaryotic genomes, intended to serve as pilot projects in the complex human genome endeavor. As could be expected, the effort has been focused on sequencing the genomes of the well-studied *E. coli* and *B. subtilis*. Yet, the fact that some mollicute genomes are about one-fifth the size of the genomes of these bacteria favors sequencing first a mycoplasma genome. In fact, the *M. pneumoniae* and *M. genitalium* genomes are currently being sequenced. Over 40% of the 800-kbp genome of *M. pneumoniae* has already been sequenced (Hilbert *et al.,* 1994), and the complete sequence of the entire genome is expected to be available in 1996 (R. Herrmann, personal communication). About 60% of the sequenced genomic regions of *M. pneumoniae* could be identified, mainly by significant sequence homology to defined genes of other bacteria and by hybridization with known gene probes. The rest of the sequences carried open reading frames of unidentified genes, presumably consisting of genes peculiar to the mycoplasma (Proft and Herrmann, 1994; Chapter B4, this volume).

In light of economization in the number of genes in mollicutes, it is somewhat surprising to find repeating genomic sequences in the minute genomes of many mollicutes. Some of these are integrated mollicute viral sequences that may be considered as genetic debris, not fulfilling any essential role (Ye *et al.,* 1994a). Others represent mobile genetic elements, such as insertion sequences and transposons, providing a source of DNA homologies for recombinational genomic reordering (Deng and McIntosh, 1994; Chapter B9, this volume). Gene fragments, such as fragments of the P1 and MgPa operons of *M. pneumoniae* and *M. genitalium,* are also found in mollicute genomes. These gene fragments are spread over the genome in multiple copies, constituting in the case of *M. pneumoniae* up to 6% of the total genomic sequences. The major pMGA adhesin gene of *M. gallisepticum* is found in multiple copies, arranged in tandem and differing from each other in minor details (Markham *et al.,* 1994). The multiple gene copies apparently function as components of an antigenic variation system, ful-

TABLE II
Surface Antigenic Variation Systems in Mollicutes

Organism	System designation	System components	Genes involved
M. hyorhinis	Vlp	Lipoproteins (four defined members)	Family of 6*vlp* genes (Yogev et al., 1991; Rosengarten et al., 1993)
M. bovis	Vsp	Lipoproteins (three defined members)	Gene family (Behrens et al., 1994)
M. gallisepticum	pMGA	Lipoprotein (p67)	Gene family (Markham et al., 1994)
		PvpA integral protein (not lipid modified)	*pvpA* gene (Yogev et al., 1994)
M. pulmonis	V-1	Lipoproteins	Gene family (Simmons et al., 1994)
U. urealyticum	MB	Lipoproteins	Gene defined (Teng et al., 1994)
M. hominis	LMP	Lipoproteins (p135; p120)	*Imp* gene family (Thorp et al., 1994; Christiansen et al., 1994)
M. fermentans		Lipoproteins (p29; p78)	Genes defined (Theiss and Wise, 1994)
M. hyopneumoniae		Lipoproteins (variable?) (p44; p50; p65; p70)	Genes not defined (Wise and Kim, 1987)
M. arthritidis		Integral proteins (not lipid modified?) p30; p36; p40	Genes not defined (Tangen et al., 1994)

filling an important role in evasion of the host immune system (see section on antigenic variation systems, this chapter). The nucleotide sequences of some of the genes and elements discussed earlier have brought up an interesting aspect that may relate to their origin. Thus, the genes for the adhesins P1, MgPa, and the antigen variable proteins Vlps and Vsps (Table II) differ significantly in their base composition from the rest of the corresponding genomes. Moreover, the *vlp* and *vsp* genes do not carry UGA codons for tryptophan, a feature characterizing the other genes in *Mycoplasma* genomes (Hilbert et al., 1994; Yogev et al., 1991), leading to the notion that these genes had originated from an unknown exogenous source via horizontal transfer.

Information is now accumulating not only on the type of genes, but also on their mode of organization in the mollicute genomes. The clustering and order of genes in the cluster may serve as a phylogenetic marker. The organization of genes at the origin of replication of mollicute genomes has attracted much attention, as this region is highly conserved in bacteria. While basically the gene order *(rnpA-rmpH-dnaA-dnaN-recF-gyrB)* has been retained in the several mollicute genomes studied so far, there are some differences, such as the lack of *recF* in *S. citri* (Ye *et al.*, 1994b) and in *M. genitalium* (Bailey and Bott, 1994), whereas in *M. hominis* and *M. capricolum* the *gyrB* gene was located in another part of the genome (Ladefoged and Christiansen, 1994; Miyata *et al.*, 1993). On the whole, gene rearrangements can be expected to occur rather frequently in the small mollicute genomes subjected to frequent genomic deletions during evolution.

The nature of the mycoplasmal DNA polymerases has presented an enigma. Early studies (reviewed in Razin, 1985) detected a single DNA polymerase in *M. orale* and *M. hyorhinis* as well as in *M. mycoides* and *U. urealyticum*, whereas *Spiroplasma* and *Acholeplasma* species appear to possess three DNA polymerases, as in *E. coli* and *B. subtilis* (Maurel *et al.*, 1989). Most intriguing was the apparent lack of exonuclease activity, considered essential in proofreading or editing of the newly synthesized DNA strands. Thus, for some time, the findings were taken to suggest that the evolution of mollicutes has wrought significant simplification of the gram-positive *pol* family, reducing it from three enzymes with exonuclease activity to a single enzyme devoid of exonuclease activity in *Mycoplasma* and *Ureaplasma* species (Razin, 1985). Barnes *et al.* (1994) suggest that this may not be so, as *M. pulmonis* was found to possess DNA polymerase III (Pol III), the enzyme essential for replicative DNA synthesis, displaying the size, primary structure, and exonuclease activity of typical gram-positive Pol III. They have also found evidence for a second DNA polymerase in *M. pulmonis*, possibly corresponding to the previously described mycoplasmal DNA polymerases, indicating that the DNA polymerase composition of *Mycoplasma* is more complex than previously thought (Barnes *et al.*, 1994).

The RecA protein constitutes another essential component in genomic DNA synthesis, enabling recombinatorial DNA repair and homologous recombination. The *recA* genes of several mollicutes have been cloned and sequenced, and the deduced amino acid sequences of the RecA polypeptides exhibited significant similarity to RecA's of other eubacteria (King *et al.*, 1994). The presence of *recA* genes and the apparent inability to engineer knock-out mutations in genomes as small as those of mollicutes suggest that RecA is critical for mollicute survival.

Protein Synthesis Machinery

The protein synthesis machinery of mollicutes resembles that of eubacteria with some modifications, in line with gene economization (Razin, 1985). Best characterized are the mycoplasmal rRNA genes. These genes are highly conserved and serve as the most important phylogenetic markers (Chapter E3, this volume). Moreover, cloned rRNA genes have been applied as efficient probes in detection and identification of mollicutes (Chapters E4–E6, this volume; Chapters A2–A4, Vol. II).

Organization of the mollicute rRNA genes generally follows the typical eubacterial order: 16S–23S–5S, functioning as an operon (Razin, 1985). In some porcine mycoplasmas the 5S rRNA genes are separated from the 16S–23S genes (Stemke et al., 1994), whereas in *M. gallisepticum* the 16S rRNA gene is separated from the 23S–5S genes in one of the rRNA gene sets of this mollicute (Chen and Finch, 1989). Mollicute genomes carry only one or two rRNA gene sets (Table I), whereas *Clostridium innocuum* and *Clostridium ramosum,* phylogenetically closest to mollicutes, have five and four rRNA gene sets, respectively (Bove, 1993). Interestingly, all of the 28 phytoplasmas examined by Schneider and Seemüller (1994), representing five primary taxonomic clusters, carry two rRNA gene sets, resembling in this respect their phylogenetic relatives, the acholeplasmas. Another feature of interest concerns the presence of tRNA genes in the intergenic region between the 16S and 23S rRNA genes. All of the seven rRNA operons of *E. coli* carry two tRNA genes in this spacer region. No tRNA genes could be found in the 16S–23S spacer region in mollicutes (Razin, 1985), except for phytoplasmas, where a single tRNA^{Ile} was located in this region (Kirkpatrick et al., 1994) and in *Acholeplasma laidlawii,* where the spacer region of one of the two rRNA operons carries tRNA^{Ile} and tRNA^{Ala} genes (Nakagawa et al., 1992).

The number of tRNA genes in mollicutes is kept to a minimum, with very few gene duplicates (Table I). Accordingly, the number of anticodons in *M. capricolum* is 28, not much higher than in mitochondria and close to the essential minimum for translation of all the amino acid codons by wobbling (Bove, 1993). A peculiar feature of mycoplasmal tRNAs is that they contain significantly fewer modified nucleosides than do other eubacterial tRNAs. Thus only 13 types of modified nucleosides are found in the total tRNAs of *M. capricolum,* compared to 23 types in *E. coli* (Bove, 1993). It thus appears that during the reductive evolution of the mollicutes, genes for many tRNAs, as well as genes for enzymes involved in tRNA nucleoside modification, were deleted.

Mollicutes, having genomes with a very low G+C (Table I), have been under strong A–T pressure throughout their evolution. This situation has resulted in codon usage favoring synonymous codons with A and T, in particular in the wobble (3′) position. Codons rich in G+C, such as CUC (leucine) and CGG (arginine), have not been found in *M. capricolum* (Oba et al., 1991; Maniloff et al., 1994). The use of UGA as a tryptophan codon instead of a stop codon by mollicutes is one of their most prominent features. As seen in Table I, not all mollicutes share this property. Thus, the phylogenetically early acholeplasmas and phytoplasmas (Toth et al., 1994) use the conventional UGG codon for tryptophan, carrying the usual tRNA^{Trp} (CCA) and retaining UGA as a stop codon. The phylogenetically later spiroplasmas, mycoplasmas, ureaplasmas, and mesoplasmas have evolved a new tryptophan codon, UGA, and a new tRNA to read UGA as tryptophan. The spiroplasmas use both UGG and UGA and have the corresponding two tRNAs: tRNA^{Trp} (CCA) and tRNA^{Trp} (UCA). The phylogenetically more recent mycoplasmas, such as *M. pneumoniae* and *M. genitalium*, have lost the tRNA^{Trp} (CCA) and kept only tRNA^{Trp} (UCA) which is able to read both UGG and UGA through wobble pairing (Bove, 1993). The use of UGA as a tryptophan codon has been applied as a taxonomic tool supporting, for example, the exclusion of the non-sterol-requiring mesoplasmas from acholeplasmas (Tully *et al.*, 1993; Chapter E2, this volume). From the practical point of view, the use of UGA as a tryptophan codon imposes a serious restriction on the expression of cloned mollicute genes in *E. coli*. Since *E. coli* regards UGA as a stop codon, translation of a mycoplasmal message in *E. coli* will stop where originally there should be tryptophan so that mycoplasmal proteins expressed in *E. coli* may be truncated. Ways to overcome this difficulty are described in Chapters B6 and B10 of this volume.

Resistance of mycoplasmal RNA polymerases to rifampicin and streptolydigin constitutes another feature distinguishing mollicutes from other eubacteria (Table I). The target of these antibiotics is the β subunit of the enzyme. Differences in the amino acid composition at certain regions of the β subunit are presumably responsible for the resistance of the mycoplasmal enzyme to the antibiotics, but these have not been defined as yet (Laigret and Bove, 1994).

Transcription signals in mollicutes, including the characteristic -10 and -35 regions in the promoters and the hairpin structures followed by stretches of U's typical of eubacterial terminators, resemble those of eubacteria (Bove, 1993). Shine–Dalgarno sequences upstream to the ATG start codon, serving in the initiation of mRNA translation, have been identified upstream to mollicute genes, although in some cases typical Shine–Dalgarno sequences could not be detected (Loechel *et al.*, 1991; Markham *et al.*, 1993).

Gene Transfer

Genetic studies in mollicutes have been hampered by the paucity of selectable markers and gene transfer systems (Dybvig, 1990). The lack of a cell wall in mollicutes would be expected to facilitate the introduction of exogenous DNA into the cells. In fact, the exchange of chromosomal DNA during direct contact of mycoplasma cells, probably by fusion of the cell membranes at the zone of contact, has been reported by Barroso and Labarere (1988), and conjugative transposition of transposon Tn*916* from *Streptococcus faecalis* to *M. hominis*, by a spontaneous mating process, was described by Roberts and Kenny (1987). However, the DNA transfer efficiency in these cases was rather low. Increased transformation and transfection efficiencies have been achieved in the presence of polyethylene glycol (PEG), and application of the electroporation procedure has generally proved to be more effective than PEG (Chapters B6 and B7, this volume). By using these transformation enhancers, stable integration of transposons of gram-positive bacteria (such as Tn*916* and Tn*4001*) into mollicute genomes has been achieved (Cao *et al.*, 1994).

Having efficient transfection procedures at hand, major efforts have been directed at developing vectors for cloning and expressing foreign genes in mollicutes. Transfection experiments with the replicative form (RF) of the *S. citri* virus SpV1, carrying as an insert a segment of the adhesin P1 gene of *M. pneumoniae*, resulted in the expression of this part of P1 in the spiroplasma. In this case, the presence of seven UGA codons in the P1 gene insert did not interfere with expression as *S. citri* can read UGA as tryptophan (Marais *et al.*, 1993). Yet, the SpV1-RF/*S. citri* cloning system is deficient as it suffers from the rapid loss of the cloned DNA insert (Chapter B6, this volume). The expression of foreign genes could be achieved in *A. laidlawii*, a mollicute that uses the conventional UGG codon for tryptophan. Thus, a recombinant *Lactobacillus lactis* plasmid pNZ18, carrying the α-amylase gene of *Bacillus licheniformis*, replicated stably in *A. laidlawii*, and the amylase gene was expressed in the mycoplasma (Jarhede *et al.*, 1994b). In addition, *A. laidlawii* could express the *S. citri* spiralin gene inserted into the pNZ18 plasmid, although the expressed spiralin, an acylated membrane protein in *S. citri*, was not acylated and remained with its uncleaved signal peptide in the cytoplasm of *A. laidlawii*. Chapter B6 (this volume) describes another approach directed at the construction of shuttle vectors capable of replication both in *S. citri* and in *E. coli*. The vectors, constructed by insertion of the *S. citri* origin of replication, *oriC*, and the *tetM* determinant into the colE1-derived plasmids, could be shuttled from *E. coli* to *S. citri* and back to *E. coli* (Renaudin *et al.*, 1994; Ye *et al.*, 1994b).

On the whole, it appears that although not all the problems involved in the development of effective shuttle vectors have been solved, the goal of efficient gene transfer among mollicutes is within reach, so that analysis of genetic mutations resulting from gene transfer among mutant strains can begin (Bove, 1993).

Mycoplasma Membrane

Lacking a cell wall and intracytoplasmic membranes, the mollicutes have only one type of membrane, the plasma membrane. The ease with which this membrane can be isolated and the ability to introduce controlled alterations in its composition have made mycoplasma membranes effective tools in biomembrane research (Razin, 1993). A multiauthored treatise devoted to the various aspects of mycoplasma membrane research has been published (Rottem and Kahane, 1993). Of the mycoplasma membrane components, membrane lipoproteins have attracted most of the attention since the 1980s, as many of these were shown to constitute the major components of mycoplasmal antigenic variation systems. Since the first report on acylated membrane proteins in mollicutes (Dahl et al., 1983), a wealth of information has been accumulating on the wide distribution in mollicutes of acylated membrane proteins carrying covalently bound fatty acids, mostly palmitic and myristic acids (Table II; see also Chapter C2, this volume).

The abundance of lipoproteins in mycoplasma membranes is remarkable in contrast to the limited number of lipoproteins found in other eubacteria. Thus, Thirkell et al. (1991) showed that about 25% of the *U. urealyticum* proteins were labeled by [^{3}H]palmitate and could be partitioned into the Triton X-114 phase by the procedure described in Chapter C3 of this volume. *Acholeplasma laidlawii* has about three times more acylated membrane proteins than *L. lactis* and *B. subtilis* (Jarhede et al., 1994a). Whether the abundance of membrane lipoproteins in mollicutes is associated with the lack of a cell wall is a moot point, but being dominant antigens, probably forming a coat on the cell surface, endows them with properties fulfilled by cell wall components in other bacteria.

Antigenic Variation Systems

The mycoplasma plasma membrane, being exposed to the external environment, is the cell organelle that comes into contact with components of the host immune system. Lacking the protection of a rigid cell wall, mycoplasmas are particularly sensitive to growth inhibition and lysis by antibodies and complement. Yet, despite this marked sensitivity, mycoplasma infections are usually chronic in nature, indicating the frequent failure of the host defense mechanisms to eradicate the parasites. It appears that many pathogenic mycoplasmas possess rather sophisticated mechanisms for rapid adaptation to changing microenvironments. Reversible switching of the expression and modification of major membrane protein antigens on the mycoplasma surface provide an effective escape from rapid destruction by the host immune system (Wise et al., 1992).

Table II presents mycoplasmal antigenic variation systems studied thus far. The data enable some generalizations to be made: (1) Most of the systems consist of lipoproteins exposed on the external cell surface and anchored to the membrane via the acyl chains. (2) The lipoproteins, functioning as dominant cell antigens, contain repetitive sequences that can spontaneously vary in size through changes in the number of these motifs. Moreover, the lipoproteins vary not only in size, but may be subjected to "on" and "off" expression states. (3) Cloning and sequencing of the lipoprotein genes revealed that the encoded proteins consist of three regions: the N-terminal leader region, an intermediate domain, and the C-terminal domain comprising reiterated sequences arranged in tandem. Deletion of the leader sequence on processing of the protein leaves a cysteine residue at the N terminus free for acylation, preferentially by palmitic and myristic acids. The resulting lipid-modified protein is anchored to the outer surface of the membrane through its lipid moiety. (4) In most cases, each mycoplasma strain carries a family of lipoproteins, encoded by multiple variant genes in several versions. The genes are expressed in various combinations and in varying lengths on the surface of the organism.

Lipoproteins may not be the only variable mycoplasma membrane proteins exposed on the cell surface. Thus, the *M. gallisepticum* PvpA is a non-lipid-modified, integral phase-variable membrane protein (Yogev *et al.*, 1994). Nevertheless, this mycoplasma carries a highly immunogenic lipoprotein, pMGA, undergoing high-frequency phase variation spontaneously and independently (Markham *et al.*, 1993, 1994). Solid experimental evidence exists for the occurrence of antigenic variation *in vivo*. Thus, multiple isolates of *M. hominis* obtained from the same joint of a septic arthritis patient during a 6-year period exhibited antigenic variation, although the strains were indistinguishable by restriction endonuclease analysis (Olson *et al.*, 1991). Experimental infections of mice with *M. pulmonis* (Talkington *et al.*, 1989) and rats with *M. arthritidis* (Droesse *et al.*, 1994) indicate that high-frequency antigenic phase variation occurs *in vivo*.

Does antigenic variability affect strain identification and classification? The answer appears to be negative as the variations are limited to a rather restricted number of protein antigens, encoded by a few genes. Therefore, the overall profiles of cell proteins and restriction patterns of genomic DNA do not reveal significant changes among the variants (Olson *et al.*, 1991; Rosengarten *et al.*, 1994), enabling the use of these molecular tools in strain identification.

Metabolism

Based on their ability to metabolize carbohydrates, the mollicutes are divided into fermentative and nonfermentative organisms. Members of the fermentative group produce acids from carbohydrates, decreasing the pH of the growth medium. Most of the nonfermentative mollicutes, and some fermentative species, possess the arginine dihydrolase pathway. Arginine hydrolysis by this pathway results in the production of ornithine, ATP, CO_2, and ammonia, raising the pH of the culture medium (Razin, 1978). Some mycoplasmas, such as *M. agalactiae, M. bovigenitalium,* and *M. bovis* (see Table I in Appendix, Vol. II), do not metabolize sugars or arginine, but are capable of oxidation of organic acids (lactate, pyruvate) to acetate and CO_2 (Miles *et al.*, 1994).

The mollicutes are generally facultative anaerobes, although the growth of some, such as *M. mycoides, M. pneumoniae,* and *M. hyorhinis,* is improved by aeration. The *Anaeroplasma* and *Asteroleplasma* species are obligate anaerobes and are very sensitive to oxygen. All the mollicutes examined so far have truncated respiratory systems. They lack a complete tricarboxylic acid cycle and have no quinones and cytochromes, ruling out oxidative phosphorylation as an ATP-generating system (for references see Razin, 1978, 1991; Miles, 1992; Pollack et al., 1995).

The urea-splitting ureaplasmas present a special case. Neither glycolysis nor arginine dihydrolase-ATP or acetate kinase-ATP-generating pathways could be detected in these organisms. The unique dependence of ureaplasmas on urea for growth and the presence of a potent urease in these organisms have led to the hypothesis that intracellular urea hydrolysis may be coupled to ATP synthesis through a chemiosmotic type of mechanism (Razin, 1978). Experimental support for the generation of a transmembrane potential, with resultant ATP synthesis through a ureaplasmal F_oF_1-type ATPase, has been provided by Romano et al., (1980) and extended and confirmed by Smith et al. (1993).

The extensive studies on the metabolism of mollicutes have been reviewed by Pollack et al. (1995) summarizing the information in the form of a large metabolic map, linking about 135 enzymatic activities detected in the cytoplasmic fraction of mollicutes. Based on the assumption of one gene per enzyme, the enzymatic activities defined so far were estimated to represent about 70% of the cytoplasmic proteins, bringing us closer to the goal of molecular definition of the components of the minimum self-replicating organism (Morowitz, 1984). The many phenotypic characteristics represented by the various metabolic activities of mollicutes are valuable markers in constructing the taxonomy of this group and should also be taken into account when considering the phylogeny of mollicutes (Pollack et al., 1995).

Taxonomy and Phylogeny

There is a consensus among bacterial taxonomists that the complete sequences of bacterial genomes will form the basis for phylogeny and, ultimately, taxonomy. Yet, as long as complete genomic sequences are not available, current bacterial taxonomy relies on the combination of phenotypic characteristics and on phylogenetic data based on partial genomic sequences, mostly those of the conserved ribosomal RNA genes.

According to 16S rRNA sequences (Weisburg et al., 1989; Maniloff, 1992), the mollicutes are divided into five phylogenetic units (clades). Acholeplasmas and anaeroplasmas are considered as the earliest mollicutes to have evolved by degenerative (regressive) evolution from their gram-positive bacterial ancestors. The spiroplasmas evolved by an early splitting of the acholeplasmal branch, and the mycoplasmas and ureaplasmas are thought to have spiroplasmal ancestors.

An important feature of mollicute phylogeny proposed by Woese (1987) is the rapid pace of their evolution, explaining the marked genotypic and phenotypic variability characterizing the mollicutes as a group.

A most important contribution of 16S rRNA gene sequences to mollicute phylogeny and taxonomy has been the placing of the uncultured phytoplasmas, as a distinct monophyletic clade, within Mollicutes, closely related to the acholeplasmas. The 16S rRNA gene sequences delineated within the phytoplasma clade, 11 or 12 distinct subclades, generally in accord with the phytoplasma strain clusters established by DNA homology, restriction enzyme analyses of the 16S rRNA genes, and the sequences of the 16S/23S intergenic spacer regions (Gunderson et al., 1994; Seemüller et al., 1994; Kirkpatrick et al., 1994; see Chapter E6, this volume). The taxonomic implications of these studies are that the phytoplasmas should be distinguished at the minimal taxonomic level of a genus and that each phytoplasma subclade should represent at least a distinct species. Nevertheless, in the absence of the phenotypic markers used to classify mollicutes (Chapter E2, this volume), taxonomic affiliations cannot be resolved at the present time in the conventional way. Thus, a provisional classification of the uncultured phytoplasmas may be introduced by applying the "candidatus" category proposed for classification of uncultured bacteria (International Committee, 1994). The phytoplasma case demonstrates well the revolution occurring in bacterial taxonomy, in which molecular data suffice for laying the phylogenetic and taxonomic basis for classification of a group of organisms for which very few phenotypic characteristics are available.

The present tendency to depend on direct genomic analysis, made available by the dramatic advancements of molecular genetic methodology, has put aside further development of classical taxonomic tools based on determination of nutritional requirements and enzymatic activities (Razin, 1992b). Even electrophoretic cell protein profiles are now used less extensively as taxonomic tools than in the 1970s and in the early 1980s, the reason being that these profiles reflect the expression of specific genes and are thus liable to changes. It is now possible to clone and sequence the genes themselves and thus have a more direct comparative measure.

The weight to be given to cholesterol requirement in mollicute taxonomy illustrates well the problem of using phenotypic characteristics as taxonomic tools. The cholesterol requirement has been considered for a long time as a major criterion in establishing high taxonomic groupings within the Mollicutes (see Chapter E2, this volume). Recent findings appear to weaken the high status of cholesterol requirement in mollicute classification. Thus, the non-cholesterol-requiring mesoplasmas, considered previously as acholeplasmas, were shown to possess molecular properties very different from those of the classic acholeplasmas, requiring their taxonomic separation (Tully et al., 1993). In addition, several *Spiroplasma* species have been shown to grow in the absence of cholesterol, raising doubts about the validity of the use of the cholesterol requirement as a sole definitive criterion in the determination of higher taxa of the class Mollicutes (Bove et al., 1994).

Mechanisms of Pathogenicity

Most mycoplasmas adhere tenaciously to the epithelial linings of the respiratory or urogenital tract, rarely invading tissues. Hence, they may be considered as surface parasites. However, with the increasing incidence of immunocompromised patients (due to AIDS, organ transplantation, etc.) evidence is accumulating for invasion of tissues and the intracellular location of some mycoplasmas, notably *M. fermentans* and *M. penetrans* (see Chapters A6 and A7, this volume). Extragenital infections by urogenital mollicutes are rather common in neonates and immunosuppressed and/or hypogammaglobulinemic patients (Meyer and Clough, 1993).

The molecular basis of mycoplasma pathogenicity remains largely elusive. Potent toxins have not been associated with mycoplasmas. The mildly toxic by-products of mycoplasma metabolism, such as hydrogen peroxide and superoxide radicals, have been incriminated as causing oxidative damage to host cell membranes (Almagor et al., 1986; Razin, 1991). It has been proposed that the intimate contact of the wall-less mycoplasmas with the host cell membrane may result in local, perhaps transient, fusion of the two membranes or in exchange of membrane components and direct "injection" of the mycoplasma cytoplasmic content, including hydrolytic enzymes, into the host cell cytoplasm. Thus, the potent nucleases of mollicutes (see Chapter D5, this volume) combined with superoxide radicals may be responsible for clastogenic effects (Stewart et al., 1994), and phospholipase A_2 of ureaplasmas was suggested to disturb prostaglandin production by substrate inhibition (Kim et al., 1994), possibly contributing to the pathogenesis of ureaplasmas in pregnancy. The potential for mycoplasma fusion with host cells is supported by a variety of experimental systems (see Chapter C4, this volume) and by the demonstration of a marked fusiogenic potential of a peculiar membrane lipid of *M. fermentans* strain incognitus (Salman et al., 1994), in line with the intracellular location of this mycoplasma in tissues of AIDS patients (Chapter A7, this volume).

Adhesion to Host Cells

The adhesion of mollicutes to host cells is a prerequisite for colonization by the parasite and for infection. The loss of adhesion capacity by mutation results in loss of infectivity, and reversion to the cytadhering phenotype is accompanied by regaining infectivity and virulence. The best defined mycoplasmal adhesins—membrane components responsible for adhesion—are those of *M. pneumoniae* and *M. genitalium*. Detailed information on the genes and the molecular properties of the encoded adhesin proteins of these mycoplasmas (P1, MgPa, and P30) can be found in Razin and Jacobs (1992) and in Baseman (1993). The topography of the adhesin molecules embedded in the mycoplasma membrane, as predicted by computer analysis of amino acid sequences, has been supplemented by epitope mapping, using monoclonal antibodies acting on adhesin molecules *in situ* or on synthetic oligopeptides corresponding to defined segments of the adhesin molecules (see Chapter F3, this volume). In this way, the presumed tridimensional conformation of the adhesins in the membrane, and the regions in the adhesin molecules acting directly in adhesion, could be proposed (Razin and Jacobs, 1992; Opitz and Jacobs, 1992).

Although the just-mentioned adhesins fulfill the major role in cytadhesion, the process appears to be multifactorial, involving in addition to the major adhesins, a number of "accessory" membrane proteins. These accessory proteins act in concert with cytoskeletal elements to facilitate the lateral movement and concentration of the adhesin molecules at the attachment tip organelle, an organelle characterizing these and several other cytadhering mollicutes. Much of the effort, forming part of the *M. pneumoniae* genome project, has been directed to the molecular definition of the attachment organelle components forming part of the mollicute cytoskeleton (Triton shell). Some of the genes and the encoded proteins of the *M. pneumoniae* Triton shell (30-kDa; 40-kDa; 90-kDa; p65; p200; HMW1; HMW3) have been identified and characterized. Interestingly, they are proline rich and exhibit repeat sequences and other motifs characteristic of eukaryotic cytoskeletal proteins. Efforts are being made to localize these proteins in the membrane, looking for the interactions of the proteins with each other to form the cytoskeletal network (Layh-Schmitt and Herrmann, 1994; Proft *et al.*, 1994; see Chapter A8, this volume).

As was stated earlier, most attention has been given to the adhesins and cytoskeletal elements of *M. pneumoniae* and *M. genitalium*. The isolation of *M. pirum* from the blood of AIDS patients has attracted attention to this mycoplasma, also characterized by a tip organelle. A gene encoding a protein, showing about 26% amino acid homology with the *M. pneumoniae* P1 adhesin, exhibiting also a proline-rich region, has been cloned from *M. pirum* (Tham *et al.*, 1994). The question of whether or not this protein acts as an adhesin is still open. Lipid-modified membrane proteins, probably acting as adhesins, have been characterized in another human mycoplasma: *M. hominis* (Henrich *et al.*, 1993, 1994). The notion that the same lipid-modified proteins responsible for the antigenic variation phenomenon act also as adhesins appears plausible, preliminary data in support of this direction are available (Sachse, 1994; Zhang *et al.*, 1994a).

The receptors on host cell membranes responsible for mycoplasma attachment, identified so far, are mostly sialoglycoconjugates and sulfated glycolipids (reviewed in Razin and Jacobs, 1992; see also Zhang et al., 1994b).

Activation of Host Immune System

Host immune reactions appear to play a major role in pathogenesis of mycoplasma infections. A variety of mollicutes have been shown to induce mitogenic stimulation of human and animal lymphocytes, acting mostly as polyclonal B-cell activators (Chapter F7, this volume). Mycoplasmal membrane components, such as lipoproteins of *M. fermentans* and *M. penetrans* (Feng and Lo, 1994), appear to act as B-cell mitogens, stimulating the proliferation of B lymphocytes. Numerous studies indicate that the interaction of mycoplasmas with macrophages and monocytes induces the expression and secretion of major proinflammatory cytokines, mostly tumor necrosis factor α (TNF-α), interleukin 1β (IL-1β), IL-6, and interferon-γ (IFN-γ) (Kostyal et al., 1994; Pietsch et al., 1994; Herbelin et al., 1994; Brenner et al., 1994; see also Chapter F5, this volume).

Cytokine induction by mycoplasma cells, membranes, and isolated membrane components, such as lipoproteins, has been demonstrated *in vitro*, using monocyte and lymphocyte cell culture systems, and *in vivo* (Nishimoto et al., 1994). Of great interest is the demonstration by Pietsch et al. (1994) of increased expression of TNF-α, IFN-γ, IL-1β, and IL-6 in mice experimentally infected with *M. pneumoniae*. Expression was greater in the lung than in the spleen, attesting to the rapid accumulation of lymphocytes in the infected site. Expression during reinfection of the mice of TNF-α and IL-6 was 10-fold, and of IFN-γ 50-fold, higher than during the primary challenge. These results suggest that pathogenesis of *M. pneumoniae* may be associated with the elevated expression of proinflammatory cytokines, in line with the hypothesis that host immune reactions are responsible to a large extent for the respiratory symptoms in *M. pneumoniae* pneumonia. Symptoms in secondary infections are accordingly more intense (Pietsch and Jacobs, 1993).

Mycoplasma membrane lipoproteins and certain lipids have been incriminated as factors inducing cytokine secretion (Herbelin et al., 1994; Salman et al., 1994). An amphipathic molecule with a lipid moiety, carrying fatty acids in ester linkage and a polyol moiety of unknown character, was isolated from *M. fermentans* membranes and named MDHM. This component is a potent murine and human macrophage and monocyte activator, inducing the release of TNF-α, IL-1, IL-6, arachidonate metabolites, and nitric oxide. Purified MDHM was at least as potent as bacterial lipopolysaccharide (Mühlradt and Frisch, 1994; Quentmeier et al., 1994; Chapter F5, this volume).

Perhaps the most potent mycoplasmal activator of the immune system is the

M. arthritidis cytoplasmic component named MAM or MAS. This soluble mitogen activates T cells by a unique pathway, characteristic of the recently described superantigen class. The gene for MAM has been cloned and sequenced, encoding for a protein of a molecular weight of 25,193, different from all other bacterial superantigens defined thus far (Cole and Atkin, 1991; see Chapter F6). Superantigens stimulate T cell V_β specifically by crosslinking the major histocompatibility complex class II (MHC II) with the V_β chain on the T cell receptor (TCR). MAM is specific for the 1-E α chain in the MHC II of mice. Human HLA-DR was postulated as being the counterpart of the murine 1-E α. Rink *et al.* (1994) showed that MAM is specific for distinct HLA-DR phenotypes and that these phenotypes stimulated different V_β-subtype T cells. Thus, MAM serves as a TCR V_β-selective T-cell mitogen, inducing massive T cell proliferation and proinflammatory cytokine secretion, leading to clinical syndromes characterized by shock and immunosuppression (Atkin *et al.*, 1994; Friedman *et al.*, 1994).

The great interest in MAM stems from the possibility that, being a superantigen, it may provide a link between antecedent infection and the subsequent emergence of autoimmunity. Postinfection sequelae, affecting a variety of organs, have been described in mycoplasma infections, particularly those caused by *M. pneumoniae* (Jacobs, 1991). It has been hypothesized that superantigens may play a role in autoimmune diseases by activating preexisting anti-self T cell clones or by polyclonal B cell activation, leading to the production of autoantibodies. The finding that MAM can trigger autoimmune disease in mice (Cole and Griffiths, 1993) and the presence of MAM-reactive T cells in the joints of rheumatoid arthritis patients (Friedman *et al.*, 1994) suggests that a search for MAM-like molecules in human autoimmune diseases should be pursued.

References

Almagor, M., Kahane, I., Gilon, C., and Yatziv, S. (1986). Protective effects of the glutathione redox cycle and vitamin E on cultured fibroblasts infected by *Mycoplasma pneumoniae*. *Infect. Immun.* **52**, 240–244.

Atkin C. L., Oliphant, A., Wei, S., Pole, A., Manohar, M., Sawitzke, A., Knudtson, K., and Cole, B. C. (1994). The *Mycoplasma arthritidis* superantigen MAM: Purification, sequencing, and characterization. *IOM Lett.* **3**, 686–687.

Bailey, C. C., and Bott, K. F. (1994). An unusual gene containing a *dna*J N-terminal box flanks the putative origin of replication of *Mycoplasma genitalium*. *J. Bacteriol.* **176**, 5814–5819.

Barnes, M., Tarantino, P. M., Jr., Spacciapoli, P., Brown, N. C., Yu, H., and Dybvig, K. (1994). DNA polymerase III of *Mycoplasma pulmonis:* Isolation and characterization of the enzyme and its structural gene, *pol* C. *Mol. Microbiol.* **13**, 843–854.

Barroso, G., and Labarère, J. (1988). Chromosomal gene transfer in *Spiroplasma citri*. *Science* **241**, 959–961.

Baseman, J. B. (1993). The cytadhesins of *Mycoplasma pneumoniae* and *M. genitalium*. *In* "Mycoplasma Cell Membranes" (S. Rottem and I. Kahane, eds.), Subcellular Biochemistry Vol. **20**, pp. 243–259. Plenum, New York.

Beherens, A., Heller, M., Kirchhoff, H., Yogev, D., and Rosengarten, R. (1994). A family of size-variant membrane surface lipoprotein antigens (Vsps) of *Mycoplasma bovis*. *Infect. Immun.* **62**, 5075–5084.

Bove, J. M. (1993). Molecular features of mollicutes. *Clin. Infect. Dis.* **17**(Suppl. 1), S10–S31.

Bove, J. M., Carle, P., Tully, J. G., Whitcomb, R. F., and Laigret, F. (1994). Sterol nonrequiring spiroplasmas: Do they deserve a new genus? *IOM Lett.* **3**, 449–450.

Brenner, T., Yamin, A., and Gallily, R. (1994). Mycoplasma triggering of nitric oxide production by central nervous system glial cells and its inhibition by glucocorticoids. *Brain Res.* **641**, 51–56.

Cao, J., Kapke, P. A., and Minion, F. C. (1994). Transformation of *Mycoplasma gallisepticum* with Tn916, Tn4001, and integrative plasmid vectors. *J. Bacteriol.* **176**, 4459–4462.

Carle, P., Laigret, F., Tully, J. G., and Bove, J. M. (1995). Heterogeneity of genome sizes within the genus *Spiroplasma*. *Int. J. Syst. Bacteriol.* **45**, 178–181.

Chen, X., and Finch, L. R. (1989). Novel arrangement of rRNA genes in *Mycoplasma gallisepticum*: Separation of the 16S gene of one set from the 23S and 5S genes. *J. Bacteriol.* **171**, 2876–2878.

Christiansen, G., Mathiesen, S. L., Nyvold, C., and Birkelund, S. (1994). Analysis of a *Mycoplasma hominis* membrane protein, P120. *FEMS Microbiol. Lett.* **121**, 121–128.

Cole, B. C., and Atkin, C. L. (1991). The *Mycoplasma arthritidis* T-cell mitogen, MAM: A model superantigen. *Immunol. Today* **12**, 271–276.

Cole, B. C., and Griffiths, M. M. (1993). Triggering and exacerbation of autoimmune arthritis by the *Mycoplasma arthritidis* superantigen MAM. *Arthritis Rheum.* **36**, 994–1002.

Dahl, C. E., Dahl, J.S., and Bloch, K. (1983). Proteolipid formation in *Mycoplasma capricolum*: Influence of cholesterol on unsaturated fatty acid acylation of membrane proteins. *J. Biol. Chem.* **258**, 11814–11818.

Deng, G., and McIntosh, M. A. (1994). An amplifiable DNA region from *Mycoplasma hyorhinis* genome. *J. Bacteriol.* **176**, 5929–5937.

Droesse, M., Tangen, G., Gummelt, I., Schmidt, R., Runge, M., and Kirchhoff, H. (1994). *Mycoplasma arthritidis* surface antigen variation *in vivo*. *IOM Lett.* **3**, 549–550.

Dybvig, K. (1990). Mycoplasmal genetics. *Annu. Rev. Microbiol.* **44**, 81–104.

Feng, S.-H., and Lo, S.-C. (1994). Induced mouse spleen B-cell proliferation and secretion of immunoglobulin by lipid-associated membrane proteins of *Mycoplasma fermentans* incognitus and *Mycoplasma penetrans*. *Infect. Immun.* **62**, 3916–3921.

Friedman, S. M., Li, Y., Zagon, G., Tumang, J. R., Sun, G.-R., and Crow, M. K. (1994). A potential role for microbiol superantigens in autoimmune disease. *IOM Lett.* **3**, 688–689.

Gundersen, D. E., Lee, I.-M., Rehner, S. A., Davis, R. E., and Kingsbury, D. T. (1994). Phylogeny of mycoplasmalike organisms (phytoplasmas): A basis for their classification. *J. Bacteriol.* **176**, 5244–5254.

Henrich, B., Berns, G., Kamla, V., and Hadding, U. (1994). Characterization of a 64 kDa membrane protein of *Mycoplasma hominis*. *IOM Lett.* **3**, 671–672.

Henrich, B., Feldman, R.-C., and Hadding, U. (1993). Cytadhesins of *Mycoplasma hominis*. *Infect. Immun.* **61**, 2945–2951.

Herbelin, A., Ruuth, E., Delorme, D., Michel-Herbelin, C., and Praz, F. (1994). *Mycoplasma arginini* TUH-14 membrane lipoproteins induce production of interleukin-1, interleukin-6, and tumor necrosis factor alpha by human monocytes. *Infect. Immun.* **62**, 4690–4694.

Hilbert, H., Himmelreich, R., Leibfried, U., Pirkl, E., Plagens, H., Proft, T., and Herrmann, R. (1994). DNA sequence analysis of the complete *Mycoplasma pneumoniae* genome and identification of genes and gene products. *IOM Lett.* **3**, 338.

International Committee on Systematic Bacteriology: Subcommittee on the Taxonomy of Mollicutes (1994). Minutes of the Interim Meeting, Bordeaux. *Int. J. Syst. Bacteriol.* **45,** 415–417.

Jacobs, E. (1991). *Mycoplasma pneumoniae* virulence factors and the immune response. *Rev. Med. Microbiol.* **2,** 83–90.

Jarhede, T., Boyer, M., Gustafsson, M., Sjöström, M., Tegman, V., and Wieslander, Å. (1994a). Signal peptides from mycoplasmas differ from those of other bacteria. *IOM Lett.* **3,** 129–130.

Jarhede, T., Le Hénaff, M., and Wieslander, Å. (1994b). Expression of spiralin and alpha-amylase genes in *Acholeplasma laidlawii*. *IOM Lett.* **3,** 581–582.

Jensen, J. S., Hansen, H. T., and Lind, K. (1994). Isolation of *Mycoplasma genitalium* strains from the male urethra. *IOM Lett.* **3,** 143–144.

Kahane, I., and Adoni, A., eds. (1993). "Rapid Diagnosis of Mycoplasmas." Plenum, New York.

Kim, J. J. Quinn, P. A., and Fortier, M. A. (1994). *Ureaplasma diversum* infection in vitro alters prostaglandin E_2 and prostaglandin F_{2a} production by bovine endometrial cells without affecting cell viability. *Infect. Immun.* **62,** 1528–1533.

King, K. W., Woodard, A., and Dybvig, K. (1994). Cloning and characterization of the *rec*A genes from *Mycoplasma pulmonis* and *M. mycoides* subspecies *mycoides*. *Gene* **139,** 111–115.

Kirkpatrick, B., Smart, C., Gardner, S., Gao, J.-L., Ahrens, U., Maurer, R., Schneider, B., Lorenz, K.-H., Seemüller, E., Harrison, N., Namba, S., and Daire, X. (1994). Phylogenetic relationships of plant pathogenic MLOs established by 16S/23S rDNA spacer sequences. *IOM Lett.* **3,** 228–229.

Kostyal, D. A., Butler, G. H., and Beezhold, D. H. (1994). A 48-kilodalton *Mycoplasma fermentans* membrane protein induces cytokine secretion by human monocytes. *Infect. Immun.* **62,** 3793–3800.

Ladefoged, S. A., and Christiansen, G. (1994). Sequencing analysis reveals a unique gene organization in the *gyr*B region of *Mycoplasma hominis*. *J. Bacteriol.* **176,** 5835–5842.

Laigret, F., and Bove, J. M. (1994). Rifampicin insensitivity of mollicutes: Sequence analysis of RNA polymerase β-subunit gene (*rpo*B) of *Spiroplasma citri*. *IOM Lett.* **3,** 602–603.

Layh-Schmitt, G., and Herrmann, R. (1994). Spatial arrangement of gene products of the P1 operon in the membrane of *Mycoplasma pneumoniae*. *Infect. Immun.* **62,** 974–979.

Lo, S.-C., Hayes, M. M., Tully, J. G., Wang, R. Y.-H., Kotani, H., Pierce, P. F., Rose, D. L., and Shih, J. W.-K. (1992) *Mycoplasma penetrans* sp. nov., from the urogenital tract of patients with AIDS. *Int. J. Syst. Bacteriol.* **42,** 357–364.

Loechel, S., Inamine, J. M., and Hu, P.C. (1991). A novel translation initiation region from *Mycoplasma genitalium* that functions in *Escherichia coli*. *Nucleic Acids Res.* **19,** 6905–6911.

Maniloff, J. (1992). Phylogeny of mycoplasmas. In "Mycoplasmas, Molecular Biology and Pathogenesis" (J. Maniloff, R. N. McElhaney, L. R. Finch, and J. B. Baseman, eds.), pp. 549–559. Am. Soc. Microbiol., Washington, DC.

Maniloff, J., Kampo, G. J., and Dascher, C. C. (1994). Sequence analysis of a temperate phage: Mycoplasma virus L2. *Gene* **141,** 1–8.

Maniloff, J., McElhaney, R. N., Finch, L. R., and Baseman, J. B., eds. (1992). "Mycoplasmas, Molecular Biology and Pathogenesis." Am. Soc. Microbiol., Washington, DC.

Marais, A., Bove, J. M., Dallo, S. F., Baseman, J. B., and Renaudin, J. (1993). Expression in *Spiroplasma citri* of an epitope carried on the G fragment of the cytadhesin P1 gene from *Mycoplasma pneumoniae*. *J. Bacteriol.* **175,** 2783–2787.

Markham, P. F., Glew, M. D., Whithear, K. G., and Walker, I. D. (1993). Molecular cloning of a member of the gene family that encodes pMGA, a hemagglutinin of *Mycoplasma gallisepticum*. *Infect. Immun.* **61,** 903–909.

Markham, P. F., Glew, M. D., Sykes, J. E., Bowden, T. R., Pollocks, T. D., Browning, G. F., Whithear, K. G., and Walker, I. D. (1994). The organisation of the multigene family which encodes the major cell surface protein, pMGA, of *Mycoplasma gallisepticum*. *FEBS Lett.* **352,** 347–352.

Maurel, D., Charron, A., and Bébéar, C. (1989). Mollicutes DNA polymerases: Characterization of a single enzyme from *Mycoplasma mycoides* and *Ureaplasma urealyticum* and of three enzymes from *Acholeplasma laidlawii. Res. Microbiol.* **140,** 191-205.
Meyer, R. D., and Clough, W. (1993). Extragenital *Mycoplasma hominis* infections in adults: Emphasis on immunosuppression. *Clin. Infect. Dis.* (Suppl. 1) **17,** S243-S249.
Miles, R. J. (1992). Catabolism in mollicutes. *J. Gen. Microbiol.* **138,** 1773-1783.
Miles, R. J., Taylor, R. R., Abu-Groun, E. A. M., and Alimohammadi, A. (1994). Diversity of energy-yielding metabolism in *Mycoplasma* spp. *IOM Lett.* **3,** 165-166.
Miyata, M., Sano, K.-I., Okada, R., and Fukumura, T. (1993). Mapping of replication initiation site in *Mycoplasma capricolum* genome by two-dimensional gel-electrophoretic analysis. *Nucleic Acids Res.* **21,** 4816-4823.
Morowitz, H. J. (1984). The completeness of molecular biology. *Isr. J. Med. Sci.* **20,** 750-753.
Mühlradt, P. F., and Frisch, M. (1994). Purification and partial biochemical characterization of a *Mycoplasma fermentans*-derived substance that activates macrophages to release nitric oxide, tumor necrosis factor, and interleukin-6. *Infect. Immun.* **62,** 3801-3807.
Nakagawa, T., Uemori, T., Asada, K., Kato, I., and Harasawa, R. (1992). *Acholeplasma laidlawii* has tRNA genes in the 16S-23S spacer of the rRNA operon. *J. Bacteriol.* **174,** 8163-8165.
Neimark, H. C., Leach, R., Mitchelmore, D., and Lange, C. (1994). Grey lung disease agent, an uncultured wall-less prokaryote, appears to be a new mycoplasma species. *IOM Lett.* **3,** 486-487.
Nishimoto, M., Akashi, A., Kuwano, K., Tseng, C.-C., Ohizumi, K., and Arai, S. (1994). Gene expression of tumor necrosis factor alpha and interferon gamma in the lungs of *Mycoplasma pulmonis*-infected mice. *Microbiol. Immunol.* **38,** 345-352.
Oba, T., Andachi, Y., Muto, A., and Osawa, S. (1991). CGG: An unassigned or nonsense codon in *Mycoplasma capricolum. Proc. Natl. Acad. Sci. USA* **88,** 921-925.
Olson, L. D., Renshaw, C. A., Shane, S. W., and Barile, M. F. (1991). Successive synovial *Mycoplasma hominis* isolates exhibit apparent antigenic variation. *Infect. Immun.* **59,** 3327-3329.
Opitz, O., and Jacobs, E. (1992). Adherence epitopes of *Mycoplasma genitalium* adhesin. *J. Gen. Microbiol.* **138,** 1785-1790.
Pietsch, K., and Jacobs, E. (1993). Characterization of the cellular response of spleen cells in BALB/c mice inoculated with *Mycoplasma pneumoniae* or the P1 protein. *Med. Microbiol. Immunol.* **182,** 77-85.
Pietsch, K., Ehlers, S., and Jacobs, E. (1994). Cytokine gene expression in the lungs of BALB/c mice during primary and secondary intranasal infection with *Mycoplasma pneumonia. Microbiology* **140,** 2043-2048.
Pollack, J. D., Williams, M. V., and McElhaney, R. N. (1995). The comparative metabolism of the mollicutes. Submitted for publication.
Proft, T., and Herrmann, R. (1994). Identification and characterization of hitherto unknown *Mycoplasma pneumoniae* proteins. *Mol. Microbiol.* **13,** 337-348.
Proft, T., Hilbert, H., and Herrmann, R. (1994). Identification and characterization of two *Mycoplasma pneumoniae* proteins located in the cytoskeleton-like Triton shell. *IOM Lett.* **3,** 676-677.
Quentmeier, H., Schumann-Kindel, G., Mühlradt, P. F., and Drexler, H. G. (1994). Induction of proto-oncogene and cytokine expression in human peripheral blood monocytes and monocytic cell line THP-1 after stimulation with mycoplasma-derived material MDHM. *Leukemia Res.* **18,** 319-325.
Razin, S. (1978). The mycoplasmas. *Microbiol. Rev.* **42,** 414-470.
Razin, S. (1985). Molecular biology and genetics of mycoplasmas *(Mollicutes). Microbiol. Rev.* **49,** 419-455.
Razin, S. (1991). The genera *Mycoplasma, Ureaplasma, Acholeplasma, Anaeroplasma,* and *Aster-*

oleplasma. In "The Prokaryotes" (A. Balows, H. G. Truper, M. Dworkin, W. Harder, and K.-H. Schleifer, eds.), 2nd Ed., Vol. 2, pp. 1937–1959. Springer-Verlag, New York.

Razin, S. (1992a). Peculiar properties of mycoplasmas: The smallest self-replicating prokaryotes. *FEMS Microbiol. Lett.* **100,** 423–432.

Razin, S. (1992b). Mycoplasma taxonomy and ecology. *In* "Mycoplasmas, Molecular Biology and Pathogenesis" (J. Maniloff, R. N. McElhaney, L. R. Finch, and J. B. Baseman, eds.), pp. 3–22. Am. Soc. Microbiol., Washington, DC.

Razin, S. (1993). Mycoplasma membranes as models in membrane research. *In* "Mycoplasma Cell Membranes" (S. Rottem and I. Kahane, eds.), Subcellular Biochemistry Vol. 20, pp. 1–28. Plenum, New York.

Razin, S., and Barile, M. F., eds. (1985). "The Mycoplasmas," vol. IV. Academic Press, Orlando, FL.

Razin, S., and Jacobs, E. (1992). Mycoplasma adhesion. *J. Gen. Microbiol.* **138,** 407–422.

Renaudin, J., Marais, A., Verdin, E., Duret, S., Laigret, F., and Bove, J. M. (1994). *Spiroplasma citri-oriC* plasmids: Expression of the *Spiroplasma phoeniceum* spiralin in *S. citri. IOM Lett.* **3,** 583–584.

Rink, L., Nicklas, W., Alvarez-Ossorio, L., Koester, M., and Kirchner, H. (1994). Differential induction of tumor necrosis factor alpha in murine and human leukocytes by *Mycoplasma arthritidis*-derived superantigen. *Infect. Immun.* **62,** 462–467.

Roberts, M. C., and Kenny, G. E. (1987). Conjugal transfer of transposon Tn916 from *Streptococcus faecalis* to *Mycoplasma hominis. J. Bacteriol.* **169,** 3836–3839.

Romano, N., Tolone, G., Ajello, F., and La Licata, R. (1980). Adenosine 5'-triphosphate synthesis induced by urea hydrolysis in *Ureaplasma urealyticum. J. Bacteriol.* **144,** 830–832.

Rosengarten, R., Behrens, A., Stetefeld, A., Heller, M., Ahrens, M., Sachse, K., Yogev, D., and Kirchhoff, H. (1994). Antigen heterogeneity among isolates of *Mycoplasma bovis* is generated by high-frequency variation of diverse membrane surface proteins. *Infect. Immun.* **62,** 5066–5074.

Rosengarten, R., Theiss, P. M., Yogev, D., and Wise, K. S. (1993). Antigenic variation in *Mycoplasma hyorhinis:* Increased repertoire of variable lipoproteins expanding surface diversity and structural complexity. *Infect. Immun.* **61,** 2224–2228.

Rottem, S., and Kahane, I., eds. (1993). "Mycoplasma Cell Membranes," Subcellular Biochemistry Vol. 20. Plenum, New York.

Sachse, K. (1994). Characteristics of *Mycoplasma bovis* cytadherence. *IOM Lett.* **3,** 431–432.

Salman, M. Deutsch, I., Tarshis, M., Naot, Y., and Rottem, S. (1994). Membrane lipids of *Mycoplasma fermentans. FEMS Microbiol. Lett.* **123,** 255–260.

Schneider, B., and Seemüller, E. (1994). Presence of two sets of ribosomal genes in phytopathogenic mollicutes. *Appl. Environ. Microbiol.* **60,** 3409–3412.

Seemüller, E., Schneider, B., Mäurer, R., Ahrens, U., Daire, X., Kison, H., Lorenz, K.-H., Hoffmann, A., Firrao, G., Avinet, L., and Stackebrandt, E. (1994). Phylogenetic classification of plant-pathogenic mycoplasmas by sequence analysis of 16S rDNA. *IOM Lett.* **3,** 224–225.

Simmons, W., Cao, Z., Glass, J., Watson, H. L., and Cassell, G. H. (1994). Comparison of the V-1 genes from two *Mycoplasma pulmonis* variants that differ with respect to disease potential. *IOM Lett.* **3,** 557–558.

Smith, D. G. E., Russell, W. C., Ingledew, W. J., and Thirkell, D. (1993). Hydrolysis of urea by *Ureaplasma urealyticum* generates a transmembrane potential with resultant ATP synthesis. *J. Bacteriol.* **175,** 3253–3258.

Stemke, G. W., Huang, Y., Laigret, F., and Bove, J. M. (1994). Cloning the ribosomal RNA operons of *Mycoplasma flocculare* and comparison with those of *Mycoplasma hyopneumoniae. Microbiology* **140,** 857–860.

Stewart, S. D., Watson, H. L., and Cassell, G. H. (1994). Investigation of the clastogenic potential of *Ureaplasma urealyticum* on human leukocytes. *IOM Lett.* **3,** 662–663.

Talkington, D. F., Fallon, M. T., Watson, H. L., Thorp, R. K., and Cassell, G. H. (1989).

Mycoplasma pulmonis V-1 surface protein variation: Occurrence *in vivo* and association with lung lesions. *Microbial Pathogen.* **7,** 429–436.
Tangen, G., Drösse, M., Struckmann-Möhrle, A. M., Körbis, M., and Kirchhoff, H. (1994). Investigation of the structure and function of variable surface antigens of *Mycoplasma arthritidis*. *IOM Lett.* **3,** 577–578.
Teng, L.-J., Zheng, X., Glass, J. I., Watson, H. L., Tsai, J., and Cassell, G. H. (1994). *Ureaplasma urealyticum* biovar specificity and diversity are encoded in multiple-banded antigen gene. *J. Clin. Microbiol.* **32,** 1464–1469.
Tham, T. N., Ferris, S., Bahraoui, E., Canarelli, S., Montagnier, L., and Blanchard, A. (1994). Molecular characterization of the P1-like adhesin gene from *Mycoplasma pirum*. *J. Bacteriol.* **176,** 781–788.
Theiss, P., and Wise, K. (1994). P29 and P78: Structural and genetic analysis of two phase variant surface lipoproteins of *Mycoplasma fermentans*. *IOM Lett.* **3,** 561–562.
Thirkell, D., Myles, A. D., and Russell, W. C. (1991). Palmitoylated proteins in *Ureaplasma urealyticum*. *Infect. Immun.* **59,** 781–784.
Thorp, L., Ladefoged, S., Birkelund, S., and Christiansen, G. (1994). Analysis of antibody-induced mutants of *Mycoplasma hominis* PG21. *IOM Lett.* **3,** 563–564.
Toth, K. F., Harrison, N., and Sears, B. B. (1994). Phylogenetic relationships among members of the class *Mollicutes* deduced from *rps*3 gene sequences. *Int. J. Syst. Bacteriol.* **44,** 119–124.
Tully, J. G., Bové, J. G., Laigret, F., and Whitcomb, R. F. (1993). Revised taxonomy of the class *Mollicutes*: Proposed elevation of a monophyletic cluster of arthropod-associated mollicutes to ordinal rank (*Entomoplasmatales* ord. nov.), with provision for familial rank to separate species with nonhelical morphology (*Entomoplasmataceae* fam. nov.) from helical species (*Spiroplasmataceae*), and amended descriptions of the order *Mycoplasmatales*, family *Mycoplasmataceae*. *Int. J. Syst. Bacteriol.* **43,** 378–385.
Tully, J. G., and Whitcomb, R. F. (1991). The genus *Spiroplasma*. *In* "The Prokaryotes" (A. Balows, H. G. Truper, M. Dworkin, W. Harder, and K.-H. Schleifer, eds.), 2nd Ed., Vol. 2, pp. 1960–1980. Springer-Verlag, New York.
Weisburg, W. G., Tully, J. G., Rose, D. L., Petzel, J. P., Oyaizu, H., Yang, D., Mandelco, L., Sechrest, J., Lawrence, T. G., Van Etten, J., Maniloff, J., and Woese, C. R. (1989). A phylogenetic analysis of the mycoplasmas: Basis for their classification. *J. Bacteriol.* **171,** 6455–6467.
Whitcomb, R. F., and Tully, J. G., eds. (1989). "The Mycoplasmas," vol. V. Academic Press, San Diego.
Wise, K. S., and Kim, M. F. (1987). Major membrane surface proteins of *Mycoplasma hyopneumoniae* selectively modified by covalently bound lipid. *J. Bacteriol.* **169,** 5546–5555.
Wise, K. S., Yogev, D., and Rosengarten, R. (1992). Antigenic variation. *In* "Mycoplasmas, Molecular Biology and Pathogenesis" (J. Maniloff, R. N. McElhaney, L. R. Finch, and J. B. Baseman, eds.), pp. 473–489. Am. Soc. Microbiol., Washington, DC.
Woese, C. R. (1987). Bacterial evolution. *Microbiol. Rev.* **51,** 221–271.
Ye, F., Laigret, F., and Bové, J. M. (1994a) A physical and genomic map of the prokaryote *Spiroplasma melliferum* and its comparison with the *Spiroplasma citri* map. *C. R. Acad. Sci. Paris* **317,** 392–398.
Ye, F., Renaudin, J., Bové, J.-M., and Laigret, F. (1994b). Cloning and sequencing of the replication origin (*ori*C) of the *Spiroplasma citri* chromosome and construction of autonomously replicating artificial plasmids. *Curr. Microbiol.* **29,** 23–29.
Yogev, D., Rosengarten, R., Watson-McKown, R., and Wise, K. S. (1991). Molecular basis of *Mycoplasma* surface antigenic variation: A novel set of divergent genes undergo spontaneous mutation of periodic coding regions and 5' regulatory sequences. *EMBO J.* **10,** 4069–4079.
Yogev, D., Menaker, D., Struzberg, K., Levisohn, S., Kirchhoff, H., Hinz, K.-H., and Rosen-

garten, R. (1994). A surface epitope undergoing high-frequency phase variation is shared by *Mycoplasma gallisepticum* and *Mycoplasma bovis*. *Infect. Immun.* **62,** 4962–4968.

Zhang, Q., Young, T. F., and Ross, R. F. (1994a). Identification and characterization of *Mycoplasma hyopneumoniae* adhesins. *IOM Lett.* **3,** 433–434.

Zhang, Q., Young, T. F., and Ross, R. F. (1994b). Glycolipid receptors for attachment of *Mycoplasma hyopneumoniae* to procine respiratory ciliated cells. *Infect. Immun.* **62,** 4367–4373.

SECTION A
Cultivation and Morphology

A1

INTRODUCTORY REMARKS
Joseph G. Tully

While impressive developments have occurred in the knowledge of the molecular biology and genetics of mollicutes, such fundamental information as the nutritional needs and the ability to grow these organisms in the laboratory still remains an important aspect in many subdisciplines within the field of mollicute research. Isolation and cultivation of previously unknown mollicutes since the early 1980s have certainly opened up new vistas in molecular and membrane biology (including cell adhesins), as well as new concepts regarding the ultrastructure of mollicutes and the invasiveness and pathogenicity of these organisms.

Isolation of mollicutes from clinical specimens is dependent on a number of critical conditions. Some of these relate to host-derived factors, such as the presence of specific antibody or antibiotics in the specimen, and to the occurrence of inhibitory components present in many vertebrate cells and tissues (cytokines, lysolecithins, etc.) and in plant or insect homogenates. Other crucial factors governing the successful isolation of mollicutes relate to the application of improved culture techniques in the laboratory and to adequate attention to good quality control procedures in evaluating culture media. A more extensive discussion of the possible role these factors play in primary isolation of mollicutes has been presented earlier (Taylor-Robinson and Chen, 1983; Tully, 1983a,b, 1985).

Most of the newly reported mollicutes discovered since the early 1980s have come from a variety of individual hosts known from past experiences to be colonized with these organisms (man, other animals, plants, and insects). From these observations it seems reasonable to draw at least two conclusions: that novel approaches in culture techniques will eventually lead to discovery of new mollicutes and that we still have a very incomplete knowledge of the mollicute flora of any one type of host. In addition, rapidly evolving technologies in detection and definition of the still uncultivable mollicutes by molecular techniques will certainly stimulate attempts to apply new culture medium formulations and procedures to the isolation and identification of such fastidious organisms.

Many of the current culture media formulations for mollicutes are based entirely (or with only modest modifications) on the medium described initially by Derrick Edward (1947). However, beginning in the late 1960s and early 1970s, this formulation was reported to be inadequate in meeting the growth requirements of the then newly described ureaplasmas and for a number of new and fastidious porcine mycoplasmas. Several major modifications in the Edward medium were successful in the cultivation of these organisms (see Freundt, 1983; Shepard, 1983). Also, in the early 1970s, helical mollicutes (spiroplasmas) were discovered in various plant diseases and in insect hosts. Some of the new spiroplasmas were successfully grown on the Edward formulation when sorbitol was added to raise the osmotic pressure of the medium (Bové and Saillard, 1979). However, more intensive efforts were required to develop new culture media to meet the needs of other more fastidious spiroplasmas from arthropod hosts. These efforts culminated in the development of the medium formulation designated SP-4. The reported application of this medium for primary isolation and maintenance of a variety of new and fastidious mollicutes since 1979 has reaffirmed its overall value in meeting the nutritional needs of many different mollicutes. Chapter A2 in this series gives an updated version of the SP-4 formulation, including some suggestive preparative techniques and several formula modifications for special circumstances.

Chapter A3 discusses some aspects of the cultivation of spiroplasmas in conventional undefined media formulations and efforts toward development of a chemically defined medium for these organisms. The initial attempts to cultivate several less fastidious spiroplasmas on a defined medium were successful and this work has given a certain insight into key nutritional requirements of these particular mollicutes. However, several spiroplasmas (from *Drosophila* and beetles) were extremely fastidious and resisted vigorous attempts to grow them in artificial media. Chapter A4 outlines and defines the successful application of insect cell cultures to cultivation of these and other fastidious spiroplasmas. Further efforts to adapt fastidious spiroplasmas to a chemically defined medium, to determine essential nutrients, and to explore other possible hosts with a variety of insect cell lines should significantly extend our knowledge of the ecology of the spiroplasmas.

Accurate quantitative evaluation of the growth of mollicutes has been one of the persistent technical problems in many areas of mollicute research. The shape and size of the mollicute cell, their slow growth rate, the complex medium required, and other factors have created problems in assessing growth by procedures utilized for conventional bacteria. A technique employing ATP-dependent luminometry for measuring the growth of mollicutes is described in Chapter A5. The procedure has been useful in establishing growth curves for very fastidious mollicutes, such as the ureaplasmas and several porcine mycoplasmas, and for evaluating sensitivity to antimicrobial agents.

The last three contributions in this section are directed to new information on the morphology of mollicutes, particularly to techniques that can establish location and interaction with host tissue or define surface components of mollicutes that often play an important role in their attachment to host tissues. Although the ability of many mollicutes to attach to a broad range of eukaryotic cell types is now well known, investigators have long argued over the question of whether mollicutes are capable of intracellular invasion of such cells. The methodology described in Chapter A6, which involves inoculation of eukaryotic cell monolayers with a test mollicute, appropriate preparation of tissue cells for ultrastructural examination, and specific immunocytochemical staining procedures on sectioned cells, provides an important technique in establishing the intracellular location of mollicutes.

The specific identification of mollicute species in diseased organs or tissues, outside of conventional culture techniques, has long been a difficult technical feat. The authors of Chapter A7 describe a series of approaches to the specific identification of mycoplasmas in such diseased tissues, using *Mycoplasma fermentans* as an example. The methodology, which involves immunohistochemical techniques with monoclonal antibodies and localization of organisms with mycoplasma-specific DNA in *in situ* hybridization, has yielded substantial evidence for the occurrence of this mollicute in postmortem tissues of patients with fatal infections. The techniques outlined here should have wide application to other natural and experimental infections with mollicutes and where information on tissue localization and organ involvement is critically needed. Finally, some unique surface protein complexes appear to play an important role in the cytadherence and the eventual pathogenicity of mollicutes. This association has stimulated the development of techniques that can localize and identify such subcellular structures on the mollicute surface. Chapter A8 outlines procedures for the application of both radioimmunoprecipitation and immunoelectron microscopy to the visualization of such substances on the mollicute cell.

References

Bové, J. M., and Saillard, C. (1979). Cell biology of spiroplasmas. *In* "The Mycoplasmas" (R. F. Whitcomb and J. G. Tully, eds.), Vol. III, pp. 83–153. Academic Press, New York.

Edward, D. G. ff. (1947). A selective medium for pleuropneumonia-like organisms. *J. Gen. Microbiol.* **1,** 238–243.

Freundt, E. A. (1983). Culture media for classic mycoplasmas. *In* "Methods in Mycoplasmology" (S. Razin and J. G. Tully, eds.), Vol. 1, pp. 127–135. Academic Press, New York.

Shepard, M. C. (1983). Culture media for ureaplasmas. *In* "Methods in Mycoplasmology" (S. Razin and J. G. Tully, eds.), Vol. 1, pp. 137–146. Academic Press, New York.

Taylor-Robinson, D., and Chen, T. A. (1983). Growth inhibitory factors in animal and plant tissues. *In* "Methods in Mycoplasmology" (S. Razin and J. G. Tully, eds.), Vol. 1, pp. 109–114. Academic Press, New York.

Tully, J. G. (1983a). General cultivation techniques for mycoplasmas and spiroplasmas. *In* "Methods in Mycoplasmology" (S. Razin and J. G. Tully, eds.), Vol. 1, pp. 99–102. Academic Press, New York.

Tully, J. G. (1983b). Sterility and quality control of mycoplasma culture media. *In* "Methods in Mycoplasmology" (S. Razin and J. G. Tully, eds.), Vol. 1, pp. 121–125. Academic Press, New York.

Tully, J. G. (1985). Newly discovered mollicutes. *In* "The Mycoplasmas" (S. Razin and M. F. Barile, eds.), Vol. IV, pp. 1–26. Academic Press, New York.

A2
CULTURE MEDIUM FORMULATION FOR PRIMARY ISOLATION AND MAINTENANCE OF MOLLICUTES
Joseph G. Tully

General Introduction

In 1976, a new tick-derived spiroplasma (SMCA organism, later designated *Spiroplasma mirum*) was identified in chick embryo-passaged material (Tully *et al.*, 1977). The inability to grow this particular spiroplasma in conventional culture medium for a variety of mollicutes and other spiroplasmas prompted R. F. Whitcomb to develop a number of new medium formulations. Most of the preparations fabricated were based on the concept of combining conventional mollicute culture medium components with supplements employed in cultivation of invertebrate tissue cell lines. From this collection, one formulation (SP-4) proved to be superior in cultivating not only the SMCA spiroplasma but a second, very fastidious, tick-derived spiroplasma, the Y32 organism (Tully *et al.*, 1977, 1981a). In a subsequent comparison of the SP-4 formulation and conventional Edward-type medium for isolation of mollicutes of human origin, SP-4 medium enhanced the primary isolation of *Mycoplasma pneumoniae* from about 400 throat specimens by 30–35% (Tully *et al.*, 1979). More recently, SP-4 medium has supported the primary isolation of several new mollicutes of human origin, such as *M. genitalium* (Tully *et al.*, 1981b) and *M. penetrans* (Lo *et al.*, 1991), and has provided enhanced isolation of other mollicutes of human origin, particularly *M. fermentans* (Dawson *et al.*, 1993).

The SP-4 formulation has also proven to be useful in the primary isolation and maintenance of most mollicutes of animal or plant/insect origin (Whitcomb,

1983) and for the isolation of mollicutes from infected eukaryotic tissue cell culture. In general, the medium supports the growth of most mycoplasmas, acholeplasmas, mesoplasmas, entomoplasmas, and spiroplasmas. However, SP-4 medium does not appear to provide the necessary nutrition for isolation or growth of the following *Mycoplasma* species (for recommended medium formulation for each, see Freundt, 1983): *Mycoplasma hyopneumoniae* and *M. flocculare* strains grow best on Friis medium whereas *M. synoviae* grows best on Frey's medium. Members of the genus *Anaeroplasma* grow best on media described by Robinson (1983). Although *Ureaplasma* species have been successfully maintained on SP-4 with a 1% ultrapure urea supplement, along with lowering of medium pH to 6.0 (D. L. Rose and J. G. Tully, unpublished data, 1985), it has not been established that this modified SP-4 is comparable to conventional ureaplasma media (Shepard, 1983) in primary isolation.

Components of SP-4 Medium

Mycoplasma broth base 11458 (Becton-Dickinson, Baltimore, MD)
Peptone, Bacto- (Difco Laboratories, Detroit, MI)
Tryptone, Bacto- (Difco)
Yeast extract 25% aqueous solution, heat sterilized (available from GIBCO, Grand Island, NY)
Tissue culture supplement, CMRL 1066 (10×), with glutamine and without sodium bicarbonate, sterile (GIBCO)
Yeastolate, Bacto- (Difco), 4% aqueous solution, heat sterilized
Serum, fetal bovine (Hyclone Laboratories, Logan, UT); heat inactivate in 100-ml volumes at 56° for 1 hour.
Agar, Noble (Difco)
Penicillin G, sodium (Lilly, Indianapolis, IN): for injection, 1,000,000 U/vial. For stock aqueous solution, add 10 ml heat-sterilized deionized water to vial. Final concentration, 100,000 U/ml.
Ampicillin (Polycillin N, Bristol Labs, Syracuse, NY); 500 mg/vial. Stock solution, add 2.5 ml heat-sterilized water to 500-mg vial. Final concentration, 200 mg/ml.
Polymyxin B sulfate (Calbiochem, San Diego, CA); 1,000,000 U/vial. Stock solution: add 10 ml heat-sterilized deionized water. Final concentration is 100,000 U/ml.
Glucose, 50% aqueous stock solution. Filter sterilize.
Arginine, 42% aqueous stock solution. Filter sterilize.
Phenol red, water soluble. (Chemalog, South Plainfield, NJ). Add 0.5 g to 500 ml of deionized water. Final concentration, 0.1%.

Comment: These recommended components gave the best results in our laboratory. However, homologous components obtained from different manufacturers may be adequate after appropriate quality control tests are performed.

Formulation of SP-4 Medium

Compounding the SP-4 is most conveniently divided into preparation of a heat-sterilized base medium component to which a series of sterile supplements are added.

Base Medium Ingredients (Amount for Final 400 ml Volume of Broth)

Mycoplasma broth base	1.4 g
Tryptone	4.0 g
Peptone	2.13 g
Deionized water	270 ml
Phenol red solution (0.1% aqueous solution)	8 ml

Heat to dissolve ingredients and adjust to pH 7.8 (about 3–4 ml 1 N NaOH. Sterilize at 121°C for 20 minutes.
For solid medium, add 0.8 g purified agar for each 100-ml volume of broth and omit phenol red indicator.

Sterile Supplements

CMRL 1066 tissue culture medium with glutamine (10×)	20 ml
Glucose (50% aqueous solution)	4 ml
Fresh yeast extract (15% aqueous solution)	14 ml
Yeastolate (4% aqueous solution)	20 ml
Fetal bovine serum (heat inactivate at 56°C for 1 hour)	68 ml
Penicillin (100,000 U/ml)	2 ml

Final pH of complete medium should be 7.6–7.8

Recommended Preparative Techniques

1. All supplements are added as sterile solutions, and filtration of supplements into final medium should be performed in a clean laminar flow hood with a

working surface recently washed down with 70% ethanol. It is advisable to pool all sterile supplements in a clean, sterile 125-ml serum bottle, using a sterile 100-ml graduated cyclinder and sterile pipettes.

2. The pooled supplements are then filter sterilized into the cooled base medium. If three or more 400-ml volumes of medium are being prepared at the same time, it is more useful to employ a 47-mm stainless-steel pressure filter (Model 4280, Gelman Sciences, Inc., Ann Arbor, MI). With one to two bottles of media, a syringe and a 47-mm plastic filter arrangement (Gelman Sciences or Millipore Filter Corp., New Bedford, MA) with hand pressure are more convenient.

3. The suggested optimal membrane filter pack arrangement in either the metal pressure filter or the plastic filter is as follows (starting at the bottom of the filter): a 0.4- or 0.45-µm filter; a nylon spacer screen; a 0.8-µm filter; a nylon spacer screen; a 1.0-µm (1000 nm) filter; and a prefiltration pad. Polycarbonate membrane filters from either Nucleopore, Inc. (Pleasanton, CA) or Poretics Corp. (Livermore, CA) have optimum flow rates. The metal or plastic filter holders with the just-mentioned filter pack should be sterilized at 121°C for 25–30 minutes.

4. It is advisable to have two or three of these sterile filter arrangements ready in case problems arise in filtering particular lots of supplements. Although some yeast extract preparations may have more particulate material than normal and may soon plug any filter arrangement, it is usually not necessary to centrifuge the yeast extract prior to filtration.

5. For hand pressure filtration, use a clean (but not necessarily sterile) 60-ml plastic syringe. Remove the plunger from the syringe and connect the syringe barrel to the sterile membrane filter holder. Place the filter holder on top of the opened bottle of sterile base medium, and then pour about 50–60 ml of the sterile supplements into the syringe barrel. Place the syringe plunger into the barrel and apply gentle hand pressure to the plunger as the material in the syringe passes through the filter. When this material has been filtered, leave the filter holder on the bottle of base medium and remove the syringe from the filter holder. With the syringe disconnected, the plunger can be withdrawn and the empty syringe barrel can be reconnected to the filter holder. Pour the remaining supplement into the syringe barrel, replace the plunger in the syringe, and complete the supplement filtration.

6. For the air pressure membrane filter, remove the metal screw cap and insert an extra (nonsterile) prefiltration pad into the filter holder. Using an inverted plastic syringe plunger, gently tap the filter pad down and over the sterile filter pack. Pour pooled supplements into the filter holder, tighten the metal cap, and allow some time for the supplements to wet the prefilter pad. Gently apply air pressure to start filtration. In most instances it is not necessary to exceed 8–10 lbs of air pressure, as excess pressure may break membrane filters.

Preparation of Solid Medium

1. Prepare the base medium as described earlier, including the addition of Noble agar prior to heat sterilization. For conventional mollicutes, 3.2 g of agar is added to the base medium (final concentration 0.8%). To obtain individual colonies of motile spiroplasmas on SP-4 agar, 9.0 g of agar is added to the base medium (final concentration 2.25%).

2. Heat sterilize the base medium as recommended earlier. Directly after sterilization, place the melted agar base in a 56°C waterbath. Measure and pool sterile medium supplements and filter through a hand pressure filter pack into a sterile, 125-ml serum bottle. Place the bottle of sterile supplements in the 56°C waterbath.

3. Hold both agar base and supplements at this temperature for 1–2 hours to equilibrate. When ready to pour plates, remove agar base and bottle of supplements to laminar flow hood and pour the hot supplements into the bottle of heated agar base. Quickly mix the two components by gentle swirling (without inducing air bubbles) and pour plates before the agar medium solidifies. An automatic dispensing apparatus (Unispense, Wheaton, Millville, NJ), delivering a volume of 7 ml into individual 60-mm sterile plastic dishes, provides a convenient means for rapid delivery of the melted agar.

SP-4 Medium with Extra Bacterial Inhibitors

1. Clinical specimens, such as sputum, throat washings, abscess material, urethral or vaginal swabs, and urine, frequently contain large numbers of bacteria. Since these organisms may well overgrow the low concentrations of penicillin in the SP-4 formulation, it is useful to supplement the medium with additional inhibitors.

2. To control gram-positive bacteria, add 4 ml of the penicillin stock solution to 400-ml volumes of broth, which increases the final concentration in the medium to 1000 U/ml.

3. To control gram-negative bacteria, add 2 ml of the Polymyxin B stock solution to 400-ml volumes of the broth, giving a final concentration of 500 U/ml.

SP-4 Medium for Isolation and Growth of Arginine-Hydrolyzing Mollicutes

1. Isolation and enhanced growth of arginine-hydrolyzing mollicutes can be enhanced by the additional supplement of arginine to SP-4 broth. There are small

quantities of arginine in the CMRL supplement, but these quantities may not be sufficient to promote growth of some mollicutes.

2. Add 2 ml of the sterile, 42% stock arginine solution to the pooled additives to be filtered into the final SP-4 broth. There is an advantage to observing an alkaline shift in the SP-4 broth from the growth of arginine-hydrolyzing mollicutes. It is, therefore, necessary to add less 1 N NaOH (0.5 to 1.0 ml instead of 3 ml) to the base medium. The final pH of the SP-4 broth should in this case be around 7.0 to 7.2.

Quality Control of SP-4 Medium

1. For the preparation of reliable culture media for mollicutes, it is absolutely crucial to carry out preliminary quality control tests on various lots of both fresh yeast extract and fetal bovine serum—the two most common components that may vary in their ability to support growth. Unsuitable lots of dehydrated base broth are rarely encountered, but have been reported. An extensive description of quality control tests for mollicutes culture media has been provided (Tully and Rose, 1983).

2. SP-4 base broth in a final 400-ml volume is prepared. Heat briefly to dissolve ingredients. The base medium, prior to sterilization, contains about 275 ml. This volume is divided into four equal quantities (about 68–69 ml each) and each aliquot is placed in a 125-ml serum bottle. After heat sterilization, each base medium is supplemented with about 32 ml of the filter-sterilized additives, as follows: 17 ml of fetal bovine serum; 5 ml CMRL 1066; 5 ml yeastolate solution; 2.5 ml fresh yeast extract; 1 ml glucose solution; and 0.5 ml penicillin solution. One bottle of complete SP-4 medium should be reserved for the current lot of fresh yeast extract and fetal bovine serum that serves as control preparation. The three remaining bottles can contain various test lots of serum or yeast extract to be compared to control broth.

3. Each of the individual test broth preparations and the control medium are distributed into a series of 10 1-dram (4 ml) vials, in a volume of 1.8 ml/vial. Label the control broth series and the individual series of test lots of serum or yeast extract with the appropriate lot number. The mollicute test organism (see later), as a frozen stock, is thawed and 0.2 ml of inoculum is added to the first vial in each medium series, giving a 1:10 dilution. Serially dilute the inoculum in each series by passing 0.2-ml volumes through the 10th vial.

4. Incubate each of the series of vials, including several uninoculated vials of each test medium, at the appropriate temperature. At 2- to 3-day intervals, record the growth, as measured by pH and/or turbidity changes.

5. Test organism: The choice of the strain or species of mollicute to be used in the quality control test is sometimes dependent on the organism the SP-4 broth is intended to cultivate. However, the most useful test organism is a low passage fastidious mollicute, such as the G37 strain of *M. genitalium,* the PI 1428 strain of *M. pneumoniae,* or the Y32 strain of *Spiroplasma ixodetis.*

6. If the SP-4 medium prepared with the test and control ingredients is able to sustain growth of the test mollicute through seven to nine 10-fold dilutions, then the medium would appear to be of satisfactory quality. However, if the test organism shows growth through only four to five 10-fold dilutions and the control medium growth is through seven to nine 10-fold dilutions, then the test lot of serum or yeast extract is unsuitable.

References

Dawson, M. S., Hayes, M. M., Wang, R. Y.-H., Armstrong, D., Budzko, D. B., Kundsin, R. B., and Lo, S.-C. (1993). Detection and isolation of *Mycoplasma fermentans* from urine of HIV positive patients with AIDS. *Arch. Pathol. Lab. Med.* **117,** 511–514.

Freundt, E. A. (1983). Culture media for classic mycoplasmas. *In* "Methods in Mycoplasmology" (S. Razin and J. G. Tully, eds.), Vol. 1, pp. 127–135. Academic Press, New York.

Lo, S.-C., Hayes, M. M., Wang, R. Y.-H., Pierce, P. F., Kotani, H., and Shih, J. W.-K. (1991). Newly discovered mycoplasma isolated from patients infected with HIV. *Lancet* **338,** 1415–1418.

Robinson, I. M. (1983). Culture media for anaeroplasmas. *In* "Methods in Mycoplasmology" (S. Razin and J. G. Tully, eds.), Vol. 1, pp. 159–162. Academic Press, New York.

Shepard, M. C. (1983). Culture media for ureaplasmas. *In* "Methods in Mycoplasmology" (S. Razin and J. G. Tully, eds.), Vol. 1, pp. 137–146. Academic Press, New York.

Tully, J. G., and Rose, D. L. (1983). Sterility and quality control of mycoplasma culture media. *In* "Methods in Mycoplasmology" (S. Razin and J. G. Tully, eds.), Vol. 1, pp. 121–125. Academic Press, New York.

Tully, J. G., Rose, D. L., Whitcomb, R. F., and Wenzel, R. P. (1979). Enhanced isolation of *Mycoplasma pneumoniae* from throat washings with a newly modified culture medium. *J. Infect. Dis.* **139,** 478–482.

Tully, J. G., Rose, D. L., Yunker, C. E., Cory, J., Whitcomb, R. F., and Williamson, D. L. (1981a). Helical mycoplasmas (spiroplasmas) from *Ixodes* ticks. *Science* **212,** 1043–1045.

Tully, J. G., Taylor-Robinson, D., Cole, R. M., and Rose, D. L. (1981b). A newly discovered mycoplasma in the human urogenital tract. *Lancet* **1,** 1288–1291.

Tully, J. G., Whitcomb, R. F., Clark, H. F., and Williamson, D. L. (1977). Pathogenic mycoplasmas: Cultivation and vertebrate pathogenicity of a new spiroplasma. *Science* **195,** 892–894.

Whitcomb, R. F. (1983). Culture media for spiroplasmas. *In* "Methods in Mycoplasmology" (S. Razin and J. G. Tully, eds.), Vol. 1, pp. 147–158. Academic Press, New York.

A3

CULTIVATION OF SPIROPLASMAS IN UNDEFINED AND DEFINED MEDIA

Kevin J. Hackett and Robert F. Whitcomb

Introduction

Because spiroplasmas differ in their ability to grow in medium formulations, strategies for their cultivation have varied (Whitcomb, 1983). The first sustained cultivation of a spiroplasma *(Spiroplasma citri)* was achieved in a relatively simple medium (BSR), similar to media subsequently found suitable for cultivating easy-to-grow spiroplasmas. This success was followed by cultivation of the corn stunt spiroplasma *(S. kunkelii)*, for which more complex media (e.g., M1D) or modified versions of simple media [C3-G (Liao and Chen, 1977)], were required. In 1977, a very fastidious organism *(S. mirum)* was cultivated in the SP-4 medium. Two additional spiroplasmas, one (the group XX CPBS) associated with the Colorado potato beetle and another (the group II SRO) that causes sex ratio distortion in *Drosophila* flies, proved to be intractable to isolation by standard procedures. These organisms were eventually isolated in coculture with insect cells (see Chapter A4, this volume); in addition, CPBS could be isolated under anaerobic conditions in insect cell-free media.

Certain fast-growing spiroplasmas and one fastidious spiroplasma grow in chemically defined media (Chang, 1989), facilitating nutritional analysis and development of mutants for molecular studies. *S. clarkii* strain CN-5, an easily cultured, fast-growing organism, has been cultured in a very simple defined medium containing bovine serum albumin (BSA) and (after a 4-year period of adaptation) seven organic nutrients (medium H-3). In contrast, the more fastidious *S. mirum* SMCA has been cultivated only in a very complex defined medium (H-1) that has the key membrane component sphingomyelin (Table III).

Spiroplasma media have been modified not only according to the target organism but also for their intended application. Whereas media used for primary isolation usually contain a rich mixture of nutrients, many fast growing, insect-associated spiroplasmas can be routinely maintained in simplified undefined media.

In our experience, some general principles apply: (i) Additional amino acids, particularly glutamine, improve growth titers, up to two or more log units in some cases. (ii) Some spiroplasmas may be able to satisfy their amino acid requirements by proteolytic digestion of BSA. (iii) Culture of more fastidious spiroplasmas may be aided by anaerobic conditions or cell culture (see Chapter A4, this volume). (iv) Serum batch, or lipid mixture (for defined media), can be critical to success. (v) Inorganic salts are probably unimportant, within reasonable physiological concentrations. It is possible that sodium can be eliminated entirely. (vi) Trace amounts of some components may occur as contaminants in medium components or glassware. For example, for some spiroplasmas, the amount of phosphate contaminating glassware or organic nutrients is sufficient for spiroplasma growth. (vii) Organic acids are not required, but may enhance growth of some spiroplasmas. (viii) Some spiroplasmas may benefit from riboflavin and niacin (Chang, 1989). (ix) Media made with a combination of adenosine, guanosine, cytidine, and thymidine support better growth than those with fewer nucleosides. (x) Sorbitol works as an excellent osmolyte.

Media Components

Spiroplasma media contain not only standard components that have been used for conventional mycoplasma media, but other supplements as well. For example, many are based on vertebrate [CMRL-1066 (Parker *et al.*, 1957)] or invertebrate [Schneider's *Drosophila* medium (Schneider, 1966)] tissue culture media. Media employed for vertebrate cells are especially rich in nucleic acid precursors and cofactors.

All media contain a mixture of salts that reflects, theoretically, the salt balance in tissue fluids that the cells encounter *in vivo*. One organic compound a α-ketoglutaric acid, has been shown to stimulate the growth of some spiroplasmas (Whitcomb, 1983).

Spiroplasma media invariably contain carbohydrates; no spiroplasmas have been found to lack a pathway for glucose fermentation, and, often, other sugars can be metabolized. Sugars (e.g., sucrose and glucose) and sugar alcohols (e.g., sorbitol and mannitol) have been used to adjust media osmolarity, with 300–700 mOsm generally optimal (Whitcomb, 1983).

Spiroplasmas respond favorably to high amino acid concentrations; hence, the success in using protein digests (peptone and tryptone) and yeastolate (which also supplies a variety of other metabolites). Mycoplasma broth base has been used in most spiroplasma media, although it is sometimes replaced by beef heart infusion broth alone. Batches of all such components, particularly broth base supplements, should be pretested to assure growth of the spiroplasma being studied.

Since, with a few exceptions (Rose *et al.*, 1993), spiroplasmas require sterols and none appear to be harmed by its presence, a source of cholesterol is always used in isolation media. Cholesterol and other lipid ingredients have been conventionally supplied to mollicutes by means of animal sera, e.g., fetal bovine serum. Agamma preparations are preferred or the sera should be heat treated at 56°C for 30–60 minutes (see Table I). For nutritional characterization studies, bovine serum fraction (1–5%) or a defined lipid fraction can be substituted for serum. A lipid supplement of cholesterol, palmitic acid, oleic acid, Tween 40, and Tween 80, with BSA as a fatty acid detoxicant, fulfills the minimal needs of many fast-growing spiroplasmas in defined media (Chang, 1989).

Buffers, pH indicators, and antibiotics are often included. Phenol red is often used as a pH indicator. Because spiroplasmas ferment glucose and other components, leading to rapid acidification of media and cessation of growth, buffers such as HEPES have been incorporated into some media. Spiroplasmas, like other mollicutes, have absolute resistance to penicillin and its derivatives, and this antibiotic is commonly incorporated into spiroplasma media. To avoid filtration during primary isolation, an antibiotic cocktail has been used for isolation of some spiroplasmas (Grulet *et al.*, 1993).

General Approaches to Medium Fabrication

Primary Isolation

Rich media, e.g., M1D and SP-4, are generally chosen for primary isolation since simple media are apt to be suboptimal, particularly if spiroplasma cells are at low concentrations. M1D medium is often used for isolation of insect-associated spiroplasmas. Medium SP-4 (containing the component CMRL-1066) is preferred for primary isolation of tick-associated spiroplasmas but supports growth of most spiroplasmas (exceptions include the sex ratio and Colorado potato beetle spiroplasmas). Medium SP-4 also permits primary isolation and growth of most *Mycoplasma, Mesoplasma, Entomoplasma,* and *Acholeplasma* species (see Chapter A2, this volume, and Chapter D4, Vol. II).

TABLE I

Spiroplasma Culture Media[a]

Component	M1D[b]	SP-4[k]	C3-G[l]
PPLO broth base[d]	2.0 g A	1.0 g A	4.5 g A
Tryptone[e]	1.0 g A	3.0 g A	
Peptone[e]	0.8 g A	1.6 g A	
Glucose	0.1 g A		
Fructose	0.1 g A		
Sucrose	1.0 g A		36.0 g A
Sorbitol	7.5 g A		
Schneider's *Drosophila* medium[f]	160 ml B		
CMRL-1066 (10X), with glutamine, without NaHCO$_3$[g]		15 ml B	
Fetal bovine serum[h]	50 ml C	50 ml C	
Horse serum[h]			60 ml B
Fresh yeast extract (25%)[i]	10 ml D[c]	10.5 ml D	
Yeastolate (2%)[j]		30 ml E	
Glucose (50%)		3 ml F	[m]
Phenol red (0.5%)	1.2 ml E	1.2 ml G	1.6 ml C
K-Penicillin G (10^5 U/ml)	2.5 ml F	3.0 ml H	2.5 ml D

[a] Amounts followed by the same letter are included in the same stock solution; total volume 300 ml.

[b] Williamson and Whitcomb (1975) and Whitcomb (1983).

[c] Present in M1A, but not in M1D.

[d] For example, Mycoplasma broth base, Baltimore Biological Laboratories, Cockeysville, Maryland; Difco is used for C3-G.

[e] For example, Oxoid Tryptone, Bacteriological Peptone.

[f] Grand Island Biological Co.; pre (e.g., Millipore AP25, AP15)- and final (e.g., Millipore 1.2, 0.45, and 0.22 μm) filtered.

[g] Grand Island Biological Co.

[h] For example, Gibco-BRL, HyClone: Heat inactivated at 56°C for 30 minutes in 100-ml (or 60 min in 500-ml) volumes with swirling agitation every 5 to 10 minutes and filtration (pre- and final filtration as for Schneider's *Drosophila* medium).

[i] For example, GIBCO, Flow Laboratories: Prefilter.

[j] For example, Difco: Prefilter.

[k] Tully *et al.* (1977) and Whitcomb (1983).

[l] Liao and Chen (1977) and Whitcomb (1983).

[m] May be needed for culture of some spiroplasmas.

Maintenance

Once spiroplasmas have been isolated, they can often be routinely maintained or grown in batch culture (e.g., for serological or nucleic acid studies) in simplified undefined media. In many cases, especially with spiroplasmas transmitted via flowers, very simple media may suffice for maintenance, e.g., medium C3-G (Table I). However, if a single medium is to be chosen for routine maintenance of a wide array of spiroplasmas from various groups or subgroups, we suggest medium M1D or SP-4; many spiroplasmas grow poorly or not at all in simpler alternatives.

Solid Media

Choice of a proper solid medium is dependent on the spiroplasma strain being studied (Whitcomb, 1983). Spiroplasmas readily form satellite colonies as they move from the primary focus of the colony through the agar to adjacent sites. Giant colonies as large as 4 mm in diameter, with no discrete central area, may be formed. Typical fried-egg colony morphology is observed with spiroplasmas only under poor growth conditions (e.g., reduced serum) or when organisms are nonmotile. An effective strategy for obtaining discrete colonies of highly motile spiroplasmas is the incorporation of high concentrations (2.25%) of agar (Tully et al., 1982). Agars of lesser purity than Noble agar (Difco) should not be used in solid media for spiroplasmas. Concentrations of Noble agar as low as 0.8% may suffice for culture of some spiroplasmas, but 1.6% agar is best for routine work.

Defined Media

Defined media (Chang, 1989) typically contain inorganic salts, organic acids, nucleosides, carbohydrates, amino acids, vitamins, cofactors, lipids, and buffer. A major problem is provision of long-chain fatty acids in a nontoxic, but readily available, form. This has generally been achieved by complexing fatty acids with BSA. If BSA is used, it should be treated to remove contaminant fatty acids and cholesterol (methods in Rodwell, 1983) since even the so-called fatty acid-free preparations of BSA may contain substantial quantities of fatty acids and cholesterol.

Medium Formulations

Formulations of the commonly used media are described next. Undefined media components are detailed in Table I and the concentrations of nutrients in these media are summarized in Table II. The components of defined media and their concentrations are given in Table III.

TABLE II
Components of Typical Undefined Media

Component	M1D[a]	SP-4[b]	BSR[c]	C-3G[d]
Inorganic salts (g/liter)				
$CaCl_2$		100[e]		
$CaCl_2 \cdot H_2O$	425[f]			
KCl	850[f]	200[e]		
$MgSO_4 \cdot 7H_2O$	1970[f]	100[e]		
NaCl	2800[f,g]	4230[e,g]		3570[h]
$NaHCO_3$	210[f]			
$NaH_2PO_4 \cdot H_2O$		70[e]		
$Na_2HPO_4 \cdot 7H_2O$	700[f]			
KH_2PO_4	240[f]			
Amino acids (mg/liter)				
L-Alanine		12[e]		
β-Alanine	270[f]			
L-Arginine–HCl	210[f]	35[e]		
L-Aspartic acid	210[f]	15[e]		
L-Cysteine	30[f]			
L-Cysteine–HCl·H_2O		130[e]		
L-Cystine	10[f]	10[e]		
Glutamic acid	420[f]	40[e]		
L-Glutamine	960[f]	50[e]		
Glycine	130[f]	25[e]		
L-Histidine	210[f]			
L-Histidine–HCl·H_2O		10[e]		
L-Hydroxyproline		5[e]		
L-Isoleucine	80[f]	10[e]		
L-Leucine	80[f]	30[e]		
L-Lysine	880[f]	35[e]		
L-Lysine–HCl	880[f]	35[e]		
L-Methionine	420[f]	7[e]		
L-Phenylalanine	80[f]	12[e]		
L-Proline	910[f]	20[e]		
L-Serine	130[f]	12[e]		
L-Threonine	190[f]	15[e]		
L-Tryptophan	50[f]	5[e]		
L-Tyrosine	270[f]	20[e]		
L-Valine	160[f]	12[e]		
Organic acids (g/liter)				
Fumaric acid	50[f]			
α-Ketoglutaric acid	110[f]			
Malic acid	50[f]			
Succinic acid	50[f]			
Carbohydrates (g/liter)				
Fructose	330		1000	
Glucose	1330[f]	5500[e]	1000	±5000

TABLE II (continued)

Component	MID[a]	SP-4[b]	BSR[c]	C-3G[d]
Sodium glucuronate		2.1[e]		
Sorbitol	25,000		70,000	
Sucrose	3330		10,000	120,000
Trehalose	100[f]			
Vitamins, reducing agents, and growth factors (mg/liter)				
Ascorbic acid		25.0[e]		
p-Aminobenzoic acid		0.025[e]		
Biotin		0.005[e]		
Calcium pantothenate		0.005[e]		
Choline chloride		0.25[e]		
Folic acid		0.005[e]		
Glutathione (reduced)		5.0[e]		
myo-Inositol		0.025[e]		
Nicotinamide		0.012[e]		
Nicotinic acid		0.012[e]		
Pyridoxal–HCl		0.012[e]		
Pyridoxine–HCl		0.012[e]		
Riboflavin		0.005[e]		
Thiamine–HCl		0.005[e]		
Cofactors (mg/liter)				
Cocarboxylase		0.5[e]		
Coenzyme A		1.25[e]		
FAD[i]		0.5[e]		
NAD[j]		3.5[e]		
NADP[k]		0.5[e]		
Nucleic acid precursors (nucleosides and nucleotides) (mg/liter)				
Thymidine		5.0[e]		
2'-Deoxyadenosine		5.0[e]		
2'-Deoxyguanosine		5.0[e]		
2'-Deoxycytidine–HCl		5.0[e]		
5-Methyldeoxycytidine		0.05[e]		
Uridine 5'-triphosphate		0.5[e]		
Lipids, lipid precursors, and lipid carriers (mg/liter)				
Cholesterol		0.1[e]		
Acetate, Na–3H$_2$O		42.0[e]		
Tween 80 (ml/liter)		2.5[e]		
Undefined components (g/liter)				
Serum, fetal bovine (%)	17	17	20	20
Beef extract	1050[g]	500[g]		
Brain heart infusion	700[g]	330[g]	22,500	7140[h]
Peptone	2660	5300		4290[h]
Tryptone	5770[g]	11,170[g]		
Fresh yeast extract, 25% (ml)	[l]	35		
Yeastolate	1050[g]	2500[e,g]		

(*continues*)

TABLE II (continued)

Component	M1D[a]	SP-4[b]	BSR[c]	C-3G[d]
Other (mg/liter)				
Ethanol (ml)		8.0[e]		
Penicillin G (10^3 U/ml)	+	+	+	+
Phenol red, 0.2% (15 ml/l)	+	+	+	+
pH	7.4	7.4	7.4	7.4
Osmolarity (mOsm)	500	330	700	700

[a] Williamson and Whitcomb (1975) and Whitcomb (1983).
[b] Tully et al. (1977) and Whitcomb (1983).
[c] Updated SMC medium; Saglio et al. (1971) and Whitcomb (1983).
[d] Liao and Chen (1977) and Whitcomb (1983).
[e] Partly or entirely from CMRL 1066 Medium: Parker et al. (1957).
[f] Partly or entirely from Schneider's *Drosophila* Medium: Schneider (1966).
[g] Some from BBL Mycoplasma Broth Base.
[h] Some from Difco PPLO Broth Base.
[i] Flavine adenine dinucleotide.
[j] Nicotinamide adenine dinucleotide.
[k] Nicotinamide adenine dinucleotide phosphate.
[l] 33 ml in M1A.

Liquid Media

MEDIUM M1D (TOTAL VOLUME, 300 ML) (WHITCOMB, 1983)

While stirring, dissolve autoclavable stock solution A components (Table I) in 70 ml water and adjust to pH 7.8 with 1 N NaOH. Autoclave for 30 minutes at 15–17 lb/in^2. Cool to room temperature. Add nonautoclavable components: Solutions B,C,E,F. Adjust with 1 N NaOH or KOH to desired pH, e.g., 7.4, and bring to 300 ml with water. If needed, adjust osmolarity with sorbitol to approximately 500 mOsm. Filter through 0.2-μm membranes. For multiliter batches, the medium can be pre(Millipore AP filters)- and final (1.2-, 0.45-, and 0.22-μm sandwich) filtered under pressure (e.g., that supplied by a sterile filtration apparatus linked to a peristaltic pump). Centrifugation of the medium for 50 minutes at 12,000 rpm (GSA rotor) in 250-ml centrifuge bottles is very useful for removing precipitates prior to filtration; this eliminates the need for prefiltering and for filtration of serum and Schneider's *Drosophila* medium.

MEDIUM SP-4 (TOTAL VOLUME, 300 ML) (WHITCOMB, 1983;
SEE ALSO CHAPTER A2, THIS VOLUME)

While stirring, dissolve autoclavable stock solution A components (Table I) in 197 ml water and adjust to pH 7.8 with 1 N NaOH. Autoclave and cool. Add nonautoclavable components: Solutions B–H. Adjust pH, media volume, and osmolarity, and filter (as for M1D).

MEDIUM C3-G (TOTAL VOLUME, 300 ML) (WHITCOMB, 1983)

Stock solution A components (Table I) are dissolved in 220 ml water, autoclaved, and cooled (as earlier). Add solutions B–D and filter. Medium BSR is an alternate simplified undefined medium (Whitcomb, 1983; Table II).

OTHER UNDEFINED MEDIA

Other media include T. B. Clark's SM-1 medium, based on Singh's mosquito culture medium (Whitcomb, 1983). Standard mycoplasma medium (HSI) (Whitcomb, 1983) has been used for tests of glucose and arginine metabolism. In this case, the horse serum component of HSI is replaced by 1% bovine serum fraction (Difco).

Solid Media (Total Volume, 300 ml)

Typically, solid media are made as follows: To the autoclavable fraction (Table I), add 2.4 to 6.8 g Noble agar (Difco) (depending on the spiroplasma; 1.6 g is used in HSI medium); sterilize as for liquid media. Equilibrate the autoclavable fraction and all other fractions to 56°C. Rapidly mix all components, adding penicillin last, and pour plates immediately.

An alternate method is to autoclave a $2\times$ solution of agar, cool to 56°C, and add this (1:1) to 56°C-equilibrated $2\times$ M1D medium.

DEFINED MEDIA (TOTAL VOLUME, 100 ML)

Prepare serum substitute: To 2.0 ml of warm (25–30°C) 100% ethanol, add, while stirring, appropriate amounts of (see Note below) fatty acids, cholesterol, phospholipids, and Tween supplements. While stirring, add 0.4 ml of this preparation dropwise to 19.6 ml of 6% BSA [essentially fatty acid-free (Sigma) or lipid extracted (Rodwell, 1983)] and adjust to pH 7.4 with 1 N NaOH or KOH. Filter sterilize and store at 26°C. [Note: For example (for 100 ml of medium), to achieve a final concentration of 20 mg of cholesterol/liter medium, add 10 mg of cholesterol to the 2 ml of ethanol solvent; 0.4 ml of lipid solution will then contain 2 mg of cholesterol.]

Prepare basal fraction: To 70 ml of warm (25–30°C) distilled water, add in sequence, while stirring, appropriate amounts of HEPES buffer, amino acids, organic acids, inorganic salts, nucleic acid precursors, cofactors, carbohydrates, vitamins and reducing agents, phenol red, and penicillin. [Constituents, particularly vitamins, can be added as stocks, e.g., 3% amino acid solutions. Other components, e.g., guanosine, may require solubilization in base or acid.] Titrate to pH 7.4 with 1 N NaOH or KOH and bring to 80 ml with water.

For the final medium, combine 20 ml of serum substitute dropwise to 80 ml of the basal fraction; sterilize by filtration.

TABLE III

Components in Typical Defined Media

Component	LD82[a]	CC-494[b]	CC-494N[c]	H-1[d]	H-3[e]
Buffer (g/liter)					
HEPES[f]	17	11.25	15	16	16
Inorganic salts (g/liter)					
$CaCl_2$	0.2	0.46			
$CaCl_2 \cdot H_2O$				0.095	
KCl		1.15		0.18	
$MgSO_4 \cdot 7H_2O$	0.8	2.18		0.095	
NaCl	4.6–6.6	2.1–8.4	6.48	2.35	
$NaHCO_3$		1.20			
$NaH_2PO_4 \cdot H_2O$		0.06		0.285	
Na_2HPO_4		0.07			
KH_2PO_4	0.4	0.03	0.04	0.285	g
Organic acids (g/liter)					
Sodium acetate·$3H_2O$		75			
α-Ketoglutaric acid	0.4	0.3		0.29	
Oxalacetic acid				0.19	
Pyruvic acid	0.4	0.3		0.19	
Carbohydrates (g/liter)					
Fructose	4.0			0.9	
Glucose	1.0	8.0	8.0	4.8	42.5
Sodium glucuronate		0.004		0.002	
Sorbitol (for osmolarity)				26.4	21
Sucrose		35.0 (for osmolarity)		1.9	
Trehalose				0.95	
Xylose				100	
Amino acids (mg/liter)					
L-Alanine	400	270	360	380	
β-Alanine	100			95	
L-Arginine	600	270	360	1900	
L-Asparagine	600	270	360	550	
L-Aspartate, K	1000	210	280	950	
L-Cysteine–HCl	600	210	280	550	
L-Glutamate, K	1800	450	600	1400	
L-Glutamine	600	630	1200	1700	7610
Glycine	400	135	180	380	
L-Histidine	400	225	300	760	
L-Hydroxyproline	200			450	
L-Isoleucine	400	90	120	450	
L-Leucine–HCl	1000	90	120	900	
L-Lysine	400	900	1200	380	
L-Methionine	400	450	600	380	
L-Phenylalanine–HCl	500	90	120	950	
L-Proline	1000	900	1200	950	
L-Serine	400	135	180	380	

TABLE III (*continued*)

Component	LD82[a]	CC-494[b]	CC-494N[c]	H-1[d]	H-3[e]
L-Threonine	200	180	240	200	
L-Tryptophan, K	200	54	72	380	
L-Tyrosine–HCl	200	270	360	380	
L-Valine	400	180	240	380	
Nucleic acid precursors (nucleosides, nucleotides, and sugars) (mg/liter)					
Adenosine	45	36	48	45	280
Guanosine	30	36	48	45	420
Inosine	30			45	
Cytidine	30	36	48	45	
Thymidine	30	36	48	45	510
Uridine	30	36	48	45	
2'-Deoxyadenosine		36		45	
2'-Deoxyguanosine		36		45	
2'-Deoxycytidine		36		45	
5-Methylcytosine	405				
5-Methyldeoxycytidine		1.8		2.0	
Uridine 5'-triphosphate	2.0	3.6		4.5	
Inosine monophosphate				2.0	
D-(−)-Ribose		4.5		9.0	
2'Deoxy-D-ribose		4.5		9.0	
Vitamins, reducing agents, and growth factors (mg/liter)					
Ascorbic acid		45		30	
p-Aminobenzoic acid	2.0	0.045		0.45	
Biotin	1.0	0.009		0.45	
Calcium pantothenate	2.0	0.009		0.85	
Choline chloride	1.5	0.45		0.65	
Folic acid	2.0	0.009		0.45	
Glutathione (reduced)	60	9.0		25	
myo-Inositol	1.5	0.045		0.65	
Nicotinamide	2.0	0.023		0.45	
Nicotinic acid	2.0	0.023	0.036	0.45	
Pyridoxal–HCl	1.0	0.023		0.45	
Pyridoxal phosphate	1.0				
Pyridoxine–HCl	2.0	0.023		0.85	
Riboflavin	2.0	0.009	0.014	1.3	
Riboflavin 5-phosphate	0.5				
Thiamine–HCl	1.0	0.009		0.45	
Thiamine pyrophosphate	2.0				
Cofactors (mg/liter)					
Cocarboxylase		0.90		4.5	
Coenzyme A	2.5	1.13		4.5	
FAD[h]	3.0	1.80		4.5	
NAD	2.5			4.5	
NADP[i]	2.5	3.6		4.5	

(*continues*)

TABLE III (continued)

Component	LD82[a]	CC-494[b]	CC-494N[c]	H-1[d]	H-3[e]
Lipids and lipid carriers (mg/liter)					
Cholesterol	9.8	18.50	18.50	20	20
Palmitic acid	14.8	12.30	12.30	10	[k]
Oleic acid		9.20	9.20	10	10
Linoleic acid	16.5				
Phosphatidic acid (egg yolk)	4.9				
Lysophosphatidylcholine (egg yolk)	4.2				
Lysophosphatidylcholine (soybean)	4.2				
Phosphatidylcholine (egg yolk)	7.0				
Sphingomyelin				20	
Tween 40 (ml/liter)		0.095	0.095	0.1	
Tween 80 (ml/liter)	0.148	0.095	0.095	0.1	
Glycerol (m/liter)	0.24				
DL-Glycerophosphate	400				
Albumin, bovine serum[j] (g/liter)	9.0	12.0	12.0	11.8	11.7
Polyamine (mg/liter)					
Spermidine	1000				
Other (mg/liter)					
Ethanol (100%, as lipid solvent) (ml)				0.04	0.04
Penicillin G (U/ml)				10^3	10^3
Phenol red (0.2%) (ml/liter)		15	15	10	20
1 N NaOH (m/liter)				19	
1 N KOH (ml/liter)					10
pH	7.5	7.5		7.35	7.30
Osmolarity (mOsm)		420–620		440	510

[a] Lee and Davis (1983).
[b] Chang and Chen (1982) and Chang (1989).
[c] Chang (1989).
[d] Hackett et al. (1987).
[e] Hackett et al. (1994).
[f] N-(2-Hydroxyethyl)piperazine-N'-(2-ethanesulfonic acid).
[g] Inorganic salts (particularly PO_4) may be needed if not present in sufficient quantities as media component "contaminants."
[h] Flavine adenine dinucleotide.
[i] Nicotinamide adenine dinucleotide phosphate.
[j] Essentially fatty acid free (Sigma) or lipid extracted (Rodwell, 1983).
[k] May be present in sufficient quantities in BSA.

Acknowledgment

Supported in part by Binational Agricultural Research and Development Grant No. US-1902-90R.

References

Chang, C.-J. (1989). Nutrition and cultivation of spiroplasmas. *In* "The Mycoplasmas" (R. F. Whitcomb and J. G. Tully, eds.), Vol. 5, pp. 201–241. Academic Press, New York.

Chang, C.-J., and Chen, T.-A. (1982). Spiroplasmas: Cultivation in chemically defined medium. *Science* **215**, 1121–1122.

Grulet, O., Humphrey-Smith, I., Sunyach, C., LeGoff, F., and Chastel, C. (1993). 'Spiromed': a rapid and inexpensive *Spiroplasma* isolation technique. *J. Microbiol. Methods* **17**, 123–128.

Hackett, K. J., Ginsberg, A., Rottem, S., Henegar, R. B., and Whitcomb, R. F. (1987). A defined medium for a fastidious spiroplasma. *Science* **237**, 525–527.

Hackett, K. J., Hackett, R. H., Clark, E. A., Gasparich, G. E., Pollack, J. D., and Whitcomb, R. F. (1994). Development of the first completely defined medium for a spiroplasma, *Spiroplasma clarkii* strain CN-5. *IOM Lett.* **3**, 446–447.

Lee, I.-M., and Davis, R. E. (1983). Chemically defined medium for cultivation of several epiphytic and phytopathogenic spiroplasmas. *Appl. Environ. Microbiol.* **46**, 1247–1251.

Liao, C. H., and Chen, T. A. (1977). Culture of corn stunt spiroplasma in a simple medium. *Phytopathology* **67**, 802–808.

Parker, R. C., Castor, L. N., and McCulloch, E. A. (1957). Altered cell strains in continuous culture: A general survey. Special Publications, N. Y. Academy of Sciences, Vol. 5, pp. 303–313.

Rodwell, A. (1983). Defined and partly defined media. *In* "Methods in Mycoplasmology" (S. Razin and J. G. Tully, eds.), Vol. 1, pp. 163–172. Academic Press, New York.

Rose, D. L., Tully, J. G., Bové, J. M., and Whitcomb, R. F. (1993). A test for measuring growth responses of mollicutes to serum and polyoxyethylene sorbitan. *Int. J. Syst. Bacteriol.* **43**, 527–532.

Saglio, P., LaFlèche, D., Bonissol, C., and Bové, J. M. (1971). Isolement, culture et observation au microscope électronique des structures de type mycoplasme associées à la maladie du Stubborn des agrumes et leur comparaison avec les structures observées dans le cas de la maladie du Greening des agrumes. *Physiol. Veg.* **9**, 569–582.

Schneider, I. (1966). Histology of the larval eye-antennal discs and cephalic ganglia of *Drosophila* cultured *in vitro*. *J. Embryol. Exp. Morphol.* **15**, 271–279.

Tully, J. G., Whitcomb, R. F., Clark, H. F., and Williamson, D. L. (1977). Pathogenic mycoplasmas: Cultivation and vertebrate pathogenicity of a new spiroplasma. *Science* **195**, 892–894.

Tully, J. G., Whitcomb, R. F., Rose, D. L., and Bové, J. M. (1982). *Spiroplasma mirum*, a new species from the rabbit tick *(Haemaphysalis leporispalustris)*. *Int. J. Syst. Bacteriol.* **32**, 92–100.

Whitcomb, R. F. (1983). Culture media for spiroplasmas. *In* "Methods in Mycoplasmology" (S. Razin and J. G. Tully, eds.), Vol. 1, pp. 147–158. Academic Press, New York.

Williamson, D. L., and Whitcomb, R. F. (1975). Plant mycoplasmas: a cultivable spiroplasma causes corn stunt disease. *Science* **188**, 1018–1020.

A4
INSECT CELL CULTURE APPROACHES IN CULTIVATING SPIROPLASMAS
Kevin J. Hackett and Dwight E. Lynn

Introduction

The rationale for using coculture to cultivate spiroplasmas is based on the observation that mollicutes are often isolated as contaminants in mammalian cell culture and evidence that cells have population-dependent requirements for metabolites which they synthesize, a discovery that gave rise to the use of feeder layers for primary cell culture.

Traditionally, mollicute–mammalian cell coculture systems have been used to elucidate pathogenic effects of mollicutes on cell lines. These studies demonstrated the importance of keeping cell lines axenic. In addition, mammalian cell lines have been used to isolate spiroplasmas and as model systems for studying pathogenic effects of spiroplasmas on eukaryotic cells. Similarly, invertebrate cell lines have been employed for studies of spiroplasma pathogenicity (McGarrity and Williamson, 1989), and, of importance to insect pathologists, insect cell lines have been used to coculture previously unculturable spiroplasmas (Hackett *et al.*, 1986), including the Colorado potato beetle spiroplasma (CPBS) and the sex ratio organism (SRO), a transovarially transmitted spiroplasma that causes loss of male progeny in infected females. Subsequently, Ueda and colleagues (McGarrity and Williamson, 1989) grew SROs in organ cultures of *Drosophila* embryos, and Williamson *et al.* (1989) obtained primary isolates of SROs in coculture with the yeast *Rhodotorula rubra*.

Materials

Cell Lines

Insect cell lines can be obtained from other researchers [the cabbage looper cell line, IPLB-TN-R^2, and the southern corn rootworm line, IPLB-DU182E, used in our studies can be obtained from us; for lists of other available cell lines, consult Hink and Bezanson (1985), Hink and Hall (1989), and the American Type Culture Collection (Gaithersburg, MD)] or by primary isolation, described herein. Although cells derived from a spiroplasma's insect host or a related host may be preferred, spiroplasmas have been isolated by using cells from an insect order different from the normal host (Hackett et al., 1986).

Media

The medium must be compatible with the cell line. Some media are listed in Table I and others are described in the Hink references cited earlier.

Culture Materials and Conditions

Culture plates: flasks or 24- to 96-well cell culture plates
Microscope: dark-field for spiroplasmas; inverted phase-contrast for monitoring cell cultures
Incubator: Cultures are usually maintained at 26°–28°C, but they will grow at ambient (23°C) temperature

Procedures

Culture of Insect Cells

PRIMARY CULTURE

The basic procedure for initiating primary cultures from insects is shown in Fig. 1 (based on Lynn, 1989). A novice to cell culture should obtain and maintain a previously developed cell line. This will provide experience necessary for handling cultures and recognizing healthy cells. Perhaps the salient point to consider in attempting to establish new cell lines is that patience is a virtue. While cell lines are now routinely started in less than 2 months, willingness to wait for the cells to grow may be the most important factor for success.

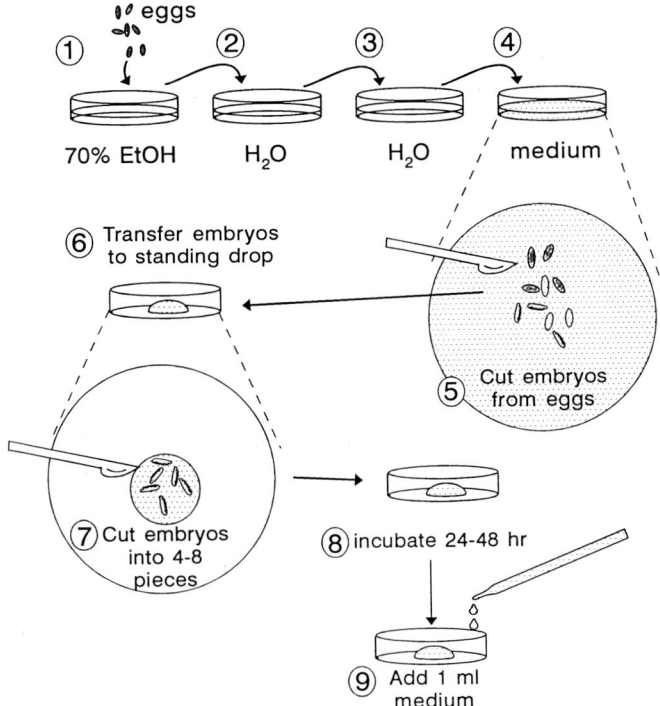

Fig. 1. For preparing primary insect cell cultures: (1) Disinfect eggs (or whole adult or immature insects if other tissue is desired) in 70% ethanol for 5–10 minutes. If eggs or insects are from a particularly dirty environment, pretreat with 0.05% sodium hypochlorite (=1% Clorox) containing a detergent (e.g., 1% Triton X-100) for 5–15 minutes before transferring to 70% ethanol. (2) Rinse in sterile distilled water for 5 minutes. (3) Transfer to fresh sterile distilled water. Hold excess material at this point while proceeding to step 4 with some of the material. (4) Transfer to a tissue culture medium containing gentamicin (50 µg/ml). (5) Place a microscalpel and fine-pointed forceps in 70% ethanol. Ignite alcohol in a flame (do not hold instrument in flame, simply ignite and hold at an angle to allow alcohol to burn off). Use the forceps and scalpel to remove embryos from eggs. (6) Transfer embryos to a 35-mm tissue culture petri dish containing a standing drop (0.1–0.2 ml) of medium with gentamicin. (7) Cut embryos into four to eight pieces with microscalpel. (8) Seal petri dish lid to bottom by stretching Parafilm around the edge of the dish and incubate at 27°C in a humidified chamber made from a tightly sealed container (such as Tupperware) holding a small beaker of water. (9) After 24–48 hours, remove Parafilm and add 1.0 ml additional medium containing gentamicin. Reseal with a fresh piece of Parafilm and return to humidified container in incubator. At 7- to 14-day intervals, add fresh medium to culture. When the dish becomes too full for easy handling (about 3 ml), remove all but 0.5 ml. If there are many cells in the removed medium, centrifuge at low speed (50–100 g, 5–10 minutes), resuspend the cells in 0.5 ml fresh medium, and return to petri dish. When the original dish contains many cells (>80% confluence), transfers can be made to a new dish or tissue culture flask.

MEDIA FORMULATION

Media formulation is an important consideration. Insect cell culture media historically have been based on hemolymph characteristics of the insect of interest. Characteristics of insect hemolymph from the viewpoint of spiroplasma habitat have been noted (Hackett and Clark, 1989). Once continuous cell lines are available, media formulations can be improved by adjusting component types and concentrations and determining the effect on cell growth.

For cell culture, three basic media formulations exist: those for Lepidoptera, generally based on Grace's (1962) medium; for Diptera, with Schneider's (1966) *Drosophila* medium being typical; and for Homoptera, which often have very simple compositions (Chiu and Black, 1967). These media have also been used to establish cell lines from insects in other orders.

In developing a new medium, a few factors are probably most important. First is the amount and ratio of inorganic ions, which vary considerably among insect orders. Also, the pH and osmotic pressure of media can have substantial impact on cell survival.

Four media commonly used in our laboratories are described in Table I. Several of these and other insect cell culture media are available commercially (e.g., GIBCO, Grand Island, NY; Sigma, St. Louis, MO; JRH Biosciences, Lenexa, KS). Media prepared in the laboratory should be made with tissue culture grade chemicals and water (TC-H_2O, such as that obtained from a deionization system, e.g., MilliQ, Millipore, Bedford, MA).

MAINTENANCE OF CELL LINES

Many insect cell lines can be maintained on a weekly subculture interval, and most insect cells grow in media buffered with phosphates, eliminating the need for a CO_2 incubator. Optimal growth usually occurs around 27°C in a temperature-controlled incubator, although most insect cells also will grow adequately at room temperature (around 23°C in modern laboratories).

Subculture (passaging) methods of insect cell lines will depend on how firmly the cells attach to the substrate. Many insect cells are loosely attached and can be subcultured by gently rocking or shaking the culture flask before transfer to fresh medium. For more firmly attached cells, cooling the culture by refrigeration (4°C) for 15–20 minutes will reduce cell adhesion by depolymerizing cell microtubules (which are involved in attachment). Cells are then detached by firmly tapping the culture or by vigorous flushing with medium from a pipette. Although this process results in some cell damage, cool temperatures also reduce the activity of lytic enzymes released into the culture medium. In the event that the cells are so firmly attached that cooling and vigorous flushing/tapping do not detach cells, either scraping or enzymatic treatment may be necessary. To scrape cells, a rubber policeman or a specially designed cell scraper is pulled across the

TABLE I
Insect Cell Culture Media Components[a]

Component	Modified Grace's[b]	DCCM[c]	Schneider's[d]	Chiu and Black's[e]
KCl	2240 A	2600 A	450 B	800 A
$NaH_2PO_4·2H_2O$	1140 A	1160 A		
$MgSO_4·7H_2O$	2780 A	1880 A	3700 B	850 A
$MgCl·6H_2O$	2280 A			
NaCl			2100 B	500 A
$NaHCO_3$			400 B	350 B
KH_2PO_4			450 B	300 B
Na_2HPO_4			2700 B	
$CaCl_2·2H_2O$	1320 B	60 B	800 A	400 A
$ZnCl_2$		0.04 C		
$MnCl_2·4H_2O$		0.02 C		
$CuCl_2·2H_2O$		0.2 C		
$(NH_4)Mo_7O_{24}·4H_2O$		0.04 C		
$CoCl_2·6H_2O$		0.05 C		
$FeSO_4·7H_2O$		0.55 D		
L-Aspartic acid		0.36 D		
Glucose	700 C	5000 E	2000 B	
Fructose	400 C			
Sucrose	26680 C			
Maltose		1000 E		
Trehalose			2000 B	
α-Ketoglutaric acid	370 D		350 C	
Fumaric acid	55 D		350 C	
Malic acid	670 D		600 C	
Succinic acid	60 D		60 C	
Thiamine–HCl	0.02 E	0.16 F		
Riboflavin	0.02 E	0.16 F		
Ca pantothenate	0.02 E	0.16 F		
Pyridoxine–HCl	0.02 E	0.4 F		
p-Aminobenzoic acid	0.02 E	0.32 F		
Folic acid	0.02 E	1.2 F		
Niacin	0.02 E	0.16 F		
Isoinositol	0.02 E	10 F		
Biotin	0.01 E	0.16 F		
Cyanocobalamin		1.0 F		
Choline chloride	0.20 E			
β-Alanine	200 F		500 D	
L-Alanine	225 F			
L-Arginine–HCl	700 F	800 G	600 D	
L-Aspartic acid	350 F	1000 G	400 D	
L-Asparagine	350 F	1300 G		

(*continues*)

TABLE I (continued)

Component	Modified Grace's[b]	DCCM[c]	Schneider's[d]	Chiu and Black's[e]
L-Cysteine			60 D	
L-Glutamic acid	600 F	1300 G	800 D	
L-Glutamine	600 F	1000 G	1800 D	
Glycine	650 F	400 G	250 D	
L-Histidine	2500 F	200 G	400 D	
Hydroxy-L-proline		800 G		
L-Isoleucine	50 F	500 G	150 D	
L-Leucine	75 F	400 G	150 D	
L-Lysine–HCl	625 F	700 G	1650 D	
L-Methionine	50 F	1000 G	150 D	
L-Proline	350 F	600 G	1700 D	
L-Phenylalanine	150 F	1000 G		
DL-Serine	1100 F	600 G	250 D	
L-Threonine	175 F	200 G	350 D	
L-Tryptophan	100 F	100 G	100 D	
L-Valine	100 F	500 G	300 D	
L-Cystine	25 G	100 H	20 D	
L-Tyrosine–HCl	50 G	300 H	500 D	
Phytone peptone		750 I		
Liver digest		2600 I		
Lactalbumin hydrolyzate	3000 H			6500 A
TC yeastolate	3000 H	5000 I	2000 D	5000 A
Glycerol (50% in H₂O)		3.2 J		
L-Glutamine		40 K		

[a] mg/liter; amounts followed by the same letter are part of the same stock solution. All solutions are made in TC–H₂O. Unless otherwise stated, stock solutions are filter sterilized (0.2 μm) and stored at 4°C. While antibiotics are useful for primary culture, good laboratory practice dictates that they not be used in routine maintenance of cell lines.

[b] Make a 500× stock of vitamins (Solution E). Mix peptides (Solution H) as a 20× stock solution and sterilize by autoclaving. Mix salts (Solution A) in 200 ml and CaCl₂ (Solution B) in 75 ml water. Mix sugars (Solution C) in 200 ml, organic acids (Solution D) in 200 ml, and amino acids (Solution F) in 300 ml water, except cystine and tyrosine (Solution G) which are dissolved in a small volume of 1.0 N HCl. Combine Solutions A, C, D, E, F, and G while stirring. Slowly add CaCl₂ (Solution B) while rapidly stirring. Slowly adjust the pH to 6.5 with 0.1 N KOH while rapidly stirring to avoid precipitation. Adjust volume to 1 liter and sterilize. Osmotic pressure should be about 350 mOsm. Aseptically add 50 ml of Solution H. Add fetal bovine serum to the final combined preparation at 5–10% (v/v). Note: modified Grace's (= Hink's TNM-FH) is available from commercial sources (GIBCO, Grand Island, NY; JRH Biosciences, Lenexa, KS; and Sigma, St. Louis, MO).

[c] Make a 1000× stock of trace minerals (Solution C), a 1500× stock solution of iron with aspartic acid (Solution D), and a 500× stock of vitamins (Solution F). Mix peptides (Solution I) as a 20× stock solution and sterilize by autoclaving. Make a 50% solution of glycerol (Solution J) and autoclave. Make a 50× stock solution of glutamine (Solution K). Mix salts (Solution A) in 200 ml,

cell surface. The cells are then dispersed into small clumps or single cells prior to transfer to fresh medium.

Trypsin is the most commonly utilized enzyme for detaching cells for subculture. Typically, the cell monolayer is rinsed with a buffer and a dilute solution of trypsin is added (we use 50 μg VMF trypsin, Worthington Biochemical Corp., Freehold, NJ, per milliliter of buffer for our *Diabrotica undecimpunctata* cell lines). The culture is either held at room temperature or warmed in an incubator for a few minutes until the cells begin to detach. At that time, fresh medium containing serum (which has anti-trypsin activity) is added to the culture, and the suspended cells are transferred to a sterile centrifuge tube and centrifuged at low speed (50–100 g) for 5–15 minutes. The cell pellet is resuspended in fresh medium and transferred to a new culture flash.

Split ratio (the number of new cultures that can be obtained from a mature parent culture) depends on a cell line's growth rate, metabolic needs, and susceptibility to injury during transfer. Slower growing cells may only be subcultured at a 1:2 split ratio every other week, whereas fast growing cells can be split 1:30 or more. Some cells cannot be split at low densities because autocrine growth factors become too dilute. If large numbers of cells are damaged during subculturing, higher initial seeding densities are necessary.

Coculture of Spiroplasmas with Insect Cells

Spiroplasmas can be isolated as described in Hackett and Clark (1989), with the technique being dependent on the location (gut or hemolymph) of the spi-

---

$CaCl_2$ (Solution B) in 75 ml, and sugars (Solution E) in 200 ml water. Dissolve amino acids (Solution G) in 300 ml water, except for cystine and tyrosine which are dissolved in a small volume of 1.0 N HCl (Solution H). Combine solutions A, E, G, and H while stirring. Slowly add Solution B while stirring rapidly. Add 1 ml Solution C, 0.67 ml Solution D, 2 ml Solution F, and 3.2 ml Solution J from stock solutions. Slowly adjust the pH to 6.5 with 0.1 N KOH while rapidly stirring to avoid precipitation. Adjust volume to 1 liter and sterilize. Aseptically add 50 ml stock Solution I and fetal bovine serum to a final concentration of 10% (v/v). Adjust osmotic pressure to 400 mOsm (for *Diabrotica* cells) with 20% (w/v) sterile (autoclaved) mannitol. Add 2 ml glutamine (Solution K stock) per 100 ml before the first use of each bottle of medium.

d Dissolve $CaCl_2$ in 50 ml (Solution A) and components of Solution B in 250 ml, Solution C in 100 ml, and Solution D in 500 ml water. Dissolve cystine and tyrosine in a small volume of 10% NaOH then add to Solution D. Combine Solutions B, C, and D while mixing continuously. Adjust pH to 6.7 (±0.05) with 0.1 N NaOH. Slowly add Solution A while stirring rapidly. Adjust volume to 1 liter and sterilize. Add fetal bovine serum shortly before use to a final concentration of 10–15% (v/v). The original formula for Schneider's medium has 500 U per liter each of penicillin and streptomycin.

e Dissolve Solution A components in 300 ml and Solution B components in 500 ml water. Mix the two fractions and adjust pH to pH 6.43 and volume to 1 liter (pH 6.43 and osmolarity 360 mOsm). Sterilize. Add fetal bovine serum to a final concentration of 7.5–20% (v/v). The original formula for Chiu and Black's medium has 100,000 U penicillin, 100 mg streptomycin, 50 mg neomycin, and 2.5 mg Fungizone per liter.

roplasma in the host. Although some researchers prefer to provide as high an inoculum as possible, this may add inhibitors from the host insect or plant; it is therefore advisable to culture various dilutions of the primary specimen. Also, care should be taken to place spiroplasmas in cell culture as soon as they are removed from the host, especially since some fastidious spiroplasmas are oxygen sensitive. In general, freshly prepared (immediately passaged) insect cell cultures support better spiroplasma growth than 3-day-old insect cell cultures.

Spiroplasma cultures should be observed at 1- to 2-day intervals, particularly during early cultivation in insect cells. The method of sampling may be important because oxygen-sensitive spiroplasmas may be damaged by exposure to oxygen during pipetting, whereas increased aeration could enhance growth of other spiroplasmas.

Microscopic examination by dark-field techniques (1250×) is critical in determining when to passage spiroplasmas in insect culture. Cultures showing 10^7 to 10^9 spiroplasmas per ml can generally be passaged 1:10, twice per week, although care should be taken to avoid passaging cultures to very low titer (e.g., less than 10^5 spiroplasmas per ml). If the number of spiroplasmas in cell cultures is too low to be passaged, fresh medium (1:4) is added every 2–3 days for the first 1–3 weeks, or until the insect cell cultures become overgrown. Chances of success are probably increased by using both passage and refeeding methods simultaneously with different cell lines. Morphologic alterations in the spiroplasma signal an adverse environment. Unhealthy signs include the formation of blebs (deformations) or knots (cells bent back on themselves) in the spiroplasma helix; spiroplasma cell lengthening without division (lack of short three to six coil forms); loss of helicity, motility, or increased rigidity; loss of refractility (long thread-like forms); or (possibly) stickiness to the microscope slide. Formation of spiroplasma clumps may indicate exhausted media or a high titer of aging spiroplasmas. Care should be taken not to confuse insect cell filaments with spiroplasmas; uninoculated insect cell controls should always be included in cultivation attempts.

If spiroplasmas appear to be unhealthy, alternate cell lines or media formulations should be tried. To develop new cell line–medium combinations, a useful strategy is to add small amounts of medium supplement (1:3 to 1:5) to existing cell cultures, so as to allow adaptation of the insect cells to a new environment. It is conceivable that some cell lines will have to be cultivated in entirely new media before they are suitable for spiroplasma isolation.

After cocultured spiroplasmas have been passaged 5–15 times, they may be adaptable to insect cell-free aerobic culture; however, success may not be achieved for 100 passages or more. Anaerobic conditions may facilitate adaptation or, in some cases, may permit direct spiroplasma isolation without coculture.

Discussion

For culture of CPBS, all cell line–medium combinations tested were adequate. Subsequent primary isolation in cell-free medium under anaerobic conditions (Anaerobic GasPak, BRL Corp., Cockeysville, MD) and studies of the cell supplied growth factor(s) suggested that insect cells were either regulating redox potential or providing carbon dioxide to the spiroplasmas. In contrast (Hackett et al., 1986), the SRO spiroplasma was only isolated in a lepidopteran cell line with a medium (H-2) specially designed for the purpose [which combined one part each of modified Grace's and DCCM media (Table I) with two parts of traditional spiroplasma medium M1D] (see formula in Chapter A3, this volume). Since subsequent isolation of SRO in coculture with yeast cells (Williamson et al., 1989) suggests some SRO metabolic flexibility, it is likely that other cell lines and a suitable medium would also be acceptable. However, this spiroplasma could not be isolated under anaerobic conditions in cell-free media.

During coculture, or culture in general, important spiroplasma attributes may be lost. For example, the cultured DW-1 strain of SRO was, like the insect-derived spiroplasma, male lethal, but lost the ability to be transovarially transmitted on repeated *in vitro* passage (Williamson et al., 1989). Other multi-passaged spiroplasmas have lost extrachromosomal elements. It is therefore important to conduct pathogenicity tests as soon as possible and to freeze ($-70°C$) aliquots of cultured organisms throughout the cultivation process for preservation of biological characteristics.

Acknowledgment

Supported in part by Binational Agricultural Research and Development Grant No. US-1902-90R.

References

Chiu, R. J., and Black, L. M. (1967). Monolayer cultures of insect cell lines and their inoculation with a plant virus. *Nature* **215**, 1076–1078.

Grace, T. D. C. (1962). Establishment of four strains of cells from insect tissue grown in vitro. *Nature* **195**, 788–789.

Hackett, K. J., and Clark, T. B. (1989). Ecology of spiroplasmas. In "The Mycoplasmas" (R. F. Whitcomb and J. G. Tully, eds.), Vol. 5, pp. 113–200. Academic Press, New York.

Hackett, K. J., Lynn, D. E., Williamson, D. L., Ginsberg, A. S., and Whitcomb, R. F. (1986). Cultivation of the *Drosophila* sex-ratio spiroplasma. *Science* **232**, 1253–1255.

Hink, W. F., and Bezanson, D. R. (1985). Invertebrate cell culture media and cell lines. *Tech. Life Sci.* **C2**, 1–30.

Hink, W. F., and Hall, R. L. (1989). Recently established invertebrate cell lines. *In* "Invertebrate Cell System Applications" (J. Mitsuhashi, ed.), Vol. 2, pp. 269–293. CRC Press, Boca Raton, FL.

Lynn, D. E. (1989). Methods for the development of cell lines from insects. *J. Tissue Culture Meth.* **12,** 23–29.

McGarrity, G. J., and Williamson, D. L. (1989). Spiroplasma pathogenicity *in vivo* and *in vitro. In* "The Mycoplasmas" (R. F. Whitcomb and J. G. Tully, eds.), Vol. 5, pp. 365–392. Academic Press, New York.

Schneider, I. (1966). Histology of the larval eye-antennal discs and cephalic ganglia of *Drosophila* cultured *in vitro. J. Embryol. Exp. Morphhol.* **15,** 271–279.

Williamson, D. L., Hackett, K. J., Wagner, A. G., and Cohen, A. J. (1989). Pathogenicity of cultivated *Drosophila willistoni* spiroplasmas. *Curr. Microbiol.* **19,** 53–56.

A5

MEASUREMENT OF MOLLICUTE GROWTH BY ATP-DEPENDENT LUMINOMETRY

Janet A. Robertson and Gerald W. Stemke

Introduction

Determination of cell populations of mollicutes presents a number of technical obstacles. First, the small cellular dimensions of mollicutes preclude total cell counts by light microscopy. Second, many species form microcolonies in broth or adhere to the surface of culture vessels, resulting in underestimated viable cell counts. Third, because of fastidious growth requirements, certain species grow poorly on agar or when highly diluted in liquid medium, leading to further underestimation of cell numbers (Stemke and Robertson, 1982). If solid medium is used, the microscopy required for the enumeration of colonies introduces another possible source of error. Relative to most common bacteria, mollicutes have long generation times; depending on the species, from several days to several weeks of incubation are required to obtain the results of viable cell counts.

Mollicute cells have notably smaller cellular mass and their cultures produce smaller populations than do other free-living prokaryotes. When turbidity is detected, it may not indicate cell mass but the precipitation of medium components due to pH changes resulting from substrate utilization (Stemke and Robertson, 1990). Even nephelometry techniques are insensitive for measuring the growth of many mollicute species. Furthermore, in the highly supplemented media used to cultivate mollicutes, cellular proteins may constitute such a small fraction of the total protein that protein measurements must be excluded as a valid index of cell numbers.

ATP is the "currency of life." For unicellular organisms under equivalent physiological conditions, the amount of ATP present usually correlates well with cell mass (Chapelle and Levin, 1968) and thus should be unaffected by cellular aggregation. One of the most sensitive methods for detecting and quantifying ATP is the light-emitting reaction of luciferin catalyzed by the firefly *(Photinus pyralis)* enzyme luciferase:

$$\text{Luciferin} + \text{ATP} + \text{O}_2 \xrightarrow[\text{luciferase}]{\text{Mg}^{+2}} \text{AMP} + \text{oxidized luciferin} + \text{light}. \quad (1)$$

Emitted light is detected and quantified by use of a luminometer which combines photodetection with the instrumentation to integrate output and express it as relative light units (RLUs) (Chapelle and Levin, 1968). Because ATP is intracellular, cells must be lysed quickly and in a manner which allows the light-emitting reaction to occur. Proprietary releasing agents are now used for this purpose. Luciferin is available as a synthesized chemical. Because many factors can modify light output and detection, the assay is usually standardized by the addition of a known amount of ATP to the reaction mixture. ATP luminometry is highly appropriate for determining the growth response of mollicutes. ATP, AMP, and ADP levels and adenylated energy pools of mollicutes were first determined using early luciferase methodology (Saglio *et al.,* 1979) and other means of ATP quantification (Beaman and Pollack, 1981, 1983). Present-day ATP luminometry uses simpler procedures and provides immediate results.

Materials

Liquid medium appropriate for growth of the test species is required for cultures, for medium controls, and for sample dilution. It is filtered through sterile 0.22-μm filters to reduce particulate matter and is then preincubated overnight prior to inoculation to reduce endogenous ATP, i.e., background RLUs, to a stable level.

Sterile, pyrogen-free water is used for ATP standards and for rinsing apparatus.

Kits of test reagents, developed initially for medical analyses, are now also available for broad industrial and environmental applications. Some of these are appropriate for measuring growth of mollicutes. Reagents are also available individually. The cost for reagents is approximately $1.50–2.00 per assay. Nucleotide-releasing agents have been designed specifically for somatic or for bacterial cells. We obtained the best results with the nucleotide-releasing agent for bacterial cells (NRB; Lumac B. V., Landgraaf, The Netherlands); NRB is stable at 20°C. Lumac reagents are available in the United States through Integrated BIOSolutions, Inc. (Princeton, NJ). Other manufacturers of reagents are Amersham, Ltd. (UK), Analytical Luminescence, Bio-Orbit (Finland), Biotrace, Ltd., (UK), Celsis, Ltd. (UK), and Sigma Chemical (St. Louis, MO).

Store, prepare, and use lyophilized reagents as indicated by the manufacturer. Reconstitute ATP standard to 5μg per ml; it can be frozen in aliquots at ~ −70°C. Although this solution is relatively stable for up to 12 hours on ice, it should not be refrozen.

Luciferin–luciferase reagent (Lumit PM, Lumac) that has been reconstituted in HEPES buffer (25 mM; pH 7.75) is also stable for about 12 hours on ice. Leftover reagent can be frozen once at −20°C but repeated freezing and thawing causes significant loss of activity.

Fluid for washing/decontaminating luminometer
Culture vessels
Record sheets
Luminometer: Many instruments are available; some have semiautomatic features. Most of our work has been done with Biocounter Model 2010A (Lumac).
Water bath set at desired growth temperature
Thermometer
Disposable test cuvettes for luminometer with rack
Micropipettors and sterile tips
Small ice bucket for ATP standard and enzyme reagent
Vortex mixer
Stopwatch
Calculator or computer

Procedure

1. Warm luminometer for the recommended time before use. Set preestablished testing parameters. We integrated the RLUs emitted over the 300- to 900-nm range of the instrument for a 10-second period at ambient temperature (~22°C).

2. Wash decontaminating solution from the enzyme delivery tube by rinsing the tube thoroughly with sterile water. Remove residual water with air under pressures. Put enzyme reagent reservoir on ice. Pump reagent through tubing until no air bubbles are present. Verify that the delivered aliquot has the desired

volume. Enzyme activity deteriorates at room temperature. If more than 30 minutes elapses between assays, return the enzyme in tubing to the reservoir, rinse tubing with sterile water, and reload the enzyme when the next assays are to be conducted.

3. Introduce 100 μl of releasing agent into a test cuvette. This can be done before each assay by injecting the agent through one channel of the luminometer or before starting the assays by preparing many cuvettes by a pipettor. We find the latter more convenient.

4. Assay procedure is as follows: to releasing agent in test cuvette (step 3), add 100 μl test sample, start stopwatch, mix for 3 seconds with vortex mixer, and then insert the cuvette into the luminometer. When 10 seconds have elapsed, inject 100 μl enzyme reagent into the cuvette. The number of RLUs indicated on the digital display can be recorded manually or by computer.

5. Determine RLUs for 100 μl of water, preincubated broth, and a series of ATP standards (5, 1, and 0.05 μg per ml water). ATP dilutions in water should show a dose response. Repeat ATP assays in broth. To obtain control values, subtract the RLUs of the diluent from the RLUs of the diluent plus ATP. If absolute rather than relative numbers of RLUs are required, use the RLUs obtained for one concentration of ATP standard to determine the correction factor *(CF)* for the RLU obtained for that experiment:

$$CF = \frac{[\text{RLU of water} + \text{ATP}] - [\text{RLU of water alone}]}{[\text{RLU of broth} + \text{ATP}] - [\text{RLU of broth alone}]}. \quad (2)$$

When internal controls are used, the amount of ATP to be added to test samples should produce RLUs within the range of the RLUs of the test samples.

6. Determine RLUs of test samples, correcting for background by subtracting the RLUs of the medium control or for quenching by applying the correction factor. When the RLUs of a test sample exceed the range of reliable readings for the instrument, dilute that sample appropriately in preincubated broth and retest. Because of the close agreement between replicate assays, single assays may suffice for preliminary experiments; we use the mean of replicate assays for final experiments.

7. For long experiments, such as growth curve determinations, use freshly prepared internal ATP standards to monitor the effects of time and temperature on enzyme activity. Compare the activity of old and new bottles of enzyme reagent as each new bottle is introduced.

8. When the RLUs at several points in exponential growth have been determined, calculate the generation time. To express the effect of potentially stimulatory or inhibitory substances added to the growth medium, insert generation time *(GT)* into the following equation:

$$\text{Percentage effect} = \frac{GT \text{ under basal conditions} - GT \text{ under modified conditions}}{GT \text{ under basal conditions}}$$
$$\times 100\%. \tag{3}$$

Acceptance of luminometry data may be improved by correlating RLUs with the most recognized method of viable cell estimation for the species under study. With this information, providing that the quenching effect of the medium is known, the ATP per viable unit can be estimated.

9. When the assays are completed, clean tubing with antibacterial wash. Leave wash in the tubing during short-term storage.

Discussion

Before undertaking luminometry assays, determine background ATP, the quenching effects of the test medium, the correction factor, and the amount of ATP appropriate for internal controls. Knowledge of the quenching effects of particular medium supplements and the growth stimulators or inhibitors under study may lead to modifications that increase assay sensitivity. For instance, certain deeply colored or turbid media can introduce a high level of quenching. We have found that color changes in media containing the phenol red indicator (red to yellow) caused little variation, whereas those in media containing bromothymol blue (yellow to green-blue) required exact pH (i.e., color) controls. The latter proved so awkward that we eliminated the indicator from the broth medium, substituting direct pH readings for rough growth estimates prior to luminometry.

The extent of ATP release from cells is an important variable in ATP luminometry. Closely related Enterobacteriaceae vary in their susceptibility to one releasing agent; for certain species, ATP release was increased by sonication (Selan et al., 1992). Thus, both the choice of releasing agent and the manner of its use should be considered. Despite chemical similarities of membranes bounding eukaryotic and mollicute cells, we found that releasing agents designed for use with animal cells did not disrupt mollicutes as efficiently as an agent designed for walled bacteria. Despite this response, differential lysis of mammalian and prokaryotic cells within a single sample may not be applicable to mollicutes.

The temperature of the growth medium in the water bath should be ascertained before experiments are initiated. The time that the cultures are taken out of incubation temperature for sampling should be minimized. Sterile technique during sampling is necessary to exclude contamination by water-borne organisms. Establish culture purity before, during, and at the end of each experiment. Sudden large increases in RLUs during the experiment may indicate contamination of the culture, the medium used as diluent, or of the reagent delivery lines. These lines should be replaced periodically or when they become blocked.

Using ATP luminometry we have been able to detect ca. 10^4 color change units$_{50}$ (CCU$_{50}$) mollicute cells per 100-μl test sample. Under optimum growth conditions, individual *Ureaplasma urealyticum* cells contained ca. 5×10^{-18} mol ATP (Stemler *et al.*, 1987) compared with ca. 2×10^{-17} mol ATP per cell for *Acholeplasma laidlawii* (Beaman and Pollack, 1981). We used luminometry to establish growth curves for the extremely slow-growing porcine species *Mycoplasma hyopneumoniae* and *Mycoplasma flocculare* (*GT* of about 10 hours) and to detect logarithmic phase growth in cultures of the faster-growing human urogenital inhabitant *U. urealyticum* (*GT* of about 60–90 minutes). Ureaplasma cultures reach relatively low maximal titers (10^7–10^8) and die rapidly. The availability of reliable growth curves allows us to harvest cells when maximum titers were reached but before death phase events have compromised the integrity of the cells or the condition of their DNA.

We also have used luminometry to establish the relative inhibition of ureaplasma serovars by Mn^{+2} (Stemler *et al.*, 1987) and to determine minimal inhibitory concentrations of antibiotics. A stable compound may be added to the test medium, the test and control media are then inoculated, and subsequent growth is monitored. When a less stable compound is under examination, we add it to the culture only after exponential growth has been detected. This approach also allows us better control over sampling times, an important consideration when examining species with long generation times. Because of continued utilization and hydrolytic decomposition, the concentration of the cellular pool of ATP decreases rapidly upon cell death (Saglio *et al.*, 1979; Stemler *et al.*, 1987; Stemke and Robertson, 1990). Thus ATP luminometry can provide a means of examining not only growth but also the decline phase of mollicute cultures. It should also prove useful in human and veterinary diagnostic mycoplasmology for monitoring cultures of blood or other important clinical specimens for the presence of mollicutes.

We consider ATP luminometry an underexploited but extremely useful tool in the study of fastidious organisms. It provides accurate, immediate, and economical estimations of growth without the tedium associated with other available methodologies.

Acknowledgments

The adaptation of this method to mollicutes was supported by an equipment grant from the Alberta Heritage Foundation for Medical Research (J.A.R.) and by operating grants from the Medical

Research Council of Canada (J.A.R.) and Farming for the Future Program of the Agricultural Research Council of Alberta. Karen Coppola of Integrated BIOsolutions supplied valuable information.

References

Beaman, K. D., and Pollack, J. D. (1981). Adenylate energy charge in *Acholeplasma laidlawii*. *J. Bacteriol.* **146,** 1055–1058.

Beaman, K. D., and Pollack, J. D. (1983). Synthesis of adenylate nucleotides by mollicutes (mycoplasmas). *J. Gen. Microbiol.* **129,** 3103–3110.

Chapelle, E. W., and Levin, G. U. (1968). Use of the firefly bioluminescent reaction for rapid detection and counting of bacteria. *Biochem. Med.* **2,** 41–52.

Saglio, P. H. M., Daniels, M. J., and Pradet, A. (1979). ATP and energy charge as criteria of growth and metabolic activity of mollicutes: Application to *Spiroplasma citri*. *J. Gen. Microbiol.* **110,** 13–20.

Selan, L., Berlutti, F., Passariello, C., Thaller, M. C., and Renzini, G. (1992). Reliability of bioluminescence ATP assay for detection of bacteria. *J. Clin. Microbiol.* **30,** 1739–1742.

Stemke, G. W., and Robertson, J. A. (1982). Comparison of two methods for enumeration of mycoplasmas. *J. Clin. Microbiol.* **16,** 959–961.

Stemke, G. W., and Robertson, J. A. (1990). The growth response of *Mycoplasma hyopneumoniae* and *Mycoplasma flocculare* based upon ATP-dependent luminometry. *Vet. Microbiol.* **24,** 135–142.

Stemler, M. E., Stemke, G. W., and Robertson, J. A. (1987). ATP measurements obtained by luminometry provide rapid estimation of *Ureaplasma urealyticum* growth. *J. Clin. Microbiol.* **25,** 427–429.

A6

INTRACELLULAR LOCATION OF MYCOPLASMAS
David Taylor-Robinson

Introduction

The question of whether mycoplasmas enter eukaryotic cells has taxed investigators for many years. Early observations (see Taylor-Robinson et al., 1991, 1993) made by examination of infected eukaryotic cell cultures after staining with vital dyes were interpreted as being suggestive of an intracellular location for the mycoplasmal organisms. In each case, staining revealed mycoplasmal organisms that covered the cells; few or no organisms were seen in the intercellular spaces. Thus, in retrospect, none of these observations was sufficient to confirm whether the organisms, which appeared to be intracytoplasmic, had gained entrance or were simply located on the surface of the cells.

As in the case of the aforementioned observations, scanning electron microscopy, although providing indisputable evidence for the association of mycoplasmas with the surface of cells (Phillips, 1978), is not a technique that can be used to demonstrate cellular invasion. On the other hand, sectioning of cells with subsequent electron microscopic examination is likely to be of more value and is an approach that was taken by a number of investigators throughout the 1960s (see Taylor-Robinson et al., 1991, 1993), again in the 1980s (Larsson and Brunk, 1981; Meloni et al., 1981; Araake et al., 1984), and more recently (Lo et al., 1989, 1993; Lo, 1992; Mernaugh et al., 1993; Stadtländer et al., 1993; Jensen et al., 1994). There seems little question that mycoplasmas are taken up by polymorphonuclear leukocytes and macrophages (Zucker-Franklin et al., 1966; Webster et al., 1988), but whether they enter epithelial cells has been less easy to resolve. Indeed, Zucker-Franklin and colleagues (1966) believed that it was not possible to determine whether mycoplasmas that were seen within vacuoles in the cytoplasm were truly intracellular or whether they were within crypts, i.e., invaginations of the cell surface, and were, therefore, extracellular.

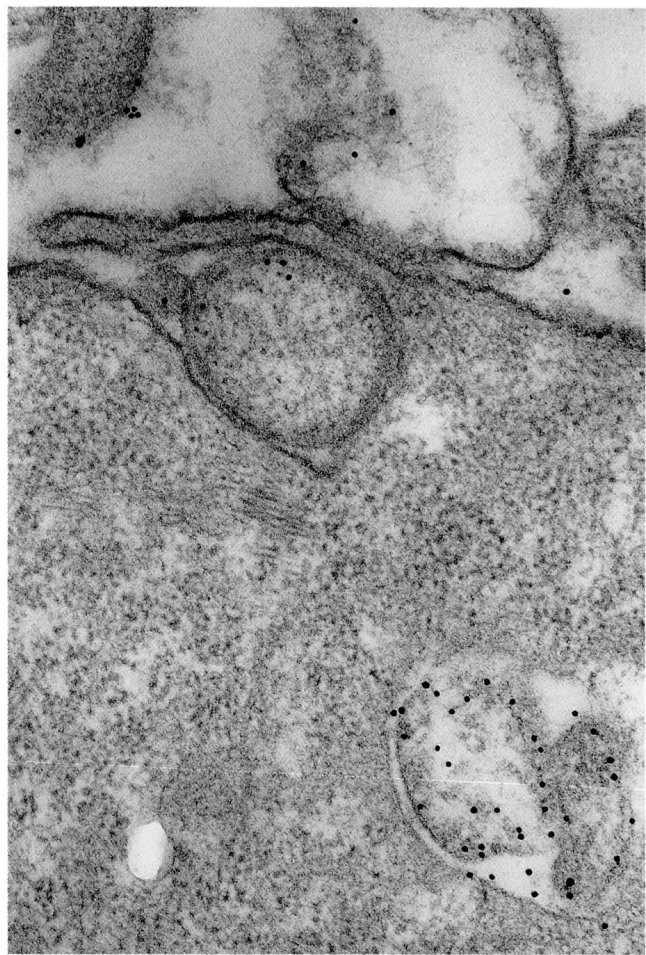

Fig. 1. Gold-labeled *M. fermentans* organisms within a HeLa cell vacuole. The membrane of the vacuole is not stained by ruthenium red, but the extracellular mycoplasma, within an invagination of the HeLa cell, and the HeLa cell membrane are stained. (Magnification: ×50,000). From Taylor-Robinson et al. (1993), *Clin. Infect. Dis.* **17,** Suppl. 1, 302–304. Copyright 1993 The University of Chicago Press. Reprinted with permission of The University of Chicago Press.

Resolution of the problem is possible only by sectioning mycoplasma-infected eukaryotic cells and undertaking transmission electron microscopy in experiments in which (i) the eukaryotic cell surface is stained so that an invagination of the surface, which otherwise may masquerade as an intracellular vacuole, may be easily appreciated and (ii) the mycoplasmas are labeled by some means so that their nature and position are readily identified. These criteria were fulfilled during the course of experiments with HeLa cells to determine whether *Mycoplasma fermentans* (Fig. 1) and *M. hominis* became intracellular (Taylor-Robinson et al., 1991, 1993). Details of the approach are presented here.

Materials

Eukaryotic Cells and Media

The goal is to produce a monolayer of eukaryotic cells in culture for subsequent mycoplasmal inoculation. The cells may constitute a primary culture or a continuous cell line, e.g., HeLa cells. It is important that they have been tested to exclude mycoplasmal contamination. Then cells (1-ml volume) suspended at a concentration of about 10^4 per ml in growth medium (see below) are seeded onto glass coverslips of 13-mm diameter contained in flat-bottomed plastic bottles of 5-ml capacity. The values for cell concentration and coverslip size are those used by the author but, of course, others may be used. The cell cultures are incubated at 37°C in an atmosphere of 5% CO_2 in air. Confluency depends on the original concentration of cells, their type, and rate of multiplication, but confluent monolayers of cells should occur within a day or so. This is checked microscopically and the medium is then replaced by 1 ml of maintenance medium before the cultures are inoculated with mycoplasmas.

The composition of growth media varies but the following is an example of medium that may be used: minimal essential medium containing 10% (v/v) fetal calf serum, and final concentrations of 3% sodium bicarbonate, 1% glutamine, 1% HEPES buffer, 0.3% sodium hydroxide, and 100 IU penicillin per ml. Maintenance medium is of the same composition except that it contains only 2% fetal calf serum.

Mycoplasma Cultivation and Antiserum

The mycoplasma under consideration may be grown in one of a variety of media, the compositions of which have been described in detail before (Freundt, 1983; Whitcomb, 1983). Specific mycoplasmal antiserum, prepared in rabbits or other animal species, or a monoclonal antibody is required.

Procedure

Inoculation of Eukaryotic Cell Cultures

The cell monolayer cultures are each inoculated with a large number of mycoplasmal organisms, e.g., 10^6-10^7 colony-forming or color-changing units. These cultures and uninoculated controls are incubated at 37°C and may be examined at various time intervals thereafter; for example, daily for 7 days. On each occasion, the number of viable mycoplasmas in an aliquot of supernatant medium may be estimated, but this is not an essential feature of the experiment. It is important, however, at each time interval to wash two or more cell monolayers and fix them for electron microscopy as described next.

Fixation and Embedding of Infected Eukaryotic Cells

The maintenance medium is removed and the cell monolayers are washed three times with 0.01 M phosphate-buffered saline "A" containing NaCl, KCl, Na_2HPO_4, and KH_2PO_4 (PBSA; pH 7.4) to remove mycoplasmas that have bound nonspecifically to the cell surface. Half of the cultures should be fixed in 3% paraformaldehyde in 0.1 M phosphate buffer, pH 7.4, for 30 minutes at room temperature. Cultures processed 1 and 2 days after inoculation are not fixed. The cells are scraped off the coverslips and are processed in microcentrifuge tubes. The samples are quenched with 0.5 M ammonium chloride in 0.1 M phosphate buffer for 30 minutes and are processed using the method of progressive lowering of temperature (Roth *et al.*, 1981). The cells are dehydrated in 30 and 50% methanol at 4°C, 80 and 90% methanol at -20°C, and infiltrated with Lowicryl K4M resin (Agar Scientific Ltd., Stanstead, UK) before final embedding in Lowicryl K4M and polymerization by long-wavelength ultraviolet light (365 nm) in microcentrifuge tubes at -30°C.

The other half of the cell cultures should be fixed in 3% glutaraldehyde in 0.1 M sodium cacodylate buffer, pH 7.4, containing 0.1% ruthenium red for 2 hours at room temperature. The cells are postfixed in 1% osmium tetroxide in cacodylate buffer containing ruthenium red for 2 hours at room temperature, scraped off the coverslips, and processed in microcentrifuge tubes. Dehydration through a graded series of acetone is followed by infiltration with Araldite resin and final polymerization at 60°C for 24 hours.

Ultrathin sections of all the polymerized blocks are cut on an ultramicrotome and collected on carbon/Formvar films on 200 mesh copper grids.

Immunocytochemistry

Ultrathin sections of cells embedded in Araldite at daily intervals after mycoplasmal inoculation of the cultures are immunolabeled, as are cells that have been embedded in Lowicryl resin after 3 and 4 days. Sections of the former are incubated with 10% hydrogen peroxide (H_2O_2) in distilled water for 10 minutes at room temperature (Baskin *et al.*, 1979) and are washed in distilled water before undertaking the immunocytochemical procedures.

The immunolabeling method is similar for both the H_2O_2-treated Araldite sections and the Lowicryl K4M sections. All procedures are carried out with 20 µl of each reagent. After preincubation of the sections with 5% bovine serum albumin (BSA, Sigma) in PBSA, they are treated for 1 hour at room temperature with specific mycoplasmal antiserum that is diluted to yield the optimal labeling density.

After washing with BSA/PBSA, the Araldite and Lowicryl sections are treated for 1 hour at room temperature with a dilution of protein A/gold in BSA/PBSA to give optimal labeling density without nonspecific background labeling. Conjugation of protein A (Sigma Chemical Co., Poole, Dorset, UK) with colloidal gold (Aurobeads G15, RPN_{475}, Amersham International, Amersham, Bucks, UK) may be undertaken using the method of Roth (1989). After further washing in BSA/PBSA and stream-washing in distilled water, the sections are air-dried and stained with aqueous uranyl acetate and lead citrate.

Control sections are included with every immunolabeling experiment. Uninfected cells are treated with the relevant mycoplasmal antiserum, and serum from the animal species prior to immunization is included at a dilution equivalent to that of the antiserum.

Interpretation of Results

If eukaryotic cells infected with mycoplasmas are embedded in Araldite and fixed with glutaraldehyde and osmium tetroxide in the presence of ruthenium red, the mucopolysaccharide surface components of both the eukaryotic and prokaryotic mycoplasmal cells will be stained. Ruthenium red reacts with anionic sites on the cell surface but does not penetrate cells. Thus, mycoplasmal organisms that remain exterior to the eukaryotic cell will be stained whereas those that have entered will not be stained. Furthermore, an invagination of the eukaryotic cell surface, which appears as an intracellular vacuole because of the way the section has been cut, will be stained; that is, the wall of the false vacuole will be stained. In contrast, the wall of a vacuole that is truly intracellular will be unstained (Fig. 1). This should help in deciding whether mycoplasmas are situated internally or externally. In addition, the presence of gold particles overlying or close to a mycoplasma-like organism following the use of specific mycoplasmal antiserum and gold labeling provides confident identification (Fig. 1) in a situation that might otherwise remain obscure.

Discussion

Attachment of certain mycoplasmas to eukaryotic cells through a specific adhesion process is a well-known phenomenon (Razin, 1985), but what happens to mycoplasmas thereafter had been unclear, although it had been widely considered that they remain extracellular. Observations on the probable intracellular location of mycoplasmas in patients with AIDS (acquired immunodeficiency syndrome) (Lo et al., 1989), which were identified later as *M. fermentans*, stimulated renewed interest in the phenomenon and culminated in experiments which proved beyond doubt that mycoplasmas could gain entry to eukaryotic epithelial cells (Fig. 1) (Taylor-Robinson et al., 1991, 1993).

Several investigators (Lo, 1992; Lo et al., 1993; Mernaugh et al., 1993; Stadtländer et al., 1993; Jensen et al., 1994) have provided electron microscope illustrations of mycoplasmas (*M. penetrans, M. genitalium*, and *M. fermentans*) apparently within eukaryotic cells. Where cells contain many mycoplasmal organisms, where the organisms are intracytoplasmic and not within a vacuole, or where they are close to the nucleus and distant from the outer cell membrane, it seems churlish to suggest that they are, in fact, extracellular and that they only appear to be intracellular. Indeed, in hindsight, some of the electron microscopic observations of early workers in the field who were cautious in the interpretation of their findings were probably illustrative of an intracellular mycoplasmal existence. Nevertheless, proof of such an existence only comes from the procedures described here.

The mechanism of cell entry is unclear and may not be the same for all mycoplasma–cell interactions. The close association of *M. genitalium* with clathrin-coated pits of human lung fibroblasts (Mernaugh et al., 1993) and organisms closely surrounded by a membrane within these cells suggests the manner in which they have entered. Furthermore, in a situation where there are many organisms *(M. penetrans)*, each closely surrounded by a membrane (Lo, 1992), it would seem untenable to argue that this points to multiple cell invaginations with each organism remaining extracellular.

Whether mycoplasmas within the cell remain viable and whether they multiply are issues that also have not been resolved. Multiplication would, of course, indicate viability. If attempts are made to assess the number and viability of intracellular organisms by undertaking viable counts of mycoplasmas in washed, infected cell cultures before and after the cells have been disintegrated to release the organisms, it should be remembered that adherent extracellular organisms could vitiate the results. Treatment of the cell cultures with an antibiotic that has mycoplasmacidal activity but does not penetrate (Mazzali and Taylor-Robinson, 1971) and/or with a specific mycoplasmal antiserum could be helpful. However, such an experiment needs to be undertaken in conjunction with electron microscopic observations. Larger numbers of mycoplasmal organisms seen by electron microscopy within cells in the later specimens of a sequential series than in the earlier specimens would suggest multiplication. However, larger numbers could simply be a reflection of a greater uptake of organisms because of an increasing number in the extracellular environment. Again, treatment of infected eukaryotic cell cultures with specific mycoplasmal antiserum to inhibit extracellular multiplication could be useful in resolving the problem.

References

Araake, M., Yayoshi, M., and Yoshioka, M. (1984). Electron microscopic studies on the attachment of *Mycoplasma pulmonis* to mouse synovial cells cultured *in vitro*. *Microbiol. Immunol.* **28**, 379–384.
Baskin, D. G., Erlandsen, S. L., and Parsons, J. A. (1979). Influence of hydrogen peroxide or alcoholic sodium hydroxide on the immunocytochemical detection of growth hormone and prolactin after osmium fixation. *J. Histochem. Cytochem.* **27**, 1290–1292.
Freundt, E. A. (1983). Culture media for classic mycoplasmas. *In* "Methods in Mycoplasmology" (S. Razin and J. G. Tully, eds.), Vol. 1., pp. 127–135. Academic Press, New York.
Jensen, J. S., Blom, J., and Lind, K. (1994). Intracellular location of *Mycoplasma genitalium* in cultured Vero cells as demonstrated by electron microscopy. *Int. J. Exp. Pathol.* **75**, 91–98.
Larsson, E., and Brunk, U. T. (1981). TEM and SEM findings in cat fibroblasts cultivated *in vitro* with and without mycoplasma. *Acta Pathol. Microbiol. Scand. A*, **89**, 9–15.
Lo, S. C. (1992). Mycoplasmas and AIDS. *In* "Mycoplasmas: Molecular Biology and Pathogenesis" (J. Maniloff, R. N. McElhaney, L. R. Finch, and J. B. Baseman, eds.), pp. 525–545. Am. Soc. Microbiol., Washington, DC.
Lo, S. C., Dawson, M. S., Wong, D. M., Newton, P. B., III, Sonoda, M. A., Engler, W. F., Wang, R. Y.-H., Shih, J. W.-K. Alter, H. J., and Wear, D. J. (1989). Identification of *Mycoplasma incognitus* infection in patients with AIDS: An immunohistochemical, *in situ* hybridization and ultrastructural study. *Am. J. Trop. Med. Hyg.* **41**, 601–616.
Lo, S. C., Hayes, M. M., Kotani, H., Pierce, P. F., Wear, D. J., Newton, P. B., III, Tully, J. G., and Shih, J. W.-K. (1993). Adhesion onto and invasion into mammalian cells by *Mycoplasma penetrans*: A newly isolated mycoplasma from patients with AIDS. *Mod. Pathol.* **6**, 276–280.
Mazzali, R., and Taylor-Robinson, D. (1971). The behaviour of T-mycoplasmas in tissue culture. *J. Med. Microbiol.* **4**, 125–138.
Meloni, G. A., Bertoloni, G., Busolo, F., and Conventi, L. (1981). Localization of mycoplasma in cultured mammalian cells. *Microbiologica* **4**, 197–203.
Mernaugh, G. R., Dallo, S. F., Holt, S. C., and Baseman, J. B. (1993). Properties of adhering and non-adhering populations of *Mycoplasma genitalium*. *Clin. Infect. Dis.* **17**, Suppl. 1, 69–78.
Phillips, D. M. (1978). Detection of mycoplasma contamination of cell cultures by electron microscopy. *In* "Mycoplasma Infection of Cell Cultures" (G. J. McGarrity, D. G. Murphy, and W. W. Nichols, eds.), pp. 105–118. Plenum, New York.
Razin, S. (1985). Mycoplasma adherence. *In* "Mycoplasmas" (S. Razin and M. Barile, eds.), Vol. 4., pp. 161–202. Academic Press, New York.
Roth, J. (1989). The colloidal gold marker system for light and electron microscopic cytochemistry. *In* "Techniques in Immunocytochemistry" (G. R. Bullock and P. Petrusz, eds.), Vol. 2., pp. 217–284. Academic Press, New York/London.
Roth, J., Bendayan, M., Carlemalm, E., Villiger, W., and Garavito, M., (1981). Enhancement of

structural preservation and immunocytochemical staining in low temperature embedded pancreatic tissue. *J. Histochem. Cytochem.* **29,** 663–671.

Stadtländer, C. T. K.-H., Watson, H. L., Simecka, J. W., and Cassell, G. H. (1993). Cytopathogenicity of *Mycoplasma fermentans* (including strain incognitus). *Clin. Infect. Dis.* **17,** Suppl. 1, 289–301.

Taylor-Robinson, D., Davies, H. A., Sarathchandra, P., and Furr, P. M. (1991). Intracellular location of mycoplasmas in cultured cells demonstrated by immunocytochemistry and electron microscopy. *Int. J. Exp. Pathol.* **72,** 705–714.

Taylor-Robinson, D., Sarathchandra, P., and Furr, P. M. (1993). *Mycoplasma fermentans*–HeLa cell interactions. *Clin. Infect. Dis.* **17,** Suppl. 1, 302–304.

Webster, A. D. B., Furr, P. M., Hughes-Jones, N. C., Gorick, B. D., and Taylor-Robinson, D. (1988). Critical dependence on antibody for defence against mycoplasmas. *Clin. Exp. Immunol.* **71,** 383–387.

Whitcomb, R. F. (1983). Culture media for spiroplasmas. *In* "Methods in Mycoplasmology" (S. Razin and J. G. Tully, eds.), Vol. 1., pp. 147–158. Academic Press, New York.

Zucker-Franklin, D., Davidson, M., and Thomas, L. (1966). The interaction of mycoplasmas with mammalian cells. I. HeLa cells, neutrophils, and eosinophils. *J. Exp. Med.* **124,** 521–532.

A7

LOCALIZATION OF MYCOPLASMAS IN TISSUES

Douglas J. Wear and Shyh-Ching Lo

Introduction

Purpose

In the study of various infectious diseases, it is important that the causative agent(s) in the diseased tissues or organs with prominent functional abnormality be identified. However, in mycoplasmal infections, isolation of fastidious mycoplasmas from clinical specimens by current culture systems is rarely successful. Conventional histopathological techniques for detecting bacteria do not stain the wall-free mycoplasmas, and electron microscopy (EM), examining only small areas of tissue, has a major sampling problem. Thus, techniques are needed that can help effectively localize specific areas in diseased tissues markedly infected by the organism.

Rationale

Our laboratory has focused more extensively in *Mycoplasma fermentans* infection which may be systemic (especially in immune-compromised patients), whereas other mycoplasma infections like those caused by *M. pneumoniae* are normally limited to one specific organ. Therefore, most of our experience presented here is based on studies of *M. fermentans*. Although mycoplasmas are prokaryotic organisms like bacteria, infection with *M. fermentans* does not induce prominent inflammatory responses in the infected tissues (Lo *et al.*,

1989a,b). Since the mycoplasma itself causes an atypical immune response in the infected hosts, characteristic histopathological features of bacterial infections, such as neutrophilic responses, are often missing. Apparent parenchymal cell damage and cytopathological changes found in diseased organs lacking the associated acute form of tissue inflammatory response become a clue to a possible *M. fermentans* infection. In the lung, *M. fermentans* infections may lead to increased numbers of chronic inflammatory cells in interstitium and intraalveolar spaces. *M. fermentans* infection is related to enlarged ballooned alveolar macrophages and hyalin membrane formation (Lo et al., 1993). In the liver, spleen, and lymph nodes, large areas of parenchyma die, leaving ghost-like cells (Lo et al., 1989a, 1991). These areas of infarct-like necrosis may be surrounded by a granulomatous wall with epithelioid cells. To localize specific areas with concentrated *M. fermentans* infection in these tissues, we use monoclonal antibodies and immunohistochemical methods to detect *M. fermentans*-specific antigens. EM study of the areas highly positive for the mycoplasmal antigens provides supportive evidence by localizing mycoplasmas in cells from the stained area (Bauer et al., 1991; Lo et al., 1989b, 1991, 1993). Immunohistologic location of positive cells and areas with marked mycoplasmal-specific antigen, plus confirmation by electron microscopy, may be sufficient for general diagnostic purposes. For research purposes, however, localization of specific areas highly positive for mycoplasma-specific DNA by *in situ* hybridization can also be performed to provide additional confirmation based on a distinct diagnostic principle (Lo et al., 1989a).

Materials and Reagents

Immunohistochemistry

Phosphate-buffered saline (PBS; GIBCO, Grand Island, NY)
10% bovine serum albumin (BSA; Sigma Chemical Co., St. Louis, MO) in PBS
5% fetal bovine serum (FBS; Sigma Chemical Co.) in PBS
1% albumin in PBS
Primary antibodies: Mouse monoclonal antibody (MAbs) to *M. fermentans* (1:600 dilution in 5% FBS) (G2 9H8 [IgG$_1$/k], 8H 12 [IgM/k]; originally raised in our laboratory (Bauer et al., 1991; Lo et al., 1989b, 1991, 1993) or monoclonal antibodies to *M. fermentans* lipid-associated membrane proteins (Wise et al., 1993)
Nonspecific mouse MAbs (IgM, MOPC 104E, and IgG$_{2b}$/k, MOPC 141) (Sigma Chemical Co.) or MAb (ascites) raised specifically against herpes virus (IgG$_1$, MCA 255, clone R1) (Whittaker Bioproducts, Walkersville, MD)

Secondary antibodies: Biotinylated horse anti-mouse IgG or biotinylated goat anti-mouse IgM (Vector Laboratories, Burlingame, CA)
Avidin–biotinylated peroxidase complex (ABC) reagent (Vector Labs)
Freshly prepared diaminobenzidine (Sigma Chemical Co.) solution with 0.2% H_2O_2
Hematoxylin counterstaining solution

Electron Microscopy

Xylene
Graded Ethanol (90, 70, 50, 30, and 10%)
PBS
2% glutaraldehyde
Dalton's fixative (1% osmium tetroxide, 4% potassium dichromate, 3.4% sodium chloride, pH 7.2)

In Situ Hybridization

Kodak Photo-Flo 200
3% Elmer's glue
Graded ethanol
PBS
Proteinase K (1 mg/ml in PBS) at 37°C
Hybridization mixture: 50% formamide, 10% (w/v) dextran sulfate, 0.6 M NaCl, 0.01 M Tris, pH 7.5, 0.5 mM EDTA, 1 mg/ml BSA, 0.02% (w/v) Ficoll, 0.02% (w/v) polyvinylpyrrolidone, 0.2 mg/ml sheared denatures salmon sperm DNA, and 0.5 mM dithiothreitol
^{35}S-labeled DNA probe (5–10 × 10^3 cpm/μl)
Doubly siliconized coverslips
Rubber cement
70 and 95% ethanol containing 0.3 M ammonium acetate
Kodak NTB3 nuclear track emulsion
0.6 M ammonium acetate
Hematoxylin counterstaining solution
M. fermentans probe [2.2-kb cloned fragment of M. fermentans DNA carries insertion-like genetic element (Hu et al., 1990) inserted into M13 mp 19 cloning vector (BRL) that has been radiolabeled with ^{35}S-labeled α-deoxyadenosine triphosphate by the chain elongation method (Messing, 1983) using the M13 universal sequencing primer (17-mer, Pharmacia) and the Klenow fragment of DNA polymerase I (BRL) (Klenow and Henningsen, 1970)]

Control DNA (M13 mp 18 or M13 mp 19 containing no insert of mycoplasmal DNA) which was labeled in same manner as *M. fermentans* probe

Procedure

Immunohistochemistry

1. Deparaffinize tissue sections by xylene and rehydrate tissue through a graded ethanol series.
2. Incubate the deparaffinized slides in 3% H_2O_2–methanol solution for 30 minutes.
3. Transfer to 10% BSA in PBS for 30 minutes to block nonspecific binding of immunoglobulins.
4. Rinse briefly with PBS.
5. Cover tissue in each slide with mouse monoclonal antibody (MAbs) (1:600 dilution in 5% FBS in PBS). Note: For controls use nonspecific mouse MAbs.
6. Refrigerate slides overnight.
7. Return to room temperature for a 1-hour incubation at room temperature.
8. Rinse extensively with 1% albumin in PBS.
9. Cover slides with secondary antisera (specificity should match the isotype of primary antibody) and incubate the slides at room temperature for 1 hour.
10. Rinse extensively with PBS.
11. Cover slides with avidin–biotinylated peroxidase complex, incubate at room temperature for 1 hour.
12. Rinse extensively with PBS.
13. Develop color reaction with diaminobenzidine and H_2O_2 substrate.
14. Counterstain with hematoxylin solution.

Electron Microscopy

1. Tissue (1–2 mm in diameter) is punched out of the paraffin block after the exact area is identified.
2. Deparaffinize tissues by hourly changes of xylene for at least three times.
3. Tissues are kept in xylene overnight and are washed in a solution containing half xylene and half ethanol the next morning.
4. Keep tissues in absolute ethanol overnight to remove xylene completely.

5. Rehydrate tissues through a graded ethanol series (90, 70, 50, 30, and 10%) for 5 minutes each change.
6. Tissues are rinsed in distilled water and placed in PBS for 30 minutes.
7. Tissues are fixed in 2% glutaraldehyde overnight.
8. Tissues are placed in Dalton's fixatives for 1 hour and processed as regular EM specimens.

In Situ Hybridization

1. Tissues are mounted on slides precoated with Kodak Photo-Flo 200 (Syrjanen and Syrjanen, 1986).
2. Formalin-fixed, paraffin-embedded tissues are serially cut 5 μm thick and are floated in a 47°C water bath. The slide is dipped in 3% Elmer's glue just prior to mounting the tissue section.
3. Air dry sections and bake overnight in a 60°C oven.
4. Deparaffinize tissue in xylene.
5. Rehydrate tissues through graded ethanol and PBS.
6. Air dry tissue sections.
7. Digest sections with proteinase K at 37°C for 15 minutes.
8. Stop proteinase reaction by washing in PBS.
9. Dehydrate through a graded ethanol series, as described earlier.
10. Air dry.
11. Add 10 μl of hybridization mixture plus ^{35}S-labeled probe ($5-10 \times 10^3$ cpm/μl) to each specimen.
12. Cover with a doubly siliconized coverslip.
13. Seal with rubber cement and allow cement to dry.
14. Denature tissue and probe DNA by heating the slides on a metal tray on top of a heating block (adjusted to 110°C) for 7 minutes.
15. Cool tray rapidly on ice.
16. Hybridize tissue sections with radiolabeled probes at 42°C in a humidified chamber for 40–45 hours.
17. Remove coverslips after hybridization and wash slides as described (Syrjanen and Syrjanen, 1986).
18. Dehydrate through 70 and 95% ethanol containing 0.3 M ammonium acetate.
19. Air dry.
20. Dip in Kodak NTB3 nuclear track emulsion diluted with an equal volume of 0.6 M ammonium acetate.
21. Expose at 4°C for 2–7 days.
22. Counterstain with hematoxylin solution after the emulsion is developed.

Discussion

The ability to recognize histopathological changes suggesting *M. fermentans* infections during routine pathological examination of diseased tissues may be the first indication to scientists and clinicians. This indication, when coupled with supporting clinical history, should trigger the series of tests described in this chapter, including immunohistochemistry, *in situ* hybridization, and EM to confirm the presence of *M. fermentans* in the tissues examined. In studies of diseased tissues by immunohistochemistry, it is important to include all the crucial controls such as antiserum or ascitic monoclonal antibody raised against nonmycoplasma antigens to assure that the reaction produced by MAbs to *M. fermentans* is specific. Histiocytes, macrophages, and many degenerating cells in diseased tissues may have endogenous peroxidase activity, which cannot be completely blocked by pretreatment with H_2O_2, and are also more likely to nonspecifically pick up immunoglobulins. Both conditions will produce a positive color reaction in diaminobenzidine and H_2O_2 substrate. We have found that *M. fermentans*-specific monoclonal antibodies stain ballooned alveolar cells and foamy macrophages in the infected lung. Intracellular granular structures stain positive in degenerating alveolar cells and type II pneumocytes which may slough into the alveolar space. In addition, some liver parenchymal cells, adrenal medullary cells, brain neuroglial cells, lymphocytes, or histiocytes in the lymph nodes, as well as glomerular endothelial and epithelial cells, capillary basement membrane, tubular epithelial cells, intratubular casts, and mononuclear interstitial cells in kidney of patients with AIDS, are stained positively. In some acute cases with fulminant *M. fermentans* infections, reactive mononuclear cells in the spleen, liver, and lungs, especially in cells in advancing margins of prominent necrosis in these organs, are most likely to be positive.

Similarly, in the *in situ* hybridization studies using *M. fermentans*-specific DNA probes, controls using vector DNA without insert of mycoplasmal DNA, or other nonspecific DNA, are important. Degenerating cells, granulocytes, and eosinophils in particular often have higher affinity of nonspecific binding to DNA probes. The stringency of washing after tissue hybridization with DNA probes is crucial, and the parallel experiment using a non-mycoplasma-specific DNA probe(s) has to be included as a necessary control. It may take a more extensive infection by the mycoplasma in tissues to be clearly identified by *in situ* hybridization. This can be more easily done in the cases of fulminant mycoplasmal infection with prominent tissue necrosis. In the spleen and lung, strongly labeled clusters of cells are found at the margin of necrosis corresponding to the area of cells both immunohistochemically positive and containing mycoplasma-like particles identified by EM.

Using EM to identify mycoplasma-like structures in tissues without any other supporting techniques can be very difficult. The organism's pleomorphism both in size and in shape without a unique "mycoplasma-specific" structure makes the organisms essentially indistinguishable from many cellular processes and cytoplasmic fragments commonly found in degenerating diseased tissues. We normally will not make a diagnosis on the basis of a few mycoplasma-like particles found in the tissue. Identification of clusters of characteristic organisms is required to make a positive diagnosis in addition to positive immunohistochemical findings. The technique of immunohistochemistry using MAbs to *M. fermentans* is very helpful in localizing specific areas for EM examination. Adequate EM experience in identification of various mycoplasmas in cultures and in infected human and animal tissues is also considered crucial in these studies.

References

Bauer, F. A., Wear, D. J., Angritt, P., and Lo, S.-C. (1991). *Mycoplasma fermentans* (incognitus strain) infection in the kidneys of patients with acquired immunodeficiency syndrome and associated nephropathy: A light microscopic, immunohistochemical and ultrastructural study. *Hum. Pathol.* **22**, 63–69.

Hu, W. S., Wang, R. Y.-H., Liou, R.-S., Shih, J. W.-K., and Lo, S.-C. (1990). Identification of an insertion-sequence-like genetic element in the newly recognized human pathogen *Mycoplasma incognitus*. *Gene* **93**, 67–72.

Klenow, H., and Henningsen I. (1970). Selective elimination of the exonuclease activity of the deoxyribonucleic acid polymerase from *Escherichia coli* B by limited proteolysis. *Proc. Natl. Acad. Sci. USA* **65**, 168–172.

Lo, S.-C., Dawson, M. S., Newton, P. B., Sonoda, M. A., Shih, J. W.-K., Engler, W. F., Wang, R. Y.-H., and Wear, D. J. (1989a). Association of the virus-like infectious agent originally reported in patients with AIDS with acute fatal disease in previously healthy non-AIDS patients. *Am. J. Trop. Med. Hyg.* **41**, 364–376.

Lo, S.-C., Dawson, M. S., Wong, D. M., Newton, P. B., Sonoda, M. A., Engler, W. F., Wang, R. Y.-H., Shih, J. W.-K., Alter, H. J., and Wear, D. J. (1989b). Identification of *Mycoplasma incognitus* infection in patients with AIDS: An immunohistochemical *in situ* hybridization and ultrastructural study. *Am. J. Trop. Med. Hyg.* **41**, 601–616.

Lo., S.-C., Buchholz, C. L., Wear, D. J., Hohm, R. C., and Marty, A. M. (1991). Histopathology and doxycycline treatment in a previously healthy non-AIDS patient systemically infected by *Mycoplasma fermentans* (incognitus strain). *Mod. Pathol.* **6**, 750–754.

Lo., S.-C., Wear, D. J., Green, S. L., and Jones, P. G., and Legier, J. F. (1993). Adult respiratory distress syndrome with or without systemic disease associated with infections due to *Mycoplasma fermentans*. *Clin. Infect. Dis.* **17**(Suppl. 1), S259–263.

Messing, J. (1983). New M13 vectors for cloning. *In* "Methods in Enzymology" (R. Wu, L. Grossman, and K. Moldave, eds.), Vol. 101, pp. 20–78. Academic Press, San Diego.

Syrjanen, S., and Syrjanen, K. (1986). An improved *in situ* DNA hybridization protocol for detection of human papillomavirus (HPV) DNA sequences in paraffin-embedded biopsies. *J. Virol. Methods* **14**, 293–304.

Wise, K. S., Kim, M. F., Theiss, P. M., and Lo, S.-C. (1993). A family of strain-variant surface lipoproteins of *Mycoplasma fermentans*. *Infect. Immun.* **61**, 3327–3333.

A8

LOCALIZATION OF ANTIGENS ON MYCOPLASMA CELL SURFACE AND TIP STRUCTURES

Duncan C. Krause and Marla K. Stevens

Introduction

Numerous improvements in techniques for resolving complex mixtures of proteins, as well as the emergence of new technologies for protein analysis, have been seen over the last twenty years. The application of techniques such as one- and two-dimensional polyacrylamide gel electrophoresis, Western immunoblotting, and microsequencing to the study of the basic biology of the mollicutes has led to the identification of a number of specific proteins which are associated with particular mycoplasma activities, most notably cytadherence and motility. Furthermore, the list of such proteins will undoubtedly continue to grow as techniques employing recombinant DNA technology are creatively applied to the identification of specific gene products (Sperker et al., 1991).

Certain mycoplasma species possess specialized structures, such as the differentiated tip "organelles" seen in *Mycoplasma pneumoniae* and *Mycoplasma gallisepticum* (Boatman, 1979). Furthermore, these mycoplasmas have a detergent-insoluble filamentous protein network (Triton shell) that is reminiscent of cytoskeletons in eukaryotic cells. These structures are thought to participate in cellular processes that include motility, cell division, and adherence to host cell surfaces. As specific proteins are linked to one or more of these activities or are identified as probable cytoskeletal elements, it is often informative to evaluate their subcellular location in order to clarify protein function. If specific antibodies are available for those proteins, two useful approaches can be applied toward that end. Whole cell radioimmunoprecipitation (Hansen *et al.*, 1981) permits the identification of proteins having epitopes that are accessible to antibody binding on the mycoplasma cell surface. Immunoelectron microscopy can be applied to whole cells, thin sections, or Triton shells (Stevens and Krause, 1991, 1992) in order to visualize the subcellular distribution of proteins in the mycoplasma cell.

Materials

For Immunoelectron Microscopy

Formvar–carbon-coated 400-mesh nickel electron microscopy (EM) grids
Needle-point EM grid forceps
TBS–BSA [0.2 M Tris–HCl (pH 8.2), 0.8% NaCl, 1% EM-grade (globulin-free) bovine serum albumin; Sigma, St. Louis, MO]
Hayflick growth medium (Hayflick, 1965)
3cm³ Luer-lok syringe with 25-gauge needle
1.2-μm pore-size Acrodisc filter (Gelman Sciences, Ann Arbor, MI)
10 × 60-mm sterile plastic petri dishes
Poly(L-lysine), 0.1% in H_2O (Sigma)
N-2 Hydroxyethylpiperazine-N-2′-ethanesulfonic acid (HEPES; 25 mM) buffer, pH 7.4, 0.8% NaCl (HEPES–NaCl)
Glutaraldehyde (EM grade)
Paraformaldehyde (EM grade)
Picric acid (EM grade)
Sodium cacodylate
Nonfat dry milk
5- or 10-nm colloidal gold-conjugated secondary antibody (specificity dictated by species in which primary antibody was prepared) (Amersham, Arlington Heights, IL)
Moisture chamber (sealable container with wet paper towels to maintain humidity)
Squeeze bottles
Tris–NaCl [20 mM Tris–HCl (pH 7.5), 150 mM NaCl]
Triton X-100, 10% (v/v) in H_2O
Phenylmethylsulfonyl fluoride (PMSF) 100 mM in 2-propanol
Osmium tetroxide, 0.1% in H_2O
Aqueous lead citrate, 4.4%
Uranyl acetate, 2% in H_2O and filtered

Critical point drier
Absolute ethanol
EM grid storage case

For Whole Cell Radioimmunoprecipitation

Hayflick growth medium (Hayflick, 1965)
20 mM phosphate-buffered saline, pH 7.2 (PBS)
Incubator, 37°C
Refrigerated centrifuge capable of 12,000 g
Oak Ridge centrifuge tubes
Sterile disposable cell scraper
Hanks' balanced salt solution (HBSS)
Dialyzed horse serum
[^{35}S] Methionine
Disposable vinyl gloves
Plastic-backed absorbent paper
Trizma base
Sodium deoxycholate
Sodium dodecyl sulfate (SDS)
Triton X-100
Tetrasodium ethylenediaminetetraacetic acid (EDTA)
PMSF, 100 mM in 2-proponal
Beckman Optima TL100 tabletop ultracentrifuge with TLA100.3 fixed angle rotor and Delrin adapters for 1.5-ml capacity microcentrifuge tubes (Beckman, Palo Alto, CA) (or standard ultracentrifuge with rotor and adapters to allow 45,000 g with volumes less than 1.0 ml)
Specific antiserum
Rocker platform
Cold room or chromatography refrigerator (4°C)
Cowan I strain of *Staphylococcus aureus* (Staph A), prepared according to Kessler (1976)
Microcentrifuge
1.5-ml capacity microfuge tubes
Hot plate for boiling water bath
2-Mercaptoethanol
Glycerol
Ammonium persulfate
TEMED
Acrylamide and bisacrylamide
Glycine

En³Hance (New England Nuclear, Boston, MA)
Polyacrylamide gel electrophoresis apparatus with power supply
Gel dryer with cold trap and vacuum pump
X-ray cassette
X-ray film (e.g., X-O-Mat XAR or XRP film, Eastman Kodak Co., Rochester, NY)
−80°C freezer

Procedure

Immunoelectron Microscopy (Stevens and Krause, 1991, 1992)

1. Wild-type *M. pneumoniae* cells attach readily to EM grids during culture, whereas nonadhering variants do not. To promote attachment of the variants, Formvar–carbon-coated, 300- or 400-mesh nickel grids are incubated under a drop of 0.1% poly(L-lysine) for 1–2 hours at room temperature in a moisture chamber (small sealable plastic container large enough to hold a small petri dish surrounded by water-saturated tissue paper). Note the difference in appearance between the coated and uncoated surfaces of the grids at the outset so that the proper grid orientation can be maintained throughout the procedure. Using filter forceps, gently grasp the grids by the edge and wash under a stream of distilled water dispensed from a squeeze bottle with a fine tip. Blot the edge of the grids on tissue paper, air dry, transfer to a sterile plastic petri dish, and UV sterilize.

2. Log-phase cultures of mycoplasmas are suspended in fresh culture medium, passed through a 25-gauge needle repeatedly to disperse clumped organisms, and filtered using an Acrodisc filter having a 1.2-μm pore size to remove remaining clumps. Passing the mycoplasmas through a 25-gauge needle poses a risk of generating an aerosol. This step should be carried out gently and with care in a biological safety cabinet. Finally, a drop of the mycoplasma suspension is incubated on top of the grids from 1 to 4 hours at 37°C.

3. Remove the grids and wash gently with HEPES–NaCl using a squeeze bottle and blot dry at the edge of the grid as in step 1. Do not allow the grids to dry completely! Immediately proceed to fixation or Triton extraction, depending on whether one wishes to probe whole cells or Triton shells, respectively.

4. For Triton extraction, place grids in Tris–NaCl–2% Triton X-100 in a petri dish and incubate the dish floating in a 37°C water bath for 30 minutes. PMSF (1 mM) can be included here to reduce protease activity. Gently rock the dish periodically for efficient solubilization. Carefully remove the grids, wash as before with HEPES–NaCl, and fix in 1.5% glutaraldehyde in HEPES–NaCl for 30 minutes at 4°C.

For immunogold labeling of whole cells, incubate the grids in 1% glutaraldehyde–1% paraformaldehyde–0.1% picric acid in 0.1–0.2 M sodium cacodylate (pH 7.2) for 30 minutes at 4°C. Note that the fixatives and cacodylate are toxic and should be handled with care and disposed of appropriately.

5. After fixation, wash the grids well with HEPES–NaCl. Grids can be stored in buffer at 4°C for several days before labeling. Before labeling with antibody, block grids with 5% nonfat dry milk in TBS–BSA for 30 minutes at room temperature.

6. Incubate the grids overnight at 4°C in a moisture chamber under a drop (15–20 μl) of the primary antibody diluted in TBS–BSA. Several factors must be determined empirically at this point for each serum. These include the antibody dilution (we have used 1:100 to 1:1000 generally), whether immunoglobulins must first be purified by protein A–Sepharose chromatography, and whether affinity purification of the antibodies is necessary (Stevens and Krause, 1992). We have taken two approaches with good success: protein A–Sepharose but no affinity purification, or no protein A–Sepharose but affinity purification.

7. The grids are then washed well with TBS–BSA and incubated in a drop (15–20 μl) of the colloidal gold-conjugated secondary antibody for 1 hour at room temperature. If the primary antibody was elicited in rabbits, the secondary antibody might be goat anti-rabbit IgG conjugated to 10 nm of colloidal gold. The concentration of the secondary antibody may have to be determined empirically. We routinely use 1:10 or 1:20 dilutions. Wash the grids as before with TBS–BSA followed by distilled water, blot dry by touching the edge of the grid with a bit of filter paper, and air dry unless processing for poststaining, in which case grids should be kept in H_2O.

8. For immunogold-labeled whole cells, counterstain the grids with 2% aqueous uranyl acetate for 8 minutes, rinse with water, followed by 4.4% lead citrate for 12 minutes. Rinse grids with H_2O and air dry. For immunogold-labeled Triton shells, postfix in 0.1% osmium tetroxide in water for 10 minutes at room temperature, rinse with water, stain with filtered aqueous uranyl acetate for 10 minutes at room temperature, rinse with H_2O, dehydrate in ethanol, critical-point dry, and shadow with a thin coat of platinum according to standard EM techniques.

9. It is best to use a proper EM grid holder for the storage of grids. This helps ensure that samples will be protected and not mixed. We routinely examine samples using a JEOLCx 100 transmission electron microscope at 80 keV.

Whole Cell Radioimmunoprecipitation (Leith et al., 1983)

1. Note: the investigator should be properly trained in the safe handling and disposal of radioactive materials before attempting this procedure. Cultures of *M. pneumoniae* are incubated in 50 ml Hayflick medium in 160-cm^2 sterile

disposable plastic tissue culture flasks at 37°C until the phenol red pH indicator becomes red-orange. The Hayflick medium is removed, the monolayer is washed with 2 × 10 ml PBS, and 25 ml HBSS containing 25 mM HEPES (pH 7.4–7.6) and 10% dialyzed horse serum is added to each flask. [^{35}S]Methionine (250 μCi) is carefully added to the HEPES-buffered HBSS, and incubation is continued for an additional 4–6 hours at 37°C. Care must be taken at this point to keep the flask level in order that the entire monolayer will remain bathed in the labeling medium. At the end of this incubation, the labeling medium is carefully transferred to an appropriate container for sterilization and disposal of the radioisotope, and 25 ml of fresh Hayflick medium is added to each flask. The incubation is continued for 1–2 hours at 37°C, after which the monolayers are washed three times with PBS, scraped from the plastic surface into 10 ml PBS using a sterile disposable cell scraper, and harvested by centrifugation at 20,000 g for 30 minutes at 4°C.

2. The mycoplasma pellet is thoroughly resuspended in 1 ml cold PBS. This is routinely done in a biological safety cabinet, and care is taken to avoid generating an aerosol. Generally this involves repeated pipetting with a 1000-μl capacity micropipettor to dissociate clumped mycoplasmas. This step requires time and patience, as the sample should be chilled on ice during the resuspension, and the process must be done gently to minimize cell damage. Once the resuspension is complete, remove a 50-μl aliquot and combine with 50 μl of 2× SP buffer [0.2 M Tris–HCl, pH 6.8, 4% (w/v) SDS, 40% (v/v) glycerol, 4% (w/v) 2-mercaptoethanol, 0.04% (w/v) bromphenol blue]. Boil for 4 minutes, then store at −20°C for polyacrylamide gel electrophoresis (PAGE). This sample will provide the total mycoplasma protein profile.

3. Divide the remaining mycoplasma suspension into up to six equal aliquots in 1.5-ml capacity microfuge tubes. The number will depend on how many sera are to be tested. Plan on at least two controls: a serum-free control and a normal serum control (preferably serum collected prior to immunization from the same animals from which the immune serum was collected). The sera to be tested should each have been heated to 56°C for 30 minutes to inactivate complement. Add 20 μl of a heat-inactivated serum sample per tube of mycoplasma suspension, then incubate at 4°C with continuous gentle rocking for 90 minutes.

4. Centrifuge the mycoplasma suspensions for 5 minutes in a microcentrifuge. Carefully decant the supernatant and drain each tube well by inverting on tissue paper. Resuspend each pellet in 200 μl PBS, centrifuge as before, and repeat the wash step once. This removes antibodies not bound to the mycoplasma cell surface.

5. Resuspend each pellet thoroughly in 500 μl TDSET [10 mM Tris-HCl, 10 mM tetrasodium EDTA, 1% (w/v) Triton X-100, 0.2% (w/v) sodium deoxycholate, 0.1% (w/v) SDS (pH 7.8)] containing 1 mM PMSF. (Note: PMSF is extremely toxic. Care should be taken in preparing and handling PMSF.) PMSF is a protease inhibitor. For some mycoplasma species it may be necessary to

include additional protease inhibitors to minimize the degradation of protein antigens.

6. The samples are incubated for 1 hour at 37°C to allow solubilization of the mycoplasmas. Tap the sides of the tubes periodically to mix the contents and promote more complete solubilization. Centrifuge the samples at 45,000 g for 30–60 minutes to remove incompletely solubilized material. A Beckman Optima TL100 tabletop ultracentrifuge with the TLA-100.3 fixed angle rotor and Beckman 1.5-ml polyallomer microcentrifuge tubes with Delrin adapters works well for this purpose. Carefully collect the supernatants from each sample and transfer to fresh centrifuge tubes.

7. Add 100 μl of Cowan I strain of *S. aureus* (Staph A) which has been formalin treated according to Kessler (1976), washed twice with 20 vol of TDSET, and resuspended to a final concentration of 10% in TDSET. Incubate at 4°C for 2 hours on a rocker platform. Immune complexes will bind to the *S. aureus* protein A during this incubation. Centrifuge the samples in a microcentrifuge for 2 minutes. Carefully remove the supernatent fluid from each tube using a micropipettor and resuspend the Staph A in 1.0 ml TDSET. Centrifuge as before and discard the TDSET wash. Repeat this wash step three more times. After the final wash and centrifugation, suspend the Staph A pellet in 50 μl 1 × SP buffer. Place the samples in a boiling water bath for 4 minutes, then centrifuge for 2 minutes in a microcentrifuge. Transfer each supernatant to a fresh tube and discard the pellets. Store the samples at −20°C until ready for SDS–PAGE.

8. The choice of conditions for analysis of immunoprecipitated proteins is dictated by the size of the proteins to be resolved. We routinely use the procedure of Laemmli (1970) for SDS–PAGE, with an acrylamide concentration of 5 or 10% in the separating gel to resolve proteins 80,000–250,000 Da or 25,000–200,000 Da, respectively. Following electrophoresis the gel is treated overnight in fixative (45% methanol–10% glacial acetic acid) and is processed for fluorography, e.g., using En³Hance (New England Nuclear), according to the manufacturer's instructions. After drying on a standard gel dryer, the gel is placed in an X-ray cassette. In a darkroom, a suitable X-ray film is placed on top of the gel. We generally clip one corner of the film to correspond to the upper right corner of the gel so that the film can be oriented properly later. The cassette is closed and placed in a −80°C freezer, generally for 4–14 days, after which the film is removed and developed.

Discussion

The immunoelectron microscopy protocol described in this chapter has been used successfully to localize the cytadherence accessory proteins HMW1,

HMW3, and HMW4 in *M. pneumoniae*. It has been our experience that confidence in the significance of findings from this approach can only come when sufficient samples have been examined and every logical control has been included, such as preimmune and serum-free controls. There are several points in particular that are worth noting and/or reemphasizing. These are detailed next.

Nickel grids are preferable for such studies, as these are less likely to react with components of the growth medium. The coating of the grids with Formvar and carbon enhances surface strength for subsequent processing. This coating has a shelf life of approximately 8 weeks, after which the grids can be cleaned and recoated.

The use of poly(L-lysine) makes it possible to compare wild-type and non-adhering variants cultured directly on EM grids. However, in each case one should establish whether the labeling pattern differs for wild-type mycoplasmas grown on grids with and without poly(L-lysine). In the referenced studies (Stevens and Krause, 1991, 1992), the only difference noted was an increase in colloidal gold background with the poly(L-lysine) pretreatment of the grids.

The optimum length and conditions for incubation of the mycoplasma cultures with the grids should be established experimentally. We generally place one to three grids per well in 24-well tissue culture dishes or in 10×60-mm petri dishes. Mycoplasma densities are generally low enough that no pH change is observed during the incubation. However, in our experience probing *M. pneumoniae* with antibodies to the HMW proteins, the background labeling (gold particles not associated with the cells) increases with incubation times greater than 2 hours. Although this may reflect a natural process of protein release, this has not been determined.

Once the mycoplasmas have been cultured on the grids, the grids should not be allowed to dry completely until after poststaining. However, it is necessary to remove any liquid that remains with the grids prior to each step in the process by holding strips of filter paper to the edge of the grids. Because many of the incubations are done in a single drop of reagent, residual liquid associated with the grids could alter the reagent concentration and introduce variability into the protocol. Buffers used at this stage are chosen based in part on their compatibility (e.g., lack of reactivity) with the EM fixatives.

The fixation process chemically alters cellular constituents. Generally, the more effective the fixative is in preserving ultrastructural detail, the greater the degree of chemical alterations. This becomes problematic when one wishes to probe fixed samples with antibodies, as too much fixation disrupts epitope characteristics and decreases the antibody recognition. Thus, a balance must be achieved between retaining antigenicity while preserving ultrastructural detail. As the nature of antigenic determinants will vary with the protein of interest (and its cognate antiserum), optimum conditions for fixation may have to be established experimentally. Furthermore, such conditions may vary further between mycoplasma species, for example, with respect to the osmolality of the fixative. This point has been addressed previously by Boatman (1979).

Following fixation, the samples must be treated with a blocking agent, which occupies reactive sites from the fixative so that antibody molecules will not stain the grid nonspecifically (in much the same way that Western immunoblots must be blocked to prevent nonspecific antibody binding to the membrane). If unexpectedly high backgrounds are observed, it may be necessary to test other blocking agents. For example, omit the nonfat dry milk and increase the albumin concentration, as there may be antibody cross-reactivity with determinants in milk proteins. Alternatively, include serum from the same species in which the secondary antibody was produced, at a concentration of 1–10%. In any case, however, globulin-free (EM-grade) BSA is recommended over less expensive forms of BSA commercially available, otherwise unacceptable background levels may result.

One must also consider the possibility that the fixation process has exposed epitopes that might have otherwise been buried beneath the lipid bilayer. This possibility can be evaluated by comparing immunogold staining with pre- and postfixation labeling. The whole cell radioimmunoprecipitation procedure offers an alternative strategy in evaluating antibody accessibility in the absence of fixation.

While immunoelectron microscopy can suggest the spatial distribution of specific proteins in the mycoplasma cell, this technique is not without its weaknesses. For example, the technique is only as specific as the serum, and the ability to establish specificity is limited by the techniques available. Just because a polyclonal "monospecific" serum reacts with a single protein band by immunoblot does not rule out the possibility that the labeling pattern seen by immunoelectron microscopy is not due to reactivity to nonproteinaceous components of the mycoplasma cell. Monoclonal antibodies provide only partial relief from this dilemma, as quite often there is a tradeoff of sensitivity for specificity, given that the monoclonal antibody will bind a single epitope, while the polyclonal antibodies may recognize several antigenic determinants. This dilemma underscores the value of mutant strains that lack the protein of interest in establishing the significance of labeling pattern by immunoelectron microscopy.

The whole cell radioimmunoprecipitation technique does not permit visualization of spatial distribution but, when properly executed, can provide convincing evidence of surface exposure of specific proteins. Several points are worth noting in the experimental protocol, however. The first concern must be artifactual precipitation of "internal" proteins due to cell damage. Mycoplasma suspensions should be kept cold at all times from the point of harvest until the actual solubilization of the mycoplasmas with detergent. In addition, we recommend that all possible sources of complement (including possibly the culture medium) be eliminated by heating as indicated. If one can include antibodies to a cytoplasmic protein as a control, this possibility can be addressed.

The second major source of concern is the efficiency of solubilization by the detergent. The TDSET formulation works well with *M. pneumoniae*, based on a comparison of radioactivity in the solubilized mycoplasma preparations before and after the high-speed centrifugation to remove insoluble material. This parameter should be evaluated when applying the technique to other mycoplasma species to confirm the effectiveness of the detergents.

Some mycoplasma proteins bind to the Staph A immunosorbent in the absence of antibody. It is important to include an antibody-free control sample along with a normal serum control. This provides the necessary profiles for identification of proteins specifically precipitated by antibody due to surface accessibility.

Acknowledgments

The development of these techniques was supported by Public Health Service Research Grant AI23362 from the National Institute of Allergy and Infectious Diseases and by a faculty research grant from the University of Georgia Research Foundation. We thank M. Farmer and C. Kelloes in particular for their technical assistance.

References

Boatman, E. S. (1979). Morphology and ultrastructure of the Mycoplasmatales. In "The Mycoplasmas" (M. F. Barile and S. Razin, eds.), Vol. 1, pp. 63–102. Academic Press, New York.

Hansen, E. J., Frisch, C. F., and Johnston, K. H. (1981). Detection of antibody-accessible proteins on the cell surface of *Haemophilus influenzae* type b. *Infect. Immun.* **33**, 950–953.

Hayflick, L. (1965). Tissue cultures and mycoplasmas. *Tex. Rep. Biol. Med.* **23**(Suppl. 1), 285–303.

Kessler, S. (1976). Cell membrane antigen isolation with the staphylococcal protein A–antibody adsorbent. *J. Immunol.* **117**, 1482–1490.

Laemmli, U. K. (1970). Cleavage of structural proteins during the assembly of the head of bacteriophage T4. *Nature (London)* **227**, 680-685.

Leith, D. K., Trevino, L. B., Tully, J. G., Senterfit, L. B., and Baseman, J. G. (1983). Host discrimination of *Mycoplasma pneumoniae* proteinaceous immunogens. *J. Exp. Med.* **157**, 502–514.

Sperker, B., Hu, P.-C., and Herrmann, R. (1991). Identification of gene products of the P1 operon of *Mycoplasma pneumoniae*. *Mol. Microbiol.* **5**, 299–306.

Stevens, M. K., and Krause, D. C. (1991). Localization of the *Mycoplasma pneumoniae* cytadherence-accessory proteins HMW1 and HMW4 in the cytoskeletonlike Triton shell. *J. Bacteriol.* **173**, 1041–1050.

Stevens, M. K., and Krause, D. C. (1992). *Mycoplasma pneumoniae* cytadherence phase-variable protein HMW3 is a component of the attachment organelle. *J. Bacteriol.* **174**, 4265–4274.

SECTION B
Genome Characterization and Genetics

B1

INTRODUCTORY REMARKS
Shmuel Razin

The impressive advances in our knowledge of the molecular biology and genetics of mycoplasmas (for reviews see Razin, 1985; Dybvig, 1990; Maniloff et al., 1992; Bové, 1993) were made possible by application of the rapidly developing molecular genetics methodology. Although the techniques for mycoplasmal DNA isolation and purification described earlier (Carle et al., 1983) have basically remained the same, significant progress took place in the development of methods for the enrichment, isolation, and amplification of the DNA of the unculturable plant and insect mycoplasma-like organisms (MLOs, phytoplasmas), described in Chapter B2. These technical developments enabled the recent breakthrough in the molecular characterization, taxonomy, and phylogeny of the MLOs and in the diagnosis of the infections caused by them (Chapter E6, this volume; Chapters D11 and E9 in Vol. II).

The introduction of pulse-field gel electrophoresis (PFGE) essentially replaced the previous methods for determining mycoplasmal genome size described by Carle and Bové (1983). The results obtained by PFGE (Chapter B3) have laid to rest the dogma, dominating since the 1970s, that there is a large gap between the genome size of *Acholeplasma* and *Spiroplasma* species and that of *Mycoplasma* and *Urealplasma* species. Furthermore, PFGE, because of its capacity to separate and define relatively large chromosomal fragments, has become useful in the construction of chromosomal restriction maps, a prelude to mycoplasmal genome sequencing and genetic mapping. The complete sequencing of a mycoplasma genome now underway (Proft and Herrmann, 1994) constitutes the first major step in the complete deciphering and molecular characterization of the machinery of a living cell (Morowitz, 1984). The mycoplasma genome sequencing project has already provided valuable information on mycoplasmal genes and their products. Chapter B4 describes in detail the strategy, the various approaches, and the methodology recommended for constructing mycoplasmal physical and genetic maps.

Demonstration of methylated cytosine and/or adenine residues in the DNA of mollicutes has indicated that mycoplasmas, like other prokaryotes, possess the restriction modification system (Razin and Razin, 1980). The possibility of using specific methylation sites in mycoplasma chromosomes as taxonomic markers has been raised (Razin, 1992). DNA methylation should also be taken into consideration in genome sequencing studies and in any procedure where restriction enzymes are to be employed. The methods recommended for DNA methylation analysis are detailed in Chapter B8.

Introduction of foreign DNA into the mycoplasma cell constitutes an essential step in their genetic manipulation. Although mycoplasmas lack the cell wall barrier, their transformation by exogenous DNA is not as easy as could be expected. The two currently available procedures of choice are polyethylene glycol (PEG)-mediated transformation and electroporation. The factors to be considered for optimal transformation by these approaches are described and discussed in detail in Chapter B7. High-frequency genomic rearrangements in mollicutes have gained much attention as possible mechanisms for antigenic variation (see Chapter C3, this volume) and for the ability of mycoplasmas to respond to environmental changes, including those presented by the host. Methods to identify and define the chromosomal sequences that undergo rearrangements are detailed in Chapter B9.

While very few new mycoplasma viruses and extrachromosomal elements have been discovered since the early 1980s, new strategies for the characterization of viral genomes and extrachromosmal elements and their distinction from each other have been devised and are described in Chapter B5. The mounting interest in mycoplasma viruses and plasmids is driven by their potential use as vectors of mycoplasmal genes. The application of recombinant DNA technology in studies of expression of mycoplasmal genes by the classical approach, including their cloning and expression in *Escherichia coli* cells, is limited by the use in most established mollicute species of UGA as a tryptophan codon instead of its conventional role as a stop codon. Thus, expression of mycoplasmal genes coding for proteins may be truncated in *E. coli*. To overcome this difficulty, several approaches are described in this section: (1) Use of *E. coli* strains carrying a UGA opal suppressor, recognizing UGA as tryptophan; (2) site-specific mutagenesis changing UGA in the mycoplasmal coding sequences to UGG (Chapter B10); and (3) cloning the genes to be expressed in *Spiroplasma citri*— an organism recognizing UGA as tryptophan—using as vectors the replicative form of the spiroplasmavirus Spv1 or an artificial plasmid constructed so that its replication is mediated by the chromosomal replication origin (*ori C*) of *S. citri* (Chapter B6).

References

Bové, J. M. (1993). Molecular features of mollicutes. *Clin. Infect. Dis.* **17** (Suppl. 1), S10–S31.
Carle, P., and Bové, J. M. (1983). Genome size determination. *In* "Methods in Mycoplasmology" (S. Razin and J. G. Tully, eds.), Vol. 1, pp. 309–311. Academic Press, New York.
Carle, P., Saillard, C., and Bové, J. M. (1983). DNA extraction and purification. *In* "Methods in Mycoplasmology" (S. Razin and J. G. Tully, eds.), Vol. 1, pp. 295–299. Academic Press, New York.
Dybvig, K. (1990). Mycoplasmal genetics. *Annu. Rev. Microbiol.* **44**, 81–104.
Maniloff, J., McElhaney, R. N., Finch, L. R., and Baseman, J. B., eds. (1992). "Mycoplasmas: Molecular Biology and Pathogenesis." Am. Soc. Microbiol., Washington, DC.
Morowitz, H. J. (1984). The completeness of molecular biology. *Isr. J. Med. Sci.* **20**, 750–753.
Proft, T., and Herrmann, R. (1994). Identification and characterization of hitherto unknown *Mycoplasma pneumoniae* proteins. *Mol. Microbiol.* **13**, 337–348.
Razin, S. (1985). Molecular biology and genetics of mycoplasmas (Mollicutes). *Microbiol. Rev.* **49**, 419–455.
Razin, S. (1992). Mycoplasma taxonomy and ecology. *In* "Mycoplasmas: Molecular Biology and Pathogenesis" (J. Maniloff, R. N. McElhaney, L. R. Finch, and J. B. Baseman, eds.) pp. 3–22. Am. Soc. Microbiol., Washington, DC.
Razin, A., and Razin, S. (1980). Methylated bases in mycoplasmal DNA. *Nucleic Acids Res.* **8**, 1383–1390.

B2

ISOLATION OF MYCOPLASMA-LIKE ORGANISM DNA FROM PLANT AND INSECT HOSTS

Bruce C. Kirkpatrick, Nigel A. Harrison,
Ing-Ming Lee, Harold Neimark, and
Barbara B. Sears

Introduction

Because mycoplasma-like organisms (MLOs or phytoplasmas) are obligate pathogens of plants and insects, isolation of MLO DNA is considerably more challenging than isolating DNA from culturable microorganisms. This chapter presents five methods that have been successfully used to isolate and separate MLO from plant or insect host DNA. The specific method used depends on the desired purity of the DNA preparation, i.e., total nucleic acid fractions are usually adequate for hybridization analyses, whereas isolation of the entire **MLO** genome by pulsed-field gel electrophoresis may be desired for constructing genomic libraries. Space limitations preclude discussions on isolating MLO plasmids but this topic is comparatively straightforward and it is discussed elsewhere (Kuske and Kirkpatrick, 1990). All procedures will be presented for MLO-infected plant materials; however, the same reagents and protocols can be used with MLO-infected insects.

A. Isolation of Total DNA and RNA from MLO-Infected Hosts

This procedure works well for extracting total nucleic acids from both herbaceous and woody plants as well as insects. The following method is essentially the same as the CTAB extraction procedure described by Doyle and Doyle (1990).

Materials

1 to 2 g of MLO-infected plants
50-ml polypropylene centrifuge tubes
Mortar and pestle
Centrifuge and rotor for 50-ml tubes
65°C hot water bath
Liquid nitrogen or powdered dry ice
Chloroform:octanol (25:1,v/v)
2-Propanol
TE buffer (10 mM Tris + 1 mM EDTA, pH 8.0)
5 M ammonium acetate
95% (v/v) ethanol
2% CTAB buffer (for 100 ml):

2% cetyltrimethylammonium bromide (CTAB)	2 g
1.4 M sodium chloride	8.2 g
20 mM EDTA, pH 8.0	3 ml of 0.5 M stock
100 mM Tris, pH 8.0	10 ml of 1 M stock
Sterile H$_2$O	to 100 ml

To facilitate preparation of the buffer, mix the Tris and the EDTA solutions with approximately 60 ml of water and heat to approximately 80°C. Add CTAB and NaCl, stir until dissolved, and then add water to 100 ml.

Procedure

1. Preheat 12 ml of CTAB buffer to 60°C in a 50-ml polypropylene tube placed in a rack in a water bath. It is convenient to process six to eight samples at a time.
2. Grind 1.5 to 2.0 g of fresh, nonnecrotic tissue (leaves, midribs, or even bark scrapings from small twigs) in liquid nitrogen or powdered dry ice, using a mortar and pestle. It helps to cut the tissues up into small pieces with scissors before grinding.

3. Immediately transfer the powdered plant tissue directly into the preheated buffer and swirl the tube to mix. Place samples in 60°C water bath. As each sample is ground up and added to the water bath, go back and gently swirl each of the preceding samples that are incubating.

4. Incubate the samples at 60°C for 25 to 30 minutes each. Start a timer after the first sample is ground and added to the buffer; it takes about 5 minutes to grind and dispense each sample. Thus by the time the last sample is finished it will be time to add chloroform:octanol to the first sample.

5. After approximately 30 minutes of incubation, remove the first tube from the water bath, place it in a rack in the hood, and let it cool for 2 to 3 minutes. Add 8 ml of chloroform:octanol (or isoamyl alcohol) (25:1). Gently swirl the tube to initially mix, then place a rubber stopper over the top of the tube and invert it several times. **Be careful not to let the chloroform vapor pressure cause the sample to spill.** Process the remaining samples in the same manner; perform the extractions and incubate the samples at room temperature.

6. Transfer the tubes to a centrifuge rotor and spin at 7000 rpm for 8 to 10 minutes at room temperature. Carefully remove the top aqueous layer and transfer it to a clean 30-ml glass Corex tube.

7. Add 8 ml ($\frac{2}{3}$ volume) of ice-cold 2-propanol to the sample. Mix thoroughly and leave at room temperature for 5 to 10 minutes. If the samples are left longer, they may turn brown. Strands of DNA or a dense precipitate should form.

8. The DNA is then collected in one of two ways:

   a. If strands of DNA are visible after the incubation period, then it is best to collect the DNA by inserting a glass disposable pipette into the solution and gently stirring it such that the strands of DNA are spooled onto the end of the pipette. **This method only works if *strands* of DNA are visible in the tube.** Transfer the spooled DNA to a tube containing approximately 4 ml of 80% ethanol and 10 mM ammonium acetate. Incubate the DNA for 1 hour or longer in the 80% ethanol. After this washing step the DNA can be collected by centrifugation at 2000 g for 5 minutes. Pour off the ethanol, invert the tube, and let it dry thoroughly. Dissolve the DNA in 1 to 3 ml of TE buffer, depending on the amount of DNA present.

   b. If no DNA strands appear following 2-propanol precipitation, then the DNA should be collected in the following manner. Centrifuge the nucleic acid/2-propanol solution at 5000 rpm for 10 minutes. Carefully pour off the liquid. Wash the pellet with 80% ethanol, centrifuge, and dry the DNA as previously described. Resuspend the nucleic acid smear in 0.75 to 1.2 ml of TE. If the liquid looks cloudy then transfer it to a microfuge tube and spin it for 4 minutes at room temperature. Transfer the clear supernatant to a fresh microfuge tube and discard the pellet.

9. Add $\frac{2}{3}$ volume of 5 M ammonium acetate to the sample, then add 2 volumes of ice-cold, 95% ethanol and incubate at -20°C for 1 to 2 hours. Collect the DNA

by centrifugation, wash with 80% ethanol, dry completely, and resuspend the DNA in TE as described in step 8.

10. Electrophorese approximately 15 µl of the sample in a horizontal electrophoresis gel. Depending on the species and tissue, there should be somewhere around 50 to 100 ng of DNA plus a generous amount of rRNA.

Discussion

This procedure has worked well with both fresh, air-dried, and freeze-dried tissues from a variety of herbaceous and woody plants and generally yields DNA that is pure enough to be digested with restriction enzymes. If contaminants are present that inhibit certain restriction enzymes, then the sample can be further extracted with phenol:chloroform (1:1), then chloroform only, and ethanol precipitated. The procedure of Doyle and Doyle (1990) goes on to treat the DNA with RNase A and to precipitate with ammonium acetate and ethanol. If the DNA is used as a sample for hybridization analysis, then it seems to be completely adequate to stop after step 10.

B. MLO Enrichment by Differential Centrifugation

This procedure utilizes differential centrifugation to produce a MLO-enriched pellet (Kirkpatrick *et al.*, 1987). The low-speed clarification eliminates most host nuclear DNA and many of the carbohydrates, tannins, and polyphenolic materials that can inhibit DNA modification enzymes are discarded in the supernatant following the high-speed centrifugation run. Although less DNA is obtained than in the total nucleic acid extraction procedure described earlier, a greater percentage of DNA is from the MLO (Ahrens and Seemuller, 1992).

Materials

Same materials as listed earlier except no liquid nitrogen is needed.
Fine glass beads or sand for facilitating the grinding of plant tissue
A standard refrigerated centrifuge such as a Beckman SuperSpeed, Model RC5B
15-ml Corex centrifuge tubes plus rubber adapters for rotor
A container of ice that can hold 8 to 12 mortars

MLO grinding buffer (MGB): For 100 ml:
$K_2HPO_4 \cdot 3H_2O$	2.17 g
KH_2PO_4	0.41 g
Sucrose	10.0 g
Bovine serum albumin, Fraction V	0.15 g

Polyvinylpyrrolidone (molecular weight 10,000)	2.0 g
Ascorbic acid	0.53 g
Sterile distilled H$_2$O	to 100 ml

Stock buffer (without ascorbic acid) can be stored at 4 or -20°C. Just prior to use add the ascorbic acid and adjust to pH 7.6 with 1M NaOH.

Procedure

1. Cut up 1.5 g of leaf midribs, roots, or phloem scrapings in small pieces and place in a cold mortar containing 8 ml of cold MGB. Incubate on ice for 10–20 minutes.
2. Grind thoroughly with a cold pestle; the addition of a small amount of glass beads or sand will facilitate grinding. Add another 5 ml of fresh buffer to the mortar and grind again until the tissue is completely broken up.
3. Transfer the homogenate to a cold 15-ml glass Corex tube, and keep the tubes on ice until all the samples are ground. Place the tubes in a cold rotor (Beckman SS34 or SA600) and centrifuge at 4500 rpm (approximately 2,000 g) for 5 minutes at 4°C. Carefully transfer the supernatant to a clean, cold 15-ml Corex tube.
4. Centrifuge the supernatant at 12,000 rpm (approximately 20,000 g) for 25 to 30 minutes at 4°C. Carefully discard the supernatant, and drain the tubes thoroughly for 1–2 minutes at room temperature.
5. Resuspend pellets in 2 ml of CTAB buffer at room temperature (see Method A). Vortex the tubes to resuspend the pellet.
6. Place the tubes in a 60°C water bath and incubate for 20 to 30 minutes. Extract nucleic acids following steps 5 through 10 in Method A except that less reagents are used, i.e., 2 ml of chloroform is used in step 5, a 15-ml Corex tube is used in step 6, and 1.33 ml of 2-propanol is used in step 7. DNA strands are generally NOT visible from an MLO-enriched pellet so collect the DNA by centrifugation following step 8b.

C. Preparation of MLO-Enriched Fractions Derived from Isolated Sieve Elements

This method further enriches for MLOs by using macerating enzymes to soften MLO-infected plant tissues and facilitates the dissection and removal of pathogen-infected phloem elements (Lee and Davis, 1983). Isolated sieve elements are then ground in a buffer, MLOs are collected by differential centrifugation, and DNA is extracted as in Method B.

Materials

Cellulase R-10, macerozyme R-10 (Yakult Pharmaceutical Industry Co., Nishinomiya, Japan) or Macerase (Calbiochem, La Jolla, Ca)
Phosphate-buffered saline (PBS) buffer
Polyvinylpyrrolidone (PVP-40)
A conical tissue homogenizer and a Tenbroeck tissue homogenizer
Enzyme solution for 100 ml:

Cellulase R-10	0.8%
Macerozyme R-10 (or Macerase)	0.4%
$CaCl_2$	1.0 mM
PVP-40	0.5%
Mannitol	0.6 M

Filter through a 0.45-μm membrane filter.

Suspending medium for 1 liter:

Mannitol	0.5 M
HEPES	30.0 mM
PVP-40	0.1%

Adjust the solution to pH 7.0 and autoclave it.

Procedure

1. Collect 50–100 g of young periwinkle leaves showing early stage symptoms and surface sterilize with 20% Clorox (approximately 1% sodium hypoclorite) in a large beaker for 5 minutes. Rinse leaves twice in sterile distilled water.

2. Strip midribs of leaves longitudinally with sharp forceps in such a way that only central portions containing the most vascular tissue (greenish) are removed. Make sure that cortex tissues on both sides of the midribs are removed.

3. Transfer the stripped vascular tissues aseptically into petri plates containing the maceration enzyme solution.

4. Incubate the petri plates overnight in the dark at 4°C.

5. Transfer each of the partially digested vascular bundles (translucent in appearance) to a petri plate containing suspending medium. With forceps, gently separate the phloem tissues (green parts of the bundles), consisting mainly of layers of sieve elements and incompletely digested phloem parenchyma cells, from xylem tissue and other cell debris and transfer the phloem tissue to a separate petri plate containing fresh suspending medium (25 ml for phloem tissues prepared from 50 g of leaves.).

6. Release MLOs and plant cell organelles into the suspending medium by

gently rupturing the sieve elements with glass tissue homogenizers, first with a conical tissue homogenizer to separate the sieve elements and subsequently with a Tenbroeck tissue homogenizer (four to six strokes) to break sieve elements and release MLOs.

7. Enrich for MLOs by differential centrifugation and extract nucleic acids from the resulting pellet as described in Method B. A slightly different protocol for extracting DNA is described by Lee and Davis (1988).

D. Isolation of MLO DNA Using Bisbenzimide/Cesium Chloride Gradients

This procedure utilizes bisbenzimide, a DNA intercalating agent that preferentially binds to A:T-rich DNA, to separate MLO DNA from plant or insect DNA in a cesium chloride density gradient. The bisbenzimide acts as a "float" to decrease the density of the MLO DNA which then forms a distinct upper band in the gradient. The MLO DNA can be recovered by puncturing the side of the centrifuge tube. The resulting DNA is highly enriched for MLO DNA. This method has been used to recover MLO DNA from a variety of infected plants and insect vectors (Harrison *et al.*, 1991; Kollar *et al.*, 1990; Sears *et al.*, 1989).

Materials

Cesium chloride
TE buffer (10 mM Tris + 1 mM EDTA)
Bisbenzimide solution (10 mg/ml in H_2O), store at 4°C in the dark)
Ultracentrifuge
Vertical rotor equivalent to a Beckman 65.2 VTI
5.1 ml (13×51 mm) Quick Seal centrifuge tubes (Beckman No.342412)
Hand-held long-wave-length (366 nm) ultraviolet (UV) light
2-Propanol
16- or 18-gauge hypodermic needle
1-ml disposable plastic syringe
50 to 200 µg of DNA from MLO-infected plant or insect
50 to 200 µg of healthy host DNA to use as a control

Procedure

1. Dissolve 4.55 g of cesium chloride in 3.0 ml of TE. Transfer this solution to a 5.1-ml Quick Seal centrifuge tube. Place tube in a rack for support.

2. Separately suspend 50 to 200 µg of DNA from the healthy or MLO-infected host in 1 ml of TE. Add the DNA solution to the cesium chloride centrifuge tube. Mix thoroughly.

3. Add additional TE to bring the volume near the top of the centrifuge tube (leave enough volume to add bisbenzimide to the top of the gradient). Check the weight of the tubes and adjust to within 0.1 g.

4. Carefully add 65 µl of 10 mg/ml bisbenzimide stock solution to the top of each centrifuge tube. Do *not* mix the tube!

5. Seal the tubes and place in the rotor. Centrifuge at 45,000 to 50,000 rpm for 24 hours at 24°C. Do *not* use the centrifuge brake to slow the rotor at the end of the run.

6. Carefully remove the tubes from the rotor and place them in a test tube rack. Use the long-wavelength UV light to illuminate the tubes from behind. The DNA will be seen as discrete light blue bands in the gradient. Gradients containing MLO DNA should have an unique, upper band of MLO DNA that is not found in healthy controls.

7. Carefully insert a hypodermic needle in the very top of the gradient tube to release negative pressure developed during the centrifugation run. Be sure to do this *before* you remove the DNA bands.

8. Remove the MLO DNA band by inserting another hypodermic needle/syringe through the side of the centrifuge tube just below the band. Slowly withdraw the band to minimize mixing of the gradient.

9. Bisbenzimide can be removed by extracting the recovered band with NaCl-saturated 2-propanol (add 25 ml of NaCl-saturated water to 200 ml of 2-propanol, use the upper 2-propanol band to extract the sample). However, we have also omitted this extraction step and recovered perfectly usable DNA following the precipitation step described later.

10. Dilute the DNA sample with 2.5 volumes of distilled H_2O, add 0.8 volume of ice-cold 2-propanol, and mix thoroughly. Incubate at -20°C for several hours, and collect the DNA by centrifuging at 8000 rpm for 15 minutes at 4°C. Gently rinse the inside of the tube with 85% ethanol. Invert the tube to drain thoroughly but do not let the DNA dry out completely at this step.

11. Resuspend the DNA in 1.5 ml of TE, add 1 ml of 5 M ammonium acetate and 5 ml of 95% ethanol, and incubate at -20°C for several hours. Collect DNA by centrifugation, wash with 85% ethanol, let the tube dry completely, and resuspend the DNA smear in approximately 1 ml of TE.

Discussion

Some plant species, such as periwinkle, may contain more than one DNA band after centrifugation. For this reason it is imperative to run a healthy control.

If healthy host DNA bands are near the MLO DNA band, it may be necessary to recentrifuge the recovered MLO band to further purify it. If this is necessary, then the recovered band can be added to a fresh gradient prepared as described in steps 1 through 3; however, it is *not* necessary to add bisbenzimide to the second gradient. Centrifuge as previously described and remove the upper MLO DNA band. We have used this procedure to isolate MLO DNA successfully from low-titer woody plant hosts such as coconut, pear, and peach. It may be necessary to run several gradients of the infected woody plant host DNA to collect a reasonable amount of MLO DNA. A second centrifugation of the pooled upper bands is essential to produce a discrete MLO band from these low-titer woody plant hosts.

E. Isolation of MLO Chromosomes Using Pulsed-Field Gel Electrophoresis

All of the proceding DNA isolation procedures yield products that are contaminated with plant or insect host DNA. The following method uses pulsed-field gel electrophoresis (PFGE) to separate physically the entire MLO genome away from contaminating host DNA. The MLO chromosome is resolved as a single band in an agarose gel; the MLO chromosome can be removed from the gel using agarose-digesting enzymes or it may be directly digested with restriction endonucleases.

Materials

MLO grinding buffer (see procedure B)
High-titer MLO-infected plant or insect hosts (see discussion)
Low-speed, refrigerated centrifuge and rotor (such as Beckman SS34 or SA600)
15-ml Corex centrifuge tubes and rubber adapters for rotor
Tris/sucrose (TS) buffer (20 mM Tris, pH 8.0, + 10% sucrose)
InCert grade (FMC) agarose for preparing PFGE blocks
SeaKem GTG or SeaKem Gold agarose (FMC) for the primary PFGE gel
2× TES buffer (0.2 M Tris, 0.2 M NaCl, 20 mM EDTA, pH 8.0)
Lysis buffer (0.5 M EDTA, 1% SDS)
Proteinase K solution in sterile water (20 mg/ml)
Storage buffer (0.5 M EDTA, 1% sarcosine, pH 8.0)
1X Tris, borate, EDTA (TBE) electrophoresis buffer
42 and 52°C water baths
A plastic mold for casting PFGE blocks. The rectangular casting hole measures approximately 1.3 mm wide × 3.5 mm long × 4.0 mm deep and holds

approximately 25 µl when full. Our mold was made by precisely cutting several slots on the flat edge of a 4-mm-thick piece of plastic. After the slots were cut, a second flat piece of 4-mm-thick plastic was glued over the edge where the slots were cut, thus forming a 3.5 × 1.3-mm rectangular hole. Tape is then applied to the bottom of the plastic mold so that it can be filled with the agarose sample, thus forming the PFGE block.

A ^{137}Cs gamma-ray source or a rare cutting restriction endonuclease, such as *Not*I
A pulsed-field gel electrophoresis system such as the CHEF-DR III (Bio-Rad)

Procedure

1. Select 1.5 to 2 g of symptomatic, high-titer MLO-infected herbaceous plant material. Leaf midribs, entire young leaves, or roots seem to be the best choices for most plant species.

2. Process the tissue to obtain a MLO-enriched pellet as previously described in section B of this chapter. The final high-speed, MLO-enriched pellet is then suspended in 10 ml of TS buffer.

3. Repeat the low-speed clarification, transfer the supernatant to a clean 15-ml Corex tube, and centrifuge at 12,000 rpm for 25 minutes. Discard the supernatant, briefly drain the tube, and then place it in an ice bucket, being careful that the MLO-enriched pellet is slightly covered with a film of TS buffer.

4. Add 80 µl of TS buffer and gently but thoroughly suspend the pellet using a wide-bore plastic pipette tip. Keep the tube on ice until just before preparing the PFGE block.

Proceed to steps 5–7 *before* grinding the tissue.

5. Prepare a solution containing 1% (10 mg/ml) InCert grade agarose in 2× TES buffer. Heat to approximately 90°C to dissolve the agarose, then transfer it to the 52°C bath to keep it melted.

6. Dispense approximately 2 ml of the lysis buffer into small glass vials and incubate in the 52°C water bath. The SDS will precipitate out of the buffer at room temperature.

7. Cover the bottom of the plastic mold with a piece of masking tape and place the mold face up on ice, making sure it is as flat as possible.

8. Transfer the tube containing the MLO suspension and a second corex tube holding approximately 0.5 ml of the 1% InCert agarose solution to the 42°C water bath. Wait approximately 1 minute and then quickly add 80 µl of the 1% agarose solution to the MLO suspension and mix thoroughly.

9. Rapidly dispense a 25-µl aliquot of the agarose/MLO suspension into each well in the mold. There should be enough material to cast six 25-µl blocks.

Incubate the blocks on ice for approximately 10 minutes. The blocks should solidify completely.

10. Carefully remove the tape from the bottom of the mold so that the blocks can be pushed out of the mold. We fabricated a small plastic plunger that is the size of the block (i.e., 1.3 × 3.5 mm) to push the PFGE block out of the mold and into the small glass vials that contain the lysis buffer. Up to six blocks can be incubated in each 2-ml lysis buffer vial.

11. Add 50 µl of the 20-mg/ml proteinase K stock solution to each vial and mix thoroughly. Incubate the PFGE blocks in the lysis solution for 24 to 30 hours at 52°C. Gently swirl the vials occasionally to keep the blocks from sticking to each other. After approximately 1 day of incubation, remove the old lysis buffer and replace it with new lysis/proteinase K buffer. Incubate for another 24 to 30 hours. The edges of the block will begin to clear as the MLO cells and host membranes are dissolved by the SDS and proteinase K. If the blocks have not completely cleared, i.e., they are still opaque in the center after 3 days of incubation, then change the lysis buffer a third time and incubate for another 24 hours (this third incubation is generally *not* necessary).

12. After the blocks have cleared (they may still be green or a light tan color but they will not longer be opaque), remove the lysis solution and replace it with storage buffer (0.5 M EDTA, 1% sarcosine). Incubate at 52°C for approximately 1 hour to equilibrate the blocks with the buffer. The vials containing the blocks can now be stored at 4°C until used (at least 6 months).

13. Prior to their use the blocks are equilibrated with TE at least three times for 20 minutes per incubation.

14. The MLO chromosomes immobilized within the PFGE block are then linearized by one of the following methods:

    a. The blocks are transferred to a polypropylene tube and irradiated with a gamma-ray source for a total exposure of approximately 12,000 rads (for procedural details see Neimark and Lange, 1990).

    b. Alternatively, the chromosomes can be linearized by digestion with a rare cutting restriction endonuclease such as *Not*I. Prior to overnight digestion with the enzyme, the blocks must be equilibrated with the appropriate restriction enzyme buffer. Three changes of restriction enzyme buffer (100 µl of restriction enzyme buffer/25 µl block for 30 minutes) are usually sufficient.

15. The blocks are then embedded in a PFGE resolving gel (1% SeaKem Gold or GTG agarose dissolved in 0.5× TBE electrophoresis buffer) following the manufacturer's protocol. The gel is run at 60-second switching times using a maximum of 160 mA for 18 hours. The gel is stained with ethidium bromide (1 µg/ml), and the MLO chromosomal band is visualized by UV illumination at 300 nm.

16. The MLO band can be excised from the gel and recovered from the

agarose matrix using the GeneClean DNA purification kit (Bio101) or gelase (EpiCentre Technologies) following the manufacturer's protocols.

Discussion

The most important factor that influences the isolation of MLO chromosomes that can be visualized by staining with ethidium bromide is the concentration of the MLO in the host tissue. For this reason it is imperative to use fully symptomatic plants that are still in good condition, i.e., not necrotic. The concentration of many MLOs, especially the decline inducing MLOs, is higher in the roots than in the shoots; however, the concentration of other MLOs, such as aster yellows, is highest in newly infected shoots. For many MLOs the relative concentration of MLOs is much higher in insect than in plant hosts and this basic protocol can be easily modified to isolate MLO chromosomes from infected insects (Neimark and Kirkpatrick, 1993).

References

Ahrens, U., and Seemuller, E. (1992). Detection of DNA of plant pathogenic mycoplasma-like organisms by a polymerase chain reaction that amplifies a sequence of the 16S rRNA gene. *Phytopathology* **82**, 828–832.
Doyle, J. J., and Doyle, J. L. (1990). Isolation of plant DNA from fresh tissue. *Focus* **12**, 13–15.
Harrison, N. A., Tsai, J. H., Bourne, C. M., and Richardson, P. A. (1991). Molecular cloning and detection of chromosomal and extrachromosomal DNA of mycoplasma-like organisms associated with witches'-broom disease of pigeon pea in Florida. *Mol. Plant–Microbe Interact.* **3**, 300–307.
Kirkpatrick, B. C., Stenger, D. C., Morris, T. J., and Purcell, A. H. (1987). Cloning and detection of DNA from a nonculturable plant pathogenic mycoplasma-like organism. *Science* **238**, 197–200.
Kollar, A., Seemuller, E., Bonnet, F., Saillard, C., and Bové, J. M. (1990). Isolation of DNA from various plant pathogenic mycoplasmalike organisms from infected plants. *Phytopathology* **80**, 233–237.
Kuske, C. R., and Kirkpatrick, B. C. (1990). Identification and characterization of plasmids from the western aster yellows mycoplasma-like organism. *J. Bacteriol.* **172**, 1628–1633.
Lee, I.-M., and Davis, R. E. (1983). Phloem-limited prokaryotes in sieve elements isolated by enzyme treatment of diseased plant tissues. *Phytopathology* **73**, 1540–1543.
Lee, I.-M., and Davis, R. E. (1988). Detection and investigation of genetic relatedness among aster yellows and other mycoplasmalike organisms by using cloned DNA and RNA probes. *Mol. Plant–Microbe Interact.* **1**, 303–310.
Neimark, H. C., and Kirkpatrick, B. C. (1993). Isolation and characterization of full-length chromosomes from non-culturable plant pathogenic mycoplasma-like organisms. *Mol. Microbiol.* **7**, 21–28.

Neimark, H. C., and Lange, C. S. (1990). Pulse-field electrophoresis indicates full-length mycoplasma chromosomes vary widely in size. *Nucleic Acids Res.* **18,** 5443–5448.

Sears, B. B., Lim, P. O., Holland, N., Kirkpatrick, B. C., and Klomparens, K. L. (1989). Isolation and characterization of DNA from a mycoplasma-like organism. *Mol. Plant–Microbe Interact.* **2,** 175–180.

B3

MOLLICUTE CHROMOSOME SIZE DETERMINATION AND CHARACTERIZATION OF CHROMOSOMES FROM UNCULTURED MOLLICUTES

Harold Neimark and Patricia Carle

Introduction

Chromosome size is a particularly significant character of mollicutes because the evolutionary history of these bacteria involved substantial losses of chromosomal DNA. The development of pulsed-field gel electrophoresis (PFGE) together with methods for minimizing shearing of chromosomes during manipulation (Schwartz and Cantor, 1984) has made possible the direct resolution of very large DNA molecules—as large as 10 Mb. PFGE has had a major impact on studies of genomic structure and organization. For mollicutes, results from chromosome size determinations by PFGE have required revision of the mollicute genera descriptions (see Discussion), and size determinations and physical mapping of mollicute chromosomes are contributing significantly to our understanding of the genetic processes that have shaped the mollicutes.

This chapter provides a rapid procedure as well as conventional procedures for preparing mollicute chromosomes for PFGE together with electrophoresis protocols with parameters for resolving DNA molecules in various size ranges. The rapid procedure, which utilizes gamma-irradiation to linearize the circular chromosomes of prokaryotes, also has permitted the isolation—free of host nucleic acids—of full-length chromosomes from uncultured mollicutes such as phytoplasmas and other mycoplasma-like organisms and has made these genomes available for molecular genetic studies (Neimark and Kirkpatrick, 1993; Neimark et al., 1992) (see Chapters B2 and E6 in this volume for additional molecular methods for studying phytopathogenic mollicutes). Thus, pulsed-field isolation of full-length chromosomes when combined with the polymerase chain reaction provides a powerful tool for characterizing uncultured prokaryotes and, to a large degree, accomplishes the equivalent of cloning.

Materials

Preparation of Bacteria in Agarose Gel Blocks

Gel mold
Tool with a screwdriver-like blade slightly smaller than the mold well for extruding blocks from the mold (cut from a plastic sheet or from a plastic syringe plunger)
Medical gloves
Autoclave tape
Agarose, certified for pulsed-field gel blocks (such as InCert agarose, FMC, Rockland, ME)
Micropipetters and sterile tips
Pasteur pipettes, sterile, and bulbs
Ice bucket [a metal block (e.g., taken from a dry bath heater) provides cold dry surface for chilling the mold]
Vortex mixer
Electric hot plate
Water baths, two
Refrigerated centrifuge capable of 12,000 g
Proteinase K (20 mg proteinase K/ml deionized water. Self digest at 37°C for 15 minutes. Store aliquots in screw cap vials at -20°C)
TE, pH 8 (10 mM Tris base, 1 mM disodium EDTA)
TES buffer (100 mM Tris base, 10 mM disodium EDTA, 100 mM NaCl; adjust to pH 8 with HCl)
Sarcosine lysis buffer (1% N-lauroylsarcosine, sodium salt, 0.5 M disodium EDTA, pH 9–9.5). The addition of 10 mM Tris is optional.
Sodium dodecyl sulfate (SDS) lysis buffer. Add 1 g of SDS to 100 ml of 0.5 M disodium EDTA, pH 9. This detergent lysis buffer is used instead of sarcosine for unculturable organisms to provide strong denaturing and lysing conditions. Before use, the buffer must be warmed and kept at 55°C to dissolve the SDS.

Blocks prepared with SDS can be washed with TE at 55°C and used for further procedures, but if blocks are to be stored, the SDS must be washed out and the blocks equilibrated with warm sarcosine lysis buffer to prevent SDS precipitation.
Small screw cap tubes such as cryotubes

Pulsed-Field Gel Electrophoresis

Candida albicans chromosomal DNA size markers (Clonetech, Palo Alto, CA)
Hansenula wingei chromosomal DNA size markers (Bio-Rad, Hercules, CA)
Saccharomyces cerevisiae chromosomal DNA size markers (Bio-Rad, Clonetech, and several other suppliers)
Schizosaccharomyces pombe chromosomal DNA size markers (Bio-Rad, Clonetech, and several other suppliers)
Glass rods for manipulating agarose blocks. Bent in L-shape to match the size of agarose blocks; can be made from Pasteur pipettes by sealing and bending the tips.
Microscope coverslips, sterile, for cutting agarose blocks
Agarose, ultrapure DNA grade, low EEO ($-m_r$): 0.10 to 0.15 such as Seakem LE agarose (FMC, Rockland, ME)
Ethidium bromide stock solution (10 mg/ml water)
Pulsed-field gel electrophoresis apparatus such as Geneline I or Geneline II (Beckman Instruments, Fullerton, CA) or CHEF II (Bio-Rad)
UV transilluminator, 300 nm
Polaroid MP4 camera or similar camera
Red 23A lens filter (Kodak)
Densitometer such as an Ultrascan XL laser densitometer (LKB, Bromma, Sweden)
^{137}Cs gamma-ray source (available in many large hospitals)
20× TAE buffer, pH 8 (0.2 M Tris base, 0.1 M sodium acetate, 0.01 M disodium EDTA). Dilute 1:20 for use.
5× TBE buffer, pH 8 (0.45 M Tris base, 0.45 M boric acid, 0.01 M disodium EDTA)

Southern Blot Analysis

Nitrocellulose or positively charged nylon membranes
A calibrated UV source for cross-linking DNA to nylon or 80°C oven

Procedure

Preparation of Chromosomal DNA in Agarose Blocks

Cells are embedded in agarose, lysed, and deproteinized by modifications of the methods devised by Schwartz and Cantor (1984). All solutions and containers are sterilized. The following items should be ready before beginning the procedure.

1. A suitable mollicute culture. Fifty to 100 ml of late exponential cultures of most mollicutes provides a sufficient number of cells to prepare a large number of agarose blocks. If chromosomes are to be cut with a restriction enzyme or if the effect of gamma-irradiation in producing DNA double-strand breaks is to be studied quantitatively, the culture should be treated with 80 µg chloramphenicol/ml for the last 90 minutes of incubation to reduce the number of replication forks by blocking reinitiation of DNA replication (Pyle *et al.*, 1988). Uncultured mollicutes or bacteria are concentrated from tissue preparations by differential centrifugation in a buffer that has been shown to be osmotically safe, e.g., by microscopy. TES buffer is suitable for the mouse grey lung agent (Neimark *et al.*, 1992) and TES does not appear to be harmful for preparing the agarose that is mixed with phytoplasmas suspended in Tris–sucrose (Neimark and Kirkpatrick, 1993).

2. One water bath at 55°C and another bath at 38°C (use 37°C for uncultured mollicutes).

3. Sterile TES buffer, ice-cold.

4. Molten 1 – 1.5% InCert agarose in TES in a 38°C water bath. (Although a temperature as high as 42°C appears to be tolerated by a number of culturable mollicutes, and higher temperatures allow more time for filling the mold before the agarose begins to gel, higher temperatures may contribute to lysis of some mollicutes.)

5. Gel mold with wells sealed with autoclave tape. Wear gloves. Fold a tab at one end of the tape to aid removal later.

6. Micropipetter for removing the volume of cell suspension to be embedded in agarose (the same pipetter is used for adding an equal volume of molten agarose to the cell suspension).

7. Another micropipetter with the volume preset for filling several mold wells or a Pasteur pipette and bulb for taking up the cell–agarose mixture and filling the mold.

8. Aliquots of sarcosine lysis solution containing 1 mg proteinase K/ml in 13 × 100-ml tubes in a 55°C water bath; 50 µl proteinase K stock solution per milliliter of lysis solution can be added while cells are being centrifuged (use 1.5 ml of lysis solution per five 25-µl blocks).

Wear gloves and be prepared to work rapidly so that little time elapses between the following steps.

1. Chill cultures on ice and centrifuge at 15,000 g for 20 minutes. Discard supernatant and drain tubes or bottles on sterilized filter paper. Cells can be washed once in cold sterile TES but washing is not essential. Carry out the following steps as rapidly as possible: resuspend the cell pellet in cold sterile TES at a 30- to 50-fold concentration relative to the original culture. For the Geneline II mold, resuspend cells in 3 ml TES. Cell pellets from small amounts of uncultured bacteria can be resuspended in as little as 75 μl of buffer before being mixed with an equal volume of agarose. The final suspension should be uniform, but if abundant amounts of cells are available, any stubborn small clumps can be left behind when the cells are taken up. Use a micropipetter to withdraw the volume of cells to be mixed with agarose, transfer to a 12 $\times$ 75-mm tube, and quickly bring to 38°C by swirling in a water bath. Immediately add an equal volume of agarose warmed at 38°C using the micropipette and mix the suspension at 38°C by swirling and/or by drawing up and down in the pipette to be used for loading. Quickly load the pipette without bubbles and fill mold wells. (Suspension of the cells and mixing with agarose should be carried out as rapidly as possible, but see Discussion for deliberate shearing of chromosomes to circumvent the need for gamma-irradiation.) Chill mold by laying flat on a cold dry surface for ca. 5 minutes (cool the Geneline II mold for 15 minutes).

2. Peel off tape and extrude the blocks into lysis solution containing proteinase K. Load about 5 blocks/1.5 ml of lysis solution (a suitable ratio for 25-μl blocks). Return the tubes to the 55°C water bath (25-μl blocks will begin to clear within minutes) and treat for 24–48 hours. (Treatment over a weekend or longer is not harmful as long as the tubes are sealed to prevent evaporation.)

3. Remove the lysis solution and replace with fresh sarcosine solution without proteinase K. DNA sample blocks can be stored for many months or even years at 4°C in screw cap tubes sealed with Parafilm (do not use conical bottom microfuge tubes). Samples can be shipped at room temperature in a small tube filled completely with sarcosinate solution.

Chromosomal DNA Size Markers

Saccharomyces cerevisiae chromosomes are good size markers for mycoplasmas and uncultured mollicutes. For the larger chromosomes found in acholeplasmas and spiroplasmas, *Candida albicans* chromosomes (1.01 to 3 Mb) or *Hansenula wingei* chromosomes (1.05 to 3.13 Mb) provide a useful range of DNA size markers; *Schizosaccharomyces pombe* also can be employed but has fewer size markers in the range useful for mollicutes. Commercially prepared *C. albicans* (Clonetech) and *H. wingei* (Bio-Rad) chromosomal DNA size markers are available; *S. cerevisiae* size markers are available from these and several other suppliers. Alternately, investigators can prepare their own yeast chromosome size markers from *S. cerevisiae* (see for example, Neimark and Lange, 1990 and references therein); *C. albicans* (ATCC 14053) spheroplasts can be prepared from late log cultures as described (Gorshorn *et al.*, 1992) and can then be embedded, lysed with 1% SDS, and digested with proteinase K as for wall-less bacteria.

Cleavage of Genomic DNA by Gamma-Irradiation (Neimark and Lange, 1990)

Genomic DNA agarose blocks are placed in sterile TE in 2-ml polypropylene tubes (NUNC, Denmark), cooled on ice, and irradiated with a ^{137}Cs gamma-ray source (Mark I, Model 68, 22.5 kCi Dual Source Irradiator, J. L. Shepherd and Assoc., San Fernando, CA) at a dose rate of 67.7 Gy per minute [Gy (Gray) = 100 rads; Fricke ferrous sulfate dosimetry, all doses within ± 5%]. Irradiation times are usually 1.85 minutes (periods of 0.46, 0.92, and 1.85 minutes were used corresponding to gamma-ray doses of 31, 62, and 125 Gy).

Following irradiation, blocks are used for electrophoresis or are stored for subsequent use. (Blocks may be stored refrigerated for a few days or returned to sarcosine lysis solution for long-term storage at 4°C). Wash blocks three more times (10–15 minutes each) in a few milliliters of TE, equilibrate in running buffer, drain on the tube wall, and insert blocks into gel wells against the forward wall without bubbles; seal wells using a small amount of agarose retained after pouring the gel.

Restriction Digestion of Agarose-Embedded Chromosomes

Restriction digestion is carried out by standard methods modified for "in-gel" treatment. Dilute the sarcosine solution by rinsing briefly and then washing blocks twice in TE for 10–15 minutes. Inactivate remaining proteinase K by treating blocks in 1 m*M* phenylmethylsulfonyl fluoride (PMSF) at pH 7.0–7.5 for 1 hour on ice (PMSF deteriorates rapidly in aqueous solution); repeat with fresh PMSF solution (a stock solution of 100 m*M* PMSF in 2-propanol is stable at room temperature for at least 9 months). Wash blocks twice in the appropriate restriction buffer for 15 minutes or more, transfer to clean tubes, and add fresh buffer and the restriction enzyme, selected to cut infrequently in low G+C DNA (see Table I). Higher enzyme concentrations and longer treatment times are required for digestion in agarose; treat overnight where enzyme stability permits or replenish enzyme as required. After digestion, wash blocks in TE and insert in the PF gel as described earlier.

TABLE I

RESTRICTION ENDONUCLEASES FOR CHROMOSOME ANALYSIS
OF MYCOPLASMAS AND LOW-GC-CONTENT ORGANISMS

Enzyme[a]	Recognition sequence	Stability[b]	Temperature (°C)
GC-only sequences			
*Apa*I	GGGCC'C	High	
*Bss*HII	G'CGCGC		50
*Nar*I	GG'CGGC		
*Sac*II	CCGC'GG		
*Sma*I	CCC'GGG	Low	25
ATGC sequences			
*Bam*HI	G'GATCC	Low	
*Bgl*I	GCCNNNN'NGGC	Medium	
*Kpn*I	GGTAC'C	Low	
*Mlu*I	A'CGCGT	High	
*Nhe*I	G'CTAGC		
*Nru*I	TCG'CGA	High	
*Sac*I	GAGCT'C		
*Sal*I	G'TCGAC		
*Xho*I	C'TCGAG	High	

[a] Not all of these enzymes are available certified for digestion of agarose-embedded chromosomal DNA for pulsed-field gel electrophoresis.

[b] Provided by BRL, Life Technologies, Inc., or certified for 16 hours of digestion by NEB.

Pulsed-Field Gel Electrophoresis

Prepare PF gel (retaining a small amount of agarose to seal blocks in wells) and perform pulsed-field elctrophoresis according to the equipment manufacturer's directions; it is important to observe safety precautions, recognize that circulating buffer can be electrically hazardous, and always shut off power before touching the system. Parameters for separating full-length chromosomes (or fragments) from mycoplasmas, acholeplasmas, spiroplasmas, and uncultured mollicutes from plants, insects, and other animals are provided in Table II for three commonly available PF electrophoresis systems. As can be seen, switching intervals are chosen depending on the size of the molecules to be separated. A switching interval that is too short will not resolve molecules above the size range of that interval; a switching interval that is too long will have decreased resolution for smaller molecules.

TABLE II

Protocols for Separating Full-Length Mollicute Chromosomes or DNA Fragments

	PF apparatus		
Protocol	Geneline	Geneline II	CHEF-DR II
A. Separation of DNA molecules or full-length chromosomes from ca. 200 to 1600 kbp			
Temperature	9–11°C	8°C[a]	14°C[c]
Buffer	0.25× TAE	0.25× TBE	0.5× TBE
Agarose gel	1% LE	1% LE	1% LE
Electric field	170 mA	370 mA	200 V[d]
Switching intervals and times	4 seconds, 30 minutes	60 seconds, 12 hours	60–120 seconds, 20 hours
	150 mA	120 seconds, 12 hours	
	60 seconds, 18 hours	180 seconds, 12 hours	
B. Separation of larger mollicute chromosomes such as those of acholeplasmas and spiroplasmas			
Temperature	10–14°C	8°C[b]	14°C[e]
Buffer	0.25 × TBE	0.25 × TBE	0.5 × TBE
Agarose gel	1% LE	0.8% LE	0.8% LE
Field strength	120 mA	80 mA	200 V
Switching intervals and times	1 minute, 8 hours	85 minutes, 30 hours	2 minutes, 53 seconds to
	2 minutes, 8 hours	60 minutes, 60 hours	3 minutes, 49 seconds,[d] 62 hours
	3 minutes, 8 hours	54 minutes, 54 hours	
		35 minutes, 48 hours	

[a] This Geneline II protocol resolves molecules from 200 to 2200 kbp.
[b] This Geneline II protocol separates molecules larger than 2200 kbp.
[c] PF conditions from Robertson et al. (1990).
[d] Switching times are varied linearly (ramped) from the initial to the final switch times.
[e] Optimized PFGE parameters for separating molecules in the size range 1600 to 2200 bp provided by Bio-Rad.

Gel Staining, Photography, and Densitometry

Stain gels in ethidium bromide (1 μg/ml of water) for 30 minutes and destain in water for 10 minutes (a typical gel is shown in Fig. 1). Greatly increased sensitivity can be achieved by employing nucleic acid dyes such as ethidium bromide dimer or the ultrasensitive dye SYBR Green I (Molecular Probes, Eugene, OR). SYBR Green I, although expensive, is approximately 25 times more sensitive than ethidium bromide when viewed at 254 nm with epi-illumination (Fotodyne, New Berlin, WI); even with standard 300-nm transillumination, SYBR Green I staining is approximately fivefold more sensitive than ethidium bromide.

Photograph ethidium bromide-stained gels transilluminated at 300 nm with a

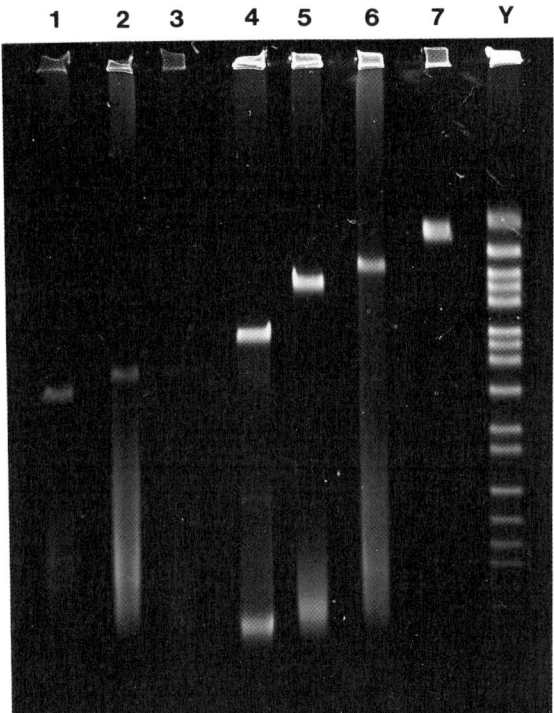

Fig. 1. Pulse-field gel electrophoresis of full-length *Mycoplasma* chromosomes made linear by gamma-irradiation. Lane 1, *M. hominis;* lane 2, *M. arginini;* lane 3, *M. orale;* lane 4, *M. pneumoniae;* lane 5, *M. pulmonis;* lane 6, *M. pullorum;* lane 7, *M. iowae;* lane Y, *Saccharomyces cerevisiae* strain YPH149 chromosome size standards. Chromosomal DNA was prepared in agarose blocks and irradiated as described under Materials and Procedure. Electrophoresis was in a 1% agarose gel at 9–11°C with a current of 150 mA and switching intervals of 60 seconds for 18 hours (preceded by an initial stage consisting of 170 mA with 4-second switching intervals for 30 minutes) in a transverse field gel electrophoresis apparatus (GeneLine, Beckman Instruments). From Neimark and Lange (1990), by permission of Oxford University Press.

red 23A filter (Kodak) using a panchromatic film (Ektapan, 10 × 12.5 cm, Kodak). Alternatively, Polaroid type 665 positive/negative film (7.3 × 9.5 cm) can be used. SYBR Green I-stained gels are photographed with a yellow filter. Special care should be taken not to distort the shape of the gel when it is positioned on the UV box.

Photographic negatives are traced with a densitometer (for example, an Ultrascan HL laser densitometer; LKB, Bromma, Sweden); negatives also can be projected with an enlarger or a slide projector. Migration distances are measured as described in Fig. 2. Chromosome sizes are estimated by comparing their

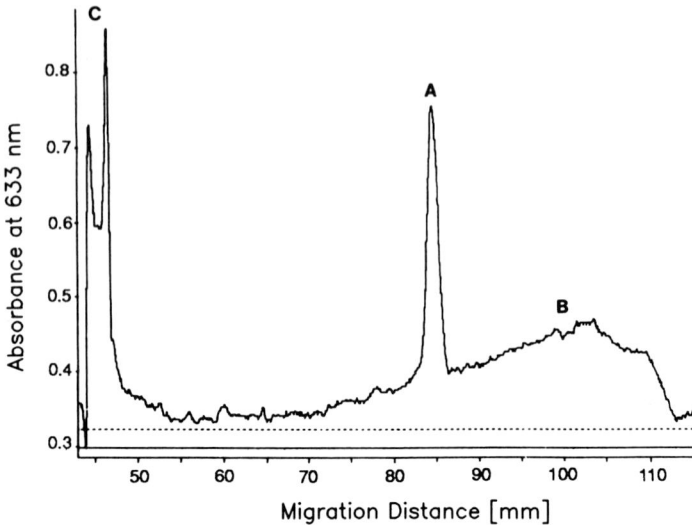

Fig. 2. Densitometer tracing of a photographic negative showing typical PFGE migration of gamma-irradiated DNA from a prokaryote (*M. hominis* PG21). (A) Trace of the band formed by full-length chromosomes which have been made linear by one double-strand break; (B) a region composed of DNA fragments of various lengths produced by multiple double-strand breaks at random sites within chromosomes; and (C) peaks produced by light scattering at the borders of the DNA sample block. Radiation dose: 125 Gy (12,500 rads). Electrophoresis was as described in Fig. 1. Migration, from left to right, is measured from the lower edge of the sample block (C) to the center of the peak of the linearized chromosome (A). From Neimark and Lange (1990), by permission of Oxford University Press.

mobilities to those of DNA size reference bands run in an adjacent lane. At least two lanes of reference marker bands should be included in gels containing multiple samples. A standard curve is produced by plotting the sizes of the reference standards versus migration distance; mobilities are then compared using the linear portion of the curve.

Southern Blot Analysis

Chromosomes resolved by PFGE are transferred to nitrocellulose or nylon membranes by standard methods (Maniatis *et al.*, 1982). To facilitate the transfer of large DNA molecules from PF gels, DNA is depurinated by treating the gel twice in $0.2\ M$ HCl for 10 minutes with gentle shaking and then rinsing briefly in distilled water before continuing with the usual denaturation and neutralization steps (denature twice in $0.5\ M$ NaOH–$1.5\ M$ NaCl for 15 minutes; rinse briefly in distilled water; neutralize twice in $0.5\ M$ Tris base in $1.5\ M$ NaCl, pH 7.0, for 15 minutes). Normally, the total UV exposure time is less than 1 minute; if several minutes of UV exposure occur, it may be necessary to take UV nicking into account and reduce the depurination time. Transfer is carried out in 10× SSC or 10× SSPE (Maniatis *et al.*, 1982). If upward blotting is used, the weight placed on the paper towel stack should not exceed 500 g to avoid compressing the gel; downward blotting may have advantages over the usual upward configuration. The transferred DNA fragments can be bound to nylon either by heating at 80°C for 45 minutes (a vacuum oven is not necessary for nylon) or by UV cross-linking. Hybridization is carried out by standard methods (Maniatis *et al.*, 1982).

Discussion

In the past, chromosome size estimates for mycoplasmas and other bacteria have been determined mainly by DNA renaturation kinetics. Results from PFGE studies have had a profound effect on the concept of chromosome size relations among mollicutes and PFGE has superseded DNA renaturation kinetics for determining chromosome size. Until recently it was thought that mollicute chromosomes fall into just two size classes: one composed of chromosomes clustering around 1600 kbp (ca. 1000 MDa) containing *Acholeplasma, Spiroplasma, Anaeroplasma,* and *Asteroleplasma* species, and the second composed of chromosomes that cluster around 760 kbp (ca. 500 MDa) containing *Mycoplasma* and *Ureaplasma* species, i.e., approximately half the size of the chromosomes in the first group. It is now clear that mollicute chromosomes span a continual range of sizes—from approximately 600 kbp to more than 2000 kbp—and although acholeplasmas and spiroplasmas have relatively large chromosomes, there is no substantial gap in chromosome size between mollicutes with large and small chromosomes (Neimark and Lange, 1990; see also Pyle *et al.*, 1988; Robertson *et al.*, 1990; Carle *et al.*, 1992). This finding has required revision of the descriptions of the mollicute genera.

PFGE combined with the preparation procedure described in this chapter, which utilizes gamma-irradiation to linearize the circular chromosomes of prokaryotes, provides a convenient and rapid method for measuring chromosome sizes and for determining whether chromosomes are circular. The procedure eliminates the need for restriction enzyme digestion and greatly reduces the manipulations and time required for preparing chromosomes for PFGE. Because gamma-irradiation produces just one sharp chromosome band in the gel, optimal pulse-switching intervals for molecules of a given size range can be determined rapidly to allow comparison of chromosome mobilities to size standards. In contrast, restriction digest fragments can span a wide range of sizes and may require several trials to determine switching intervals suitable for measuring the mobilities of all fragments. Thus, for simply determining chromosome size, this procedure is ideal, but obviously restriction digestion is required for physical mapping. The use of gamma-irradiation to produce linear chromosomes for PFGE was devised independently by Van der Bliek *et al.*, (1988).

It is possible to circumvent the need for gamma-irradiation by deliberately shearing chromosomes to introduce a single break in a sufficient number of chromosomes to produce a visible band (for example, by repeatedly pipetting chromosomes up and down in a long capillary Pasteur pipette), but such manipulation is feasible only for organisms that can be cultured.

With any pulsed-field system, one should be alert to the possible occurrence of anomalous mobility (for a discussion and references, see Neimark and Lange, 1990). Also, the possibility that the low G+C content of mollicute chromosomal DNA might cause anomalous mobility has been considered (Pyle *et al.*, 1988; Neimark and Lange, 1990). However, no correlation appears to exist between chromosomal DNA mobility and low average G+C content (Neimark and Lange, 1990). Pyle and co-workers (1988) reported that the mobilities of all restriction fragments of mycoplasma DNA chromosomes tend to be greater relative to *S. cerevisiae* when shorter pulse times are employed; this has the effect that sizes of fragments are underestimated at shorter pulse times. Gravius and colleagues (1994) reported that high-G+C-content molecules, particularly lower molecular weight fragments, migrate faster than lambda (λ) phage molecules (50% G+C) and it appears that the pulse program can affect the relative migration of G+C-rich fragments relative to lambda concatemers [Gravius *et al.* (1994) employed a 4- to 55-second pulse ramp over 28 hours and did not report results with longer pulse periods].

PFGE analysis of full-length chromosome preparations has been especially helpful in studies of uncultured bacteria. These preparations made possible the determination of phytoplasma chromosome sizes and the detection of more than one chromosome in cells concentrated from many phytoplasma "strains" (Neimark and Kirkpatrick, 1993; Neimark, Schneider, Kirkpatrick, and Seemuller, in preparation). This method also can provide clonally pure DNA for molecular genetic studies. For the latter work, the highest quality freshly deionized and filtered water should be used for making solutions which should also be filtered; added precautions include treating plastic gel trays and other items with 10% sodium hypochlorite (household bleach) for several minutes to destroy any contaminant DNA, followed by rinsing with the high quality water.

We expect that size measurements will contribute significantly to tracing chromosome evolution among the branches of the mollicutes.

Acknowledgment

This research was supported in part by a NIH Biomedical Research Support Grant to the SUNY Health Science Center at Brooklyn.

References

Carle, P., Rose, D. L., Tully, J. G., and Bové, J. M. (1992). The genome size of spiroplasmas and other mollicutes. *IOM Lett.* **2,** 232.

Gorshorn, A. K., Grindle, S. M., and Scherer, S. (1992). Gene isolation by complementation in *Candida albicans* and applications to physical and genetic mapping. *Infect. Immun.* **60,** 876–884.

Gravius, B., Cullum, J., and Hranueli, D. (1994). High G+C-content DNA markers for pulsed-field gel electrophoresis. *BioTechniques* **16,** 52.

Maniatis, T., Fritsch, E. F., and Sambrook, J. (1982). "Molecular Cloning: A Laboratory Manual." Cold Spring Harbor Laboratory Press, Cold Spring Harbor, NY.

Neimark, H., and Kirkpatrick, B. C. (1993). Isolation and characterization of full-length chromosomes from nonculturable plant-pathogenic mycoplasma-like organisms. *Mol. Microbiol.* **7,** 21–28.

Neimark, H. C., and Lange, C. S. (1990). Pulse-field electrophoresis indicates full-length mycoplasma chromosomes range widely in size. *Nucleic Acids Res.* **18,** 5443–5448.

Neimark, H. C., Leach, R., Mitchelmore, D., and Lange, C. (1992). Chromosome isolation and electron microscopic studies on an uncultured wall-less prokaryote: The Grey Lung agent. *IOM Lett.* **2,** 260.

Pyle, L. E., Corcoran, L. N., Cocks, B. G., Bergemann, A. D., Whitley, J. C., and Finch, L. R. (1988). Pulsed-field electrophoresis indicates larger-than-expected sizes for mycoplasma genomes. *Nucleic Acids Res.* **16,** 6015–6025.

Robertson, J. A., Pyle, L. E., Stemke, G. W., and Finch, L. R. (1990). Human ureaplasmas show diverse genome sizes by pulsed-field electrophoresis. *Nucleic Acids Res.* **18,** 1451–1455.

Schwartz, D. C., and Cantor, C. R., (1984). Separation of yeast chromosome-size DNAs by pulsed-field gradient gel electrophoresis. *Cell* **37,** 67–75.

Van der Bliek, A. M., Lincke, C. R., and Borst, P. (1988). Circular DNA of 3T6R50 double minute chromosomes. *Nucleic Acids Res.* **16,** 4841–4851.

B4

PHYSICAL AND GENETIC MAPPING

Thomas Proft and Richard Herrmann

General Introduction

Mollicute genomes are small in size, ranging from about 600 to 2500 kbp (Herrmann, 1992). Therefore, physical and genetic maps can be constructed with ease compared to the much larger genomes of conventional bacteria like *Escherichia coli* (4700 kbp), *Bacillus subtilis* (4200 kbp), or *Stigmatella aurantiaca* (9400 kbp).

The complete nucleotide sequence of a genome constitutes the highest possible resolution. This, of course, involves a formidable amount of work and is unnecessary for many purposes. Instead, the so-called restriction map may be satisfactory. This is the complete set of DNA restriction fragments produced by a given restriction endonuclease ordered in the same linear fashion as the fragments are aligned in the genome itself. A restriction map can be converted to a genetic map by localizing genes onto individual restriction fragments. To construct a meaningful genetic map, a restriction map with a high resolution is needed. For instance, a 1000-kbp-long genome, cut only 10 times by a restriction endonuclease, is divided into fragments too large to permit constructing reasonably precise genetic maps, since too many genes map to one fragment, and the relative order of these genes cannot be determined without additional experiments. For mollicute genomes 10- to 20-kbp-long fragments at the average would provide the desirable precision.

The classical and outdated procedure to convert a physical map into a genetic map is to map the relative positions of two or more gene loci by genetic crosses. Genes are now being localized on a physical map (restriction map) by DNA–DNA or DNA–RNA hybridization experiments with specific gene probes.

Construction of Restriction Map

To construct a detailed restriction map, one should start off with establishing a map with a relatively small number of restriction sites (10–20) and then increasing the resolution of this map gradually.

The first crucial step is the selection of restriction enzymes. A suitable combination consists of a rarely cutting endonuclease (3–6 sites in the genome) and of a more frequently cutting one (10–20 sites). Based on the G+C content and the size of the genome, enzymes could be preselected, but in practice it pays off to test a large number of various restriction endonucleases and to select not only by the absolute number of fragments, but also by the size distribution of the fragments. A set of fragments which can be clearly separated by pulsed-field gel electrophoresis (PFGE) is preferable. A prerequisite for a correct analysis is knowledge of the size of the complete genome as well as the size and the number of the individual restriction fragments. Two different approaches can be used to establish the correct order of restriction fragments: (1) one-dimensional PFGE combined with additional methods for aligning individual restriction fragments and (2) two-dimensional pulsed-field gel electrophoresis (2D-PFGE) (Bautsch, 1988; Römling *et al*, 1989).

The more elegant method is 2D-PFGE. The method consists of two sequential steps. First, a mollicute genome is partially digested by an endonuclease X and the resulting DNA fragments are separated by PFGE in the first dimension. A gel strip with the separated DNA fragments is cut off and incubated again with endonuclease X until the DNA fragments are completely digested. The fragments are then separated by a second PFGE, where the fragments migrate perpendicularly to the first direction. Figure 1 explains the principle and shows the evaluation of a simple example. Instead of using the same enzyme for a partial/complete digestion, the analysis could be done with two different enzymes, each digesting to completion. The analysis has to be repeated in the reverse order. An alternative procedure to the 2D-PFGE method is to analyze by one-dimensional PFGE the restriction endonuclease-treated genomic DNA and link the individual fragments by additional means. The following strategies are frequently used: (1) construction of a bank of linking clones, (2) double digest with different restriction enzymes and hybridization with fragment-specific gene (DNA) probes, and (3) application of the Smith–Birnstiel technique.

Linking Clones

A linking clone consists of a relatively short DNA fragment from the genomic DNA which contains *one* restriction site for the enzyme selected for construction of the restriction map. Such a linking clone aligns two adjacent DNA fragments

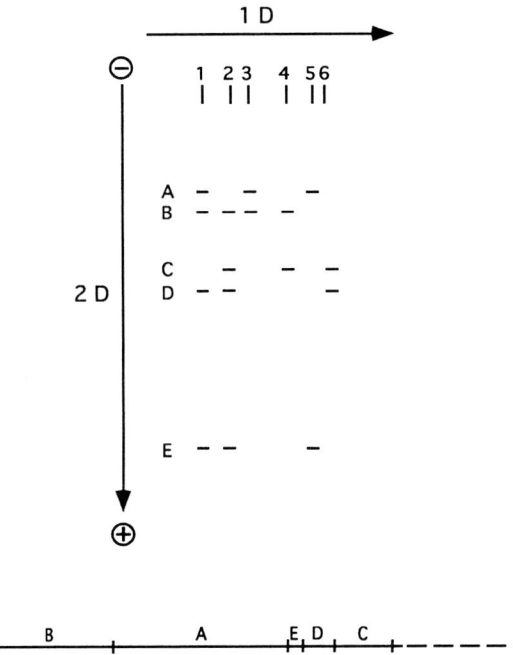

Fig. 1. Alignment of restriction fragments by 2D-PFGE. Genomic DNA was partially digested with a restriction endonuclease and the fragments were separated by PFGE (1D). The numbers (1–6) indicate the different fragments, which are products of partial digestion. After digestion to completion with the same endonuclease in the gel matrix, the sample was analyzed by a second PFGE (2D) which ran perpendicularly to the first PFGE. The alignment of the fragments A–E is shown.

(Fig. 2). Linking clones are also useful for further restriction mapping (see Smith–Birnstiel technique).

Double Digest with Different Restriction Enzymes

A double digest with two different enzymes and comparison of this gel pattern with the pattern after digestion with one enzyme only is a very helpful test for prescreening. For instance, a combination of a rarely cutting enzyme A (5–7 sites) and a more frequently cutting enzyme B (10–20 sites) shows which restriction fragments generated by endonuclease A are cut by enzyme B and vice versa (Fig. 3). Any fragment that is cut only once by the second enzyme represents a linking fragment. In the next step the large fragments generated by the rarely cutting enzyme A can be isolated, labeled with ^{32}P, and used in Southern blotting

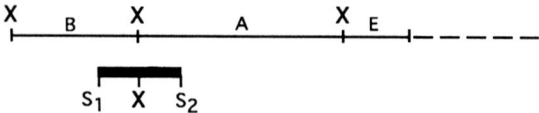

Fig. 2. Example of a linking clone. The thick bar represents a clone which links fragment B to A. X, S_1, and S_2 are restriction sites. Fragments X–S_1 and X–S_2 are specific probes for fragments B and A, respectively (see text).

experiments of genomic DNA digested by enzyme B. Results from these experiments indicate which of the fragments generated by enzyme B partially overlap with or are completely contained within the larger fragments produced by enzyme. A. Instead of isolating restriction fragments, it is also possible to use cloned DNA fragments with unique sequences for correlating restriction fragments generated by different enzymes. The advantage of the one-dimensional PFGE compared with 2D-PFGE is that the DNA bands are sharper and the calculation of DNA fragment sizes is more accurate.

Smith–Birnstiel Technique

The Smith–Birnstiel technique (Smith and Birnstiel, 1976) enables the establishment of the order of restriction fragments of any linear DNA, such as ge-

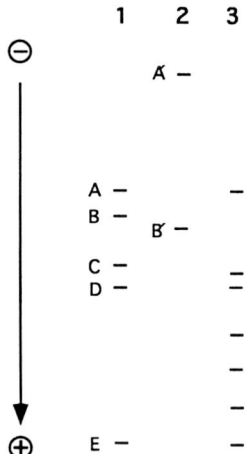

Fig. 3. Double digest with different restriction endonucleases. Lane 1, digest of a circular DNA with an endonuclease X cutting five times; lane 2, digest with an endonuclease Y cutting twice; and lane 3, double digest with endonucleases X and Y. The pattern in lane 3 shows that fragments B and C contain restriction sites for endonuclease Y.

nomic DNA fragments or linearized cosmid or plasmid DNA. This is done by partial digest of the fragment to be mapped with an appropriate endonuclease and by determining the size of all those intermediate fragments which have the same unique end (Fig. 4). In addition, it is absolutely necessary to know the size and number of all restriction fragments which occur after a complete digest. Comparing the size difference between intermediate fragments with the size of the individual restriction fragments to be mapped, one can conclude which fragment is causing the increase from one particular intermediate to the next larger one. By repeating this analysis two by two with all the neighboring intermediate fragments, it is possible to deduce the correct order of restriction fragments. This requires good size markers, preferentially homopolymers of sequenced DNA.

For labeling one end of the DNA to be mapped, any DNA probe which hybridizes between this end and the first restriction site of the endonuclease selected for constructing the map on this fragment may be used, for instance, one-half of a linking clone fragment or a specific oligonucleotide. This ensures that only those DNA products of the restriction analysis which carry the specific

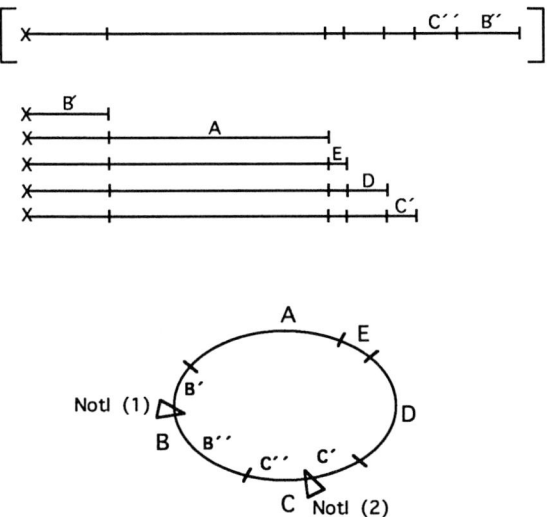

Fig. 4. Alignment of restriction fragments by the Smith–Birnstiel technique. For simplicity, only two *Not*I sites and five *Xho*I sites were drawn. Fragments B' and C' are specific for the two ends of the larger *Not*I fragment, whereas fragments B" and C" are specific for the ends of the smaller *Not*I fragment. All the intermediate fragments of the complete *Not*I digest of genomic DNA hybridizing with a labeled B' fragment are shown. Intermediates with no B' fragment will not show up under the conditions used. The map in brackets shows the alignment for the complete genome.

end show up, whereas all the other intermediates are not visible (Fig. 4). The results of such an analysis should be confirmed by repeating the experiment, but taking the other end of the large fragment to be mapped as a reference point.

Since the double-digest method described earlier is a standard method, we will only describe the procedure for the less frequently used method of linking clones and of Smith–Birnstiel.

MATERIALS

Preparation of Linking Library

Restriction enzyme buffer 10×: use the buffers recommended by the manufacturer.
Ligation buffer (T4 DNA ligase) 10×: 500 mM Tris–HCl, pH 8.0; 100 mM MgCl$_2$; 100 mM dithiothreitol (DTT); 10 mM EDTA; 10 mM ATP; 500 μg/ml bovine serum albumin (BSA)
Sodium acetate, 3 M, pH 5.2 (adjust to pH 5.2 with glacial acetic acid)
TE buffer: 10 mM Tris–HCl, pH 8; 0.1 mM EDTA
Buffered phenol: 1000 g phenol; 1.2 g hydroxyquinoline (antioxidant); 250 ml deionized H$_2$O; 62.5 ml 1 M Tris base
Solutions for plasmid isolation: (I) 50 mM glucose, 25 mM Tris–HCl, pH 8; 10 mM EDTA, pH 8; store at 4°C; (II) 0.2 N NaOH, 1% sodium dodecyl sulfate (SDS); (III) 5 M potassium acetate, 60 ml; glacial acetic acid, 11.5 ml; deionized H$_2$O, 28.5 ml; store II and III at rom temperature
Ethanol, 70% (v/v)
TBE buffer 5x for agarose gel electrophoresis: 0.45 M Tris base; 0.45 M boric acid; 0.01 M EDTA, pH 8, working solution is 0.5x TBE
Agarose: 1% low melting agarose (melting temperature 65°C) for isolation of DNA fragments; 0.5–1.5% standard agarose (melting temperature 87°C) for analytical gels
Anion-exchange columns, e.g., Quiagen (Diagen, Germany) for plasmid purification (follow the instructions of the manufacturer)
LB medium, per liter: Bacto-tryptone, 10 g, Bacto-yeast extract, 5 g; NaCl, 10 g, deionized H$_2$O, pH 7.0
LB plates: add 1.5% Bacto-agar to LB medium
Concentrations of antibiotics stock solution: tetracycline, 10 mg/ml; ampicillin, 50 mg/ml; chloramphenicol, 25 mg/ml
Plasmid vectors: numerous plasmids vectors are available; for standard cloning procedures we frequently use pUC18/19, pSP64/65, pBR322, pBluescript (Stratagene, La Jolla, CA). Do not use a vector that contains the restriction site separating the fragments to be linked. These and other plasmid vectors also

provide a convenient source for isolating antibiotic resistance genes. For detailed information concerning all aspects of molecular cloning, we recommend consulting standard books such as Sambrook et al. (1989).

Escherichia coli strains for cloning and propagation of plasmids. The strain used should meet at least the following requirements: sensitivity against various antibiotics; good growth and plating behavior; defect in general recombination and restriction modification systems; and lactose utilization (α-complementation). The frequently used *E. coli* strains are XL1-Blue, HB101, DH5, and DH5α.

Water baths
Centrifuge (refrigerated) for preparative purposes
Centrifuge (refrigerated) for Eppendorf tubes
Standard horizontal gel electrophoresis apparatus
UV transilluminator: 254 and 312 nm
Eppendorf tubes; all manipulations in volumes smaller than 0.5 ml are done in Eppendorf tubes unless otherwise stated
Micropipetters and sterile tips

Smith–Birnstiel Technique

Preparation of bacteria in agarose and lysis of cells are described in Chapter B3, this volume.

TE buffer: 10 mM Tris–HCl, pH 8; 0.1 mM EDTA. This buffer contains only 0.1 mM EDTA compared to the conventional buffer with 1 mM EDTA
TE$_{50}$: 10 mM Tris–HCl, pH 8; 50 mM EDTA
TE$_{50}$ containing 40 μg/ml phenylmethylsulfonyl fluoride (PMSF) for inactivation of proteinase K
Lysis buffer: 1% *N*-lauroylsarcosine; 0.5 M EDTA; 10 mM Tris–HCl, pH 9.5; 1 mg/ml proteinase K (stock solution 20 mg/ml in H$_2$O, store at -20°C)
Restriction enzyme buffer 10x: follow the instructions of the manufacturer
EDTA, 0.5 M (pH 8); adjust to pH 8 with sodium hydroxide
Transfer solution for capillary transfer of DNA fragments from the gel to a nylon membrane: 0.4 N NaOH, 1.5 M NaCl; 0.25 N HCl for depurinization of DNA prior to transfer
Ethidium bromide: stock solution (10 mg/ml water), working solution (1 μg/ml)
Agarose of low electroendosmosis certified for use in PFGE
Nylon membranes
Apparatus for clamped homogeneous electric field electrophoresis (CHEF)
Oven, 80°C
UV transilluminator: 254 and 312 nm

PROCEDURE

Preparation of Linking Library

For simplicity, we make the following assumptions: The genome to be mapped is about 1000 kbp in size. The restriction endonucleases used are *ApaI* for constructing the map and *Hin*dIII for preparing the linking library. *Hin*dIII cuts the genome 200–500 times.

1. Digest 5 μg of genomic DNA with nuclease *Hin*dIII in 50 μl buffer. The DNA concentration for restriction analysis should not be higher than 200 μg/ml. Follow the instructions of the manufacturer concerning buffer and temperature for the enzyme reaction. Calculate the required enzyme units to get complete digestion in a given time. Increase the calculated units by a factor of 3 to ensure that the genomic DNA is completely digested.

2. Check for completeness of digestion by taking samples (0.5 μg) after one-third, two-thirds, and at the end of the calculated reaction time and run them on standard agarose gels (0.8% agarose). The fragment pattern of the last two time points should be the same.

3. Inactivate the restriction enzyme by extraction with an equal volume of phenol/chloroform (1:1).

4. Separate the aqueous upper phase and add 0.1 volume of 3 M sodium acetate and 2.5 volumes of ethanol, keep it on ice for 20 minutes, and collect the DNA by centrifugation for 15 minutes at 4°C at 12,000 g ; wash the pellet three times with 70% (v/v) ethanol. Dissolve the DNA in TE buffer at a concentration of 200 μg/ml.

5. Ligate 0.2 μg ($\approx$0.06–0.16 pmol*) cleaved genomic DNA to 0.4 μg ($\approx$0.2 pmol) of a plasmid vector which was linearized by *Hin*dIII in a final volume of 20 μl†. The vector plasmid should not have an *ApaI* restriction site. Incubate for 12–16 hours at 16°C or at room temperature (22°C) for 2–4 hours. If the percentage of plasmids with an insert is low (<10%), increase the amount of cleaved genomic DNA by a factor of 3 to 5 in the ligation mixture.

6. Since the efficiency of the ligation reaction and the quality of competent cells vary, it is advisable to do a pilot experiment and transform competent *E. coli* cells with 5 μl of the ligation mixture. There are many different procedures for preparing competent cells; we use routinely frozen competent cells (Hanahan, 1983) from different *E. coli* strains (see Materials). Test the quality of the competent cells by transformation with 0.001 pmol plasmid for 200 μl cells. About 20,000 transformants represent a library.

* The calculation is based on the following: the molecular weight of one base pair is 660; the genome is 1000 kbp long and *Hin*dIII cuts it statistically 244 times (1000 kbp/4^6 bp), neglecting the G+C content of the genomic DNA and of the DNA sequence of the restriction site.
† Treat the linearized vector with calf intestinal alkaline phosphatase (see p. 146).

7. For transformation of 200 µl competent cells do not use more than 0.1 pmol vector DNA. Follow the standard procedure. Keep all buffers and cells at 4°C. Mix competent cells with ligated DNA, incubate for 20 minutes at 4°C, shift for 4 minutes to a 37°C water bath, and cool again on ice for a few minutes. Add 0.8 ml LB medium to the transformation mixture and incubate for 30 minutes at 37°C to permit expression of the antibiotic resistance gene. After that, dilute the cells in 9 ml of LB medium which contains the selective antibiotic and grow the bacteria for 4 more hours at 37°C with good aeration for amplification of the cloned plasmid.

8. Harvest cells by centrifugation for 10 minutes at 8000 g and isolate the plasmids by the rapid alkaline method. The final purification of the plasmid mixture could be done by an affinity column (Quiagen). Determine the plasmid concentration.

9. Digest the plasmid preparation with *Apa*I. Be aware that only those plasmids which carry a linking fragment will be digested (a *Hin*dIII fragment with an *Apa*I site). Therefore, you will not see a shift from the circular form of plasmid to the linear one if you monitor the digestion by agarose gel electrophoresis.

10. To find the linking fragments, a selectable marker, for instance, a gene coding for antibiotic resistance, will be ligated to those plasmids which have an *Apa*I site. The ligation has to be done under conditions which conserve the *Apa*I sites. Add 2 µg of plasmid DNA to 0.2 µg (≈0.2 pmol) of a gel-purified DNA fragment (assumed length 1500 bp in a final volume of 20 µl) coding for an additional selectable marker, for instance, resistance to chloramphenicol.

11. Transform competent cells again with a maximum of 5 µl ligation mixture; allow expression of the antibiotic resistance genes and plate them on an agar plate containing 100 µg/ml ampicillin and 25 µg/ml chloramphenicol. Colonies growing under these conditions (37°C up to 24 hours) should be picked up and individually grown in 2 ml of LB medium containing ampicillin (50 µg/ml) and chloramphenicol (25 µg/ml) for up to 24 hours.

12. Isolate plasmid DNA from the individual cultures by the rapid alkaline method and check for (a) the DNA fragment carrying the chloramphenicol resistance gene by *Apa*I restriction analysis and (b) the linking fragment by *Hin*dIII restriction analysis. The expected results for positive clones after restriction with *Apa*I are as follows: all the clones should have the fragment coding for the chloramphenicol resistance gene and a second fragment which consists of the vector DNA and the original cloned genomic *Hin*dIII fragment with the *Apa*I restriction site. Digestion of this second fragment with *Hin*dIII will give three fragments, one being the linearized plasmid and the two other ones being the halves of the *Apa*I-digested *Hin*dIII fragment. If one of these linking clones is used as a hybrization probe in a Southern blot analysis of *Apa*I-digested genomic DNA, only two fragments should show up. They have to be adjacent on the bacterial chromosome.

Mapping by Smith–Birnstiel Technique

The Smith–Birnstiel technique can be applied to establish the order of restriction fragments on linear DNA fragments with defined ends for instance, linearized plasmids, cosmids, genomic DNAs, or individual restriction fragments. As an example we take a genome (1000 kbp) with two *Not*I sites and 10–15 *Xho*I sites. Two linking clones for the *Not*I sites are also available. The procedure for the PFGE is more or less the same as described by Neimark and Carle (Chapter B3, this volume). For a detailed description of PFGE, see Birren and Lai (1993).

1. Prepare agarose plugs with 3×10^9 cells (equivalent to about 6 μg DNA, assuming 1.5–2 genomes per cell) in a final volume of 200 μl.

2. Transfer the agarose plugs to a sterile tube containing lysis buffer and 1 mg/ml proteinase K, and incubate for 24 hours at 50°C. The plugs should be completely covered with lysis solution; shake gently.

3. Wash the plugs twice for 10 minutes with TE, and inactivate the proteinase by incubating the plugs for 30 minutes at 50°C in TE containing PMSF at a concentration of 40 μg/ml. Repeat this step. Wash the plugs three times for 10 minutes at 50°C in TE. Store at 4°C in 10 mM Tris–HCl and 50 mM EDTA (pH 8.0).

4. Equilibrate the plugs with the appropriate restriction buffer (twice for 10 minutes). Transfer to a new sterile tube, add 40 μl of 10x restriction buffer, 160 μl TE, bovine serum albumin (DNase-free) at a final concentration of 50 μg/ml, and restriction enzyme (*Not*I) at a concentration 10 times higher than that calculated for complete digestion in a given time. Stop the reaction by replacing the restriction buffer by washing the plugs in TE twice for 10 minutes at 4°C. Store the plugs in 10 mM Tris–HCl and 50 mM EDTA (pH 8.0).

5. Cut off a 40-μl slice from the *Not*I-treated agarose plug and check for complete digestion by PFGE.

6. For partial digestion with endonuclease *Xho*I, cut the residual part of one plug into four slices (40 μl), and wash them twice for 10 minutes at 4°C with 1x restriction buffer. Set up four tubes with one plug slice in each 1.5-ml Eppendorf tube and add 8 μl 10x restriction buffer and 40 μl TE; incubate for 30 minutes at 37°C. Add then to each gel slice 40 μl 1x restriction buffer containing the enzyme *Xho*I which has been prepared in the meantime by a twofold serial dilution of *Xho*I in 40 μl *Xho*I buffer. The enzyme concentration in the starting solution should be high enough to ensure complete digestion in a given time (e.g., 2 hours).

7. Stop the reaction by adding 10 μl of 0.5 M EDTA (pH 8) and by cooling on ice.

8. Analyze the DNA by an appropriate PFGE system. That is, a system which

produces straight lanes and separates DNA fragments reproducibly by size. The contour-clamped homogeneous electric field (CHEF) technique is recommended. The PFGE conditions depend very much on the expected size range of the DNA. The CHEF system is adequate for analyzing fragments ranging between 50 and 2000 kbp in length. Follow the instruction of the manufacturer concerning pulse times.

9. Load the samples (slices containing about 1 μg DNA) on the prepared agarose gel and run the gel under conditions that are optimal for the expected size range of fragments. Load in every third well a size marker, for instance, phage λ DNA concatemers, homopolymers of any linearized plasmids or cosmids, or yeast (*Saccharomyces cerevisiae*) chromosomes. Apply the marker in two concentrations: at a higher concentration in the two outside lanes, for staining the gel with ethidium bromide; and at a lower concentration interspaced after each third lane. The less concentrated size markers will be visualized only by hybridization with labeled probes after DNA transfer from the gel to a nitrocellulose or nylon membrane. Stain the gel in ethidium bromide (1 μg/ml) for 30 minutes. The pattern of the size marker with the high concentration shows the quality of the run. The pattern of the low concentration size markers should be almost invisible under these conditions, otherwise it would exhibit strong signals after hybridization. (The concentration of size marker depends on the size range to be covered and on the extent of polymerization.) Transfer the DNA to a nylon membrane by Southern blotting. We prefer nylon because it is easier to handle than nitrocellulose and the DNA can be transferred to this membrane by a sodium hydroxide solution (1.5 M NaCl, 0.4 N NaOH). To improve the transfer of large DNA fragments, soak the gel twice for 10 minutes in 0.25 M HCl at room temperature prior to DNA transfer. The DNA is fixed to the membrane by heating at 80°C for 1 hour.

10. Prepare the two *Not*I-linking clones. Digest each clone with *Not*I and with a second endonuclease having a unique restriction site in the multiple cloning region of the plasmid. Two fragments will be obtained from each linking clone, all together four specific probes, one for each end of the two *Not*I fragments. Separate the fragments of the digested linking clone DNA by semipreparative agarose gel electrophoresis. Use low melting agarose; 1–2 μg of each fragment will be sufficient. Elute the DNA fragments from the gel by hot phenol extraction (Sambrook *et al.*, 1989).

11. Label these probes with ^{32}P by nick translation or by oligonucleotide (hexamer) randomly primed DNA synthesis (Sambrook *et al.*, 1989). For interpretation of results, see Fig. 4. Nonradioactive label, for instance, by digoxigenin containing DNA precursor, is equally useful.

12. Use these labeled probes together with the size marker-specific probes for hybridization of the Southern blot filters. For interpretation of results, see Fig. 4 and Wenzel *et al.* (1992).

By a slightly modified procedure any plasmid- or cosmid-cloned DNA fragment can be mapped. One has to be careful with the selection of vectors. They should have restriction sites for a rarely cutting enzyme (e.g., *Sfi*I and *Not*I) flanking the cloning site. The plasmids or cosmids are linearized by one of the rarely cutting enzymes and treated further as described by Wenzel *et al.* (1992).

DISCUSSION

The construction of a restriction map with a high resolution cannot be achieved by a single method only, but a combination of the four methods described earlier will always give satisfying results, for example, see Ladefoged and Christiansen (1992). An alternative method can be based on the construction of an ordered library consisting of overlapping cosmids. For a 1000-kbp genome, a set of 40–50 cosmid clones should be sufficient, as has been shown for *Mycoplasma pneumoniae* (Wenzel and Herrmann, 1989). The advantage of this approach is that one is dealing with DNAs ranging between 40 and 50 kbp in size. These can be analyzed by conventional 0.5% agarose gel electrophoresis, generated in large quantities and further subdivided into smaller restriction fragments ensuring high resolution and precise mapping. Construction of such an ordered cosmid library requires much more work during the initial phase of library establishment, but after that all further investigations are facilitated significantly. Only one obstacle to this approach remains, as there are some conflicting reports concerning the stability of cloned DNA. Bové *et al.* (1990) reported that cosmid clones with DNA inserts from *Spiroplasma citri* could not be maintained in *E. coli*. In addition, it is possible that some DNA fragments cannot be cloned in a plasmid or cosmid vector at all. However, it would be still worthwhile to have an incomplete cosmid bank covering "only" 80% of the genome since this could already facilitate constructing physical and genetic maps. Nevertheless, prior to recommending the use of cosmid gene libraries as standard tools for constructing maps, it has to be established whether the DNA of mollicutes with a G+C content below 30 mol% can be cloned and maintained in *E. coli* without problems.

Construction of Genetic Map

INTRODUCTION

To convert a physical map into a genetic map, gene-specific DNA or RNA probes are required which hybridize in Southern blotting experiments with a single fragment of the restriction map. One has to differentiate between widely distributed conserved bacterial genes and species-specific genes. It is easy to find probes for conserved genes [see below, strategies (1) and (2)]; but difficulties are caused by species-specific genes or by genes that are also present in other bacteria, but have not yet been detected and characterized. For instance, open reading frames, sometimes overlapping, can be proposed based on data from a DNA sequence analysis. If none of these frames show a significant homology to a known gene or gene product, it is difficult to pin down the frame actually used by the bacterium and to define the hitherto unknown gene. In these instances we recommend strategy (4). This procedure can also be applied if antibodies to a specific protein are the only analytical tool available. If we have information on a partial amino acid sequence of a protein, e.g., 10–15 amino acids at the N-terminal region, then strategy (3) is helpful.

The following strategies can be followed: (1) For mapping conserved genes it might be sufficient to use the cloned corresponding genes from another species (heterologous probe). Typical examples are the genes for ribosomal RNAs, the F_0F_1-ATPase, or the elongation factor EF-Tu. (2) If one is interested in a survey of a genome, random cloning and a sequencing strategy at a reasonably large scale might be the method of choice (Peterson et al., 1993). By this approach, several hundred individual clones have to be sequenced and the data analyzed by comparison to DNA and protein sequence databases. About 50% of the sequence data can be identified by significant homology. The clones serve as probes for mapping. (3) If a protein sequence is known, for instance, the N-terminal region of a purified protein, it is possible to deduce a DNA sequence and use it as a probe. Because of the degenerate code, a mixture of oligonucleotides between 15 and 25 nucleotides long may be adequate in most cases. (4) Species-specific antibodies can be used to identify fusion proteins with a mollicute-specific moiety (Proft and Herrmann, 1994). The principle of the method (Fig. 5) is to fuse an open reading frame to a gene that is expressed very well in *E. coli*. Gene fusion increases the probability that the cloned mollicute moiety is expressed and the product is maintained stably in *E. coli* so that it can be isolated in sufficient quantities for immunization of mice or rabbits to get high-titered sera. The construction of gene fusion is done either in a directed manner, based on DNA sequence data, or by shotgun cloning without any DNA sequence information. Fusion proteins recognized by the specific antibodies have a mollicute-specific protein moiety. A positive reaction is taken as proof that a mollicute DNA fragment has been fused in frame and that the expressed mollicute moiety is also expressed *in vivo* in the bacterium. The clone of the gene fusion serves as a specific probe for mapping. A fusion protein can also be used to induce the synthesis of monospecific antibodies in mice or rabbits and to localize the corresponding proteins in the mollicute subcellular fractions by Western blotting. The crucial part of these experiments is the quality of antisera used for immunoscreening. We have subfractionated bacterial cells and used the subfractions for immunization (membrane, cytosolic, Triton X-100-insoluble fraction). We received cell fraction-specific antibodies which helped us to assign gene products of open reading frames to a subcellular fraction. Again, since the strategies (1), (2), and (3) apply standard techniques, we will only describe in detail strategy (4).

Identification of Mollicute Gene Products

Principle: The construction of expression libraries from genomes or gene libraries and screening with cell fraction-specific antisera is outlined in Fig. 5.

MATERIALS

Construction of Expression Libraries from Genomes or Gene Libraries

Endonuclease *Bgl*II, endonuclease *Sau*3A, calf intestinal alkaline phosphatase (CIP), T4 DNA ligase (Boehringer Mannheim)
Buffer M 10×: 100 mM Tris–HCl, pH 7.5; 100 mM MgCl$_2$; 500 mM NaCl; 10 mM dithioerythritol (DTE)
Buffer A 10×: 330 mM Tris–acetate; 100 mM magnesium-acetate; 660 mM potassium-acetate; 5 mM DTE
CIP buffer 10×: 500 mM Tris–HCl, pH 9.0; 10 mM MgCl$_2$; 1 mM ZnCl$_2$; 10 mM spermidine
Ligation buffer (T4 DNA ligase) 10×: 500 mM Tris–HCl, pH 8.0; 100 mM MgCl$_2$; 100 mM DTT; 10 mM EDTA; 10 mM ATP; 500 μg/ml BSA
TE buffer: 10 mM Tris–HCl, pH 8.0; 0.1 mM EDTA
0.25 M EDTA; phenol; chloroform; 3 M sodium-actate, pH 5.2; ethanol; 70% (v/v) ethanol
Expression vectors: pQE13, pQE14, pQE15 (Diagen, Hilden Germany), see Fig. 6
E. coli strain M15, containing plasmid pSUPMP [pSUPMP codes for the lac-controlled UGA suppressor gene *trp*176, a chloramphenicol resistance marker, and the *lac* repressor (*lac*I gene)]
1.5-ml reaction tubes (Eppendorf), water bath (37°C)

Fig. 5. Strategy for the identification of mollicute gene products. Genomic or cloned mollicute DNA is used for construction of expression libraries in a shotgun approach. The mollicute DNA is expressed in pQE13-15 as a fusion protein with the mouse dihydrofolate reductase (DHFR) in the three possible reading frames. Six N-terminal histidine residues allow the rapid and efficient purification of the fusion protein using immobilized metal chelate affinity chromatography with Ni^{2+} cations. In combination with a suppressor plasmid (pSUPMP), the UGA stop codons (used as a tryptophan codon in many mollicutes) can be suppressed. Positive clones can be identified by screening the expression library with cell fraction-specific antisera. The purified fusion proteins are used for production or purification of monospecific antibodies.

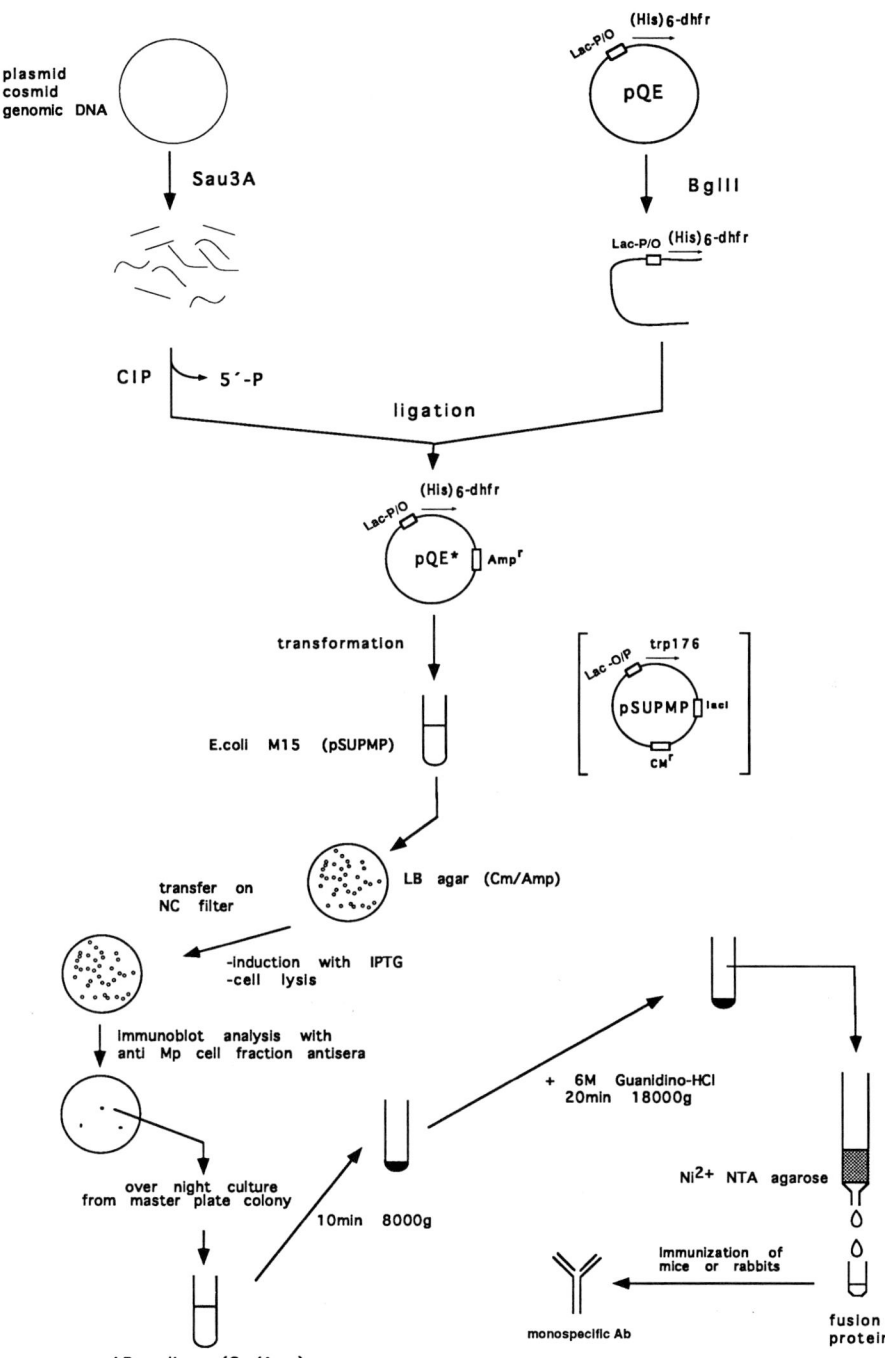

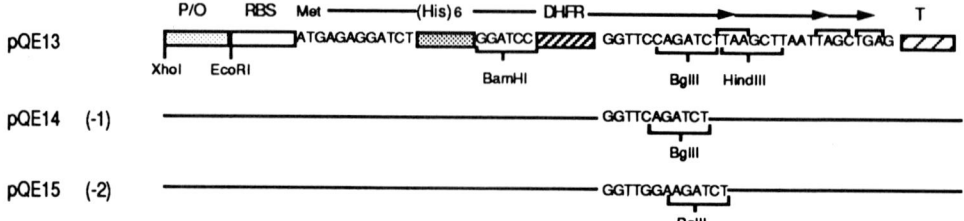

Fig. 6. Expression region and cloning site of the pQE13-15 vectors. The vectors allow the regulated expression (induction by IPTC) of a fusion protein with the mouse dihydrofolate reductase (DHFR) and six N-terminal histidine residues. The vectors differ in the number of base pairs between the *dhfr* gene and the *Bgl*II cloning sites and permit the expression of the mollicute DNA in the three possible reading frames.

Screening Expression Libraries

LB medium, per liter: Bacto-tryptone, 10 g, Bacto-yeast extract, 5 g; NaCl, 10 g; deionized H_2O, pH 7.0

LB agar plates (82 mm) containing 100 μg/ml ampicillin and 25 μg/ml chloramphenicol

LB agar plates with the antibiotics and with 5 mM isopropyl-β-D-thiogalactopyranoside (IPTG)

Nitrocellulose filters (BA85, 82 mm, Schleicher & Schuell)

Fractionation of M. pneumoniae Cells for Antisera Production

Tris buffer: 10 mM Tris–HCl, pH 7.5; 150 mM NaCl

Triton X-100 (Sigma) 10% vol/vol) solution, aprotinin (Trasylol, Boehringer Mannheim)

Branson cell disrupter B15 with a microtip, ultracentrifuge TL-100 with a TLA 100.2 rotor (Beckman), polycarbonate centrifuge tubes (Beckman)

1.5-ml Eppendorf tubes

PROCEDURE

Construction of Expression Libraries from Genomes or Gene Libraries

1. Add to reaction tube A 8 μl (1 μg) of pQE13, pQE14, and pQE15 each, 3 μl 10× buffer M, 2 μl endonuclease *Bgl*II (40 U/μl), and adjust to 30 μl with TE.

2. In reaction tube B digest 5–50 μg cloned DNA (plasmid, cosmid) or

genomic DNA with endonuclease *Sau*3A (5 U/μl). The amount of enzyme and the final volume depend on DNA concentration and the number of expected restriction fragments. However, the DNA concentration should not be higher than 200 μg/ml. Add $\frac{1}{10}$ volume of 10× buffer M to the reaction mixture.

3. Incubate both reaction tubes for 2 hours at 37°C.
4. Adjust the volume of both test tubes to 100 μl with TE and extract twice with 100 μl phenol and once with 100 μl phenol/chloroform (1:1, v/v).
5. Add 10 μl 3 *M* sodium acetate to both tubes and precipitate the DNA with 250 μl of ethanol. Wash the pellet twice with 500 μl 70% ethanol.
6. Resuspend precipitated DNA fragments (tube B) in 30–50 μl TE. Add $\frac{1}{10}$ volume 10× CIP buffer and 1 U calf intestinal phosphatase for 50 pmol 5' ends. Incubate for 30 min at 37°C.
7. Stop the reaction with $\frac{1}{10}$ volume of 250 m*M* EDTA. Extract and precipitate the DNA as described in steps 4 and 5.
8. Resuspend the DNA pellet of tube A in 30 μl TE.
9. Add 10 μl (=1 μg, =0.4 pmol) vector DNA from tube A, 1 μl (1 U‡) T4 ligase and 4 pmol DNA fragments (tube B) in a new tube, add 4 μl 10× ligase buffer, adjust to 40 μl with TE, and incubate overnight at 16°C.

E. coli M15 (pSUPMP) are transformed with the recombinant expression vectors according to the method of Hanahan (1983).

Plating Cells and Induction of Fusion Proteins

1. Plate 10, 50, and 100 μl of the bacterial suspension on LB agar plates without IPTG and incubate overnight at 37°C (the colonies should be grown to a maximum diameter of 0.2–0.5 mm).
2. Overlay those plates which show an appropriate colony number (200–500) with a nitrocellulose filter.
3. Mark the filter position in three asymmetric locations with a syringe and waterproof drawing ink.
4. Remove the filters and place them upside down on fresh agar plates containing 5 m*M* IPTG and incubate at 37°C for 4 hours. Keep the master plates at 4°C.

Colony Screening with Antisera against M. pneumoniae Cell Fractions

The lysis of the *E. coli* cells on the filters is described by Sambrook *et al.* (1989). The procedure for developing the filters is similar to that for Western blot

‡ Weiss unit.

filters. It is recommended to presaturate the antisera with *E. coli* lysate (Sambrock *et al.*, 1989). Several antisera dilutions should be tested. Positive clones for inoculation of liquid cultures can be picked from the master plate.

Fractionation of M. pneumoniae Cells for Antisera Production

STEP A. MEMBRANE AND CYTOSOLIC FRACTIONATION

1. Suspend *M. pneumoniae* cells harvested from 1 Roux flask (about 2 mg protein) in 1 ml double-distilled H_2O in a 1.5-ml reaction tube and add 10 U aprotinin.
2. Sonicate the cells under cooling conditions with 10 bursts of 15 seconds.
3. Centrifuge the suspension at 6000 g for 10 minutes at 4°C.
4. Transfer the supernatant containing the disrupted cells in a PC tube and centrifugate at 130,000 g for 1 hour at 4°C.
5. Carefully remove the supernatant containing the cytosolic proteins.
6. Store the cytosolic fraction and the pellet (membrane fraction) at -20°C.

STEP B. ISOLATION OF TRITON SHELL BY TRITON X-100 EXTRACTION

1. Suspend the *M. pneumoniae* cells in Tris buffer (10 mM Tris–HCl, pH 7.5, 150 mM NaCl) as described in step A.1 and add 10 U aprotinin.
2. Add 25 µl of 10% Triton X-100 solution and incubate on a shaker for 30 minutes at 37°C.
3. Centrifuge at 14,000 g for 30 minutes.
4. Remove the supernatant and repeat steps B.1–3 with the pellet.
5. Discard the supernatant and store the pellet (Triton shell) at -20°C.

For immunization of rabbits with cell fractions, see "Preparing the Fusion Protein for Immunization" steps 3 and 4.

Fusion Protein Purification by Immobilized Metal Chelate Affinity Chromatography (IMAC)

The method described by Hochuli *et al.* (1988) is followed.

MATERIALS

Column with 2 ml NTA-agarose (nitrilotriacetic acid-coupled Sepharose CL-6B; Diagen, Hilden)
Buffer A: 6 M guanidine hydrochloride; 0.1 M Na_3PO_4, pH 8.0

Buffer B: 8 M urea; 0.1 M Na_3PO_4; 0.01 M Tris–HCl, pH 8.0
Buffer C: 8 M urea; 0.1 M Na_3PO_4; 0.01 M Tris–HCl, pH 6.3
Buffer D: 8 M urea; 0.1 M Na_3PO_4; 0.01 M Tris–HCl, pH 5.9
Buffer E: 8 M urea; 0.1 M Na_3PO_4; 0.01 M Tris–HCl, pH 4.0
Buffer F: 6 M guanidine hydrochloride; 0.1 M Na_3PO_4, pH 2.7
Resin regeneration solutions: 0.2 M CH_3COOH; 0.05 M Na_2EDTA; 0.2 M NaOH; 0.1 M $NiSO_4 \cdot 6H_2O$
1 M IPTG solution
LB medium containing 100 μg/ml ampicillin and 25 μg/ml chloramphenicol
500-ml Erlenmeyer flask; 30-ml Corex tube; 1.5-ml Eppendorf tubes

PROCEDURE

Induction of Fusion Protein for IMAC

1. Inoculate 200 ml LB medium containing antibiotics with 10 ml of an overnight culture.
2. Incubate for about 2 hours at 37°C (until an OD_{600} of 0.7 is reached).
3. Add 20 μl of the 1 M IPTG solution and incubate for 3–5 hours at 37°C.
4. Harvest the cells by centrifugation at 8000 g for 10 minutes.

Immobilized Metal Chelate Affinity Chromatography

1. Resuspend the pellet in 20 ml buffer A and stir in room temperature for 30 minutes.
2. Centrifugate the lysate in the Corex tube at 18,000 g for 20 minutes.
3. Apply the supernatant onto the column containing 2 ml NTA-agarose charged with Ni^{2+} (see regeneration).
4. Wash the resin with 20 ml of buffer A, 10 ml of buffer B, and 15 ml of buffer C.
5. Elute the monomeric molecules with 20 ml of buffer D and collect 1-ml samples.
6. Elute the multimeric molecules with 8 ml of buffer E and collect 1-ml samples.
7. Wash the resin with 10 ml of buffer F.

Regeneration of the NTA-Resin

1. Wash the resin with 10 ml of 0.2 M CH_3COOH, 10 ml of 0.05 M Na_2EDTA, 10 ml of 0.2 M NaOH, and 10 ml of 0.2 M CH_3COOH.

2. Charge the resin with 10 ml of 0.1 M NiSO$_4$ · 6H$_2$O.
3. Wash the resin with 10 ml 0.2 M CH$_3$COOH and 20 ml of buffer A.

Preparing Fusion Protein for Immunization

1. Measure each collected fraction from IMAC at OD$_{280}$.
2. Pool the best fractions and dialyze against H$_2$O overnight at 4°C.
3. Determine the protein concentration (e.g., Lowry assay).
4. Immunize rabbits or mice with 50 μg fusion protein mixed with an appropriate adjuvant. Booster injections (two to three times); follow in intervals of 1 week for mice and 4 weeks for rabbits

Alternatively, monospecific antibodies can be produced by affinity purification of cell fraction-specific antisera. This can be achieved by incubation of the antisera with fusion protein fixed on a nitrocellulose sheet followed by elution of the monospecific antibodies with 100 mM glycine, pH 2.7. After neutralization with a $\frac{1}{10}$ volume of 1 M Tris–HCl, pH 8, the antibodies can be used for immunoblot analysis.

DNA Sequence Analysis of Vector-Fragment Transition

The recombinant expression plasmids can be purified using the rapid alkaline lysis method of Sambrook *et al.* (1989).

For sequencing of the vector-fragment transition it is necessary to synthesize an oligonucleotide as a primer that binds approximately 50 bp upstream of the cloning site. We recommend a primer with the following sequence:

5' GCGTCCTCTCTGAGGTCCAG 3'

Localization of Gene Products in Cell Fractions

MATERIALS

Tris buffer: 10 mM Tris–HCl, pH 7.5; 150 mM NaCl
1.8 M KJ solution; 0.5 M NaCl solution: 0.1 N NaOH solution; aprotinin (1 U/μl)
10% Triton X-100; 10% Triton X-114.
Sample buffer 2× : 125 mM Tris–HCl, pH 6.8; 4% SDS; 3% 2-mercaptoethanol; 0.01% bromophenol blue; 20% glycerol
50% trichloroacetic acid
1.5-ml Eppendorf tubes

PROCEDURES

Preparation of Cell Fractions and Membrane Fractionation

1. Suspend *M. pneumoniae* cells harvested from 1 Roux flask (about 2 mg protein) in 1 ml double-distilled H_2O in an Eppendorf tube.
2. Continue as described in "Fractionation of *M. pneumoniae* Cells for Antisera Production" steps A2–5.
3. Resuspend the pellet (membrane fraction) in 1 ml of 0.5 M NaCl and 10 µl of aprotinin and incubate on the shaker for 30 minutes at room temperature.
4. Centrifugate at 130,000 g for 60 minutes.
5. Carefully remove the supernatant (cytosolic contamination and weakly bound peripheral membrane proteins) and store at -20°C.
6. Resuspend the pellet in 1 ml of 0.1 N NaOH and repeat steps 3–5. The supernatant contains peripheral membrane proteins. The pellet contains more tightly bound peripheral membrane proteins, integral membrane proteins, and proteins of the cytoskeleton-like Triton shell.

Extractions Using Triton X-100 and KJ Solution

The method was described by Stevens *et al.* (1992).

1. Suspend *M. pneumoniae* cells harvested from 1 Roux flask (about 2 mg protein in 1.2 ml of Tris buffer, and add 300 µl 10% Triton X-100 and 15 µl aprotinin.
2. Continue as described in "Fractionation of *M. pneumoniae* Cells for Antisera Production" steps B2–4.
3. Resuspend the pellet (Triton shell) in 1 ml of 1.8 M KJ solution and incubate on a shaker for 30 minutes at room temperature.
4. Centrifugate at 14,000 g for 30 minutes at 4°C.
5. Carefully remove the supernatant.

Triton X-114 Extraction

The method was described by Bordier (1981) and modified by Riethman *et al.* (1987).

1. Suspend *M. pneumoniae* cells harvested from 1 Roux flask (about 2 mg protein in 1.35 ml of Tris buffer, and add 150 µl 10% Triton X-114 (three times precondensed and equilibrated with the Tris buffer) and 15 µl aprotinin.
2. Incubate on a shaker for 30 minutes at 4°C.

3. Centrifugate at 14,000 g at 4°C for 30 minutes.

4. Carefully remove the supernatant and transfer it to a new tube. The pellet can be reextracted if desired, repeating steps 1–3.

5. Incubate the supernatant for 5 minutes in a 37°C water bath. The solution will become cloudy, indicating micelle condensation.

6. Centrifuge at 14,000 g for 5 minutes at room temperature and remove the upper aqueous phase (contains the hydrophilic proteins) from the lower detergent phase (contains mainly integral membrane proteins). An additional pellet may appear during phase separation. This pellet shows a distinct protein profile and is considered as a separate fraction.

7. Both the aqueous and the detergent phases should be washed three times. Therefore, add 10% Triton X-114 to the aqueous phase and Tris buffer to the detergent phase to final concentrations of 1% TX-114 and continue as described in steps 2–6.

Localization of Proteins in Cell Fractions

The proteins recognized by antifusion protein antisera can be localized in the prepared cell fractions using immunoblotting techniques.

1. Precipitate the solubilized proteins by adding 0.15 vol. of 50% TCA to the samples, mix, and centrifugate at 14,000 g for 15 minutes.

2. Discard the supernatant and dissolve each pellet in 250 μl H_2O and 250 μl 2× sample buffer. If the color of the sample buffer changes from blue to yellow due to the remaining traces of TCA, blow some ammonium chloride vapor into the tube until the color changes back to blue.

3. Load 10-μl samples of each fraction on an SDS–polyacrylamide gel, transfer the proteins onto a nitrocellulose sheet, and screen with antifusion protein antisera of interest.

Discussion

The most convenient method for identifying conserved genes is to use the corresponding cloned genes from a phylogenetically closely related bacterial strain. Those clones are often not available. So far, most bacterial genes have been cloned from *E. coli* or *B. subtilis*. Thus, in many instances, one has to work with heterologous probes from only distantly related bacteria. An additional problem is caused by the low G+C content, lower than 30 mol%, for the majority of the mollicutes' DNAs. As a consequence, specific interaction between the DNA probe and the DNA to be identified takes place by A-T pairing contributing less to the stability of a hybrid DNA than a G-C pair. This requires working at low stringency and reduces the possibility of getting significant hybridization signals. For example, Wenzel and Herrmann (unpublished) failed to identify with a *Mycoplasma* strain PG 50-derived cloned gene, coding for the β subunit of the DNA-dependent RNA polymerase, the corresponding gene in *M. pneumoniae*, but others succeeded to do this with mycoplasmas having a low G+C content. In any case, using heterologous probes, different stringencies of hybridization conditions have to be tried. This is done most conveniently by keeping the temperature constant (37°C) and by varying the concentrations of formamide. One percent formamide decreases the melting temperature of dsDNA by about 0.5°C. In most hybridization experiments there will be more than one signal. Even with only a single signal, a few hundred nucleotides have to be sequenced for confirmation. A similar procedure has to be applied when using a set of degenerate synthetic oligonucleotides as probes.

Not much is known about the stability of mollicute DNA cloned and amplified in *E. coli*. We assume that the use of UGA as a tryptophan codon instead of a stop codon in many mollicutes has a stabilizing effect. It prevents, through premature stops, the synthesis of potentially toxic *M. pneumoniae*-coded proteins in *E. coli*. A premature translational stop is of course a disadvantage for our described method of identification of gene products, which is based on the synthesis of fusion proteins in *E. coli*. This can be overcome to some extent by introducing a suppressor gene for the UGA codon into *E. coli* and by using an *E. coli* mutant deficient in the gene for the release factor RF-2 (Smiley and Minion, 1993; Proft and Herrmann, 1994; see Chapter B11, this volume). The efficiency of suppression is about 50% for each UGA codon, e.g., if there are six UGA codons in a gene, only 1% of the gene product will be the full-length protein, even in an *E. coli* UGA suppressor strain. For standard fusion proteins, a mollicute-specific moiety of 50–100 amino acids is sufficient. This can be achieved by the suppression of one or two UGA codons, which might be present in a 150- to 300-bp-long fused DNA. If one is interested in a fully translated gene, then a cloning vector should be used which places the polyhistidine tag at the C-terminal end of the fusion protein. This permits isolation by nickel affinity chromatography of only those fusion proteins that were translated to the end.

The success of the described methods depends largely on the quality of the antisera. We had good experience with antisera directed against subcellular fractions. The fractions yielded high-titered, multispecific sera and permitted screening for many gene products with a small number of sera. An additional advantage of the antisubcellular fraction-specific antisera was that these sera enabled us to localize unknown proteins in the mycoplasma cell.

This approach will not work for proteins that are either weakly immunogenic or underrepresented in the cell fraction used for immunization. False-positive clones will show up if silent repetitive sequences are present that are not translated in the bacteria but code for the same or very similar segment of a protein, i.e., part of an *in-vivo*-translated gene.

The immunological approach can be modified by synthesizing, based on DNA sequence information, fusion proteins which could be used to induce the synthesis of monospecific antibodies. Western blots of bacterial protein extracts with these sera is another way to show the correlation between a given DNA sequence (open reading frame) and an *in vivo*-expressed bacterial protein.

References

Bautsch, W. (1988). Rapid physical mapping of the *Mycoplasma mobile* genome by two-dimensional field inversion gel electrophoresis techniques. *Nucleic Acids Res.* **16**, 11461–11467.

Birren, B., and Lai, E. (1993). "Pulsed Field Gel Electrophoresis'. A Practical Guide." Academic Press, San Diego.

Bordier, C. (1981). Phase separation of integral membrane proteins in Triton X-114 solution. *J. Biol. Chem.* **256**, 1604–1607.

Bové, J. M., Laigret, F., Finch, L. R., Carle, P., Ye, F. C., Citti, C., Saillard, C., Grau, O., Renaudin, J., Bové, C., Whitley, J., and Williamson, D. (1990). Genomes of Spiroplasmas. *IOM Lett.* **1**, 70–71.

Hanahan, D. (1983). Studies on transformation of *Escherichia coli* with plasmids. *J. Mol. Biol.* **166**, 557–580.

Herrmann, R. (1992). Genome structure and organization. *In* "Mycoplasma: Molecular Biology and Pathogenesis" (J. Maniloff, R. N. McElhaney, L. R. Finch, and J. B. Baseman, eds.), pp. 157–168. *Am Soc. Microbiol.*, Washington DC.

Hochuli, E., Bannwarth, W., Doebeli, H., Genz, R., and Stueber, D. (1988). Genetic approach to facilitate purification of recombinant proteins with a novel metal chelate adsorbent. *Bio-Technology* **11**, 1321–1325.

Hochuli, E., Doebeli, H., and Schacher, A. (1987). New metal chelate adsorbent selective for proteins and peptides containing neighbouring histidine residues. *J. Chromatogr.* **411**, 177–184.

Ladefoged, S. A., and Christiansen, G. (1992). Physical and genetic mapping of the genomes of five *Mycoplasma hominis* strains by pulsed-field gel electrophoresis. *J. Bacteriol.* **174**, 2199–2207.

Peterson, S. N., Hu, P.-C., Bott, K. F., and Hutchison, C. A., III (1993). A survey of the *Mycoplasma genitalium* genome by using random sequencing. *J. Bacteriol.* **175**, 7918–7930.

Proft, T., and Herrmann, R. (1994). Identification and characterization of hitherto unknown *Mycoplasma pneumonniae* proteins. *Mol. Microbiol.* **13**, 337–348.

Riethman, H. C., Boyer, M. J., and Wise, K. S. (1987). Triton X-114 phase fractionation of an integral membrane surface protein mediating monoclonal antibody killing of *Mycoplasma hyorhinis*. *Infect. Immun.* **55**, 1094–1100.

Römling, U., Grothues, D., Bautsch, W., and Tümmler, B. (1989). A physical genome map of *Pseudomonas aeruginosa* PAO. *EMBO J.* **8**, 4081–4089.

Sambrook, J., Fritsch, E. F., and Maniatis, T. (1989). "Molecular Cloning," 2nd Ed. *Cold Spring Harbor Laboratory Press,* Cold Spring Harbor, NY.

Smiley, B. K., and Minion, F. C. (1993). Enhanced readthrough of opal (UGA) stop codons and production of *Mycoplasma pneumoniae* P1 epitopes in *Escherichia coli*. *Gene* **134**, 33–40.

Smith, H. O., and Birnstiel, M. L. (1976). A simple method for DNA restriction site mapping. *Nucleic Acids Res.* **3,** 2387–2398.

Stevens, M. K., and Krause, D. C. (1992). *Mycoplasma pneumoniae* cytadherence phase-variable protein HMW3 is a component of the attachment organelle. *J. Bacteriol.* **174,** 4265–4274.

Wenzel, R., and Herrmann, R. (1989). Cloning of the complete *Mycoplasma pneumoniae* genome. *Nucleic Acids Res.* **17,** 7029–7043.

Wenzel, R., Pirkl, E., and Herrmann, R. (1992). Construction of an *Eco*RI restriction map of *Mycoplasma pneumoniae* and localization of selected genes. *J. Bacteriol.* **174,** 7289–7296.

B5

CHARACTERIZATION OF VIRUS GENOMES AND EXTRACHROMOSOMAL ELEMENTS
Kevin Dybvig

Introduction

Extrachromosomal nucleic acids have been reported in several Mollicutes. Analysis of most of these elements has revealed that they are the replicative form of a mycoplasma virus genome. In only a few cases have these elements proven to be cryptic plasmids.* Accordingly, when a new extrachromosomal element is first identified, it should not be assumed that it is a plasmid without seriously addressing the possibility that it is of viral origin. Electron microscopy (EM) is an important tool for detection of virus-like particles in suspect strains, using methods described previously (Cole, 1983). Another approach for virus detection is to assay for infectious particles using plaque-forming units (PFU) assays, as described in this chapter. Strategies for characterization of virus genomes and extrachromosomal elements should include a determination of whether they are DNA or RNA, single or double stranded, and linear or circular.

Materials

Centrifuge
Growth medium (broth)

---

* In this chapter, "plasmid" is the designation given to self-replicating, extrachromosomal elements (usually double-stranded, circular DNA) that are not some form of a viral genome. Distinctions between the genome of a defective virus and a plasmid are blurred.

Growth medium (agar) in 60 × 15-mm petri dishes
Reagents for nucleic acid isolation and agarose gel electrophoresis
Filters, 0.2 μM
Soft agar overlay solution: 0.7% agar in a buffer that promotes stability of the mycoplasma. Phosphate-buffered saline often works well. This solution can be stored as a liquid in a 65°C incubator.
DNase I
RNase A
S1 nuclease and reaction buffer: A typical 1× buffer for S1 consists of 200 mM NaCl, 50 mM sodium acetate (pH 4.5), 1 mM ZnSO$_4$, and 0.5% glycerol.
Exonuclease VII (Exo VII)

Procedures

Isolation of Viruses

Chapters F3–F5 in "Methods in Mycoplasmology," Vol. II, published in 1983, discussed methods for isolation and characterization of mycoplasma viruses using PFU assays. However, these earlier chapters provided methods from the perspective of an investigator already possessing a cell strain that is known to be an indicator for the virus. In practice, if the presence of a virus is suspected, based on either EM analysis or on the observation of an extrachromosomal element, it is necessary to initiate a search for an indicator strain. There are numerous factors to consider. First, the strain suspected of harboring a virus is a likely lysogen and is not a good candidate for an indicator. Second, the host range for the virus may be narrow; numerous strains of whatever species the suspect strain is a member of should be examined as possible indicators. Third, many mycoplasmal strains grow poorly; thus, obtaining well-formed cell lawns on which to observe viral plaques should not be taken for granted. Fourth, if virus replication is slow, overly dense cell lawns may completely obscure the appearance of plaques. Fortunately, most of these considerations are conveniently dealt with by using the megaplaque assay described in this chapter. It is readily adaptable for handling large numbers of candidate indicator strains and requirements concerning the density of the cell lawns are not as rigorous as for conventional PFU assays.

PREPARATION OF VIRUS STOCKS FROM SUSPECT STRAINS

Prior to examining suspect strains for the presence of infectious virus, stocks of the suspect virus that are free of host cells should be prepared. Most cells can be removed from a culture by centrifugation (e.g., 10,000 g for 10 minutes) followed by filtration of the supernatant through a 0.2 μm filter. For most viruses, the resulting filtrate can be stored indefinitely at 4°C with little loss of titer.

MEGAPLAQUE ASSAY

1. Incubate a culture of the candidate indicator strain to late-logarithmic growth phase.
2. Mix (vortex gently to avoid air bubbles), 30 to 300 μl of cell culture with 1.5 ml of soft agar overlay solution that has been precooled to 42°C.
3. Quickly pour the overlay mixture onto an agar plate that has been pre-warmed to room temperature.
4. Immediately spot 10 μl of suspect virus stock onto the center of the plate.
5. Let the plate stand for 15 minutes to allow the overlay to harden.
6. Incubate the plate at the host's optimal growth temperature.
7. Examine the plates at 24 and 48 hours incubation. The appearance of a zone of low cell density (megaplaque) around the area in which the virus stock was spotted is indicative of infectious (viral) material. Compare the plates to controls in which uninoculated growth medium, as opposed to virus stock, was spotted onto the agar overlay.

PFU ASSAY

Once a potential indicator strain has been identified using the megaplaque assay, a PFU assay should be used for verification. Because the PFU assay is based on discrete plaques, the condition of the cell lawn is much more critical than it is for the megaplaque assay. Also, the titer of the virus in the suspect stock is unknown. To visualize discrete plaques, serial dilutions (10-fold) of the virus stock should be examined.

1. Incubate a culture of the candidate indicator strain to late-logarithmic growth phase.
2. Prepare serial 10-fold dilutions of the virus stock, down to 10^{-8}.
3. To 10 μl of each virus dilution, add 30 to 300 μl of cell culture and incubate at the host's optimal growth temperature for 1 hour to allow for virus adsorption to host cells.
4. Mix (vortex gently to avoid air bubbles), the cell–virus mixture with 1.5 ml of soft agar overlay solution that has been prewarmed at 42°C.
5. Quickly pour the overlay mixture onto an agar plate that has been pre-warmed to room temperature.
6. Let the plate stand for 15 minutes to allow the overlay solution to harden.
7. Incubate the plate at the host's optimal growth temperature.
8. Examine the plates at 24 and 48 hours incubation. The appearance of small zones of low cell density (plaques) are indicative of infectious (viral) material.

VIRUS MULTIPLICATION

Once the indicator strain has been identified, it should be proven that the suspected virus can multiply in the indicator strain. If multiplication occurs, then the possibility that the plaques result from the expression of bacteriocin-like activity can be ruled out. This can be accomplished by picking plaques or megaplaques in the form of agar plugs and using them to inoculate uninfected cell cultures. The inoculated culture is then incubated at the host's optimal growth temperature, and samples are removed at various times and assayed for PFU activity. A rise in PFU during the infection cycle indicates that a virus has, indeed, been isolated. The virus can then be further characterized as described in Chapters F3–F5 of Vol. II of "Methods in Mycoplasmology."

VIRUS PURIFICATION: A REQUIREMENT FOR CHROMOSOME ANALYSIS

Assuming the search for a mycoplasma virus was initiated because of the observation of an extrachromosomal element, the investigator will likely wish to determine whether the extrachromosomal element is truly viral in origin. The isolation of the virus and the identification of an indicator strain, as described earlier, provide the needed tools. By comparing nucleic acids isolated from virus-infected cells to nucleic acids isolated from uninfected cells, it should be possible to conclude whether the presence of the extrachromosomal element is dependent on virus infection.

A potentially major undertaking is to analyze the genome of infectious virions. Virus purification can be a major hurdle until infection conditions that yield a high titer of virus are identified. Because the replicative form of viral genomes is often different from the form of the genome in virus particles, it is *essential* to purify the virus free from host cell debris. Most purification protocols involve banding the virus on either density or sedimentary velocity gradients. Some viruses, such as the P1 virus isolated from *Mycoplasma pulmonis* (Dybvig et al., 1987), are unstable in many gradient media. In particular, many viruses are unstable in CsCl. P1 virus needs exogenous protein to maintain infectivity, and its purification was made feasible by inclusion of 1% bovine serum albumin (BSA) in all buffers. Even in the presence of BSA, P1 is unstable in CsCl and sucrose solutions. Approaches for characterization of viral genomes are discussed next.

Characterization of Plasmids and Viral Genomes

Most extrachromosomal elements are molecules of covalently closed circular (CCC), double-stranded DNA (dsDNA). These elements can be isolated by using standard plasmid purification methods, such as cesium chloride gradient centrifugation. The characterization of these elements is generally routine: analysis of restriction fragments by agarose gel electrophoresis can determine the size and restriction map of the element; Southern hybridization analysis using the element as a probe can be used to determine its host range; and molecular cloning can be a prelude to DNA sequence analysis. The question considered here is how to address unusual molecules, either plasmids or viral genomes, that have characteristics different from CCC dsDNA. EM is a powerful tool for nucleic acid analysis and should be employed whenever feasible. However, even laboratories that specialize in EM often lack the required equipment and expertise to examine rotary shadowed nucleic acids. Therefore, enzymatic treatments for nucleic acids analyses are emphasized here.

DNA VS RNA

Digest a sample of the nucleic acid with RNase A (0.5 µg/ml for 30 minutes at 37°C) or DNase I (0.5 µg/ml for 30 minutes at 37°C) and analyze the digestion products by agarose gel electrophoresis. Sensitivity to these treatments will usually clearly demonstrate if the molecule is DNA or RNA. If resistance to both enzymes is observed, the nucleic acid may be double-stranded RNA (dsRNA). In the presence of salt ions that stabilize RNA base pairing, dsRNA is resistant to RNase A. By using salt-free buffers (e.g., 10 mM Tris, pH 8.0; 1 mM EDTA), dsRNA will be degraded by RNase A.

SINGLE-STRANDED VS DOUBLE-STRANDED DNA

Many viruses contain single-stranded DNA (ssDNA) genomes. A simple way to determine strandedness is to digest the DNA with S1 nuclease (0.2 units of enzyme at 37°C for 30 minutes) and analyze the digestion products by agarose gel electrophoresis. At low enzyme concentrations, S1 specifically degrades ssDNA. However, controls must be included to ensure that dsDNA is S1 resistant under the conditions employed. For an example of using S1 nuclease sensitivity to determine strandedness, the reader is referred to Dybvig *et al.* (1985).

CIRCULAR VS LINEAR DNA

Although plasmids are usually circular, linear plasmids have been described in some bacterial systems. Viruses with linear genomes are common. It is usually routine to demonstrate that a dsDNA molecule is circular because of the superhelicity of CCC dsDNA. S1 nuclease will nick CCC dsDNA, producing an open circular (OC) form, and more extensive digestion with S1 will convert the OC form to linear (L) DNA (Beard *et al.*, 1973). The CCC, OC, and L forms all have different migration properties on agarose gels. Therefore, by comparing the mobility of S1 nuclease-digested DNA on agarose gels, using either various concentrations of S1 enzyme or various incubation times to control the reaction, circular molecules can be identified as such based on the conversion of the CCC form to the OC form to the L form. DNA molecules that exhibit no change in mobility following S1 digestion are candidates for linear molecules. However, proof of linearity may be surprisingly difficult. Some linear plasmids have covalently closed "hairpin" ends. Therefore, these DNAs are resistant to digestion with exonucleases and are resistant to denaturation by routine methods. Similarly, some linear plasmids and viral genomes have protein covalently linked to one end of each strand and are consequently resistant to the activity of some exonucleases. Probably the best method to definitively show that an extrachromosomal molecule is linear is by EM.

In the case of viral, ssDNA genomes, digestion with Exo VII can be used as a simple test to determine whether the DNA is linear or circular. Exo VII specifically degrades linear, ssDNA. For an example of using Exo VII to determine circularity vs linearity of a ssDNA genome, the reader is referred to Dybvig *et al.* (1985).

Discussion

The experimental approach for isolation of mycoplasma viruses as outlined earlier is straightforward and can be conducted in virtually any laboratory. Therefore, when an extrachromosomal element has been detected, attempts at virus isolation should be considered. Obviously, failure to find a virus does not prove that no viruses are present. In particular, a virus would be hard to isolate if it were present at a very low titer. However, viruses present in low titer would not produce an extrachromosomal element that is readily detected when total genomic DNA is analyzed on ethidium bromide-stained agarose gels. Therefore, when a virus cannot be isolated, it would be appropriate to assume that any extrachromosomal element is a *bona fide* plasmid until evidence to the contrary is obtained.

It can be assumed that small extrachromosomal elements, such as the pADB201 and pKMK1 plasmids isolated from *Mycoplasma mycoides* subsp. *mycoides,* are not viral genomes because of their size. Even small viruses have genome sizes of over 4 kb, whereas pADB201 and pKMK1 are less than 2 kb (Bergemann *et al.*, 1989; King and Dybvig, 1992). The limited coding capacity of these plasmids is insufficient to encode for proteins other than those required for DNA replication. Nucleotide sequence analysis of these plasmids has confirmed that additional factors needed for virus assembly, such as capsid proteins, are not encoded. Indeed, using the methods described in this chapter, attempts to isolate a virus from the *M. mycoides* subsp. *mycoides* strain that harbors pKMK1 were unsuccessful.

Some viruses such as mycoplasma virus L2 have noncytocidal infectious cycles (Putzrath and Maniloff, 1977). Nevertheless, PFU assays are successful because virus-infected cells grow more slowly than do uninfected cells. In such cases, the titer of the virus stock is important for a successful PFU assay. Cell lawns containing no plaques have the same appearance as cell lawns containing too many plaques that form a single megaplaque covering the entire lawn. By using the megaplaque assay, the titer of the virus is less critical. Because virus is spotted onto the center of the cell lawn, the perimeter of the lawn remains uninfected. The megaplaque is confined to a small area rendering it observable regardless of the titer of the virus.

Because cell density is critical for a successful PFU assay, a range of cell concentrations should be examined when initially establishing the assay for a new virus. For species that strongly alter the pH of the medium during growth, the use of agar medium containing a pH indicator (e.g., phenol red) can be helpful. Cell lawns that show a dramatic change in pH after overnight growth are probably too dense, whereas cell lawns that fail to show a change in pH after 48 hours of incubation are usually too sparse.

A wide variety of nucleic acid structures have been described in mycoplasmas. Known examples of mycoplasma virus genomes include circular and linear forms of ssDNA and dsDNA (for a review see Maniloff, 1988). Extrachromosomal dsRNA molecules have been described in *M. pulmonis* (Dybvig *et al.*, 1987). These molecules were only observed in cells that had been infected with mycoplasma virus P1, but whether the dsRNA molecules were of viral versus cellular origin was not determined. Many mycoplasmas produce double-stranded ribonuclease (Marcus and Yoshida, 1990), suggesting that dsRNA may be frequently encountered by these organisms.

Viral genomes have a wide range of possible structures, and myriad pitfalls may confound the analysis of a newly isolated virus. As an example, the chromosome of mycoplasma virus P1 was difficult to isolate, let alone characterize. As described earlier, purification of P1 virus was problematic because infectivity was unstable in many solutions. Once the virus was purified, routine phenol extraction of nucleic acids from virions resulted in low yields. The P1 genome consists of linear dsDNA of about 11.3 kb (Zou *et al.*, 1995). Covalently attached to the 5' end of each strand is a terminal protein (TP). The presence of TP probably causes P1 DNA to partition inefficiently into the aqueous phase during phenol extraction, explaining the low yield of purified P1 DNA. In support of this explanation, protease treatment prior to phenol extraction dramatically improves the efficiency of P1 DNA isolation. As described in other bacteriophage systems, e.g., φ29 (Garcia *et al.*, 1984), the TP on P1 DNA can be removed by treatment with piperidine. Due to the presence of TP, P1 DNA is resistant to the 5'-specific λ exonuclease. P1 DNA is surprisingly hard to denature, suggesting that TP may noncovalently bind with the 3' end of each strand. The properties of P1 DNA are such that it was not until EM analysis was performed that its linearity was clearly demonstrated.

References

Beard, P., Morrow, J. F., and Berg, P. (1973). Cleavage of circular, superhelical simian virus 40 DNA to a linear duplex by S1 nuclease. *J. Virol.* **12**, 1303–1313.

Bergemann, A. D., Whitley, J. C., and Finch, L. R. (1989). Homology of mycoplasma plasmid pADB201 and staphylococcal plasmid pE194. *J. Bacteriol.* **171**, 593–595.

Cole, R. M. (1983). Virus detection by electron microscopy. *In* "Methods in Mycoplasmology" (S. Razin and J. G. Tully, eds.), Vol. **2**, pp. 407–412. Academic Press, New York.

Dybvig, K., Liss, A., Alderete, J., Cole, R. M., and Cassell, G. H. (1987). Isolation of a virus from *Mycoplasma pulmonis*. *Isr. J. Med. Sci.* **23**, 418–422.

Dybvig, K., Nowak, J. A., Sladek, T. L., and Maniloff, J. (1985). Identification of an enveloped phage, mycoplasma virus L172, that contains a 14 kilobase single-stranded DNA genome. *J. Virol.* **53**, 384–390.

Garcia, J. A., Penalva, M. A., Blanco, L., and Salas, M. (1984). Template requirements for initiation of phage φ29 DNA replication *in vitro*. *Proc. Natl. Acad. Sci. USA* **81**, 80–84.

King, K. W., and Dybvig, K. (1992). Nucleotide sequence of *Mycoplasma mycoides* subspecies *mycoides* plasmid pKMK1. *Plasmid* **28**, 86–91.

Maniloff, J. (1988). Mycoplasma viruses. *Crit. Rev. Microbiol.* **15**, 339–389.

Marcus, P. I., and Yoshida, I. (1990). Mycoplasmas produce double-stranded ribonuclease. *J. Cell. Phys.* **143**, 416–419.

Putzrath, R. M., and Maniloff, J. (1977). Growth of an enveloped mycoplasmavirus and establishment of a carrier state. *J. Virol.* **22**, 308–314.

Zou, N., Park, K., and Dybvig, K. (1995). Mycoplasma virus P1 has a linear, double-stranded DNA genome with inverted terminal repeats. *Plasmid* **33**, 41–49.

B6

PLASMID AND VIRAL VECTORS FOR GENE CLONING AND EXPRESSION IN *SPIROPLASMA CITRI*

J. Renaudin and J. M. Bové

General Introduction

Genetic analyses of bacteria are often impaired by the lack of methods for transformation and by the lack of suitable gene vectors. Indeed, the ability to transform bacteria using either naturally occurring or chemically induced competence is still the exception and not the rule.

In mollicutes, genetic studies have been hampered by the paucity of selectable markers and by the absence of gene transfer systems. In addition, the use of recombinant DNA technology to study expression of mollicute genes by cloning in *Escherichia coli* has been limited because many mollicutes (spiroplasmas, mesoplasmas, entomoplasmas, mycoplasmas, and ureaplasmas) use the codon UGA to encode for tryptophan. Cloning in a spiroplasma (or a mycoplasma) host would overcome these difficulties and would help to investigate gene expression and the function of the polypeptides they encode. Transfer of genetic material into mollicutes would also provide a way to complement mutations. Advances in the development of mollicute genetic systems have been reviewed (Dybvig, 1990).

This chapter describes two vectors used to introduce and express foreign genes in *Spiroplasma citri*. One is the replicative form of spiroplasmavirus SpV1 of *S. citri*, and the other is a plasmid whose replication is mediated by the chromosomal replication origin (*oriC*) of *S. citri*. The chapter also describes a protocol for efficient electrotransformation of *S. citri*.

Materials, Buffers, and Chemicals

Binocular microscope
Orbital shaker
Refrigerated centrifuges and rotors
Laminar flow hood
Thermal cycler
Electroporator: Gene pulser unit and pulse controller from Bio-Rad Laboratories, Inc.
Electroporation cuvettes, 0.4-cm electrode gap
Microcentrifuge
Dry incubators at 32° and 37°C
HS buffer: 8 mM HEPES (N-2-hydroxyethylpiperazine-N'-2-ethanesulfonic acid), pH 7.4, 280 mM sucrose
Laemmli solubilization buffer: 50 mM Tris–HCl, pH 6.8; 5% 2-mercaptoethanol; 2% SDS; 0.1% bromophenol blue; 10% glycerol
MOPS buffer: 40 mM MOPS [3-(N-morpholino) propanesulfonic acid]; 1 mM EDTA; 10 mM potassium acetate
PBS buffer: 140 mM NaCl; 2.7 mM KCl; 4 mM Na$_2$HPO$_4$; 1.8 mM KH$_2$PO$_4$; pH 7.2
SP4 culture medium (Whitcomb, 1983)
TEG buffer: 25 mM Tris–HCl, pH 8.3; 10 mM EDTA; 50 mM glucose
TE buffer: 10 mM Tris–HCl, pH 8; 1 mM EDTA

Procedures

Electrotransformation of Spiroplasma citri

INTRODUCTION

Electroporation refers to the process of subjecting living cells to a rapidly changing, high-strength electric field, thereby producing transient pores in their outer membranes. This process has been used to introduce DNA into eukaryotic and prokaryotic cells. Electroporation has provided a significant advance over chemical means for transforming many strains of *E. coli* and a variety of gram-negative and gram-positive species, including species of the class Mollicutes such as *Acholeplasma laidlawii*, *S. citri*, and *Mycoplasma pneumoniae*.

The protocol designed for optimal transformation of *S. citri* (Stamburski *et al.*, 1991; Gasparich *et al.*, 1993) is described next.

PROTOCOL

1. Grow spiroplasma cells in SP4 medium to late log phase [$\sim 10^9$ CFU (colony-forming units)/ml].
2. Aliquot into 1.5-ml fractions and centrifuge at 18,000 g for 15 minutes. Discard the supernatant and resuspend each pellet in 1 ml of HEPES–sucrose (HS) buffer.
3. Sediment the cells by centrifugation at 18,000 g for 15 minutes.
4. Resuspend each pellet in 0.2 ml of HS buffer, and combine cells from two pellets into an electroporation cuvette (0.4-cm electrode gap).
5. Add DNA (usually 0.1 to 1 µg in 20 µl H$_2$O), mix, and keep the cell/DNA mixture on ice for 10 minutes.
6. Switch on the gene pulser. Set the voltage on 2.5 kV, the capacitance on 3 µF, and the pulse controller on 1000 Ω.
7. Insert the cuvette in the pulser chamber and apply one single pulse. In these conditions the electric field strength is 6.25 kV/cm and the time constant is approximately 1 to 2 mseconds.
8. When spiroplasma cells are transfected with SpV1 viral DNA, plate the treated cells immediately onto a 4-hour lawn of *S. citri* R8A2 grown on SP4 solid medium, and incubate at 32°C. Viral plaques appear within 24 hours.
9. When spiroplasma cells are transformed with a plasmid containing an antibiotic resistance gene marker, add 0.8 ml of SP4 medium and incubate for 3 hours at 32°C (to allow phenotypic expression) before plating on solid SP4 medium containing the appropriate antibiotic concentration (2 µg/ml tetracycline, when the *tetM* determinant is used).

DISCUSSION

Transformation with viral DNA, i.e., transfection, was the first successful transfer of DNA in mollicutes in general and in spiroplasmas in particular. Transfection of *S. citri* by the SpV1-RF DNA was used to compare and optimize methods of transformation. In the case of *S. citri*, both polyethylene glycol (PEG)-mediated protocol and electroporation can be used. However, electroporation yields higher transfection frequency (approximately 10^{-3} transfectants per CFU) than the PEG method (approximately 10^{-4} transfectants CFU).

Use of Replicative Form of SpV1 for Gene Cloning in Spiroplasma citri

INTRODUCTION

SpV1 is a rod-shaped virus with a circular single-stranded DNA genome. The double-stranded replicative form has been used as a vector to introduce and express foreign DNA in *S. citri*. The chloramphenicol acetyltransferase (CAT) gene as well as the G fragment of the *Mycoplasma pneumoniae* adhesin P1 gene were expressed regardless of the presence of UGA codons in the reading frame (Stamburski *et al.*, 1991; Marais *et al.*, 1993; Renaudin and Bové, 1994). The SpV1 replicative form (RF) has been used for the following reasons:

1. The SpV1 genome has been fully sequenced; it contains intergenic regions in which foreign DNA can be inserted without disrupting any viral gene.

2. Because of the rod-shaped morphology of the virion, the presence of additional DNA in the viral genome does not prevent its encapsidation into viral particles.

3. The SpV1 RF can be efficiently transfected into *S. citri* (see previous section). Transfected cells produce virions, the presence of which is revealed by plaque formation, from which recombinant viral DNA can be recovered.

PROTOCOL

Purification of the SpV1 Replicative Form

1. Grow a 600-ml culture of *S. citri* R8A2 (ATCC 27556 Rockville, MD) in SP4 medium to mid-log phase ($\sim 10^8$ CFU/ml).

2. Add the viral inoculum (10^9 to 10^{10} PFU/ml) at a multiplicity of infection of approximately 0.5 to 1 and incubate at 32°C for 16 hours.

3. Harvest the infected cells (3 × 200 ml) by centrifugation at 20,000 g for 40 minutes and resuspend each pellet in 2 ml of TEG buffer. Transfer into a 15-ml centrifuge tube, then successively add 0.4 ml of 10% SDS and 0.6 ml of 5 M NaCl. Mix by inverting the tubes and keep on ice for 40 minutes.

4. Centrifuge at 20,000 g for 40 minutes and save supernatants.

5. The nucleic acids are recovered from the aqueous phase by standard phenol deproteinization and ethanol/sodium acetate precipitation.

6. The ethanol precipitate is dissolved in 2 ml TE buffer and treated with 10 μg/ml ribonuclease at 37°C for 30 minutes.

7. The SpV1 RF is further purified by ethidium bromide–cesium chloride density gradient centrifugation.

8. According to this procedure, a 600-ml culture of SpV1-infected spiroplasma cells yields approximately 200 to 250 μg of purified SpV1-RF DNA.

Insertion of Foreign DNA into Viral RF

1. The SpV1 RF (2 to 5 μg) is linearized by digestion with restriction endonuclease *Sau*3AI or *Mbo*I, which cut at a unique site in the intergenic region I3 (Fig. 1), and is 5'-dephosphorylated by treatment with alkaline phosphatase. As a result, religation of the RF can only occur when a DNA fragment with intact 5'-phosphate groups is inserted.

2. The linearized, dephosphorylated RF (20 to 50 ng) is ligated to approximately 100 ng of insert DNA with compatible ends with 1 to 10 units of T4 DNA ligase in a final volume of 10 to 20 μl.

3. Restriction of DNA, dephosphorylation, and ligation, as well as the fill-in reaction needed to insert blunt-end DNA fragments, are performed according to standard procedures (Sambrook *et al.*, 1989).

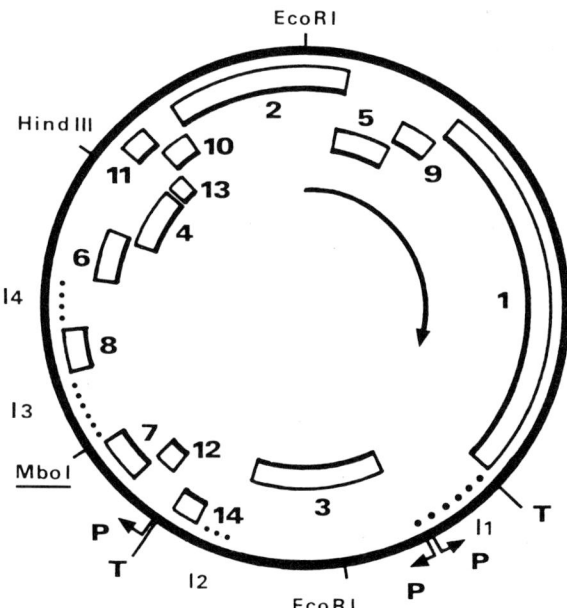

Fig. 1. Gene organization of SpV1-R8A2B (from Renaudin and Bové, 1994). Open reading frames are numbered 1 to 14. Dots indicate the intergenic regions I1 to I4. P, promoter-like sequences; T, terminator-like sequences.

Transformation of Spiroplasma citri

S. citri cells are electrotransformed with 1 to 5 µl of the ligation mixture as described in the previous section.

Detection and Characterization of the Recombinant SpV1 Clones

To detect the recombinant viral clones, *in situ* hybridization of SpV1 plaques can be performed according to the standard protocol described for bacteriophage M13 plaques. However, because of the reduced size of SpV1 plaques and the lower amount of virions produced (compared to M13), it is recommended to perform an amplification step by transferring (with toothpicks) initial plaques to a new indicator lawn on which larger plaques are formed.

The following procedure is used for rapid preparation of RF DNA from recombinant SpV1 viral clones.

1. Pick a single plaque in 1 ml of an early log phase culture of *S. citri* R8A2 and incubate at 32°C for 24 hours.

2. Use this 1-ml infected culture to inoculate a fresh 10-ml *S. citri* culture and incubate for an additional 16 hours.

3. Harvest infected cells by centrifugation at 20,000 g for 30 minutes and resuspend the pellet in 200 µl TEG buffer.

4. Successively add 50 µl of 10% SDS and 65 µl of 5 M NaCl. Keep on ice for 30 minutes.

5. After centrifugation at 20,000 g for 30 minutes, the nucleic acids are recovered from the supernatant by phenol deproteinization and ethanol/acetate precipitation. The nucleic acid precipitate is dissolved in 20 µl TE buffer.

This protocol yields enough SpV1 RF DNA to be analyzed by Southern blot hybridization according to standard procedures (Sambrook *et al.*, 1989).

Expression of Cloned DNA Insert in Spiroplasma citri

Expression in spiroplasma cells of the DNA fragment inserted in the SpV1 RF is followed by characterizing the insert-specific transcripts and the translation products. The RNA transcripts are characterized by Northern blot hybridization of total RNA extracted from *S. citri* cells infected with the recombinant SpV1 virus. Determination of the initiation and termination sites of transcription requires further characterization of the transcription products by primer extension and nuclease S1 hydrolysis techniques. Translation products are usually detected by Western blot analysis using specific antibodies and/or by their enzymatic activity. Because these techniques have been extensively described, this section only focuses on RNA isolation from spiroplasma cells, preparation of protein samples for Western blot analysis, and detection of chloramphenicol acetyltransferase (CAT) activity in spiroplasma cells carrying the CAT gene.

Total RNA Isolation

RNA is isolated from SpV1-infected *S. citri* cells in culture by using the guanidinium thiocyanate–cesium chloride method (Chirgwin *et al.*, 1979).

1. Grow a 400-ml culture of *S. citri* R8A2 to early log phase and inoculate with recombinant SpV1 virions at a multiplicity of infection of 0.5 to 1.
2. Harvest cells by centrifugation (20,000 g for 40 minutes) 16 hours after infection.
3. Disperse each pellet (corresponding to 200 ml of culture) in 3.5 ml of 10 mM Tris–HCl buffer, pH 7.4, containing 4 M guanidinium thiocyanate and 7% 2-mercaptoethanol and add sarcosyl to 2% final concentration to lyse the cells.
4. Add 1 g of CsCl to each 2.5 ml of homogenate and layer the homogenate (8 ml for two pellets) onto a 3-ml cushion of 1.033 g/ml CsCl in 10 mM EDTA (pH 7.4) ($n=1.4047$; $d=1.758$ g/cm^3) in a Beckman SW41 rotor tube.
5. Centrifuge at 180,000 g for 16 hours at 20°C. Discard the supernatant carefully and dissolve the pellet in 1 ml of 10 mM Tris–HCl buffer, pH 7.4, containing 5 mM EDTA and 1% SDS.
6. Extract once with a 4:1 mixture of chloroform and 1-butanol. Save the aqueous phase and reextract the organic phase.
7. Combine the two aqueous phases, add 0.1 volume of 3 M sodium acetate, pH 5.2, and 2.2 volumes of ethanol. Store at -20°C for a least 2 hours. Recover RNA by centrifugation.
8. Dissolve the pellet in 1 ml of sterile water and reprecipitate with ethanol in the presence of 1.25 M ammonium acetate. The final pellet is resuspended in 0.2 ml sterile water. The yield is approximately 0.5 to 2 mg of total RNA per 600 ml of *S. citri* culture.

Preparation of Protein Samples

Because spiroplasma cells are grown in a complex medium containing fetal calf serum, extensive washing of cells is needed to remove proteins of the culture medium.

1. Harvest spiroplasma cells (50 ml culture) by centrifugation at 20,000 g for 40 minutes.
2. Disperse each cell pellet (corresponding to 25 ml of culture) in 10 ml of PBS buffer and collect cells by centrifugation.
3. Wash the cells twice more as just described.
4. Resuspend the final pellet in 0.5 ml of Laemmli solubilization buffer.
5. Heat the solubilized proteins in a boiling water bath for 5 minutes. Centrifuge at 10,000 g for 10 minutes to remove the insoluble material and save supernatant.
6. Store the protein samples (0.5 to 2 mg protein per 50 ml of *S. citri* culture) at -20°C until use.

Detection of CAT Activity

The CAT gene has been used as a model gene to demonstrate the potential of the SpV1 RF for gene cloning and expression in *S. citri* (Stamburski et al., 1991). Translation of the CAT gene into a functional polypeptide was demonstrated by the presence of CAT activity in cell-free extracts of spiroplasma cells carrying the CAT gene.

Preparation of Spiroplasma Cell-Free Extracts

1. Collect spiroplasma cells ($\sim 10^9$) from 1 ml of culture by centrifugation at 15,000 g for 20 minutes in a microcentrifuge. Wash the pellet in 1.5 ml of HS buffer and resuspend the final pellet in 0.1 ml of 250 mM Tris–HCl buffer, pH 7.5.
2. Apply four cycles of freezing in liquid nitrogen and thawing at 37°C to the samples (vortex at each cycle), and remove cell debris by centrifugation at 15,000 g for 5 minutes.
3. Heat the supernatant at 65°C for 10 minutes to inactivate deacetylases, clarify by centrifugation, and store the supernatant at -20°C until use.

CAT Assay. This Assay Is Essentially According to Gorman et al. (1982).

1. Incubate 30 µl of each extract at 37°C for 2 hours with 170 µl of the following reaction mixture: 100 µl of 250 mM Tris–HCl, pH 7.5, 50 µl (1 µCi) of [^{14}C]chloramphenicol (specific activity, 60 mCi/mmol), and 20 µl of acetyl coenzyme A (3.5 mg/ml).
2. Stop the reaction and extract acetylated forms of [^{14}C]chloramphenicol by adding 1 ml of ethyl acetate.
3. Dry the organic layer (upper phase) and resuspend in 20 µl of ethyl acetate.
4. Load 5 µl of each sample on silica gel thin-layer plates and submit to ascending chromatography with a 95:5 chloroform:methanol mixture as the solvent for 90 minutes.
5. Air dry the plates and autoradiograph.

The presence of CAT activity is detected by the ability of cell-free extracts to acetylate chloramphenicol into acetylated forms (1-acetyl, 3-acetyl, and 1,3-acetyl) of chloramphenicol. When subjected to silica gel thin-layer chromatography, these acetylated forms exhibit a higher mobility than nonacetylated chloramphenicol.

The amount of acetylated chloramphenicol can be quantitatively determined by counting the radioactivity present in the corresponding spots of the silica gel plate.

Alternatively, the presence of the CAT enzyme (but not its activity) can be detected by Western immunoblotting using commercially available CAT antibodies.

DISCUSSION

The potential of the SpV1-RF/*S. citri* cloning system to express genes, namely those containing UGA codons, has been definitively demonstrated. However, some limitations have to be mentioned here. The first limitation is the lack of information on the restriction–modification system of the spiroplasmal cloning host. Our data only showed that DNA amplified in *E. coli* strains HB101 or TG1 could be cloned in *S. citri* R8A2. However, it could not be cloned when amplified in *E. coli* DH5αF', a strain which, in contrast to strains HB101 and TG1, possesses a functional *Eco*K restriction–modification system. A second limitation is the relative instability of the recombinant SpV1 RF. When the recombinant virus is propagated in the absence of selection pressure, loss of the cloned DNA insert occurs within the first 10 propagations. In the future, the SpV1-RF/*S. citri* cloning system must be improved for greater stability and increased expression of the DNA insert within the SpV1-derived vector.

Construction of Artificial Plasmids for Gene Cloning in Spiroplasma citri

INTRODUCTION

Replication of the bacterial chromosome is initiated at a well-defined region, the origin of replication (*oriC*). A remarkable feature of the replication origins is the conservation of the *dnaA* gene and the flanking noncoding regions, containing *dnaA* boxes and AT-rich repeats. A similar organization occurs in mollicutes (Fujita *et al.*, 1992). The self-replicating property of the *oriC* region of *S. citri* has been used to construct artificial plasmid vectors which replicate in this organism (Ye *et al.*, 1994).

ISOLATION OF *oriC* REGION OF *Spiroplasma citri*

Two conserved domains of the DNA protein have been identified (Ogasawara *et al.*, 1991). One is the ATP-binding site (YNPLFIYG) and the other, close to the C-terminal end, is GGRDHTTV. From these two amino acid sequences, taking into account the known *S. citri* codon usage, the sequences of the two following degenerated oligonucleotides have been deduced: primer 1, 5'-TATAATCC(TA)TT(AG)

TTTAT (TA)TATGG-3' and primer 2, 5'-(AT)AC(AT)GT(AT)GTATGATC (TC)C(TG)(AT)CC(AT)CC-3'.

These oligonucleotides are used as primers for polymerase chain reaction amplification of the *dnaA* gene of *S. citri*. The 25-μl reaction mixture contains 20 to 50 ng of *S. citri* genomic DNA, 0.2 mM MgCl$_2$, 0.05% W1 detergent, 0.1 mM of each dNTP, 0.2 mg/ml BSA, 0.08 μM of primer 1, 0.24 μM of primer 2, and 1 to 2 units of *Taq* DNA polymerase.

The amplification reaction is performed in a Perkin–Elmer Cetus thermal cycler over 36 cycles, each of 45 seconds at 92°C, 1 minute at 40°C, and 1 minute at 72°C, with an additional step of 20 minutes at 72°C for chain termination.

The amplified products are analyzed and purified by agarose gel electrophoresis. The 0.8-kbp agarose DNA band is excised, and the DNA is eluted from the gel using the GeneClean II kit (BIO 101, Inc.) procedure. To establish the restriction map of the *dnaA* gene region, the purified amplified DNA fragment is used as the probe in Southern blot hybridization of restricted genomic DNA. The restriction DNA fragments containing the *dnaA* gene are rescued from total DNA by cloning in *E. coli*. Genomic or subgenomic DNA libraries are constructed in plasmid vectors and are screened with the *dnaA* probe according to standard procedures.

In the case of *S. citri*, we have shown that a 2-kbp *Taq*I restriction fragment which contains the entire sequence of *dnaA* gene and the flanking *dnaA* boxes is capable of autonomous replication (Ye et al., 1994).

CONSTRUCTION OF PLASMID VECTORS

The 2-kbp *oriC* fragment of *S. citri* is linked to the *tetM* determinant, a 4.2-kbp *Hinc*II fragment rescued from plasmid pJI3 (Burdett et al., 1982), by ligation, then the ligation mixture is transformed into *S. citri* cells by electroporation as described earlier.

S. citri transformants are selected by plating on solid SP4 medium containing 2 μg/ml tetracycline. Under these conditions, transformed cells resistant to tetracycline form colonies within 10 days after plating.

Individual colonies are separately transferred into 2 ml SP4 liquid medium supplemented with 2 μg/ml tetracycline. During passaging, the antibiotic concentration is progressively increased from 2 to 4, 8, and 15 μg/ml.

Plasmids are subsequently purified from spiroplasma transformants by the cleared-lysate procedure described earlier for the SpV1 RF and are analyzed by agarose gel electrophoresis and Southern blot hybridization.

Alternatively, shuttle vectors that replicate both in *E. coli* and in *S. citri* are constructed by inserting the *S. citri oriC* and the *tetM* determinant in ColE1-derived plasmids (Renaudin et al., 1995). The advantage of such plasmids is that

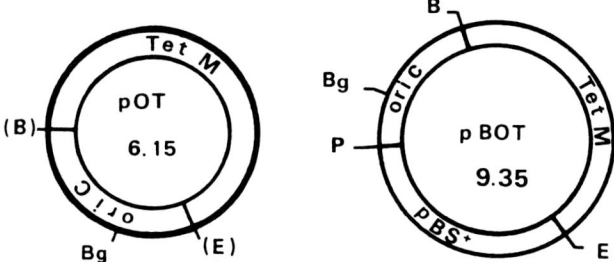

Fig. 2. Restriction maps of plasmids pOT and pBOT. Tet M, DNA fragment containing the *tetM* determinant gene. OriC, DNA fragment containing the chromosomal replication origin of *S. citri*. pBS+, ColE1-derived plasmid (Stratagene).

they can be easily constructed in *E. coli* prior to being transferred into spiroplasmas.

These original plasmid constructs (pOT and pBOT in Fig. 2) can then be engineered for various cloning purposes.

DISCUSSION

Plasmids pOT and pBOT are electrotransformed into *S. citri* at a high frequency (10^{-3} transformants/cell). However, transformation frequency is highly dependent on the spiroplasmal host strain. In the presence of selection pressure (tetracycline), the plasmids are stably maintained for more than 50 generations. In *S. citri*, plasmid pOT is maintained extrachromosomally while plasmid pBOT, which replicates both in *S. citri* and in *E. coli*, integrates into the spiroplasmal host chromosome. These two plasmids have been successfully used as vectors to introduce foreign DNA into *S. citri*. Expression of the *S. phoeniceum* spiralin gene in *S. citri* strain ASP1 was demonstrated (Renaudin et al., 1995).

It is hoped that *oriC* plasmids will be able to stably maintain larger inserts than would viral or plasmid vectors. These plasmids hold considerable promise for the development of genetic studies in spiroplasmas in that they offer a way to complement mutations.

References

Burdett, V., Inamine, J., and Rajagopalan, S. (1982). Heterogeneity of tetracycline resistance determinants in *Streptococcus*. *J. Bacteriol.* **149,** 995–1004.
Chirgwin, J. M., Przybyla, A. E., MacDonald, R. J., and Rutter, W. J. (1979). Isolation of biologically active ribonucleic acid from sources enriched in ribonuclease. *Biochemistry* **18,** 5294–5299.

Dybvig, K. (1990). Mycoplasmal genetics. *Annu. Rev. Microbiol.* **44,** 81–104.
Fujita, M. Q., Yoshikawa, H., and Ogasawara, N. (1992). Structure of the dnaA and DnaA-box region in the *Mycoplasma capricolum* chromosome: Conservation and variations in the course of evolution. *Gene* **110,** 17–23.
Gasparich, G. E., Hackett, K. J., Stamburski, C., Renaudin, J., and Bové, J. M. (1993). Optimization of methods for transfecting *Spiroplasma citri* strain R8A2 HP with the Spiroplasma virus SpV1 replicative form. *Plasmid* **29,** 193–205.
Gorman, C. M., Moffat, L. F., and Howard, B. H. (1982). Recombinant genomes which express chloramphenicol acetyltransferase in mammalian cells. *Mol. Cell. Biol.* **2,** 1044–1051.
Marais, A., Bové, J. M., Dallo, S. F., Baseman, J. B., and Renaudin, J. (1993). Expression in *Spiroplasma citri* of an epitope carried on the G fragment of the cytadhesin P1 gene from *Mycoplasma pneumoniae*. *J. Bacteriol.* **175,** 2783–2787.
Ogasawara, N., Moriya, S., and Yoshikawa, H. (1991). Initiation of chromosome replication: Structure and function of *ori*C and DnaA protein in eubacteria. *Res. Microbiol.* **142,** 851–859.
Renaudin, J., and Bové, J. M. (1994). SPV1 and SPV4, spiroplasma viruses with circular, single-stranded DNA genomes, and their contribution to the molecular biology of spiroplasmas. *Adv. Virus Res.,* **44,** 429–463.
Renaudin, J., Marais, A., Verdin, E., Duret, S., Foissac, X., Laigret, F., and Bové, J. M. (1995). Integrative and free *Spiroplasma citri oriC* plasmids: Expression of the *Spiroplasma phoeniceum* spiralin in *Spiroplasma citri*. *J. Bacteriol.,* **177,** 2870–2877.
Sambrook, J., Fritsch, F., and Maniatis, T. (1989). "Molecular Cloning: A Laboratory Manual," 2nd Ed. Cold Spring Harbor Laboratory Press, Cold Spring Harbor, NY.
Stamburski, C., Renaudin, J., and Bové, J. M. (1991). First step toward a virus-derived vector for gene cloning and expression in spiroplasmas, organisms which read UGA as a tryptophan codon: Synthesis of chloramphenicol acetyltransferase in *Spiroplasma citri*. *J. Bacteriol.* **173,** 2225–2230.
Whitcomb, R. F. (1983). Culture media for spiroplasmas. *In* "Methods in Mycoplasmology" (S. Razin and J. G. Tully, eds.), Vol. 1, pp. 147–159. Academic Press, New York.
Ye, F., Renaudin, J., Bové, J. M., and Laigret, F. (1994). Cloning and sequencing of the replication origin (*OriC*) of the *Spiroplasma citri* chromosome and construction of autonomously replicating artificial plasmids. *Curr. Microbiol.* **28,** 23–29.

B7

ARTIFICIAL TRANSFORMATION OF MOLLICUTES VIA POLYETHYLENE GLYCOL- AND ELECTROPORATION-MEDIATED METHODS

Kevin Dybvig, Gail E. Gasparich, and Kendall W. King

Introduction

Currently, there is no general method for transformation of mycoplasmas. Artificial transformation of a few species has been described using methods based on either polyethylene glycol (PEG) or electroporation. For PEG-mediated transformation, the methods are similar to protocols originally developed for transformation of *Bacillus subtilis* protoplasts (Chang and Cohen, 1979). The principle behind electroporation is the use of high-voltage electric field pulses to transiently disrupt the cell membrane allowing for the uptake of exogenous DNA. This technique has proven to be useful in the transformation of both prokaryotes and eukaryotes (Shigekawa and Dower, 1988). The method of choice for transformation of mycoplasmas, PEG-mediated versus electroporation, depends on the species in question and on the nature of the DNA molecule to be used.

Materials

PEG-Mediated Transformation

PEG 8000 (Sigma)
Plasmid or viral DNA (10–20 μg DNA is usually required)
Centrifuge
Growth media (broth and agar)

FOR *Acholeplasma laidlawii*

40% (w/v) PEG in S/T buffer (0.5 M sucrose, 0.01 M Tris, pH 6.5)
Dilution buffer: 0.01 M Tris, pH 8.0

FOR *Mycoplasma capricolum* AND *Mycoplasma mycoides*

Wash buffer: S/T
70% (w/v) PEG in 0.01 M Tris (pH 6.5)
$CaCl_2$
Yeast tRNA
Dilution buffer: S/T

FOR *Mycoplasma pulmonis*

40% (w/v) PEG in PBS (0.01 M Na_2HPO_4, 0.14 M NaCl, pH 7.3)
Yeast tRNA
Dilution buffer: PBS

FOR *Spiroplasma citri*

44% (w/v) PEG in S/T
Dilution buffer: 0.01 M Tris, pH 8.0

Electroporation

Microcentrifuge
Electroporation system (e.g., Bio-Rad Gene Pulser, including pulse controller and 0.4-cm gap cuvettes)
Electroporation buffer: 0.008 M HEPES (N-2-hydroxyethylpiperazine-N'-2-ethanesulfonic acid) buffer, pH 7.4, in 0.28 M sucrose

Growth media (broth and agar)
Plasmid or viral DNA

Procedure

PEG-Mediated Transformation

1. Grow cells to late logarithmic phase.
2. Harvest cells by centrifugation at 10,000 g for 10 minutes.
3. For *M. capricolum* and *M. mycoides,* wash once in S/T buffer.
4. Resuspend cells in growth medium (*A. laidlawii* and *S. citri*), PBS (*M. pulmonis*), or 0.1 M $CaCl_2$ (*M. capricolum* and *M. mycoides;* following suspension in $CaCl_2$, incubate cells on ice for 30 minutes).
5. To 250 µl of cell suspension, add plasmid or viral DNA plus 10 µg yeast tRNA (optional for some species) in a total volume of 50 µl. Mix.
6. Immediately add 2.0 ml of PEG solution. Mix.
7. Incubate up to 2 minutes at room temperature.
8. Add 10–15 ml dilution buffer. Mix.
9. Harvest cells by centrifugation at 12,000 g for 12 minutes.
10. Resuspend cells in growth medium.
11. Incubate for 30 minutes to 2 hours at 37°C (32°C for *S. citri*) to allow for expression of the resistance gene product (if applicable).
12. Remove a small sample of the culture for colony-forming units (CFU) determination. Assay the rest of the culture for transformants on selective medium or assay for transfectants as plaque-forming units (PFU) on lawns of susceptible host cells.

Electroporation

1. Grow host cells to mid-logarithmic phase. Assay a sample from this culture for CFU.
2. Aliquot 1.5 ml of cell culture into microcentrifuge tubes (two per reaction), and harvest cells at 18,000 g for 20 minutes at room temperature.
3. Resuspend pellets in 200 µl electroporation buffer and combine two pellets into one tube.
4. Harvest cells as described earlier and resuspend cells in 400 µl electroporation buffer.
5. Add CsCl-purified DNA to host cells in electroporation buffer (0.1 µg for

viral DNA; 10 μg for plasmid DNA) and place mixture in a 0.4-cm gap electroporation cuvette. Electroporate at 6.5 kV/cm (2.5 kV in a 0.4-cm gap cuvette) at 1000 ohm resistance and 3 μFD capacitance for a pulse of 0.5–1.5 mseconds.

6. For transfection experiments, assay the electroporation mixture for PFU using lawns of appropriate host cells. For transformation experiments involving plasmids containing antibiotic resistance determinants, add 600 μl growth medium to the electroporation mixture, incubate at the host's optimal growth temperature to allow expression of the resistance determinant, and assay for CFU on selective medium.

Discussion

PEG-mediated transformation of a mollicute was first described for transfection of *A. laidlawii* with DNA from mycoplasma virus L2 (Sladek and Maniloff, 1983). It is currently the method of choice for transformation of *M. pulmonis, M. mycoides,* and *M. capriocolum* (Dybvig and Alderete, 1988; King and Dybvig, 1991; King and Dybvig, 1994b). In particular, repeated attempts to transform *M. pulmonis* using a variety of electroporation conditions have been unsuccessful (K. Dybvig, unpublished data). In contrast, electroporation is the only method yet reported for transformation of *Mycoplasma pneumoniae* (Hedreyda *et al.,* 1993). Therefore, both the PEG method and electroporation should be considered when examining species for which transformation has not been previously characterized.

Electroporation and the PEG-mediated method usually work about equally well for transformation of *A. laidlawii* and *S. citri.* In the case of *A. laidlawii,* the most efficient transformation method depends on the DNA molecule being employed. For transfection with mycoplasma virus L1 DNA (a circular molecule of about 4.5 kb), the PEG method is about 10 times more efficient than electroporation (Lorenz *et al.,* 1988). In contrast, transfection with mycoplasma virus L3 DNA (a 39-kb linear molecule) has only been successful using electroporation (Lorenz *et al.,* 1988). Whether differences in the size of the DNA molecule and/or some other factor(s) determines which transformation method is more efficient remains to be determined. The largest DNA molecule successfully used for PEG-mediated transformation of mycoplasmas has been the Tn*916*-containing plasmid pAM120 (26 kb) (Dybvig and Cassell, 1987). In cases in which electroporation and the PEG method are equally efficient, electroporation is probably the method of choice, as the PEG method can sometimes yield variable results.

Several factors should be considered for optimal transformation using the PEG method. These include the PEG concentration, choice of buffers, concentration of the transforming DNA molecule, host cell concentration, presence of $CaCl_2$, and presence of carrier DNA or RNA. The source of the PEG is critical, and different lots from the same supplier can yield very different results. Sterilization of PEG solutions by filtration yields a different final product than does sterilization by autoclaving. In determining an optimal PEG concentration, the investigator should be consistent regarding the sterilization method. Filtration is impractical for the high concentrations of PEG used to transform *M. mycoides* and *M. capricolum*. High concentrations of PEG in solution can solidify at room temperature, and it may be necessary to briefly thaw the solution prior to transformation. The buffers used for PEG-mediated transformation are also crucial. Many mycoplasmas maintain excellent viability in PBS, and PBS is usually a good transformation buffer. However, PBS should be avoided if cells are to be transformed in the presence of $CaCl_2$ (e.g., as for transformation of *M. mycoides*) because a precipitant (probably calcium phosphate) can form that interferes with transformation efficiencies.

Factors to be considered for optimal transformation via electroporation include field strength, buffer, concentration of the transforming DNA molecule, and host cell concentration. It may also be beneficial to gently treat the host cells with trypsin immediately prior to transformation. The first reported use of electroporation with mollicutes was with *A. laidlawii* and the mycoplasma viruses L1 and L3 (Lorenz et al., 1988). To obtain an optimal efficiency of one plaque per 3×10^7 DNA molecules for the L1 virus, a brief digestion of cells with trypsin prior to electroporation was 16-fold more efficient than without trypsin treatment. Trypsin treatment may inactivate exonucleases in the medium or it may remove cell surface proteins that interfere with DNA uptake.

A protocol similar to that of Lorenz et al. (1988) has been used for transfection of *S. citri* with DNA from spiroplasma viruses SVTS2 and SpV1 (McCammon et al., 1990; Stamburski et al., 1991). The protocol for electroporation using SpV1 DNA has been optimized to obtain a maximum efficiency of 9.1×10^{-4} PFU/CFU (Gasparich et al., 1993). This protocol is described in the Procedure section of this chapter.

Regardless of the transformation method, a phenotypic marker is necessary for the selection of transformants. The capacity of different mycoplasmal viruses to form plaques on lawns of host cells has been exploited, and various antibiotic resistance determinants have also been used. The *tetM* gene, encoding resistance to tetracycline, has been shown to function in a wide variety of mycoplasmas. Other markers that have been used with success in some mycoplasmal species include genes encoding for resistance to erythromycin, gentamicin, chloramphenicol, and neomycin/kanamycin (Dybvig, 1990; King and Dybvig, 1994a; Stamburski et al., 1991). Genes encoding resistance to heavy metals may potentially serve as additional markers. The genetic marker(s) can be located on a transposon, in which case the expression of resistance is dependent on integration of the transposon into the host chromosome, or on plasmids capable of either integration or self-replication within the host. Southern hybridization analysis should be used to confirm the presence of the genetic marker in putative transformants. In most cases, antibiotic selection should be maintained to prevent loss of the marker.

References

Chang, S., and Cohen, S. N. (1979). High frequency transformation of *Bacillus subtilis* protoplasts by plasmid DNA. *Mol. Gen. Genet.* **168,** 111–115.
Dybvig, K. (1990). Mycoplasmal genetics. *Annu. Rev. Microbiol.* **44,** 81–104.
Dybvig, K., and Alderete, J. (1988). Transformation of *Mycoplasma pulmonis* and *Mycoplasma hyorhinis:* Transposition of Tn*916* and formation of cointegrate structures. *Plasmid* **20,** 33–41.
Dybvig, K., and Cassell, G. H. (1987). Transposition of gram-positive transposon Tn*916* in *Acholeplasma laidlawii* and *Mycoplasma pulmonis. Science* **235,** 1392–1394.
Gasparich, G. E., Hackett, K. J., Stamburski, C., Renaudin, J., and Bové, J. M. (1993). Optimization of methods for transfecting *Spiroplasma citri* strain R8A2 HP with the spiroplasma virus SpV1 replicative form. *Plasmid* **29,** 193–205.
Hedreyda, C. T., Lee, K. K., and Krause D. C. (1993). Transformation of *Mycoplasma pneumoniae* with Tn4001 by electroporation. *Plasmid* **30,** 170–175.
King, K. W., and Dybvig, K. (1991). Plasmid transformation of *Mycoplasma mycoides* subsp. *mycoides* is promoted by high concentrations of polyethylene glycol. *Plasmid* **26,** 108–115.
King, K. W., and Dybvig, K. (1994a). Mycoplasma cloning vectors derived from plasmid pKMK1. *Plasmid* **31,** 49–59.
King, K. W., and Dybvig, K. (1994b). Transformation of *Mycoplasma capricolum* and examination of DNA restriction-modification in *M. capricolum* and *Mycoplasma mycoides* subsp. *mycoides. Plasmid* **31,** 308–311.
Lorenz, A., Just, W., da Silva, Cardosa, M., and Klotz, G. (1988). Electroporation-mediated transfection of *Acholeplasma laidlawii* with mycoplasma virus L1 and L3 DNA. *J. Virol.* **62,** 3050–3052.
McCammon, S. L., Dally, E. L., and Davis, R. E. (1990). Electroporation and DNA methylation effects on the transfection of spiroplasma. *Zentralbl. Bakteriol. Suppl.* **20,** 60–65.
Shigekawa, K., and Dower, W. J. (1988). Electroporation of eukaryotes and prokaryotes: A general approach to the introduction of macromolecules into cells. *Biotechniques* **6,** 742–751.
Sladek, T. L., and Maniloff, J. (1983). Polyethylene glycol-dependent transfection of *Acholeplasma laidlawii* with virus L2 DNA. *J. Bacteriol.* **155,** 734–741.
Stamburski, C., Renaudin, J., and Bové, J. M. (1991). First step toward a virus-derived vector for gene cloning and expression in spiroplasmas, organisms which read UGA as a tryptophan codon: Synthesis of chloramphenicol and acetyltransferase in *Spiroplasma citri. J. Bacteriol.* **173,** 2225–2230.

B8

DNA METHYLATION ANALYSIS

Aharon Razin and Paul Renbaum

Introduction

Methylated bases have been known to exist in DNA since the early 1950s. However, only since the 1980s or so has DNA methylation been demonstrated to play a role in a variety of biological phenomena. The new discoveries of biological processes that involve DNA methylation evoked high interest in modification of DNA and, consequently, several novel methods of analyzing DNA methylation have been developed (for review see Razin and Riggs, 1980). Initially, paper chromatography and electrophoresis were used to separate the minor modified bases from the four major bases in DNA. These laborious methods were quickly replaced by more sophisticated techniques using high-performance liquid chromatography (HPLC), mass spectrometry, gas chromatography, and nearest neighbor analysis. However, all of these methods suited only the analysis of overall methylation in DNA. To analyze the methylation status of specific sites in the DNA, other methods had to be developed. Several methods are now available to study individual methylation sites in the DNA. These methods use methyl-sensitive restriction enzymes combined with gel electrophoretic separation of restriction fragments and their visualization by staining or Southern blotting. Genomic sequencing is a more sophisticated method suitable for analyzing 5-methylcytosine residues which are not part of restriction sites (Saluz and Jost, 1993). Some of the methods that are currently under extensive use are described in this chapter in detail. A general assay of methyltransferase activity is also discussed.

Materials

88% formic acid (HPLC grade)
SCX $\frac{10}{25}$ Partisil Whatman column
NH$_4$COOH (pH adjusted with glacial acetic acid)
[α-^{32}P]dNTP,S-[*methyl*-^{3}H] adenosylmethionine
DNA polymerase I
Micrococcal nuclease
Spleen phosphodiesterase
Venom phosphodiesterase
Alkaline phosphatase
Pancreatic DNase I
Sephadex G-50
Kodak chromogram sheets 13255 cellulose without fluorescent indicator
X-ray films for autoradiography
Restriction enzymes
Nitrocellulose sheets for hybridization
Gene screen nylon mesh
GF/C filters

High-Performance Liquid Chromatography

Numerous studies have used HPLC for the analysis of overall methylation levels in DNA. This method is based on the breakdown of the DNA to nucleoside monophosphates, nucleosides, or free bases (Razin and Razin, 1980). It, therefore, provides both qualitative and quantitative information on the modified bases in DNA. DNA can be digested to 5'-NMP using DNase I and venom phosphodiesterase, and the nucleotides are then converted to nucleosides by treatment with alkaline phosphatase. Alternatively, the DNA can be hydrolyzed by acid to the free base level. Because of space limitations, only the analysis of free bases will be described in detail.

Procedure

DNA samples (10–50 μg) are hydrolyzed at 180°C for 1 hour in 200 μl of 88% formic acid in sealed glass tubes. Samples are dried under a stream of nitrogen and dissolved in 50 μl of 0.1 M HCl. Samples (20 μl) are injected into a SCX $\frac{10}{25}$ Whatman column, and the bases are eluted by degassed 0.02 M NH$_4$COOH buffer (pH 2.4) at an ambient temperature and a flow rate of 40 μl/hour. Absorbance at 280 or 254 nm is recorded. A typical elution profile of authentic base markers under such conditions is presented in Fig. 1.

Nearest Neighbor Analysis

This method (Gruenbaum *et al.*, 1981) provides a quantitative estimate of methylated bases in the DNA and, at the same time, gives some indication as to the distribution of the methylated bases, at random or restricted to specific sites. According to this method, randomly nicked genomic DNA is end-labeled at the nicks with *Escherichia coli* DNA polymerase I and one of the four α-^{32}P-labeled 2'-deoxynucleoside 5'-triphosphates (dGTP, dCTP, dATP, or dTTP). The purified labeled DNA (after the removal of unreacted nucleotide) is digested by spleen phosphodiesterase and micrococcal nuclease to yield 2'-deoxynucleoside 3'-monophosphates. By this digestion, the radioactive phosphate is transferred from the incorporated nucleotide to its 5' nearest neighbor in the DNA. The labeled nucleotides are separated by two-dimensional chromatography on thin-layer cellulose sheets and autoradiographed. The relative radioactivity in the spot representing the major base and the corresponding modified base serves to quantify the degree of methylation. The method is illustrated in the flow diagram (Fig. 2) for [α-^{32}P]dGTP as the labeling nucleotide. The same procedure can be used for labeling with [α-^{32}P]dATP, [α-^{32}P]dTTP, or [α-^{32}P]dCTP.

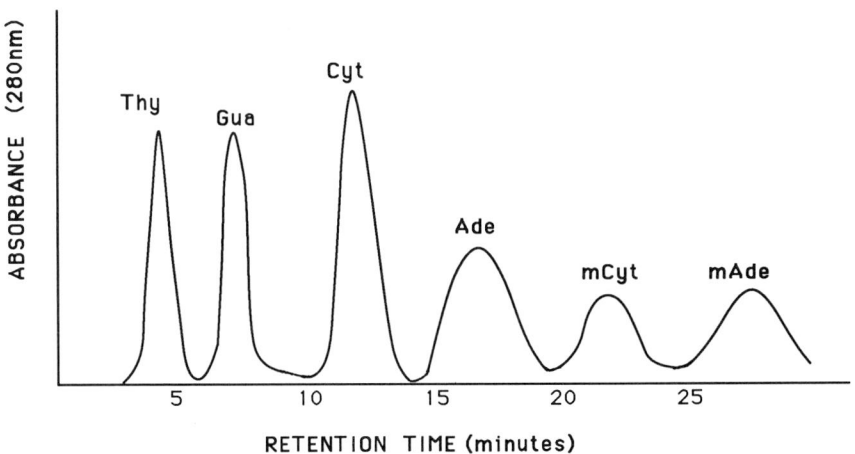

Fig. 1. Separation of authentic base markers by HPLC on a SCX $\frac{10}{25}$ column under conditions as described in the text. Thy, thymine; Gua, guanine; Cyt, cytosine; Ade, adenine; m^{5}Cyt, 5-methylcytosine; m^{6}Ade, 6-methyladenine. Quantitation of the individual bases is described in Razin and Razin (1980).

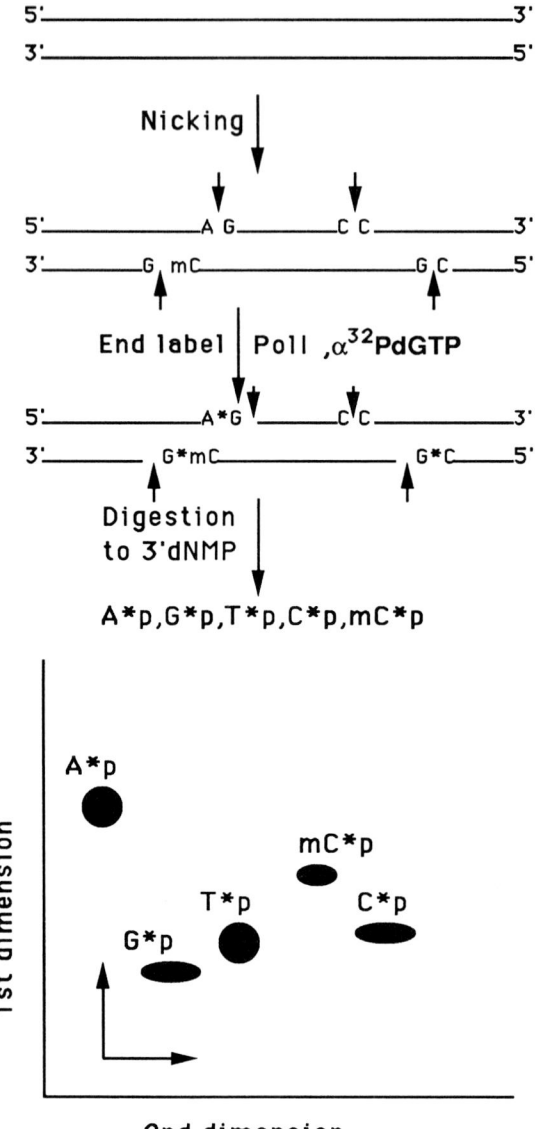

Fig. 2. Nearest neighbor analysis. Genomic DNA is nicked by DNase I (at arrows) and the nicks are end-labeled by [α-^{32}P]dGTP. The ^{32}P label is transferred to the nearest 5′ neighbor by digesting the DNA with micrococcal nuclease and spleen phosphodiesterase. The resulting ^{32}P-labeled 3′NMPs are separated by 2D thin-layer chromatography and subjected to autoradiography. The ratio of the radioactive spots mC*p to C*p represents the methylated fraction of cytosine residues in CpG. For chromatographic conditions, see text.

PROCEDURE

One-microgram DNA samples (in 1–16 μl volume) are nicked with 2μl DNase I (10 μg/ml) in a buffer containing 50 mM potassium phosphate (pH 7.2), 5 mM MgCl$_2$, and 1 mM 2-mercaptoethanol. After 15 minutes of incubation at 37°C, 7 U of *E. coli* DNA polymerase I is added to the cooled reaction mixture together with 2μl α-^{32}P-labeled dGTP, dATP, dCTP, or dTTP (3000 Ci/ml), and a 30-minute incubation is carried out at 15°C. The reaction is terminated by adding 4 μl of 0.5 M EDTA, and unreacted [α-^{32}P]dNTP is removed by Sephadex G-50 spin chromatography. The labeled DNA is digested to deoxynucleoside 3'-monophosphates by incubating in 20 μl final volume for 2 hours at 37°C with 2 μl (1.4 mg/ml) micrococcal nuclease in 25 mM Tris–HCl (pH 8.0) and 5 mM CaCl$_2$, followed by incubation for 2 hours at 37°C with 2 μl (70 U/ml) spleen phosphodiesterase. The digested DNA is applied to cellulose thin-layer sheets and subjected to two-dimensional chromatography. Chromatography in the first dimension is by isobutyric acid:H$_2$O:NH$_4$OH (132:40:2,v/v), and in the second dimension by saturated (NH$_4$)$_2$SO$_4$:1 M sodium acetate:2-propanol (160:36:4,v/v).

Restriction Enzyme Analysis

The most popular assay for methylation of specific DNA sequences is based on the digestion of genomic DNA with restriction enzymes that are sensitive to methylation at their recognition sites (Bird, 1978). The DNA digest is then subjected to electrophoresis through an agarose gel that is subsequently stained by ethidium bromide or blotted to nitrocellulose sheets. The blots are hybridized with α-^{32}P-labeled specific DNA probes and autoradiographed (see Fig. 3). Ethidium bromide-stained gels reveal the extent of digestion of the DNA by the restriction enzyme. This reflects the level of methylation at the restriction site along the entire genome. A representative analysis of adenine methylation at GATC sites in genomes of mycoplasma and spiroplasma strains is presented in the ethidium-stained gel (Fig. 4). Although this analysis is limited to the methylation status of a specific site (GATC), it shows the state of methylation of the site along the entire genome. Accordingly, the three *Mycoplasma* species DNAs are not methylated at GATC sites, whereas four *Spiroplasma* species DNAs are methylated at GATC (Table I).

Genomic DNA Sequencing

Genomic sequencing can be used to analyze the status of methylation of all cytosine residues. In the original method for genomic sequencing, the 5-methylcytosine residues resist the C reaction and, therefore, appear as a blank on a

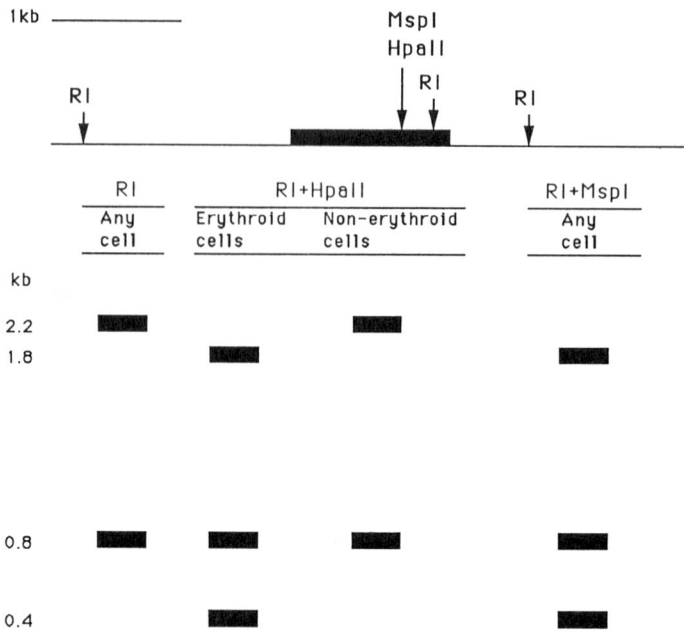

Fig. 3. Schematic presentation of a typical analysis of the status of methylation of a single site in a gene sequence. Rabbit genomic DNA was digested with *Eco*RI (RI) or with RI in combination with *Hpa*II. *Hpa*II does not cut at CCGG sites when the inner C is methylated. *Msp*I is an isoschizomer of *Hpa*II that cuts at CCGG sites regardless of the methylation status of the inner cytosine residue. The restriction digests were electrophoresed on a 0.8% agarose gel, blotted to nitrocellulose sheets, and hybridized to a labeled rabbit β-globin probe. In cells that express β-globin, the *Hpa*II site is unmethylated, whereas in cells that do not express the gene, the site is methylated. The *Eco*RI/*Msp*I digest reveals only one CCGG site in the β-globin gene region.

sequencing gel. Since this method detects each 5-methylcytosine residue in the DNA sequence, this technique complements the restriction enzyme analysis which is limited to sites for which methyl-sensitive restriction enzymes exist. Following the conventional reactions of the chemical sequencing method, the DNA fragments are subjected to electrophoresis through an 80 × 40-cm 8% acrylamide gel. The gel is electrotransferred to Gene screen nylon mesh and the nylon sheet is hybridized with a labeled sequence-specific probe. A missing band on the autoradiogram in a cytosine position reflects the presence of 5-methylcytosine in this position in the sequence. The disadvantage of this procedure is that it will not allow distinction between unmethylated and partially methylated cytosine residues in a population of DNA molecules. This drawback has been overcome by a modified method in which cytosine residues are deaminated to uracil, whereas 5-methylcytosine remains nonreactive. The sequence under in-

B8 DNA Methylation Analysis

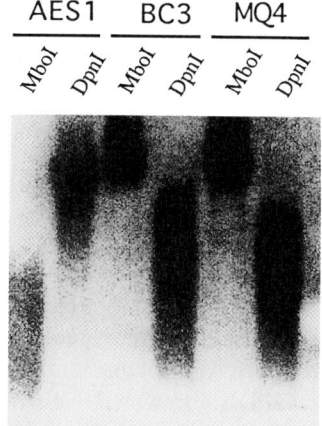

Fig. 4. Three examples of *Spiroplasma* species (strains, AES1, BC3, and MQ4) are presented. The restriction digests were electrophoresed on agarose gels and stained with ethidium bromide. *Mbo*I and *Dpn*I are isoschizomers that recognize the same sequence, GATC. *Mbo*I cuts only when the adenine residue in GATC is unmethylated on both strands. *Dpn*I cuts at this site only when adenine residues are methylated on both strands. Thus, the conclusion from this restriction pattern is that AES1 DNA is unmethylated at GATC sites, whereas the adenine residue in GATC sites in BC3 and MQ4 DNA is methylated on both strands.

vestigation is then amplified by polymerase chain reaction (PCR) with two sets of strand-specific primers to yield a pair of fragments, one from each strand, in which all uracil and thymine residues have been amplified as thymine and only

TABLE I

GATC METHYLATION IN *Mycoplasma* AND *Spiroplasma* SPECIES

Genus and species	*Mbo*I	*Dpn*I
Mycoplasma		
M. capricolum	+	−
M. gallisepticum	+	−
A. laidlawii	+	−
Spiroplasma		
S. citri (R8A2)	−	+
S. melliferum (BC-3)	−	+
S. monobiae (MQ-1)	−	+
S. velocicrescens (MQ-4)	−	+
S. floricola (BNR-1)	+	−
S. culicicula (AES-1)	+	−

5-methylcytosine residues have been amplified as cytosine. The PCR products can be sequenced directly to provide a strand-specific average sequence for the population of molecules or can be cloned and sequenced to provide methylation maps of single DNA molecules. Detailed description of the methods can be found in Saluz and Jost (1993).

DNA Methyltransferase Assay

To assay DNA methyltransferase activity, the rate of incorporation of tritium counts from S-[*methyl*-^{3}H]adenosylmethionine into unmethylated DNA is measured. The most efficient substrate for prokaryotic methyltransferases is unmethylated duplex DNA. However, comparable activity can sometimes be observed with hemimethylated DNA in which one strand of the DNA is methylated. In contrast, the mammalian methyltransferase shows two orders of magnitude higher activity with the hemimethylated substrate (Razin, 1989).

The methyltransferase assay is carried out in a 40-μl reaction mixture that contains 1 μl (1 μCi) S-[*methyl*-^{3}H]adenosylmethionine (15 Ci/mmol); 1–2 μg of DNA substrate, and cell extract or methyltransferase preparation (see below) in a buffer containing 10 mM Tris (pH 7.5), 5 mM EDTA, 5 mM 2-mercaptoethanol, and 0.1 mM phenylmethylsulfonyl fluoride. The reaction, carried out at 37°C, is terminated by the addition of a 200-μl solution of 0.5% sodium dodecyl sulfate in 0.5 M NaOH. This mixture is subsequently heated for 2 hours at 60°C and is extracted once with chloroform/isoamyl alcohol (24:1,v/v). To 150 μl of the upper phase, 10 μl of calf thymus DNA (1 mg/ml) is added as carrier and DNA is precipitated by 5 ml cold 10% trichloroacetic acid (TCA). The precipitate is filtered and washed with cold 5% TCA on GF/C filters. Filters are dried and counted in a toluene-based scintillation fluid.

Methyltransferase extract is prepared by suspending the cells in 50 mM Tris–HCl (pH 8.0), 10 mM EDTA, and 10 mM 2-mercaptoethanol, and rupturing the cells by sonication. The supernatant, following 1 hour of centrifugation at 100,000 g, is dialyzed against the suspension buffer in 50% glycerol for 18 hours. Methylase activity in this crude extract is stable for at least 4 months at -20°C.

Discussion

In light of the variety of methods for DNA methylation analysis, the selection of the technique is of critical importance. In most DNAs, only one or two types of modified bases have been found. One is methylated on the amino group at position 6 of the adenine ring (m^{6}Ade), and in the other the carbon 5 in the cytosine ring (m^{5}Cyt) is methylated. In a few organisms, 4-methylcytosine, in which carbon 4 of the cytosine base is methylated, has also been detected. A first step in establishing the existence of modified bases in DNA is to hydrolyze the DNA to the nucleotide, nucleoside, or base level and use one of the available chromatographic methods to separate the minor bases from the major bases. With these methods, the extent to which adenine or cytosine residues are modified can be estimated.

Once the presence of methylated bases in the DNA is established, the next step is to provide information on the distribution of the modified bases along the DNA. In many cases, the modified base appears in palindromic sequences that are recognized by available restriction enzymes. A few of these restriction enzymes are sensitive to methylation and, therefore, usually cut at the site when it is unmethylated. However, there are exceptions to this rule in which the restriction enzyme only cuts when the site is methylated on both strands. One such example is the restriction enzyme *DpnI* that cuts at GATC sites in which the adenine residue is methylated on both strands. Modified bases that are present in restriction sites can easily be analyzed by digestion of the genomic DNA with the appropriate methyl-sensitive restriction enzyme and by separating the restriction fragments by gel electrophoresis. In this regard, a note of caution should be made. DNA that has been cloned loses its authentic modification and acquires the methylation profile of the host. Similarly, DNA that has been amplified by PCR loses its original methylation pattern and becomes completely unmethylated.

Modified bases that occur in sequences for which methylation-sensitive restriction enzymes are not available can best be analyzed by genomic sequencing. However, the available genomic sequencing methods are only suitable for the analysis of m^{5}Cyt and are limited to regions in the DNA to which probes exist. If the overall distribution of the modified bases is sought, the nearest neighbor analysis provides answers as to whether the modified base is randomly distributed or appears in specific sequences. This method also allows a quantitative estimate of the overall methylation in specific sequences. It was successfully used in the case of mammalian DNA where the only methylated base is cytosine, present exclusively in CpG-containing sequences, but not all CpG sites are methylated. It was also used in the analysis of methylation in spiroplasmas (Nur *et al.*, 1985) and helped to discover CpG methylation in *S. monobiae* strain MQ1 (Renbaum *et al.*, 1990). It should, however, be noted that in this spiroplasma, adenine residues in GATC sites are methylated as well (see Fig. 4).

References

Bird, A. P. (1978). Use of restriction enzymes to study eukaryotic DNA methylation. II. The symmetry of methylated sites supports semiconservative copying of the methylation pattern. *J. Mol. Bio.* **118,** 49–60.

Gruenbaum, Y., Stein, R., Cedar, H., and Razin, A. (1981). Methylation of CpG sequences in eukaryotic DNA. *FEBS Lett.* **124,** 67–71.

Nur, I., Szyf, M., Razin, A., Glaser, G., Rottem, S., and Razin, S. (1985). Procaryotic and eucaryotic traits of DNA methylation in Spiroplasmas (Mycoplasmas). *J. Bacteriol.* **164,** 19–24.

Razin, A. (1989). DNA methylases. *In* "Genetic Engineering" (J. K. Setlow, ed.) Vol. 11, pp. 1–11. Plenum, New York.

Razin, A., and Razin, S. (1980). Methylated bases in mycoplasmal DNA. *Nucleic Acids Res.* **8,** 1383–1390.

Razin, A., and Riggs, A. D. (1980). DNA methylation and gene activity. *Science* **210,** 604–610.

Renbaum, P., Abrahamove, D., Fainsod, A., Wilson, G. G., Rottem, S., and Razin, A. (1990). Cloning, characterization and expression in *Escherichia coli* of the gene coding for the CpG DNA methylase from *Spiroplasma* sp. strain MQ1 (M. SssI). *Nucleic Acids Res.* **18,** 1145–1152.

Saluz, H. P., and Jost, J. P. (1993). Major techniques to study DNA methylation. *In* "DNA Methylation: Molecular Biology and Biological Significance" (J. P. Jost and H. P. Saluz, eds.), pp. 11–26. Birkhauser-Verlag, Basel.

B9

IDENTIFICATION AND CHARACTERIZATION OF GENOME REARRANGEMENTS

Bindu Bhugra and Kevin Dybvig

Introduction

How do mycoplasmas, being the smallest of the self-replicating prokaryotes, maximize their coding capacity while maintaining a genome of limited size? The answer to this question may lie in the potential to continually create new coding regions by rapidly rearranging the chromosome. For example, gene rearrangements occur in *Mycoplasma pulmonis* at a high frequency. Some of these rearrangements apparently regulate the expression of surface antigens and generate heterogeneous cell populations with varied phenotypes. Phenotypic changes mediated by high-frequency chromosomal rearrangements presumably affect the ability of the mycoplasma to respond to environmental changes, likely playing a significant role in the pathogenic potential of the organism (Dybvig, 1993). The purpose of the following method is to identify chromosomal sequences that undergo high-frequency rearrangement, with the view that these sequences will likely be key determinants to cell survival.

Materials

Reagents, equipment; and supplies needed for:

Propagation of mycoplasmas

Filter cloning (0.2-μm filters; sterile 1-ml pipettes)
DNA isolation
Agarose gel electrophoresis
Molecular cloning
Southern hybridization
DNA sequencing and sequence analysis

Procedure

The strain chosen for this analysis should be filter cloned *prior* to examining cells for chromosomal rearrangements, especially if the strain has been highly passaged since it was first isolated. The goal of these experiments is to identify DNA rearrangements that occur at a *high* frequency. Filter cloning will remove subpopulations that may have arisen via relatively rare spontaneous mutations that may have occurred at any time since the strain was originally isolated. To filter clone, cell cultures are gently passed through a 0.2-μm filter and immediately assayed for colony-forming units (CFU) (Dybvig et al., 1989). The resultant colonies are picked and propagated as described below. Because mycoplasmal genomes vary at high frequencies, any colony picked from an agar plate should *always* be given a new strain designation. Once the strain has been filter cloned, high-frequency DNA rearrangements can be identified and characterized as follows.

1. Subclone the strain by randomly selecting (from agar plates) well-separated colonies for further propagation (in broth). It is recommended that the selected colonies be derived by filter cloning methods. Because mycoplasmal colonies are small, it is advantageous to pick colonies as agar plugs. Sterile 1-ml pipettes are convenient for this purpose. Agar plugs are dispensed into 1–2 ml of broth and incubated until late-logarithmic growth phase. Each of these cultures is given a new strain designation. They are stored at -70°C in 15% glycerol and used as stocks for later experiments.

2. Chromosomal DNA is isolated from small-scale cultures (usually 30 ml) that are grown using aliquots of the stocks of each subclone prepared in step 1. The DNA isolation method described by Dybvig and Alderete (1988) works well for most mycoplasma species.

3. Chromosomal DNA from the various subclones is digested with a restriction enzyme and analyzed by agarose gel electrophoresis. It is critical to resolve as many DNA fragments as possible. Electrophoresis on 0.8% agarose gels for 12–16 hours using about 2V/cm of gel length is usually sufficient. The separated DNA fragments are visualized by staining the gel with ethidium bromide. DNA

fragments that are present in some subclones, but not in others, are referred to as restriction fragment length polymorphisms (RFLPs). DNA fragments that are present in the parent strain but not in some subclones are "precursor" fragments, whereas DNA fragments that are present in some subclones but not in the parent strain are "product" fragments.

4. Gel-purified DNA fragments, either precursors or products, corresponding to RFLPs are cloned into an *Escherichia coli* vector.

5. If a precursor fragment was cloned in step 4, Southern hybridization analysis using the cloned fragment to probe chromosomal DNA from the RFLP-containing subclone will identify the appropriate product fragment. Conversely, if a product fragment was cloned in step 4, Southern hybridization analysis will identify the precursor fragment in the parent strain.

6. The precursor/product fragments identified in step 5 are cloned.

7. The restriction maps of the cloned precursor and product fragments are compared to identify the specific region in which recombination took place to generate the RFLP. These variable regions are subcloned and their nucleotide sequences determined. These data should reveal the mechanism responsible for generation of the RFLP.

DNA sequence analysis may also reveal whether the DNA rearrangement would likely affect expression of neighboring genes. By comparing potential genes in the vicinity of the DNA rearrangement to sequences in the various nucleotide and protein databases, it may be possible to speculate on the function of these genes. If this is the case, experiments can be designed to examine the likely phenotypic consequences of the DNA rearrangement.

Discussion

Mycoplasmas generally grow as aggregates, which may contain heterogeneous populations with varied phenotypes and genotypes. Passing mycoplasmal cultures through 0.2-μm filters immediately prior to assaying for CFU provides an effective method to eliminate cell aggregates, as colonies obtained using this method should be derived from single cells. Filter cloning of cultures is not mandatory. We have had success, with or without filter cloning, in identifying RFLPs in subclones of a single strain of *M. pulmonis* (Bhugra and Dybvig, 1992).

It is sometimes necessary to propagate and analyze a fairly large number of colonies in order to identify one or more RFLP-containing subclones. Usually, analysis of 50–100 random colonies is sufficient. Serial passage of mycoplasmal cultures before subcloning generates additional heterogeneity in populations increasing the efficiency of RFLP isolation (Bhugra and Dybvig, 1992).

Restriction endonuclease digestion of the small mycoplasmal chromosome results in fewer fragments relative to most bacterial genomes. For mycoplasmas, therefore, it is relatively easy to detect RFLPs on routine agarose gels stained with ethidium bromide. Although any restriction enzyme can be used for RFLP analysis, it seems that the size distribution of *Hin*dIII cleavage products is particularly well suited for this method. A limitation of the method, though, is that many RFLPs may not be recognized due to comigration of variable DNA fragments with other DNA fragments of similar size. An attractive alternative technique to examine genomic rearrangements is the Southern cross-hybridization method of Potter and Dressler (1986).

Mechanisms for generating DNA rearrangements are diverse. In *M. pulmonis*, three distinct types of rearrangements have been identified. One is the transposition of IS*1138*, an insertion sequence element indigenous to this organism (Bhugra and Dybvig, 1993). Another is the inversion of a site-specific DNA invertible element that regulates the restriction and modification properties of the cell (Dybvig and Yu, 1994). The third involves the reassortment of genes encoding variable lipoprotein surface antigens (Bhugra and Dybvig, 1995, unpublished data). Clearly, high-frequency DNA rearrangements in *M. pulmonis* have important phenotypic consequences.

Genome rearrangements have also been described in transposon Tn*916*-containing strains of *Mycoplasma mycoides* subsp. *mycoides* (Youil and Finch, 1990). Large (several hundred kilobases) tandem duplication and/or inversion of chromosomal sequences located adjacent to Tn*916* insertion sites were noted. Perhaps, these rearrangements were linked to, or induced by, transposition of Tn*916*. Examples of large rearrangements such as these are best studied using pulsed-field gel electrophoresis rather than conventional agarose gel electrophoresis.

All of the RFLPs in *M. pulmonis* that have been examined to date are the result of DNA rearrangement. However, this does not necessarily always have to be the case. For example, a point mutation located within a recognition site for the restriction enzyme being used for this analysis will generate a RFLP. However, mutations such as this are rare, and investigators should not be overly concerned with the possibility of studying this relatively uninteresting example of an RFLP. Another scenario that could generate RFLPs in the absence of DNA rearrangements is changes in the methylation state of the chromosomal DNA. If changes in DNA methylation at specific sites happen to overlap with the recognition sequence of the restriction enzyme used for RFLP analysis, RFLPs would result. Because most DNA methylases in prokaryotes are thought to be constitutively expressed, this possibility may seem unlikely. However, as mentioned earlier, *M. pulmonis* contains a novel, phase-variable restriction and modification system. It is not yet known whether such systems are prevalent in mycoplasmas, but the possibility that some RFLPs in mycoplasmas are generated by changes in chromosomal methylation must be considered.

References

Bhugra, B., and Dybvig, K. (1992). High-frequency rearrangements in the chromosome of *Mycoplasma pulmonis* correlate with phenotypic switching. *Mol. Microbiol.* **6**, 1149–1154.

Bhugra, B., and Dybvig, K. (1993). Identification and characterization of IS*1138*, a transposable element from *Mycoplasma pulmonis* that belongs to the IS*3* family. *Mol. Microbiol.* **7**, 577–584.

Dybvig, K., and Alderete, J. (1988). Transformation of *Mycoplasma pulmonis* and *Mycoplasma hyorhinis:* Transposition of Tn*916* and formation of cointegrate structures. *Plasmid* **20**, 33–41.

Dybvig, K., Simecka, J. W., Watson, H. L., and Cassell, G. H. (1989). High-frequency variation in *Mycoplasma pulmonis* colony size. *J. Bacteriol.* **171**, 5165–5168.

Dybvig, K. (1993). DNA rearrangements and phenotypic switching in prokaryotes. *Mol. Microbiol.* **10**, 465–471.

Dybvig, K., and Yu, H. (1994). Regulation of a restriction and modification system via DNA inversion in *Mycoplasma pulmonis*. *Mol. Microbiol.* **12**, 547–560.

Potter, H., and Dressler, D. (1986). A 'Southern Cross' method for the analysis of genome organization and the localization of transcription units. *Gene* **48**, 229–239.

Youil, R., and Finch, L. R. (1990). Genome rearrangements in *Mycoplasma mycoides* subsp. *mycoides*. *IOM Lett.* **1**, 276–277.

B10

EXPRESSION OF MYCOPLASMAL GENES IN *Escherichia coli*

Paul Renbaum and Aharon Razin

Introduction

The expression of foreign genes in *Escherichia coli* is a powerful laboratory manipulation. Genes can be isolated, characterized, and expressed (or overexpressed) for biochemical studies, and their products can be purified with relative ease in large quantities. The major limitations of this system stem from differences between *E. coli* and the parent organisms. Transcriptional signals vary among organisms, and therefore genes of interest are customarily inserted under the control of an *E. coli* promoter. The most severe problems may occur when proper expression of a foreign protein requires post-translational modifications. Expression of mycoplasmal and spiroplasmal proteins (along with those of mitochondria) present an additional complication as these organisms use the universal stop codon UGA to code for tryptophan (Inamine *et al.*, 1990). Translation in *E. coli* without UGA suppression results in truncated proteins as peptides are prematurely released from ribosomes. This chapter describes the cloning and expression of mycoplasmal genes based on a strategy used in cloning the *Spiroplasma* sp. MQ1 CpG DNA methylase (Renbaum *et al.*, 1990). Many procedures used here are standard methods and similar ones can be found in Ausubel *et al.* (1994) or in Sambrook *et al.* (1989) and are only briefly described here. Procedures that are more specific to mycoplasma or have particular relevance in this chapter are given in greater detail.

Materials

E. coli strains: XL1-Blue MRF (Stratagene, Inc.); ZIP23; CJ236; BMH71-18
Plasmids: pUC19; pBluescript; pSU77
1 M NaCl
1 M Tris–HCl,pH 6.3,pH 7.5,pH 8,pH 8.3
0.5 M EDTA
10% N-laurylsarcosine (w/v)
Proteinase K (20 mg/ml)
RNase A (20 mg/ml, DNase free)
Phenol
Chloroform
Isoamyl alcohol
Restriction enzymes: *Sau*3A;*Bam*HI
T4 DNA ligase
T4 DNA polymerase
Calf intestinal alkaline phosphatase
Sucrose
38-ml Beckman SW 27 polyallomer tubes
Gradient maker
Spectrophotometer
Ethanol
Electrophoresis equipment: agarose; gel box; power supply; UV light source
IPTG (Isopropyl-β-dithiogalactopyranosidase)
X-Gal (5-bromo-4-chloro-3-indolye-β-D-galactoside, in 2% dimethylformamide (w/v)
1 M KCl
1 M MgCl$_2$
Lysozyme
10% sodium dodecyl sulfate (SDS)
3 M sodium acetate, pH 5.2
Acidic phenol, pH 4
[γ-^{32}P]ATP
Polynucleotide kinase
1 M Dithiothreitol (DTT)
20 mM dNTPs
Reverse transcriptase (Life Sciences Inc.)
Ammonium acetate
Formamide loading buffer
6% polyacrylamide denaturing gel
LB media

M9 minimal media
M9 media supplemented with 18 amino acids (excluding cysteine and methionine)
Rifampicin
[^{35}S]Methionine
2-Mercaptoethanol
Glycerol
40% polyacrylamide (w/v)
2% bisacrylamide (w/v)
10% ammonium persulfate
TEMED
Tris–glycine buffer
Methanol
Glacial acetic acid
Glycerol
Ampicillin
Chloramphenicol

Choosing Host Strain and Vector

The choice of a proper host strain can have a serious impact on the success or failure of a cloning/expression project. The host should be efficiently transformable, and Rec⁻(recombination deficient, the most commonly used mutant is *rec*A, although *recB,C,D,F*, and *J* mutants are also available). The strain should also be deficient in any restriction systems such as *hsdR, mrr, mcrA*, and *mcrBC*, which recognize and restrict foreign DNA by the absence or presence of methyl groups at particular sites. The strain XL1-Blue MRF (available from Stratagene) is highly recommended.

Another desirable trait, especially if the foreign gene product may negatively affect *E. coli* growth, is inducible expression. The simplest systems use vectors with a *lac* promoter–operator, followed by a multiple cloning site, and a portion of the *lacZ* gene. This plasmid is transformed into a host cell carrying the *lacI*^{Q} mutation, an overexpressor of the Lac repressor (often carried on an episome, F′, and may require selective pressure to prevent curing, such as growth on minimal media or antibiotics). The addition of IPTG (a lactose analog) sequesters the Lac repressor and allows transcription to proceed. Additionally, strains containing particular deletions in the *lac* operon (i.e., *lacZ*ΔM15) can be transformed with a *lacZ* vector complementing this deletion, allowing for blue–white selection of those clones with foreign DNA. Two highly proven vectors are pUC19 and pBluescript (Stratagene). pBluescript has the added advantages of a large multiple cloning region, single-strand DNA rescue, and T7 and T3 RNA polymerase promoters.

UGA Opal Suppression

Mycoplasma and *Spiroplasma* species use the universal stop codon UGA to code for tryptophan and, therefore, genes containing UGA-Trp codons, when expressed in *E. coli,* will give rise to truncated proteins. It is possible to circumvent this problem by using a tRNA *opal* suppressor that inserts tryptophan residues at UGA sites. Although such suppressors may be encoded on the chromosome of some *E. coli* strains, we know of no host strain commonly used for cloning and expression that carries an opal suppressor. However, plasmid-borne suppressors such as pSU77, encoding *trp* t176, a strong *opal* suppressor under the inducible *lac*UV5 promoter, can easily be cotransformed into almost any *E. coli* host with no detrimental effects.

Library Construction

Library construction entails the preparation of insert and vector DNAs, along with ligation and transformation into host cells. High molecular weight DNA is purified from a 1-liter mycoplasma culture. Cells are grown and harvested at 12,000 g, 4°C for 10 minutes, washed twice in 0.25 M NaCl, 25 mM Tris–HCl, pH 7.5, 5 mM EDTA, and the pellet resuspended in 10 ml 0.1 M EDTA, 1% N-laurylsarcosine (w/v); proteinase K is added to 200 μg/ml, and the suspension is incubated at 50°C for 2 hours. RNase A (DNase free) is added to 100 μg/ml, and the incubation is continued at 37°C for 2 hours. The treated lysate is extracted gently: twice with phenol/chloroform/isoamyl alcohol (25:24:1, v/v) and twice with chloroform/isoamyl alcohol (24:1). The aqueous phase is then dialyzed extensively against TE (10 mM Tris–HCl, pH 8, 1 mM EDTA) at 4°C.

The purified DNA should be partially digested with a four-base cutter restriction enzyme which yields sticky ends, to produce fragments in the 2- to 10-kb range. Restriction conditions should be empirically calibrated using twofold serial dilutions of enzyme at a set time and temperature to give the desired range of digestion products. The reaction should then be scaled up to 50–150 μg genomic DNA. Divide the reaction into three parts, with enzyme concentrations slightly above, slightly below, and equal to the empirically determined concentration (we digested *Spiroplasma* sp. MQ1 DNA with *Sau*3A, using 0.16–0.3

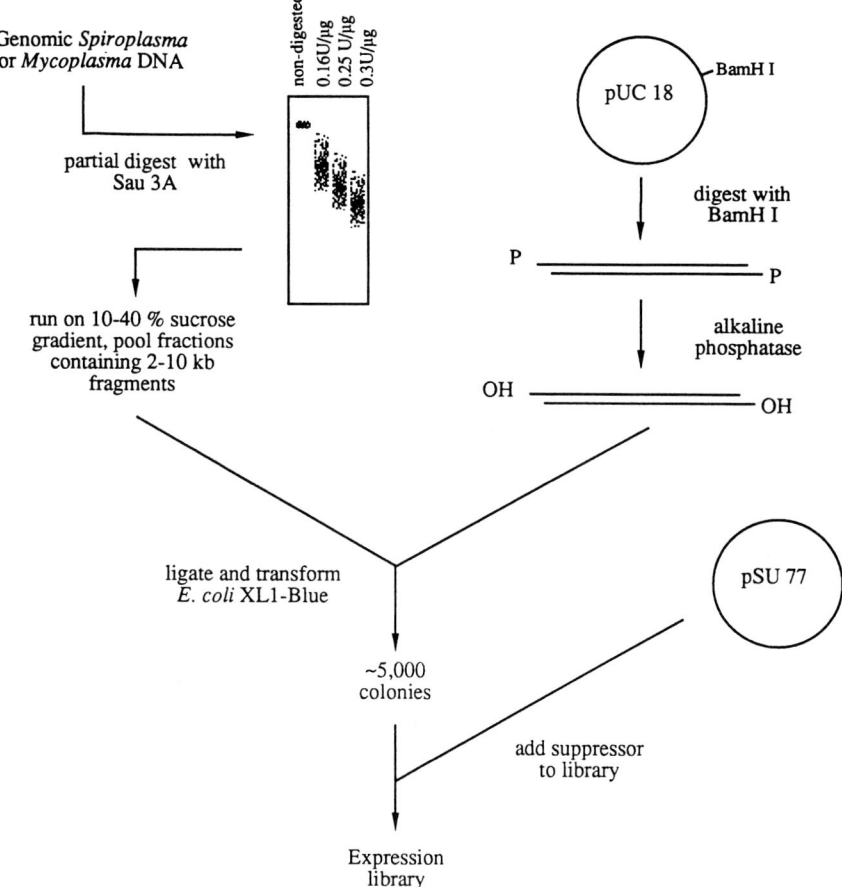

Fig. 1. Flow sheet showing the preparation of a mycoplasma or spiroplasma genomic expression library in *E. coli*.

U/µg DNA for 20 minutes at 37°C). The reaction is then immediately stopped by the addition of EDTA to 20 mM and put in a 65°C water bath for 20 minutes.

The resulting fragments are extracted with phenol/chloroform/isoamyl alcohol (25:24:1), ethanol precipitated, and run on a 10–40% sucrose density gradient to remove both the high and the very low molecular weight DNA. The gradients are prepared in 38-ml Beckman SW 27 polyallomer tubes with a standard gradient maker in a buffer of 1 M NaCl, 20 mM Tris–HCl, pH 8.0. A maximum of 50 µg DNA is applied to each gradient and centrifuged at 20,000 g for 23 hours at 20°C. One milliliter fractions are collected from the bottom of the tube, and the DNA content of each fraction is quantitated spectrophotometrically at 260 nm.

Those containing DNA are ethanol precipitated and examined by electrophoresis on a 0.7% agarose gel. Fractions containing DNA fragments in the 3- to 10-kb range are pooled and ligated into a complementary restriction site (i.e., for inserts digested with *Sau*3A use *Bam*HI) with 50 ng dephosphorylated vector, prepared with calf intestinal alkaline phosphatase, at a molar ratio of 3:1 (inserts:vector) using T4 DNA ligase. The ligated DNA is used to transform the host cells which were previously transformed with the suppressor. We recommend using the standard high efficiency protocol of Hanahan (1985) to achieve results of 10^7-10^8 transformants/µg DNA. A portion of the ligation mixture may be plated on agar freshly spread with IPTG/X-Gal (80 µl of 50 mM IPTG, 1% X-Gal) to ascertain how many transformants contain recombinant DNA. Blue colonies express the LacZ protein from the vector (a blue colony attests to the breakdown of X-Gal, a LacZ substrate), whereas white colonies indicate that foreign DNA interrupts the *lacZ* coding sequence. To determine the average insert size, restriction analysis of plasmid DNA minipreps from 10–20 white colonies should be performed.

Selection

Three methods can be employed to select clones from a library: hybridization, antibody recognition, and protocols based on enzymatic activity. While expression is not a requirement for isolation by DNA hybridization, the latter two methods depend on expression and, hence, UGA suppression. Clonal isolation based on enzymatic activity can be a long and cumbersome project, which is dependent on the size of the library and the sensitivity of the enzymatic assay. Frequently, an assay may be sensitive enough to allow examination of groups of 10–100 clones together. This, combined with the small genome size of mycoplasma (approximately 500 genes), may make it a feasible method.

Gene Analysis and Characterization

Primary analysis of a cloned gene includes restriction mapping and sequencing. Aside from showing restriction sites, a restriction map will reveal the orientation of the cloned insert in relation to the vector. If the insert is expressing protein, then identification of clones in both orientations demonstrates the ability of the insert to undergo transcription from an endogenous promoter. Sequencing provides not only a coding sequence, but a clearer picture of those signals involved in transcription and translation, including the presence of UGA codons. To fully characterize transcription signals, it is necessary to examine messenger RNA. Primer extension of mRNA molecules with reverse transcriptase provides a qualitative analysis.

RNA Extraction

An overnight *E. coli* culture (transformed with the suppressor and the gene of interest) is diluted 1:20, grown until 0.3 OD_{595}, and induced with 1 mM IPTG. The cells are harvested at 0.5 OD; 12.5 ml of culture, 10 g of crushed ice, and 1.25 mg of chloramphenicol are added; and the tube is kept at room temperature until the ice has melted. The sample is centrifuged at 2000 g for 10 minutes and the pellet is resuspended in 1 ml of buffer containing 10 mM Tris–HCl, pH 7.3, 10 mM KCl, and 5 mM $MgCl_2$. Add 150 μg lysozyme to 0.5 ml of culture and freeze at -70°C for 30 minutes. Defrost the samples in a 65°C water bath, add 35 μl 10% SDS, and incubate for 2 minutes at 65°C. Add 12.5 μl 3 M sodium acetate, pH 5.2, and extract the sample twice with 0.6 ml acidic phenol (pH 4). To the aqueous phase, add $\frac{1}{10}$ volume of 3 M sodium acetate, pH 5.2, EDTA to 1 mM, and precipitate with 2.5 volume of 100% ethanol, rinse with 70% (v/v) ethanol, dry, and dissolve the pellet in 100 μl distilled deionized water (DDW), and store at -70°C.

Primer Extension

Forty nanograms of a specific antisense primer located in the first 100–200 bp of the transcript is labeled using 50 μCi [γ-^{32}P]ATP with polynucleotide kinase. Eight nanograms of the labeled primer is mixed with 30 μg total RNA, 13 μl freshly prepared buffer (0.385 M Tris–HCl, pH 8.3, 0.535 M KCl, 38.5 mM $MgCl_2$, 38.5 mM DTT, in a total volume of 47 μl, annealed at 65°C, and moved to room temperature for 5 minutes. A 2.5-μl mix of 20 mM dNTPs and 4.5 U of reverse transcriptase (Life Sciences Inc.) is added and the reaction is incubated at 42°C for 1 hour, followed by inactivation at 75°C for 10 minutes. After treatment with 1 μg RNase for 30 minutes at 37°C, the volume is brought to 250 μl with TE, 25 μl ammonium acetate is added, and the DNA is ethanol precipitated overnight, rinsed in 70% ethanol, dried, and resuspended in 6 μl water. Formamide loading buffer (4 μl) is added, and the reaction is run on a 6% polyacrylamide denaturing gel.

Site-Directed Mutagenesis

Although the recombinant protein may be sufficiently expressed, the suppressor tRNA is constantly competing with release factor 2. The replacement of UGA codons from the coding sequence by UGG-Trp can both simplify and significantly improve protein expression. The simplest method of *in vitro* site-specific mutagenesis incorporates a primer which eliminates a unique restriction site from the plasmid in addition to the desired base change (Deng and Nickoloff, 1992; also available as a kit from Clonetech). More than one primer can be annealed simultaneously to the denatured double-stranded plasmid, and T4 DNA polymerase is used to synthesize the mutagenized strand. After the synthesis and ligation reactions, the DNA is digested with the restriction enzyme whose site was altered to remove self-annealed parental plasmids. The reaction is then transformed into *E. coli* BMH 71-18, a *mut*S host (DNA mismatch repair deficiency mutation), and is grown overnight as a batch culture. Highly purified DNA is recovered and digested with a large excess of the same enzyme to eliminate all parental plasmids and is used for a second transformation on LB plates with antibiotic. Individual colonies are examined for the absence of the altered restriction site, and the UGA-containing regions are sequenced in positive clones.

Selective Protein Labeling

When it is inconvenient or impossible to assay enzymatic activity, it is still possible to examine protein expression by selectively labeling those proteins located downstream from a T7 promoter (Clarke and Carbon, 1976). An *E. coli* host is cotransformed with (1) a plasmid containing the gene of interest subcloned downstream from a T7 RNA polymerase promoter and (2) a compatible (different origin and antibiotic) plasmid containing the genes encoding (i) T7 RNA polymerase under the control of λ P_L promoter, (ii) λ repressor c*I857* temperature-sensitive under the control of the P_{lac} promoter–operator, and (iii) the UGA suppressor (if necessary). An overnight culture, grown in LB media supplemented with the appropriate antibiotics, 1 mM IPTG, at 30°C, is diluted 1:10 in the same medium and grown to 0.4 OD, also at 30°C. One milliliter of culture is removed, washed once with M9 minimal medium, and then resuspended in M9 medium supplemented with 18 amino acids (prepare a 0.1% solution of each amino acid, excluding cysteine and methionine, dilute 1:200) and incubated for an additional 40 minutes at 30°C. The culture is shifted to 42°C for 60 minutes to inactivate the temperature-sensitive λ repressor, which in turn permits expression from the P_L promoter, inducing T7 RNA polymerase. Rifampicin is added to a final concentration of 0.25 mg/ml, and incubation is continued for 10 minutes at 42°C and is then shifted to 30°C for an additional 20 minutes. One-half milliliter of culture is removed and labeled with 10 μCi [^{35}S] methionine, incubating for 1–5 minutes at 30°C. The cells are spun down for 10 seconds, resuspended in 0.1 ml cracking buffer (60 mM Tris–HCl, pH 8, 1% 2-mercaptoethanol, 1% SDS, 10% glycerol, 0.01% bromophenol blue), and heated to 100°C for 5 minutes, and 10–20 μl is loaded onto a SDS–polyacrylamide stacking gel (4% stacking gel prepared with 0.125 M Tris–HCl, pH 6.8, and 10% separating gel prepared with 0.375 M Tris–HCl, pH 8.8). The samples are electrophoresed in 1× Tris–glycine buffer at 180 V for 1–1.5 hours. The gel is fixed in 5% methanol, 7.5% glacial acetic acid, 2% glycerol, dried, and autoradiographed.

Discussion

The cloning and expression of mycoplasmal and spiroplasmal genes are relatively straightforward. These organisms have a small genome, and the transcription signals characterized so far are readable in *E. coli* (Renbaum *et al.*, 1990; Mouches *et al.*, 1985; Glaser *et al.*, 1992). The only outstanding difficulty is the presence of UGA-Trp codons in both mycoplasmal and spiroplasmal messages. These messages absolutely require a UGA opal suppressor for proper translation. However, the necessity of a suppressor can be turned into a distinct advantage: when the gene of interest includes a UGA-directed tryptophan, controlling transcription of the suppressor will limit translation of the protein even when the recombinant gene is driven by an endogenous promoter. This can protect host cells from deleterious effects a foreign gene product may exert.

The host/suppressor combination provides a powerful system capable of expressing efficiently both mycoplasmal and spiroplasmal genes. However, it must be noted that there will be two different plasmids within each *E. coli* host cell: the suppressor plasmid and the gene of interest. Inherently, this is not a problem; both plasmids can be taken up even in the same cotransfection, provided that two conditions are met: (1) each plasmid must be selectable with a separate antibiotic and (2) replication of each plasmid must be directed by a different origin. In the protocol just described, the cloning vectors pUC19 or pBluescript (both derivatives of pBR322) are based on the pMB1 replicon and encode ampicillin resistance; whereas pSU77 (a derivative of pACYC184) is based on the p15A origin and encodes chloramphenicol resistance.

Occasionally the need arises to express three genes at a given time (i.e., the suppressor, the gene of interest under a T7 polymerase promoter, and T7 polymerase; this, with rifampicin and [^{35}S]methionine, will selectively label only the gene product under the T7 promoter; see below). It is possible, yet somewhat awkward, to use three plasmids with three different antibiotics and origins. However, since most commonly found plasmids use the pMB1 origin and therefore require subcloning into a different vector, it is most often simpler to put two of the three genes into one vector. Another possibility is to engineer a host strain that expresses at least one of the three genes. This can be done by engineering a recombinant bacteriophage λ with one of the three genes, infecting the host, and selecting a lysogen. A convenient method to verify (or select for) suppressor function is to introduce the suppressor into *E. coli* ZIP238, a strain with an *opal* mutation in the *lacZ* gene. When grown on McConkey (lactose) agar, colonies expressing the suppressor will acquire the ability to break down lactose and will appear red.

When preparing DNA fragments for a genomic library, it is advisable to avoid restriction enzymes that are sensitive to methylation such as *Hpa*II (CmCGG) or *Mbo*I (GmATC) or insensitive to heat inactivation such as *Taq*I. Partial enzymatic digestion is advantageous over using a six-base cutter to complete digestion or sonication because it allows both representation of the total genome and a high ligation efficiency.

If the gene to be cloned is expected to have possible detrimental effects on host cell growth, the ligation products can be transformed into a host without the suppressor. This will ensure no expression and therefore no negative selection. The resulting library should be collected (scraped from the agar plates) and amplified by growing the cells as a batch culture. Purify the plasmid DNA (as a batch) and use it to transform the same host cells, but this time harboring the suppressor in order to form an expression library.

To determine whether a representative library was prepared, the following formula can be employed: $N = \ln(1-P)/\ln(1-f)$, where N is the required number of different recombinants, P is the desired probability of this particular sequence to be represented (99%), and f is the fraction of the genome in a single recombinant (average insert length/total genome length approximately 10^6 bp) (Clarke and Carbon, 1976). For an expression library, it is necessary to ensure an open reading frame including the *lacZ* promoter from the vector (as it is not known whether the recombinant gene will carry a promoter that can drive transcription in *E. coli*), and therefore we must multiply N by 6, the number of possible different reading frames. Considering 5.5 kb as an average insert size, the equation becomes

$$N = \ln(1-0.99)/\ln(1-[5.5 \times 10^3/1 \times 10^6]) = 765 \times 6 = 4590 \text{ transformants.}$$

Site-specific mutagenesis should be performed with T4 DNA polymerase for the synthesis reaction rather than Klenow polymerase because the T4 enzyme lacks strand displacement activity. An alternative method to that described previously uses single-strand DNA which was grown in a *dut*⁻ (dUTPase), *ung*⁻ (uracil *N*-glycosylase) host such as *E. coli* CJ236. This bacterium will incorporate a number of uracil residues in place of thymine residues. The uracil-containing strand is used as a template for *in vitro* synthesis (with dTTP) of the complementary mutagenized strand, and the resulting double-stranded plasmid is introduced into a *ung*$^+$ host. The uracil-containing strand is degraded, leaving only the synthesized mutant strand to replicate (Kunkel, 1985).

The best indication of protein expression is enzymatic activity. Unfortunately, this may not be possible with enzymes that have no convenient assay, structural proteins, or foreign proteins which are not properly processed in *E. coli*. However, by using T7 RNA polymerase, it is possible to selectively label only those proteins from genes cloned downstream from a T7 promoter. This is accomplished by shutting down cellular transcription with rifampicin, an inhibitor of the β subunit of *E. coli* RNA polymerase. Rifampicin blocks the initiation of new cellular RNA chains without interfering with the transcription from the T7 RNA polymerase. After allowing existing transcripts to complete translation, [^{35}S]methionine is added to the culture and, following a brief incubation, the cells are harvested, and the crude extract is run on a gel. This process allows labeling of only those proteins whose message was transcribed from a T7 promoter.

References

Ausubel, E., Brent, R., Kingston, R., Moore, D., Seidman, J., Smith, J., and Struhl, K. J., eds. (1994). "Current Protocols in Molecular Biology." Wiley, New York.

Clarke, L., and Carbon, J. (1976). A colony bank containing synthetic Col E1 hybrid plasmids representative of the entire *E. coli* genome. *Cell* **9**, 91–99.

Deng, W. P., and Nickoloff, J. A. (1992). Site-directed mutagenesis of virtually any plasmid by eliminating a unique site. *Anal. Biochem.* **200**, 81–88.

Glaser, G., Hyman, H., and Razin, S. (1992). Ribosomes. *In* "Mycoplasmas, Molecular Biology and Pathogenesis" (J. Maniloff, R. N. McElhaney, L. R. Finch, and J. B. Baseman, eds.), pp. 169–177. Am. Soc. Microbiol. Washington, DC.

Hanahan, D. (1985). Techniques for transformation of *E. coli*. *In* "DNA Cloning: A Practical Approach" (D. M. Glover, ed.), Vol. I, pp. 109–135. IRL Press, Oxford.

Inamine, J., Ho, K. C., Loechel, S., and Hu, P. C. (1990). Evidence that UGA is read as a tryptophan codon rather than as a stop codon in *M. pneumoniae*, *M. genitalium*, and *M. gallisepticum*. *J. Bacteriol.* **172**, 504–506.

Kunkel, T. A. (1985). Rapid and efficient site specific mutagenesis without phenotypic selection. *Proc. Natl. Acad. Sci. USA* **82**, 488–492.

Mouches, C., Candresse, T., Barroso, G., Sailard, C., Wroblewski, H., and Bové, J. M. (1985). Gene for spiralin, the major membrane protein of the helical mollicute *Spiroplasma citri*, cloning and expression in *E. coli*. *J. Bacteriol.* **164**, 1094–1099.

Renbaum, P., Abrahamove, D., Fainsod, A., Wilson, G. G., Rottem, S., and Razin, A. (1990). Cloning, characterization, and expression in *E. coli* of the gene coding for the CpG methylase from *Spiroplasma* sp. MQ1 (M.SssI). *Nucleic Acids Res.* **18**, 1145–1152.

Sambrook, J., Fritsch, E. F., and Maniatis, T., eds. (1989). "Molecular Cloning, A Laboratory Manual." Cold Spring Harbor Press, Cold Spring Harbor, NY.

SECTION C
Membrane Characterization

C1

INTRODUCTORY REMARKS
Shmuel Razin

Procedures for isolation of mycoplasma membranes, characterization and manipulation of membrane lipid components, and electrophoretic analysis of membrane proteins, outlined in "Methods in Mycoplasmology" (Razin and Tully, 1983), are still valid as they have not been changed since the early 1980s. The recent focusing on the role of mycoplasma membrane components in pathogenicity has been associated with the development and application of new procedures directed mostly to the genetic, chemical, and immunologic characterization of membrane proteins exposed on the cell surface. Some of these proteins serve as adhesins, playing a role in mycoplasma adhesion to host tissues (Chapter F2 and F3, this volume), while others act as modulators of the immune system (Chapters F5–F9, this volume) and constitute the dominant antigens of mycoplasmas.

Mycoplasma membranes were shown to carry a high number of lipoproteins with a maximum molecular mass larger than in other bacteria. These proteins, anchored in the membrane through covalently attached acyl chains, are exposed on the cell surface and thus act as important cell antigens. Interestingly, the lipoproteins, as well as some nonmodified integral membrane proteins, exhibit antigenic phase variation, a property expected to play an important role in avoidance by the parasite of host immune responses (see the general introduction to this volume). Methods for specific isotopic labeling of covalently bound ligands of mycoplasma membrane proteins are detailed in Chapter C2, whereas methods for the selective isolation and analysis of the amphiphilic mycoplasma membrane components by Triton X-114 phase partitioning are described in Chapter C3. This chapter also includes the procedure for the assessment of expression patterns of surface epitopes in mycoplasma populations by colony immunoblotting.

This procedure is particularly useful in determining the frequency of phase variation, dual expression patterns, and distribution in a population of a particular antigenic component.

The recent demonstration of intracellular location of some pathogenic mycoplasmas (Chapters A6 and A7, this volume) has attracted more attention to the possibility of fusion of the mycoplasma membrane with that of the host cell as an initial step in penetration. In addition, the surface-exposed mycoplasma membrane may fuse with artificial lipid vesicles (liposomes) and thus serve as a means for transfecting mycoplasmas with plasmids encapsulated in the liposomes. The approaches and techniques applied in activation and evaluation of membrane fusion are described in Chapter C4.

The mycoplasma membrane, being the only barrier separating the cytoplasm from the outside environment, is the seat for the systems transporting the many exogenous nutrients required by the fastidious organisms. Some of the transport systems function as ion channels, generating electrochemical gradients that can be used for many essential functions, such as controlling cell volume, internal pH regulation and uptake of nutrients. Methods for measuring the electrochemical gradients and membrane potentials are detailed in Chapter C5, and methods for measuring ion flows and their effects on mycoplasma cell volume are described in Chapter C6.

Reference

Razin, S., and Tully, J. G., eds. (1983). "Methods in Mycoplasmology," Vol. I, Section D. Academic Press, New York.

C2

POSTTRANSLATIONAL MODIFICATION OF MEMBRANE PROTEINS

Åke Wieslander, Susanne Nyström, and Anders Dahlqvist

Introduction

The major functions of biological membranes include physical containment of cell contents, matrix for structural and enzyme proteins, storage and use of chemiosmotic energy, and communication and signaling. For the wall-less mollicutes, all these functions must reside in the single cytoplasmic membrane. This also includes proteins normally associated with the periplasmic space, cell wall, and outer membrane of other bacteria. Many of these proteins are the targets for different types of covalent modifications. The small genome size of mycoplasmas implies a restricted coding capacity, where the majority of genes are most likely constitutive, responsible for the so-called "housekeeping" functions (Wieslander *et al.*, 1992). This is supported by the constancy in protein profiles obtained by sodium dodecyl sulfate (SDS) gel analysis of cells grown under different laboratory conditions. However, it is not known whether mycoplasmas can selectively express new proteins and functions on association *in vivo* with potential hosts as has been described for several pathogenic bacteria.

Generally, the individual amino acid residues in proteins can be chemically modified *in vivo* in many different ways (Wold, 1981). Information for many modifications and processing sites reside in the amino acid sequence. This is also the case for a variety of binding motifs and domains. The recent PROSITE compilation of such sites and patterns, available in several molecular biology computer programs, lists more than 800 different patterns (Bairoch, 1993). The biological occurrence, recognition, and chemical analysis of many of these have been described (Aitken, 1990). Compared to extensively studied bacteria like *Escherichia coli*, little has been done in the field of membrane protein modification in mycoplasmas. This chapter deals with modifications of specific mycoplasma membrane proteins, established especially by metabolic labeling with radioactive isotopes.

Materials and Procedures

Table I summarizes the different modifications established for specific mycoplasma membrane proteins. It is impractical to list here all chemicals, techniques, and equipment used for the analyses of protein modification due to space limitations. The selected references shown contain sufficient descriptions and further references that can be consulted and easily followed.

Growth Media and Conditions

Essentially all studies of modification of mycoplasma membrane proteins have been performed with cells cultured normally (>20 hours) in standard mycoplasma growth media (see Chapter A2, this volume). All of these media were undefined and contain serum components, influencing markedly the amounts of isotopes needed for efficient labeling *in vivo* (see below). For most labeling studies (Table I), culture volumes of 5 to 50 ml medium are sufficient, with the lower volume being adequate for fast-growing, high-yielding species like *Acholeplasma laidlawii*. This is ≈ 1 to $20\times$ the volumes used in similar studies with the common bacteria like *E. coli* and *Bacillus* species in defined media.

Isotopes

For most commercial radioactive chemicals the ^{3}H-labeled ones are usually of substantially higher specific activity than the ^{14}C ones and are often less expensive. It is our experience that the resolution and sharpness of bands and spots in (dried) gels are sometimes decreased by the presence of various radioactive enhancers/amplifiers needed in the gel to visualize ^{3}H-labeled proteins. Such enhancers are usually not needed for molecules containing isotopes with higher radiation energy like ^{14}C and ^{35}S. For ^{3}H- and ^{14}C-labeled fatty acids (FA) we use 6 and 2 μCi per ml growth medium, respectively. This provides for good labeling of membrane proteins and for an extremely strong labeling of the lipids. One should avoid adding more than ≈ 0.1 μmol/ml (including cold FA in the medium) since the FAs may inhibit growth at high concentrations. Similar levels of radioactivity are used for the labeling of membrane proteins during growth by

TABLE I
COVALENT MODIFICATIONS OF SPECIFIC MYCOPLASMA MEMBRANE PROTEINS

Modification type	Ligand/residue	Label donor[a]	Species	Reference
Lipid modification	Fatty acids	^{3}H, ^{14}C	See Table II	Nyström et al. (1992)
	Glycerol	2-^{3}H	M. capricolum	Dahl and Dahl (1984)
	Cysteine	^{35}S	M. hyorhinis	Rosengarten and Wise (1991)
	Isoprenoid	[^{14}C]Mevalonic acid	A. laidlawii	Nyström and Wieslander (1992)
Export signal processing	Signal peptide	[^{14}C]Leu/[^{35}S]Cys	M. hyorhinis	Cleavinger et al. (1994)
Redox components	Flavin	[^{14}C]Riboflavin	A. laidlawii	Nyström et al. (1992)
	Iron	[^{59}Fe]Cl$_3$	A. laidlawii	Jägersten et al. (1982)
	Iron, copper, FMN, sulfur	CA	A. laidlawii	Reinards et al. (1981)
Protein cofactors	Phosphate	[^{32}P]Phosphate in vivo	A. laidlawii	Archer et al. (1978)
		[γ-^{32}P]ATP in vitro	M. gallisepticum	Platt et al. (1988)
		[γ-^{32}P]GTP in vitro	A. laidlawii	Dahlqvist, unpublished (1993)
		[β-^{32}P]UDP-Glc in vitro	A. laidlawii	Dahlqvist, unpublished (1993)
	Lipoic acid	[^{14}C]Octanoic acid	A. laidlawii	Wallbrandt (1992)
Monosaccharide metabolites	Glucose, fructose, mannose, glucosamine	^{3}H, ^{14}C-uniform ^{14}C ^{14}C ^{14}C	A. laidlawii	Nyström et al. (1992)

[a] Labels in ligands; CA, chemical analysis.

commercial radiolabeled amino acid mixtures. For individual amino acids occurring less frequently in the proteins, the amounts may be increased. Stock volumes can be reduced, and organic solvents removed, by evaporation under a stream of N_2 gas. Thereafter the proper growth medium is added (nonsterile) and the isotopes are dissolved by stirring. Note that medium protein components are essential as carriers for FAs. The complete medium volume is then sterilized by filtration.

Polyacrylamide Gel Systems

Several high-resolution one-dimensional (1D) systems can be used (Hames and Rickwood, 1990). In many instances these are sufficient for the visualization and enumeration of labeled membrane proteins as illustrated in Fig. 1. However, for a more detailed chemical analysis of individual proteins, these must be separated or purified by immunological, chromatographic, or other electrophoretic techniques. This is illustrated by the 2D polyacrylamide gel electrophoresis (PAGE) analysis in Fig. 2; note the crowded labeled proteins in certain regions. Proper two-dimensional gel systems are described in Hames and Rickwood (1990). Ready-made gels supplied by several companies can also be used. For

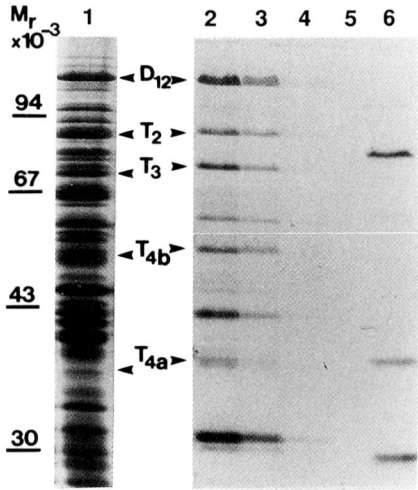

Fig. 1. Covalent modification of membrane proteins. Fluorogram of an SDS–PAGE gel with membrane proteins from *A. laidlawii* cells grown with labeled ligands. Lane 1, Coomassie-blue stained specimen; lane 2, [^{14}C]14:0; lane 3, [^{14}C]16:0; lane 4, [^{14}C]18:0; lane 5, [^{3}H]glycerol; and lane 6, [^{14}C]riboflavin. The marked proteins were identified by monospecific antibodies after immunoblotting (from Nyström et al., 1992, with permission).

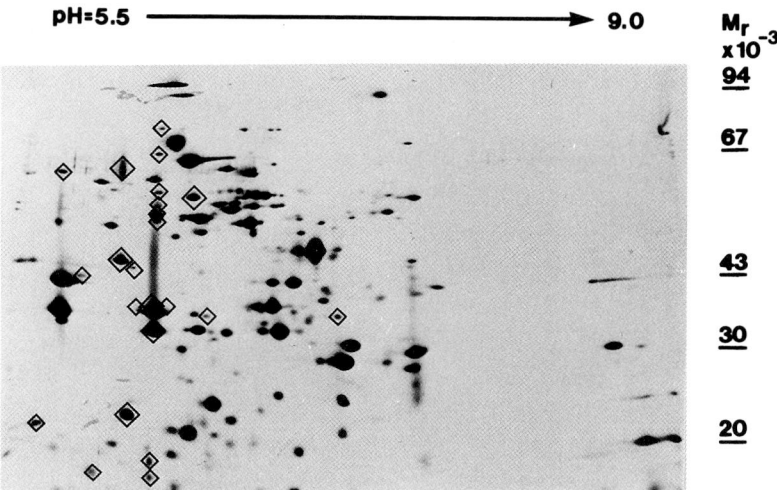

Fig. 2. Two-dimensional PAGE of *A. laidlawii* membrane proteins. Silver-stained gel of defatted membrane proteins from cells grown with [^{14}C]16:0 and analyzed by isoelectric focusing followed by gradient SDS–PAGE. Labeled proteins are boxed (from Nyström et al., 1992, with permission).

several labels dissolved or covalently incorporated into the membrane lipids, an extraction ("delipidation") of the membranes by organic solvents is advisable as it reduces the radioactive background in gels and blots (Hayashi and Wu, 1992; Nyström et al., 1992). It also enhances the separation of the proteins in the the first pH gradient dimension of two-dimensional gels (Nyström et al., 1992). A partial, but convenient, purification can be achieved by phase partitioning of solubilized proteins in the detergent Triton X-114, as originally described by Bordier (1980) and subsequently modified by others (e.g., Hooper and Turner, 1992; Nyström et al., 1992; Chapter C3, this volume). Here, delipidation interferes with detergent solubilization and should be avoided. Most mycoplasma membrane lipoproteins accumulate selectively in the bottom detergent phase on partitioning, separated from many other cellular proteins (Fig. 3).

Detection and Analysis

Most covalently bound ligands have been detected by autoradiography of PAGE gels. A convenient, fast, and inexpensive system for drying gels between sheets of transparent cellophane has been described (Wallevik et al., 1982). Radioactivity in such gels (with or without enhancer, see earlier) is detected by exposure to X-ray film (1–2 days to 3 weeks) or more rapidly by recent imaging

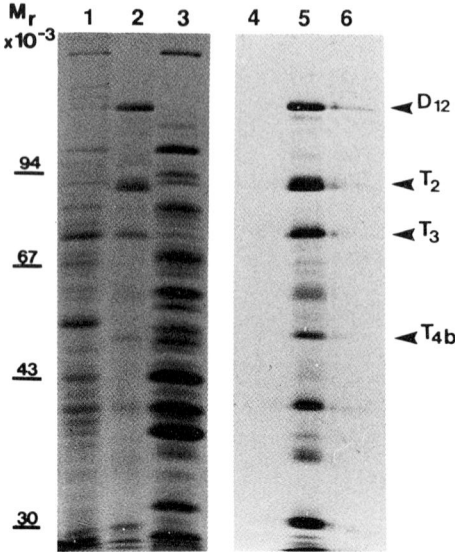

Fig. 3. Triton X-114 phase partitioning of membrane lipoproteins. SDS–PAGE gels with proteins from *A. laidlawii* cells grown with [^{14}C]16:0. Lanes 1–3, Coomassie-stained specimen; and lanes 4–6, autoradiogram of 1–3. Lanes 1 and 6, Triton X-114 insoluble (pellet) proteins; lanes 2 and 5, proteins in bottom detergent phase; and lanes 3 and 4, proteins in top aqueous phase (from Nyström *et al.*, 1992, with permission).

techniques ("direct" or "screen" equipment); the latter also enables quantification. Film-detected or stained spots and bands can be cut out from the dried gels, rehydrated, digested, analyzed chemically, and quantified by liquid scintillation counting (Nyström *et al.*, 1992).

For understanding the molecular function of modification, it is essential that *all* different covalently bound ligands, or ligand constituents, on the selected proteins, as well as the nature of bonds, are characterized chemically. This also involves the amino acid site for modification. Kinetics are also important, as certain ligands such as phosphate may have a rapid turnover and are lost during cell harvest; here labeling is only possible *in vitro* (e.g., Platt *et al.*, 1988). Different phosphate donors (ATP, GTP, or UDP-Glc) yield different labeling patterns and intensities in *A. laidlawii* (Table I). The types of bonds for fatty acid chains (Aitken, 1992; Hayashi and Wu, 1992) and phosphates (Cortay *et al.*, 1991; Hardie, 1993) can be indicated by the sensitivity of the (radioactive) ligands to release by acid or alkaline conditions. Modification sites can be localized after partial degradation of the peptide chain. Such fragments can be sequenced and also analyzed by mass spectrometry (Hooper and Turner, 1992). Phosphorylated amino acids from fully degraded proteins, or of fragments thereof, can also be analyzed (Hardie, 1993). Released lipid chains are usually characterized chemically by a combination of thin-layer chromatography (TLC) (normal/reversed phase), gas–liquid chromatography, high-performance liquid chromatography (HPLC) (reversed phase), and mass spectrometry (Aitken, 1992; Hayashi and Wu, 1992).

Discussion

It is evident from Fig. 1 that covalent modification of mycoplasma membrane proteins can be conventionally detected by one-dimensional slab gels. However, for a more thorough characterization the proteins must be further purified by various methods. This may include cloning and sequencing of the corresponding gene(s). So far, the overwhelming majority (but not all) of the envelope-associated lipoproteins in bacteria appear to be modified by fatty acid chains at the *N* terminus according to a consensus mechanism, different from the four presently known mechanisms in eukaryotes (Hooper and Turner, 1992). In bacteria, a specific target sequence specifies signal peptide cleavage and a consecutive modification of the new N-terminal cysteine residue with one ether-bonded glycerol carrying two ester-linked acyl chains and one amide-linked alkyl chain at the Cys amino group (Hayashi and Wu, 1992). Occasionally, the latter third chain is lacking. Such proteins are hydrophilic, lack transmembrane segments, and are anchored in the membrane by their three (or two) hydrocarbon chains. The number of lipoproteins in the membrane is higher and their maximum molecular mass is larger in mollicutes (Table II; Wieslander *et al.*, 1992). As in other bacteria, a preference for saturated chains can be also be noted (Table II). Furthermore, the mycoplasma proteins all seem to have isoelectric points on the acidic pH side (Nyström *et al.*, 1992).

So far, not all the features of the bacterial consensus modification mechanism (Hayashi and Wu, 1992) have been unequivocally and completely demonstrated for any mycoplasma lipoprotein. The genes for several lipoproteins have been cloned and sequenced (see, e.g., Bové *et al.*, 1993; Wise, 1993), revealing typical lipoprotein signal sequences but slightly unusual signal peptidase processing sites and no hydrophobic transmembrane segments in the mature proteins. These proteins are obviously lipid modified in their native hosts. However, a *Mycoplasma hyorhinis* protein expressed in *E. coli* could not be processed and modified (Cleavinger *et al.*, 1994), and the *Spiroplasma citri* spiralin protein could not be processed when expressed in *A. laidlawii* (Jarhede *et al.*, 1995). Glycerol incorporation has so far been demonstrated only for a subset of *M. capricolum* lipoproteins (Table II). The fraction of ester-bound chains in mycoplasmal lipoproteins appears to be larger than for the consensus modification

TABLE II
Membrane Lipoproteins in Mollicutes

Species	Molecular mass (kDa)[a]	Acyl chains[b]	Reference
A. laidlawii A	15–130	14:0 > 16:0 > 18:0 > 18:1c > 18:2c[c]	Nyström et al. (1992)
A. laidlawii B	17–100	16:0 (no 18:1c)[c]	Dahl et al. (1985)
M. arginini	14–105	16:0 > 14:0	Wieslander, unpublished (1988)
M. capricolum	15–100	16:0 > 18:1c	Dahl et al. (1983)
M. fermentans	15–150	16:0	Wise et al. (1993)
M. gallisepticum	15–116	16:0	Forsyth et al. (1992)
M. hyopneumoniae	44–65	16:0 > ? (also 18:1c)[c]	Wise and Kim (1987)
M. hyorhinis	23–120	16:0[c]	Bricker et al. (1988)
S. melliferum	15–85	16:0 > 14:0 > 18:2c > 18:0 > 18:1c	Wróblewski et al. (1989)
U. urealyticum	16–126	16:0	Thirkell et al. (1991)

[a] Range of lipoprotein molecular masses estimated from gel photos. Most species contain 15–25 lipoproteins.

[b] Fatty acids tested and preferences of incorporation. 14:0, myristic; 16:0, palmitic; 18:0, stearic; 18:1c, oleic; and 18:2c, linoleic acid. Glycerol is incorporated in some *M. capricolum* lipoproteins, but not in *A. laidlawii* or *S. melliferum* (other species not tested).

[c] Acyl chain composition verified (TLC/HPLC) after release from the proteins.

(Wieslander et al., 1992), indicating that a third, amide-linked chain may often be missing. It is doubtful if just two (or one) acyl chains are sufficient to firmly anchor a large lipoprotein (cf. Table II) to a membrane.

The unexpected finding of isoprenoid-modified proteins in *A. laidlawii*, distinct from the prokaryotic lipoproteins but common in eukaryotes (Nyström and Wieslander, 1992), emphasizes the need for further molecular studies of protein modification mechanisms in mollicutes. The ecological niches of the cell-wall-deficient mollicutes as surface parasites on eukaryotic cells may involve special adaptations of the bacterial modification consensus mechanisms, including even the use of eukaryotic host mechanisms.

Acknowledgments

Our work has been supported by the Swedish Natural Science Research Council. We thank Mrs. Karin Bjurström for secretarial assistance.

References

Aitken, A. (1990). "Identification of Protein Consensus Sequences." Ellis Horwood, Chichester.
Aitken, A. (1992). Structure determination of acylated proteins. In "Lipid Modification of Proteins: A Practical Approach" (N. M. Hooper and A. J. Turner, eds.), pp. 63–88. Oxford Univ. Press, Oxford.
Archer, D. B., Rodwell, A. W., and Rodwell, E. S. (1978). The nature and location of Acholeplasma laidlawii membrane proteins investigated by two-dimensional gel electrophoresis. Biochim. Biophys. Acta **513,** 268–283.
Bairoch, A. (1993). The PROSITE dictionary of sites and patterns in proteins: Its current status. Nucleic Acids Res. **21,** 3097–3103.
Bordier, C. (1980). Phase separation of integral membrane proteins in Triton X-114 solution. J. Biol. Chem. **256,** 1604–1607.
Bové, J. M., Foissac, X., and Saillard, C. (1993). Spiralins. In "Subcellular Biochemistry" (S. Rottem and I. Kahane, eds.), Vol. 20, pp. 203–223. Plenum, New York.
Bricker, T. M., Boyer, M. J., Keith, J., Watson-McKown, R., and Wise, K. S. (1988). Association of lipids with integral membrane surface proteins of Mycoplasma hyorhinis. Infect. Immun. **56,** 295–301.
Cleavinger, C. M., Kim, M. F., and Wise, K. S. (1994). Processing and surface presentation of the Mycoplasma hyorhinis variant lipoprotein VlpC. J. Bacteriol. **176,** 2463–2467.
Cortay, J.-C., Nègre, D., and Cozzone, A.-J. (1991). Analyzing protein phosphorylation in prokaryotes. In "Methods in Enzymology" (T. Hunter and B. M. Setton, eds.), Vol **200,** 214–227. Academic Press, New York.
Dahl, C. E., and Dahl, J. S. (1984). Phospholipids as acyl donors to membrane proteins of Mycoplasma capricolum. J. Biol. Chem. **259,** 10771–10776.
Dahl, C. E., Dahl, J. S., and Bloch, K. (1983). Proteolipid formation in Mycoplasma capricolum. J. Biol. Chem. **258,** 11814–11818.
Dahl, C. E., Sacktor, N. C., and Dahl, J. S. (1985). Acylated proteins in Acholeplasma laidlawii. J. Bacteriol. **162,** 445–447.
Forsyth, M. H., Tourtellotte, M. E., and Geary, S. J. (1992). Localization of an immonodominant 64 kDa lipoprotein (LP 64) in the membrane of Mycoplasma gallisepticum and its role in cytadherence. Mol. Microbiol. **6,** 2099–2106.
Hames, B. D., and Rickwood, D., eds. (1990). "Gel Electrophoresis of Proteins: A Practical Approach," 2nd Ed. Oxford Univ. Press, Oxford.
Hardie, D. G., ed. (1993). "Protein Phosphorylation: A Practical Approach." Oxford Univ. Press, Oxford.
Hayashi, S., and Wu, H. C. (1992). Identification and characterization of lipid-modified proteins in bacteria. In "Lipid Modification of Proteins: A Practical Approach" (N. M. Hooper and A. J. Turner, eds.), pp. 261–285. Oxford Univ. Press, Oxford.
Hooper, N. M., and Turner, A. J., eds. (1992). "Lipid Modification of Proteins: A Practical Approach." Oxford Univ. Press, Oxford.
Jägersten, C., Odelstad, L., and Johansson, K.-E. (1982). Identification of iron- and phosphorus-containing antigens of the Acholeplasma laidlawii cell membrane. FEBS Lett. **144,** 130–134.
Jarhede, T. K., Le Hénaff, M., and Wieslander, Å. (1995). Expression of foreign genes and selection of promoters in Acholeplasma laidlawii. Microbiology. (in press).
Nyström, S., Wallbrandt, P., and Wieslander, Å. (1992). Membrane protein acylation: Preference for exogenous myristic acid or endogenous saturated chains in Acholeplasma laidlawii. Eur. J. Biochem. **204,** 231–240.
Nyström, S., and Wieslander, Å. (1992). Isoprenoid modification of proteins distinct from mem-

brane acyl proteins in the prokaryote *Acholeplasma laidlawii*. *Biochim. Biophys. Acta* **1107**, 39–43.
Platt, M. W., Rottem, S., Milner, Y., Barile, M. F., Peterkofsky, A., and Reizer, A. (1988). Protein phosphorylation in *Mycoplasma gallisepticum*. *Eur. J. Biochem.* **176**, 61–67.
Reinards, R., Kubicki, J., and Ohlenbusch, H.-D. (1981). Purification and characterization of NADH oxidase from membranes of *Acholeplasma laidlawii*, a copper-containing iron-sulfur flavoprotein. *Eur. J. Biochem.* **120**, 329–337.
Rosengarten, R., and Wise, K. S. (1991). The Vlp system of *Mycoplasma hyorhinis:* Combinatorial expression of distinct size variant lipoproteins generating high-frequency surface antigenic variation. *J. Bacteriol.* **173**, 4782–4793.
Thirkell, D., Myles, A. D., and Rusell, W. C. (1991). Palmitoylated proteins in *Ureaplasma urealyticum*. *Infect. Immun.* **59**, 781–784.
Wallbrandt, P. (1992). "The Pyruvate Dehydrogenase Complex of *Acholeplasma laidlawii*." Ph. D. Thesis, Umeå University, Umeå.
Wallevik, K., Jensenius, J. C., Andersen, J., and Poulsen, A. M. (1982). A simple and reliable method for the drying of polyacrylamide slab gels. *J. Biochem. Biophys. Methods* **6**, 17–21.
Wieslander, Å., Boyer, M. J., and Wróblewski, H. (1992). Membrane protein structure. *In* "Mycoplasmas: Molecular Biology and Pathogenesis" (J. Maniloff, R. N. McElhaney, L. R. Finch, and J. B. Baseman, eds.), pp. 93–112. Am. Soc. Microbiol. Washington, DC.
Wise, K. (1993). Adaptive surface variation in mycoplasmas. *Trends Microbiol.* **1**, 59–63.
Wise, K. S., and Kim, M. F. (1987). Major membrane surface proteins of *Mycoplasma hyopneumoniae* selectively modified by covalent bound lipid. *J. Bacteriol.* **169**, 5546–5555.
Wise, K. S., Kim, M. F., Theiss, P. M., and Lo, S.-C. (1993). A family of strain surface lipoproteins of *Mycoplasma fermentans*. *Infect. Immun.* **61**, 3327–3333.
Wold, F. (1981). In vivo chemical modification of proteins (post-translational modification). *Annu. Rev. Biochem.* **50**, 783–814.
Wróblewski, H., Nyström, S., Blanchard, A., and Wieslander, Å. (1989). Topology and acylation of spiralin. *J. Bacteriol.* **171**, 5039–5047.

C3

VARIANT MEMBRANE PROTEINS

Kim S. Wise, Mary F. Kim,
and Robyn Watson-McKown

Introduction

Because of their small genome, mycoplasmas encode and express relatively few proteins compared to other eubacteria. A small proportion of mycoplasmal gene products are associated with or are transported through the single plasma membrane of these wall-less organisms. Nevertheless, membrane proteins provide vital functions for all aspects of cellular physiology that require molecular exchange across the membrane boundary. In addition, some membrane proteins contribute to the physical interface that mediates mycoplasma adaptation to host environments. Some surface proteins provide specialized functions, such as avoidance of host immune responses. Indeed, many mycoplasma species display "families" of membrane proteins that vary both in their expression pattern (phase variation) and in their structural attributes (size variation). These systems create variation among strains, as well as in isogenic populations, where high frequency switching generates complex mosaics even in clonal populations. Widespread occurrence of surface protein variation in mycoplasmas demands an efficient way to assess membrane protein profiles of these organisms.

Variable surface proteins, as well as other functional proteins associated with the single membrane, are often integral membrane proteins, as defined by their amphiphilic behavior under conditions of detergent solubilization. Many variable membrane proteins are covalently modified by lipid. The amphiphilic property of these mycoplasmal lipoproteins, and their anchorage in the membrane, is due to acyl chain modification of a N-terminal Cys residue in the processed mature product (Cleavinger *et al.*, 1994). Lipid modification renders otherwise hydrophilic surface proteins amphiphilic. Whether through lipid modification or hydrophobic (e.g., transmembrane) segments in the primary amino acid sequence, mycoplasma proteins with amphiphilic properties are, *per se*, an inherently interesting class of products.

A general method for selective isolation and analysis of amphiphilic mycoplasma membrane components has been developed based on the segregation of proteins (or other components) that associate with detergent micelles, from their hydrophilic counterparts that do not. This occurs as a phase separation during temperature-dependent condensation of the nonionic detergent Triton X-114 (TX-114). TX-114 detergent phase fractionation has been widely used since the application was reported by Bordier (1981). TX-114 phase partitioning was first reported for analysis of mycoplasma membrane proteins by Riethman *et al.* (1987) and was modified (Boyer and Wise, 1989; Bricker *et al.*, 1988; Wise and Kim, 1987) to optimize the procedure. Several variations of the technique have since been applied to diverse mycoplasma species, some of which are reviewed in Wieslander *et al.* (1992; Chapter C2 of this volume). A standard method for this procedure is presented here, along with other applications (selective metabolic labeling and colony blot immunostaining) that facilitate the characterization of variable mycoplasma membrane proteins. The use of isolated membrane proteins as specific targets for immunologic screening is also discussed.

Materials

For Triton X-114 Phase Fractionation of Mycoplasmas

TS buffer (154 mM NaCl, 10 mM Tris, pH 7.4), containing 100 mM phenylmethylsulfonyl fluoride (PMSF) or another protease inhibitor; alternative buffers such as phosphate-buffered saline (PBS; 2.7 mM KCl, 1.2 mM KH_2PO_4, 138 mM NaCl, 8.1 mM $Na_2 HPO_4$, pH 7.4) can also be used.

1 to 2 mg mycoplasma protein (Lowry *et al.*, 1951), from organisms harvested from broth culture and washed by centrifugation with TS buffer to remove medium constituents.

1.5-ml conical microcentrifuge tubes with screw cap (e.g., Fisher Scientific, Pittsburgh, PA, Catalog No. 05-664-33).

Refrigerated (4°C) microcentrifuge, capable of attaining 12,000 g without an increase of temperature during centrifugation.

Room temperature microcentrifuge equipped with a swinging bucket rotor.

37°C water bath or heating block.

Triton X-114, specially purified for membrane research (e.g., Boehringer Mannheim, Catalog No. 1033 441).

Working stock solution of "10%" Triton X-114 (refer to Procedures for method of preparation).
Rotator apparatus.
Vortex mixer.
P1000 and P200 Pipetman or equivalent pipettors.
Wide-bore pipette tips (e.g., United Laboratory Plastics, St. Louis, MO, Catalog No. UP4005).
Apparatus for sodium dodecyl sulfate–polyacrylamide gel electrophoresis (SDS–PAGE).

For Metabolic Labeling of Mycoplasmas

Modified Hayflick broth medium containing 20% horse serum (e.g., Gibco BRL Life Technologies, Inc., Gaithersberg, MD, Catalog No. 230-6050AJ).
RPMI 1640 medium without methionine, cysteine, and cystine (e.g., ICN Biomedicals, Inc., Costa Mesa, CA, Catalog No. 16-464-54).
L-[^{35}S]Cysteine, specific activity >800 Ci/mmol (e.g., ICN Biomedicals, Inc., Catalog No. 51002).
L-[^{35}S]Methionine, specific activity >1000 Ci/mmol (e.g., ICN Biomedicals, Inc., Catalog No. 51001H).
Base medium for metabolic labeling: Dialyze 20 ml of Hayflick (or other) broth medium at 4°C against five 100-ml changes of Cys- and Met-free RPMI 1640 medium or against medium deficient in other specific amino acids chosen for radiolabeling. Sterile filter the dialyzed sample and store aliquots in sterile tubes at -70°C. This base medium should be centrifuged at 12,000 g for 5 minutes after thawing to remove any particulates.
Labeling medium: To label with cysteine, supplement base medium with 0.5 to 1.0 mCi/ml L-[^{35}S]cysteine and 15 μg/ml unlabeled L-methionine. To label with L-[^{35}S]methionine, supplement base medium with 0.5 to 1.0 mCi/ml L-[^{35}S]methionine and 50 μg/ml unlabeled cysteine.
Sterile PBS.
Sterile microcentrifuge tubes.
13-ml sterile, round base polypropylene centrifuge tubes with screw caps (e.g., Sarstedt, Inc., Catalog No. 60.540).
Microcentrifuge.

For Colony Immunoblots to Detect Antigenic Variation in Mycoplasma Populations

Petri dishes containing 25 ml modified Hayflick solid medium containing 1.0% Noble agar (e.g., Difco Laboratories, Detroit, MI, Catalog No. 0142-01).

Nitrocellulose 82-mm filter circles, grade BA85, 0.45 μm pore size (e.g., Schleicher & Schuell, Keene, NH, Catalog No. 40-20440).

CO_2 incubator.

Blocking solution: TS buffer containing 3% (w/v) bovine serum albumin (BSA) (e.g., Sigma Chemical Company, St. Louis, MO, Catalog No. A-6793).

Murine monoclonal antibody (MAb) (or other Abs) to surface epitopes on mycoplasma membrane antigens.

Peroxidase-conjugated antibody (goat) against mouse immunoglobulin G (lgG) (Organon Teknika Corp., Cappel Research Products, Durham, NC, Catalog No. 55550) or against lg of appropriate species to detect other primary antibodies used.

o-Dianisidine (Sigma Chemical Company, Catalog No. D-9143) or other insoluble chromogenic peroxidase substrate of choice.

Procedures

Principles of TX-114 Phase Partitioning

The principle of phase separation using TX-114 is illustrated in Fig. 1. Briefly, at lower temperatures, the detergent takes on a micellar form characteristic of aqueous "solutions" of many detergents (Fig. 1a). Heating the sample above a temperature characteristic for a particular detergent (the "cloud point") results in microcondensation of micelles, giving the sample a cloudy appearance (Fig 1b). Centrifugation of this sample at a temperature above the cloud point sediments micelles into a physically distinct phase highly enriched in detergent (Fig 1c). This detergent (TX) phase is denser, is more viscous, and is delineated by a sharp interface with the less dense and less viscous upper aqueous (AQ) phase, which contains very little detergent (Bordier, 1981). Proteins or other compounds with amphiphilic properties are associated with micelles during initial solubilization and remain associated after phase fractionation. Proteins and other compounds with hydrophilic characteristics do not associate with micelles and remain in the aqueous phase. Therefore, any complex mixture of components that can be solubilized in TX-114 (including most components of mycoplasmas) can be separated by their phase partitioning properties. Amphiphilic (integral) proteins associated with the single membrane of mycoplasmas would be predicted to partition with the detergent phase in this system.

TX-114 is particularly suited for this technique. It is a mild nonionic detergent (with many properties similar to Triton X-100). More importantly, it has a cloud point (about 20°C) in a temperature range compatible with the integrity of biological compounds such as proteins (Bordier, 1981). Using the range of 0–4°C for

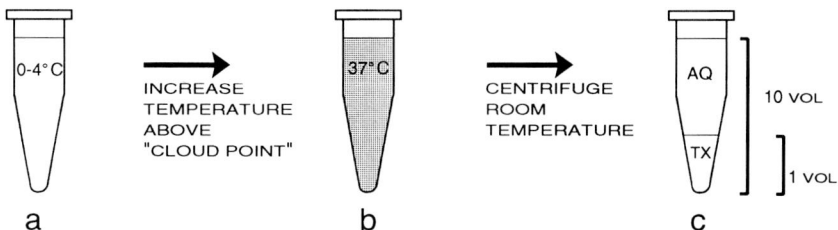

Fig. 1. Steps in TX-114 phase partitioning. Stages include initial micellar detergent solution at low temperature (a), condensation of micelles at temperatures above the cloud point (b), and the resulting phase separation yielding the heavier detergent (TX) phase and the lighter aqueous (AQ) phase (c). The preferred ratio of TX and AQ phases is indicated. This is determined by the initial TX-114 concentration, which is prepared from a stock solution empirically calibrated to meet this criterion.

solubilization in the cold, and from 25°C (or room temperature) to 37°C for phase separation, this system has many practical advantages.

Preparation of TX-114 Working Stock Solution

The concentration of TX-114 during solubilization and phase fractionation is important, although it can be varied somewhat without affecting the results of the procedure. Practical factors to consider include: (1) having a sufficient detergent concentration to solubilize reasonable quantities of mycoplasmal components, (2) establishing a volume ratio of the partitioned phases that allows efficient reextraction ("washing") of the TX phase, and (3) maintenance of a consistent volume of the TX phase during sequential washing steps.

Establishing the operational working concentration of TX-114 is somewhat empirical and is influenced by the fact that the volume ratio of the partitioned phases is dependent on the initial detergent concentration. A suitable and convenient approach is to prepare a TX-114 working stock solution at a concentration equal to that desired after phase fractionation. That in turn depends on the volume ratio desired. The conditions illustrated in Fig. 1 indicate a reasonable system for most applications. It shows a partitioning where the resulting TX phase is one-tenth the volume of the initial sample prior to phase separation. If, after removal of the AQ phase shown (Fig. 1c), the lower TX phase is resuspended in buffer to the original sample volume and the sample is repartitioned, the lower phase will again occupy one-tenth the original sample volume. A stock solution is therefore prepared at a concentration to meet these criteria.

Bordier (1981) reported the concentration of the TX phase after partitioning under similar conditions to be approximately 10–11% (w/v). However, our experience with commercial preparations of this detergent indicates that the stated concentration often cannot be used for this calculation. Instead, a simple series of test dilutions of the commercial preparation is made on a small scale, to determine which dilution will yield the desired 1 to 10 ratio of TX to AQ phase volumes after partitioning. The appropriate dilution is used to prepare a larger stock solution, which is then arbitrarily defined as the 10% working stock of TX-114. This process is performed with each new commercial preparation, using the same buffer that is to be used for sample preparation, solubilization, and partitioning. The 10% TX-114 working stock is prepared as follows:

1. Deliver increasing amounts of commercially obtained detergent to a series of chilled microcentrifuge tubes and add cold TS buffer to each tube up to a volume of 1 ml. Prepare one tube with approximately 10% (w/v) of the undiluted commercial preparation and set up the dilution series in a range that includes a twofold higher and twofold lower concentration. Amounts of detergent can be determined by weight (measured by weighing the detergent added to a tared tube) or estimated by volume. In the latter case, use a disposable wide bore pipette tip, allow sufficient time for the viscous solution to enter the tip, remove excess detergent from the outside of the tip, and be sure that all of the detergent in the tip is delivered. Maintain the tubes at 0 to 4°C and thoroughly mix these various dilutions of TX-114 on a rotator at 4°C.

2. Transfer 100 µl of each test dilution to a new tube containing 900 µl of cold TS buffer. Mix thoroughly at 0 to 4°C.

3. Transfer the tubes to 37°C, incubate for 5 minutes to condense micelles, and centrifuge in a swinging bucket rotor at room temperature (above 22°C) for 3 minutes at 8000 g.

4. Observe the series of tubes at room temperature and record which tube yielded a TX phase volume of 100 µl, using another tube filled with 100 µl of water for visual calibration. The *original* dilution of TX-114 (i.e., in step 1) corresponding to this tube is defined as 10% TX-114 (this factor can also be estimated by interpolating between the two tubes closest to the 100-µl volume of TX phase). Use this dilution factor to prepare a large volume of 10% TX-114 working stock from the commercial preparation.

5. Prepare about 25 ml of 10% TX-114 working stock solution, using reasonable precautions to minimize contamination. Weighing is the preferred method, but if a large pipette is used to measure the volume of the viscous commercial preparation, be sure that the entire volume is delivered. Mix the stock well by rotating the bottle at 4°C for several hours, and confirm the concentration by performing a phase fractionation as in steps 2 through 4. If the concentration is not suitable, repeat the calibrating dilution series with smaller increments to improve accuracy. Alternatively, small amounts of the commercial detergent preparation or buffer can be added to the working stock to correct the concentration (testing each addition, after thorough mixing, by phase fractionation). Working stock solutions are generally stable for several months at 4°C.

TX-114 Phase Partitioning of Mycoplasma Proteins

The following is a typical protocol for the separation of amphiphilic mycoplasma proteins based on the method described in Boyer and Wise (1989). All procedures are carried out in 1.5-ml microcentrifuge screw cap tubes. The protocol describes extraction in a 1-ml volume. However, the extraction volume can be scaled down to 200 µl (e.g., when organisms from a 1-ml broth culture are to be extracted) or it can be scaled up to extract organisms from larger cultures.

1. In a chilled microcentrifuge tube, suspend 1 to 2 mg of mycoplasma protein in 900 µl of cold, sterile TS buffer containing the protease inhibitor. Using a wide-bore pipette tip, add 100 µl of cold 10% TX-114 working stock solution. (This viscous solution must be allowed to completely fill the pipette tip, and excess on the tip should be removed with a tissue.) The stock TX-114 solution will settle to the bottom of the tube. Thoroughly dissolve it by vortexing or suspending with a Pipetman. Bubbles and foaming should be avoided. (For smaller scale extractions, mycoplasmas obtained from 1- to 2-ml broth cultures can be microcentrifuged, and the final, washed pellet can be resuspended in 200 µl of a cold 1% TX-114 solution, prepared by diluting the 10% stock solution 1:10 with buffer.)

2. Mix the sample well, and place the tube on a rotator at 4°C for 30 to 120 minutes. (At this step a small sample can be removed, which represents the total material, for later comparison with the fractionated material.)

3. Keep the tube on ice to remain below the cloud point and transfer it to a refrigerated microcentrifuge. Centrifuge the sample at 12,000 g for 5 minutes at 4°C, being sure that the apparatus does not heat during centrifugation. A tightly packed, insoluble pellet should be obtained under a uniform, clear supernatant. (*Note:* If the sample has warmed during centrifugation, it may yield a pellet, but under a dense and viscous condensed detergent phase that may appear cloudy. This condition is not suitable for subsequent steps. Be sure that any manipulations of the sample occur at 4°C or below.)

4. Carefully transfer the clear supernatant to a new tube, avoiding the pellet. Incubate this tube at 37°C for 5 minutes. The solution will become cloudy, indicating the condensation of detergent micelles. Centrifuge the sample for 3 minutes at 8000 g in a microcentrifuge equipped with a swinging bucket rotor at room temperature. This results in a phase separation yielding a clear, viscous lower detergent (TX) phase sharply delineated from the upper aqueous (AQ) phase (see Fig. 1c). The volume of the lower TX phase should be approximately one-tenth the total volume in the tube. If only the TX phase proteins are to be prepared, the AQ phase can be removed and discarded, and the remaining TX phase processed as described in step 6 below. However, the AQ phase is often of interest and can be processed as indicated in the following step 5.

5. To prepare the AQ phase, remove as much of this upper phase as possible, without disturbing the lower TX phase. Transfer the AQ phase to a new, chilled

tube and estimate the volume by pipetting a measured volume of water to the same level in an empty tube. "Wash" the AQ phase as follows:

    a. To the measured volume of AQ phase, add one-tenth that volume of cold 10% TX-114 working stock solution. Thoroughly mix by vortex and pipetting, and incubate 3 to 5 minutes on ice to thoroughly dissolve the detergent (the solution should appear clear).

    b. Transfer the tube to 37°C for 5 minutes to condense micelles, and microcentrifuge the resulting cloudy suspension at 8000 g for 3 minutes at room temperature to separate the phases.

    c. Transfer the upper AQ phase to a new tube. Discard the lower detergent phase. Wash the upper AQ phase three times by repeating steps 5a and 5b, each time transferring the upper AQ phase to a fresh, chilled tube. This fraction can then be analyzed by SDS–PAGE and used for several purposes. It contains only traces of detergent (Bordier, 1981).

6. To prepare the TX phase, the lower TX phase generated in step 4 should be processed as follows:

    a. After removing the upper AQ phase (see Fig. 1c), chill the tube containing the TX phase on ice. Add cold buffer to the condensed TX phase to restore to the original extraction volume (1 ml in the case described here). Resuspend and dissolve the condensed detergent thoroughly (avoiding foaming) and incubate the sample for 3 to 5 minutes on ice. This solution should be clear and uniform.

    b. Transfer the tube to 37°C and incubate for 5 minutes to condense detergent micelles. Centrifuge the cloudy sample at room temperature for 3 minutes in a swinging bucket microcentrifuge. Again, the resulting lower TX phase should be approximately one-tenth the volume in the tube. Remove and discard the upper AQ phase.

    c. Save the lower TX phase and repeat wash steps 6a and 6b three times.

    d. After the last wash, discard the upper AQ phase and resuspend the lower TX phase to the starting volume. This results in a 1% TX-114 concentration.

    e. Chill the tube thoroughly and mix the sample to completely dissolve the detergent phase. Microcentrifuge the tube at 12,000 g for 5 minutes at 4°C. Be sure the sample does not warm during centrifugation (see note in step 3). Transfer the supernatant to a fresh tube. Any insoluble material, which sometimes appears during repeated washing of the TX phase, is removed by this step and should be discarded.

    f. This supernatant contains the TX-phase proteins. It can be used at this concentration (approximately 1% TX-114) or the proteins can be concentrated by one more step of phase fractionation, where the TX phase is reconstituted with a smaller volume of buffer. The final material can be stored at -70°C and should be mixed well at 0 to 4°C after thawing.

Metabolic Labeling of Mycoplasma Membrane Proteins

The procedure for metabolic labeling of mycoplasmas is adapted from Rosengarten and Wise (1991). It has been used to label *M. hyorhinis, M. fermentans* (Wise *et al.*, 1993), and (by substituting FF medium for Hayflick medium as the base medium for metabolic labeling) *M. hyopneumoniae* (Wise and Kim, 1987).

1. Transfer 10 ml of a mid- to late-logarithmic mycoplasma broth culture to a sterile 13-ml polypropylene tube and centrifuge for 5 minutes at 12,000 g.

2. Carefully wash the resulting pellet *in situ* twice with sterile PBS, being careful to remove all medium from the walls and cap of the tube, but without dislodging the pellet.

3. Carefully, but thoroughly, resuspend the pellet in 1 ml labeling medium and incubate the sample at 37°C for 12 to 18 hours.

4. Transfer the sample to a 1.5-ml conical microcentrifuge tube and centrifuge the labeled cells for 10 minutes at 12,000 g. Wash the pellet twice by resuspending in 1 ml cold PBS and centrifuging. Suspend the final pellet in 1 ml of 1% TX-114 (prepared by diluting the 10% stock). Proceed with step 2 under TX-114 Phase Partitioning of Mycoplasma Proteins.

Colony Immunoblots to Detect Surface Epitope Phase Variation

This procedure assesses the expression patterns of surface epitopes on mycoplasma populations growing in colonies on agar medium. It involves making a "lift" of colonies onto a nitrocellulose filter and immunostaining the filter to detect antigen expression on different colonies or within the population in single colonies. This is particularly useful in determining the frequency of phase variation, dual expression patterns, or distribution in a population of a particular component.

1. Prepare standard agar plates seeded with 200–2000 (colony-forming units) of mycoplasmas and incubate until colonies are large (e.g., 5 days for most *M. hyorhinis* strains).

2. While plates are fresh and at 37°C or room temperature, carefully place a dry nitrocellulose filter disc (with the smooth side down) onto the agar surface. This can be done by touching the center of a bowed filter to the middle of the plate and slowly "rolling" the edges outward as the leading edges of the filter become wet.

3. After 5 minutes at room temperature, remove the filter by firmly grasping an edge with forceps and lifting steadily until the entire filter is free. It should not be allowed to drop back onto the agar during removal. (If colonies are to be retrieved later, sterile filters should be used, and their position should be marked

on the plate prior to removal by placing asymmetric needle holes through the paper and underlying agar.)

4. Immerse the filter in blocking solution and incubate at room temperature for a minimum of 1 hour or at 4°C overnight.

5. Immunostain the filter by standard procedures using the primary antibody to the surface antigens of interest, followed by the secondary-conjugated antibody and substrate to stain the imprint of the colony population.

6. Dry the immonostained filters and observe or photograph the filter, using a standard high-magnification stereoscope or inverted microscope, with incident lighting applied to the stained side of the filter.

Discussion

Technical Aspects of Phase Fractionation

1. Initial TX-114 extraction of mycoplasma preparations usually results in some material that cannot be solubilized, even in detergent excess or after repeated extraction. These components appear to be a selective set of proteins when analyzed by SDS–PAGE. In contrast, when detergent is limiting (e.g., when excessive amounts of mycoplasma protein are extracted in the initial solubilization), the amount of pelleted material after solubilization is much greater. This represents organisms not fully dissociated by detergent. This material shows the entire spectrum of mycoplasma components in SDS–PAGE. Nevertheless, the TX-114 soluble supernatant material obtained from either of these solubilization conditions can be used for further fractionation.

2. Some mycoplasma species or strains also tend to generate insoluble material during subsequent phase fractionation. This is deposited under the TX phase as pelleted material during each cycle of washing the TX phase and can be removed (by step 6e under TX-114 Phase Partitioning of Mycoplasma Proteins). Specific proteins may precipitate possibly due to a selective susceptibility to changing conditions during phase partitioning. An example is the P95 protein of *M. fermentans* (Fig. 2a) and other high molecular weight components of this species.

3. The number of cycles needed to wash the TX phase depends on the purity of the material required. Since the condensed TX phase (roughly 10%, w/v) still contains aqueous solvent, it also contains AQ phase components that do not associate with micelles. Each cycle of washing probably removes about 90% of the AQ phase components from the TX phase (Bordier, 1981), less the amount in the residual aqueous phase not removed from the interface after each partition-

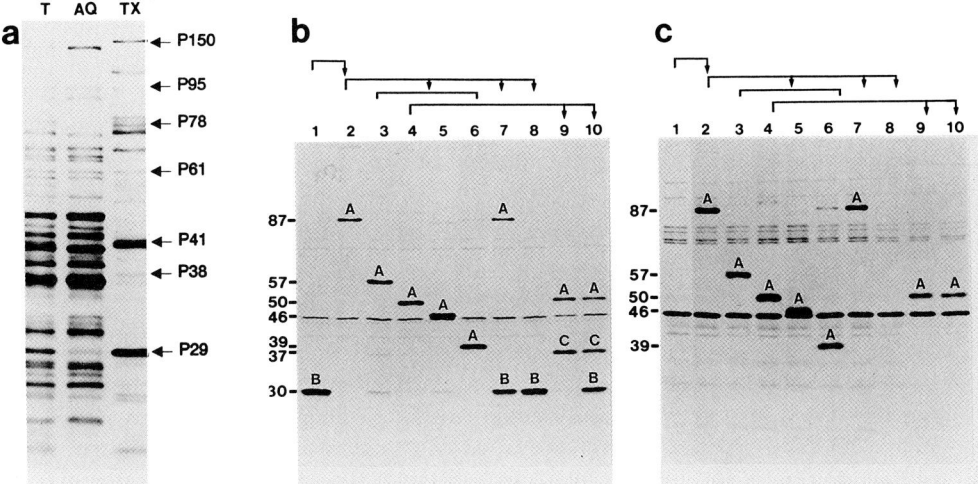

Fig. 2. TX-114 phase fractionation and comparison of variable TX phase lipoproteins from metabolically labeled mycoplasmas. Organisms were labeled with [^{35}S]Cys (a and b) or [^{35}S]Met (c). Phase fractionation was performed, and samples were analyzed by SDS–PAGE and autoradiography. (a) Total *M. fermentans* PG18 proteins (T), and the proteins partitioning into the AQ or TX phases. Arrows indicate specific TX phase surface membrane proteins further characterized elsewhere (Wise et al., 1993). (b and c) TX phase proteins prepared from clonal isolates in an isogenic lineage of *M. hyorhinis* SK76, labeled with either [^{35}S]Cys (b) or [^{35}S]Met (c). The lipoproteins VlpA, VlpB, and VlpC are indicated by corresponding letters. Sizes of the products are indicated at the left of each panel in kilodaltons. Portions of this figure are adapted and reproduced with permission: (a) from Wise et al. (1993) and (b and c) from Rosengarten and Wise (1991).

ing. For many purposes, three cycles of careful partitioning renders the TX phase sufficiently depleted of AQ phase contaminants. This includes residual medium components, the great majority of which partition with the aqueous phase (Riethman et al., 1987). However, additional cycles may be required for very sensitive systems, such as immunization with TX phase proteins. Even after several cycles of repartitioning, TX phase proteins may induce antibodies directed to specific AQ phase components due to trace AQ phase contaminants in the immunizing preparation.

4. Loading SDS–PAGE gels with large amounts of TX-114 can lead to "fanning" of channels as they run. This is more problematic if the final TX phase preparation has been adjusted to a high concentration of protein and detergent. TX phase proteins can be further concentrated and the detergent removed by precipitation. To do this, remove the upper AQ phase from the final condensed TX phase and add 9 vol of cold absolute methanol to the condensed TX phase (e.g., 900 μl methanol to 100 μl TX phase derived from a 1000-μl extraction).

Mix the sample well and place at -70°C overnight. Centrifuge the sample for 10 minutes at 12,000 g in the cold. (CAUTION: A small hole should be placed in the lid of the tube to avoid pressure buildup as the methanol warms.) After centrifugation, remove the methanol completely. This procedure extracts detergent and many lipid components and dehydrates the sample. Generally,. the precipitated protein is insoluble in aqueous buffers, but can be solubilized in other detergents or in sample buffer for SDS–PAGE analysis. The methanol-precipitated material is also compatible with further organic extraction used to analyze protein-linked lipids (Wieslander et al., 1992; see also Chapter C2, this volume).

5. Some reports have indicated forms of variable proteins in both TX and AQ phases during phase fractionation, particularly with larger size variants. The basis of this phenomenon has not been determined but might be explained by two features. First, hydrophobic moieties of proteins may interact more strongly with each other than with detergent micelles, possibly yielding an "aqueous" multimer revealing no sites for micellar interaction. Second, if partitioning is monitored by Western immunoblot procedures, degradation resulting in the removal of hydrophobic portions of the protein (e.g., lipid acylated N termini of otherwise hydrophilic lipoproteins) could yield a minor population of a nearly full-length hydrophilic product that contains the pertinent epitopes. This would be predicted to partition in the AQ phase. It is of course possible that some variable surface proteins are extrinsic and exist as nonprocessed and processed species, each with different partitioning properties.

Application of TX-114 Phase Fractionation for Analysis of Variable Lipoproteins and Detection of Specific Host Antibodies

Several experimental formats can be used to select and compare mycoplasmal variants expressing different profiles of membrane proteins. Two examples are illustrated in Fig. 2, using SDS–PAGE analysis of metabolically labeled TX-114 phase-fractionated proteins. Figure 2a compares [^{35}S]Cys-labeled proteins from whole *M. fermentans* PG18 organisms (T) with those proteins selectively partitioning into the AQ or TX phases. The TX phase products indicated by arrows have also been defined with specific antibodies by immunoblotting SDS–PAGE gels loaded with a similar panel of fractionated proteins (Wise et al., 1993). Most of the TX phase proteins indicated are also labeled with fatty acids, consistent with a lipoprotein structure. Notably, however, several TX phase proteins do not label with lipid and probably represent other forms of amphiphilic membrane proteins.

A second example (Figs. 2b and 2c) shows variation in size, expression, and differential metabolic labeling of variable lipoproteins of *M. hyorhinis*. Because these products (VlpA, VlpB, and VlpC) are amphiphilic, they are easily fractionated into the TX phase. They are readily distinguished by SDS–PAGE profiles from the relatively few other labeled proteins in the TX phase. The two panels represent TX phase proteins from exactly the same set of isogenic clonal variants of *M. hyorhinis* chosen to display different patterns of Vlp expression. Figure 2b shows the products labeled with [^{35}S]Cys, an amino acid that occurs invariably in the prokaryotic lipoprotein motif and only once in any Vlp sequence. The presence, size, and relative abundance of Vlp products can be easily detected. In Fig. 2c, TX phase profiles of the same variants are represented, but after labeling with [^{35}S]Met. In this case, only VlpA is labeled, due to the presence of one Met residue in its sequence (Rosengarten and Wise, 1991; Yogev *et al.*, 1991). VlpB and VlpC sequences contain no Met residues and are thereby readily distinguished as distinct translational products (VlpB and VlpC are also distinguished by monoclonal antibodies used in Western blots of these same panels). Interestingly, differential labeling with these two compounds shows that Met is not converted to Cys in *M. hyorhinis* under the conditions employed. TX-114 fractionation of metabolically labeled mycoplasmas may therefore reveal several features of variable amphiphilic membrane proteins. It is judicious to use multiple labels in initial studies using this approach. For example, mature VlpB and VlpC contain no Met or Leu residues, nor do they stain with Coomassie brilliant blue or some silver stain procedures. In contrast, labeling with Cys reveals all of these products. Cys labeling may be particularly suitable for expressing variable mycoplasma lipoproteins containing this residue in the typical prokaryotic processing and acylation motif (Cleavinger *et al.*, 1994; Wieslander *et al.*, 1992).

Finally, several studies have indicated that proteins and lipoproteins in the TX phase constitute major immunogenic surface components recognized by host antibodies during mycoplasmal infection and disease (e.g., Kim *et al.*, 1990; Rosengarten and Wise, 1991). Therefore, separation of TX phase proteins yields a preparation enriched in relevant antigens. These can be readily resolved and characterized by SDS–PAGE and Western blot analysis. Notably, several cross-reactions involving mycoplasmal AQ phase proteins are removed by the TX phase fractionation, thereby yielding a clearer pattern of reaction with TX phase components compared to the sometimes complex and nonspecific patterns acquired using SDS–PAGE and Western blots of whole organisms.

Colony Immunoblotting to Examine Surface Antigen Variation

1. High-resolution colony blots can reveal marked differences in the expression of specific components in isogenic populations (Fig. 3a), including the presence of radial "sectoring" in individual colonies (Fig. 3b), which is the hallmark of high-frequency phase variation. The detailed analysis of this phe-

nomenon and its molecular basis in *M. hyorhinis* are reported elsewhere (Rosengarten and Wise, 1991; Yogev *et al.*, 1991).

2. Detailed resolution of expression patterns, as well as artifacts and technical idiosyncrasies, can arise with this procedure. Some of these are illustrated in Fig. 3.

a. High-resolution blots with low background immunostaining can reveal varied patterns of antigen expression within isogenic populations (Figs. 3a and 3b). "Tails" shown in Fig. 3a result from deformation of colonies during removal of the filters. These are typically oriented in one direction of the filter and may occur on some colonies more than others.

b. Some antibodies give background staining of all colonies; however, this may still be easily distinguished from positive colonies (Fig. 3c). Figure 3c also shows the rough surface encountered on some nitrocellulose filters.

c. "Sector-like" artifacts can occur (Fig. 3d), possibly due to colony deformation during removal of the filter.

d. Some mycoplasma species, strains, isolates, or even colonal variants show marked differences in the quality of immunoblots obtained. Figure 3e shows a population expressing the antigen, but widespread damage to the colony surfaces (apparently during removal of the filter) renders a mottled pattern that is of little use for interpreting variation. The explanation of variability in blot

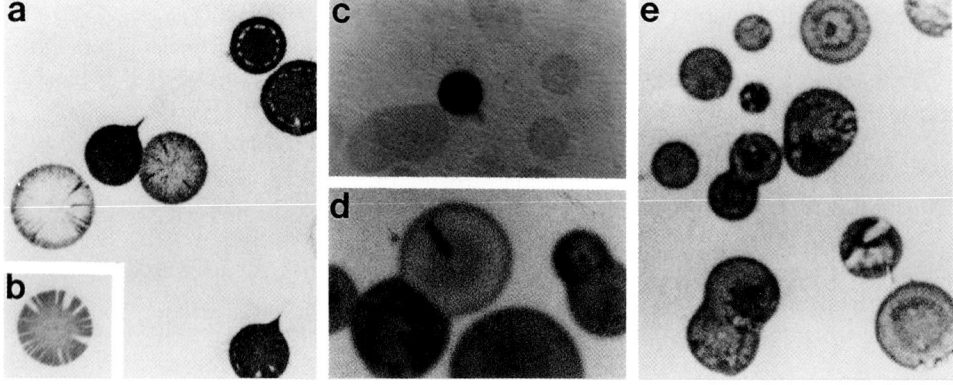

Fig. 3. Example of colony immunoblots of various mycoplasma populations stained with murine monoclonal antibodies (MAbs) to specific surface antigens. These include *M. hyorhinis* clonal populations stained with MAb to Vlp surface lipoproteins, showing sectoring, unusual ring-like staining patterns, and artifactual "tails" on two colonies (a and b); and *M. fermentans* clonal populations stained with MAbs to lipoproteins P61 (c), P150 (d), and P29 (e). All colony imprints shown range from 0.3 to 1.5 mm in diameter. MAbs and surface antigens are described in Rosengarten and Wise (1991) and Wise et al. (1993).

quality among populations is not clear, but may reflect major differences in the physical properties of colonial populations, even among clonal variants. Use of plates when colonies are old, or when desiccation has occurred, also can generate a mottled pattern of transfer.

Acknowledgments

Studies presented in this chapter were supported by DHHS Grants AI31656 and AI32219 from the National Institute of Allergy and Infectious Diseases.

References

Bordier, C. (1981). Phase separation of integral membrane proteins in Triton X-114 solution. *J. Biol. Chem.* **256**, 1604–1607.
Boyer, M. J., and Wise, K. S. (1989). Lipid-modified surface protein antigens expressing size variation within the species *Mycoplasma hyorhinis*. *Infect. Immun.* **57**, 245–254.
Bricker, T. M., Boyer, M. J., Keith, J., Watson-McKown, R., and Wise, K. S. (1988). Association of lipids with integral membrane surface proteins of *Mycoplasma hyorhinis*. *Infect. Immun.* **56**, 295–301.
Cleavinger, C. M., Kim, M. F., and Wise, K. S. (1994). Processing and surface presentation of the *Mycoplasma hyorhinis* variant lipoprotein VlpC. *J. Bacteriol.* **176**, 2463–2467.
Kim, M. F., Heidari, M. B., Stull, S. J., McIntosh, M. A., and Wise, K. S. (1990). Identification and mapping of an immunogenic region of *Mycoplasma hyopneumoniae* p65 surface lipoprotein expressed in *Escherichia coli* from a cloned genomic fragment. *Infect. Immun.* **58**, 2637–2643.
Lowry, O. H., Rosenbrough, N. J., Farr, A. L., and Randall, R. J. (1951). Protein measurement with the Folin phenol reagent. *J. Biol. Chem.* **193**, 265–275.
Riethman, H. C., Boyer, M. J., and Wise, K. S. (1987). Triton X-114 phase fractionation of an integral membrane surface protein mediating monoclonal antibody killing of *Mycoplasma hyorhinis*. *Infect. Immun.* **55**, 1094–1100.
Rosengarten, R., and Wise, K. S. (1991). The Vlp system of *Mycoplasma hyorhinis*: Combinatorial expression of distinct size variant lipoproteins generating high-frequency surface antigenic variation. *J. Bacteriol.* **173**, 4782–4793.
Wieslander, Å., Boyer, M. J., and Wróblewski, H. (1992). Membrane protein structure. In "Mycoplasmas: Molecular Biology and Pathogenesis." (J. Maniloff, R. N., McElhaney, L. R. Finch, and J. B. Baseman, eds.), pp. 93–112. Am. Soc. Microbiol., Washington, DC.
Wise, K. S., and Kim, M. F. (1987). Major membrane surface proteins of *Mycoplasma hyopneumoniae* selectively modified by covalently bound lipid. *J. Bacteriol.* **169**, 5546–5555.
Wise, K. S., Kim, M. F., Theiss, P. M., and Lo, Shyh-Ching. (1993). A family of strain-variant surface lipoproteins of *Mycoplasma fermentans*. *Infect. Immun.* **61**, 3327–3333.
Yogev, D., Rosengarten, R., Watson-McKown, R., and Wise, K. S. (1991). Molecular basis of *Mycoplasma* surface antigenic variation: A novel set of divergent genes undergo spontaneous mutation of periodic coding regions and 5' regulatory sequences. *EMBO J.* **10**, 4069–4079.

C4

MEMBRANE FUSION
Shlomo Rottem and Mark Tarshis

Introduction

The lack of a rigid cell wall and the fact that the cells are bounded by a single membrane, the cytoplasmic membrane, seem to favor fusion of mycoplasmas with host cells. Thus, membrane fusion has been suggested as a possible mechanism of mycoplasma–host cell interaction. Several studies presented an indication of a fusion process (Haberer and Frosch, 1982; Dimitrov *et al.*, 1993). Apostolov and Windsor (1975) demonstrated by ultrathin sectioning and freeze-etching the intimate contacts between erythrocytes and *Mycoplasma gallisepticum* and suggested that these areas represent a fusion process. Prakash and Gabridge (1981) observed an increased uptake of *M. pneumoniae* by 10% polyethyleneglycol (PEG)-treated fibroblasts and suggested that fusion occurred. More recently, the fusion of *M. fermentans* with cultured lymphocytes has been presented (Dimitrov *et al.*, 1993). Fusion of mycoplasma cells with lipid vesicles has been also demonstrated (Grant and McConnell, 1979). The mycoplasma–liposome fusion process has been intensively characterized (Salman *et al.*, 1991, 1993; Tarshis *et al.*, 1993) and was utilized for the transfer of plasmids encapsulated in lipid vesicles into *Spiroplasma floricola* (Salman *et al.*, 1992).

For measuring fusion, a number of sensitive assays based on fluorescence dequenching have been developed. They utilize the mixing of aqueous contents entrapped within the lipid vesicles and fusion is determined by monitoring the generation of a fluorescent product as the vesicles fuse. Another approach is based on membrane mixing. The probe is incorporated into membrane lipids and, on fusion, their surface density decreases and the fluorescence of the energy donor is dequenched. Hoekstra *et al.* (1990) have developed an assay relying on the relief of self-quenching of fluorescence of a fluorescent fatty acid, octadecylrhodamine B (R18), which is spontaneously inserted into native membranes at high concentrations. The fluorescence of R18 is dequenched as its concentration in the membrane decreases. Although the assay is prone to some artifacts, such as lipid exchange on cell–cell or small unilamellar vesicle (SUV)–cell aggregation, this probe has the advantage of not dissociating from membranes by spontaneous transfer of free monomers through the aqueous phase or by a collision-mediated transfer process.

Materials

Cell Preparations

Cells in an isotonic buffer (about 2 mg cell protein/ml)
A eukaryotic cell culture

Chemicals

Polyethylene glycol 8000 (PEG)
N-Tris(hydroxymethyl)methyl 2-aminoethanesulfonic acid (TES)
5,6-Carboxyfluorescein
Triton X-100
Octadecylrhodamine B chloride (R18, Molecular Probes, Eugene, OR)

Special Equipment

Rotary evaporator (Rotavapor, Buchi Laboratories)
Spectrofluorimeter, e.g., Perkin–Elmer LS-5B with excitation at 560 nm and emission at 590 nm
Inverted microscope
Probe sonicator, e.g., Heat Systems or Branson

Procedures

Small Unilamellar Vesicle (SUV) Preparation

1. Dissolve 5–10 mg of mycoplasma membrane lipids in 3–5 ml of chloroform in a 15-ml round-bottom centrifuge tube and evaporate the solvent to dryness under a stream of nitrogen, forming a thin lipid film on the walls of the tube.
2. Rehydrate the lipids by adding TES buffer to a final lipid concentration of 2 mg/ml and mix vigorously by a vortex mixer for 5 minutes.
3. Sonicate the milky lipid suspension for 12 minutes at 4°C under a stream of nitrogen using a sonicator operated with a 3-mm-diameter probe at 160 W.
4. The clear SUV suspension thus obtained is stored at 4°C under nitrogen and should be used within 4–5 days.

Large Unilamellar Vesicle (LUV) Preparation

1. Dissolve 5 mg of mycoplasma membrane lipids in 5 ml of chloroform in a 15-ml round-bottom centrifuge tube and evaporate the solvent to dryness under a stream of nitrogen.
2. Redissolve the dry pellet in 1.5 ml of diethyl ether and add 0.5 ml of TES buffer. Vigorously shake the two-phase mixture for 1 minute and then sonicate three times for 5 seconds at 4°C under a stream of nitrogen using a probe sonicator equipped with a 3-mm probe and operated at 20 W.
3. Evaporate the opalescent mixture obtained in a rotary evaporator at 45 rpm and 30°C under the negative pressure of 400 mm Hg until the suspension becomes a gel.
4. Briefly vortex the gel and continue to evaporate at 750 mm Hg until a homogenous suspension is obtained.
5. Dilute the resulting LUV preparation with 2 vol of TES and pass 100-μl volumes of the LUV through a Sephadex G-50 column.
6. The LUV suspension thus obtained is stored under nitrogen at 4°C and should be used within 2–3 days.

Fusion of Lipid Vesicles (SUV or LUV) with Mycoplasma Cells

A. LABELING OF LIPID VESICLES WITH R18

1. Rapidly inject 3 μl of an ethanolic solution of R18 (1 mg/ml) into 200–250 μl of a lipid vesicle dispersion (1.0 mg of lipids per ml) and incubate the mixture at 37°C for 15 minutes in the dark.

2. Remove unincorporated R18 from the lipid vesicle preparation by passing 100-μl samples through a Sephadex G-50 column and store the R18-labeled vesicles in the dark at 4°C under nitrogen until used (within 2 days of preparation).

B. FUSION OF LIPID VESICLES WITH MYCOPLASMAS

1. Mix 20 μl of R18-labeled vesicles with 180 μl of intact mycoplasma cells (2 mg of cell protein/ml) and 20 μl of 50% PEG 8000 and incubate the mixture in the dark for up to 40 minutes at 37°C.
2. Stop the reaction by the addition of 2 ml of cold PBS and measure the intensity of fluorescence dequenching in a spectrofluorimeter with excitation at 560 nm and emission at 590 nm and with splits 5 and 10 correspondingly and with correction for light scattering.
3. Add 20 μl of 20% Triton X-100 to the cuvette, mix vigorously, and repeat the measurement. The dequenching obtained in the presence of Triton X-100 is taken as 100%, representing infinite dilution of the probe. The fluorescence dequenching is calculated from the equation

$$DQ\ (\%) = [F - (F_1 \times I/I_{tr})]/[F_1 - (F_1 - I/I_{tr})],$$

where F is the fluorescence measured at the end of the incubation period; F_1 is the fluorescence after solubilization with Triton X-100; and I and I_{tr} are the fluorescence of the reaction mixture before incubation (zero time) and after solubilization with Triton X-100, respectively.

Fusion of Mycoplasmas with Eukaryotic Cells

1. Label mycoplasmas (200 μl containing 0.4 mg cell protein) by the same procedure described earlier for the labeling of lipid vesicles, except that for the removal of unincorporated R18, wash the cells twice with 2 ml of cold PBS.
2. Dilute the eukaryotic cells in PBS to a final concentration of 10^6 cells/ml. Check the viability of cells by the trypan blue exclusion test.
3. Mix 10 μl of R18-labeled mycoplasmas with 0.5 ml of the eukaryotic cell samples (5×10^5 cells) precooled to 4°C. Add PEG (5% final concentration) and incubate the mixture at 4°C for 30 minutes in the dark.
4. Sediment the cells for 2 minutes at 2000 rpm, wash the pellet gently with PBS, and resuspend in 1 ml of PBS.
5. Incubate the mixture at 37°C for 40 minutes and stop the reaction by adding 1.5 ml of cold PBS.
6. Measure the intensity of fluorescence dequenching as described earlier.

Cell Content Mixing Measurements

1. Prepare lipid vesicles as described earlier, but with TES buffer containing 40 mM 5,6-carboxyfluorescein.
2. Remove the fluorescent dye outside the vesicles by passing through a Sephadex G-50 column as described earlier.
3. Incubate 30 μl of 5,6-carboxyfluorescein-containing vesicles with 200 μl of mycoplasma cells (about 300 μg of protein) in PBS containing 5% PEG for 40 minutes at 37°C and stop the reaction by the addition of 2 ml of cold PBS.
4. Measure the fluorescence dequenching degree in a spectrofluorimeter with excitation at 490 nm and emission at 520 nm and with slits 5 and 10 correspondingly (experimental release).
5. Add 20 μl of 20% Triton X-100 to the cuvette, mix the contents, and repeat the measurements. The degree of fluorescence obtained in the presence of detergent (total release) is taken as 100% 5,6-carboxyfluorescein release. The fluorescence of SUV or LUV mixed with the cells, but without further incubation (zero time), is taken as a control.

Discussion

The extent of R18 dequenching directly correlates with the SUV/mycoplasma ratios in the medium. Maximum fluorescence is obtained with a ratio of 1:10. It is therefore recommended to use this ratio in fusion experiments. In many cases, lipid vesicle–mycoplasma fusion or mycoplasma–eukaryotic cell fusion was found to depend on PEG. PEG is a dehydrating polymer widely used for mediating cell–cell fusion in the preparation of hybridoma cells. It was suggested that PEG either alters the physical state of bulk water adjacent to the cell surface and the water of hydration of phospholipid polar groups in membranes or induces changes in the lipid phase transition, thus destabilizing bilayer backbone and, consequently, inducing fusion (Burger and Verkleij, 1990). The requirement for divalent cations (Ca^{2+} or Mg^{2+}) for lipid vesicle–mycoplasma cells or mycoplasma–eukaryotic cell fusion is expected in view of the negatively charged cell surface of mycoplamas. These cations bring the bilayers into close proximity by overcoming the repulsive hydration forces between the interacting polar head groups of negatively charged phospholipids and destabilize the attached bilayers. Cations also form salts with lipid molecules which lead to removal of water, facilitating direct molecular contacts between bilayers and lipid packing deformation.

Proteins are directly involved in the initiation of membrane fusion. They are required for specific membrane recognition and for the approach of two membranes into molecular contact and the breakdown of the hydration barrier, which, in turn, may induce a local destabilization of lipid bilayers and the formation of a fusion intermediate (Hoekstra, 1990). Fusion can result from a protein–lipid interaction, as was illustrated by fusion of viruses with liposomes (Cytovsky *et al.*, 1988). In this case, fusion occurs due to the insertion of hydrophobic peptide moieties into the interior of the bilayer of recipient membranes. Although the inhibition of fusion by proteolytic enzymes supports the idea that mycoplasmas possess protease-sensitive receptors responsible for a close contact with SUV (Salman *et al.*, 1991), the possibility cannot be excluded that, following digestion, membrane lipids tend to aggregate in protein-free membrane areas that may prevent them from fusing.

Cholesterol in mycoplasma membranes was shown to be critical to their ability to fuse with enveloped Sendai and influenza viruses (Cytovsky *et al.*, 1988), as well as for the fusion of *M. capricolum* with lipid vesicles (Salman *et al.*, 1991, 1993; Tarshis *et al.*, 1991). The cholesterol requirement for fusion in the target mycoplasma membrane could be met by a variety of planar sterols having a free β-hydroxyl group, but differing in the aliphatic side chain (Tarshis *et al.*, 1993).

References

Apostolov, K., and Windsor, G. D. (1975). The interaction of *Mycoplasma gallisepticum* with erythrocytes. 1. Morphology. *Microbios* **13**, 205–215.
Burger, K. N. J., and Verkleij, A. J. (1990). Membrane fusion. *Experientia* **46**, 631–644.
Cytovsky, V., Rottem, S., Nussbaum, O., Laster, Y., Rott, R., and Loyter, A. (1988). Animal viruses are able to fuse with prokaryotic cells: Fusion between Sendai or influenza virions and Mycoplasma. *J. Biol. Chem.* **263**, 461–467.
Dimitrov, D. S., Franzoso, G., Salman, M., Blumenthal, R., Tarshis, M., Barile, M. F., and Rottem, S. (1993). *Mycoplasma fermentans* (incognitus strain) cells are able to fuse with T-lymphocytes. *Clin. Infect. Dis.* **17**, S305–S308.
Grant, C. W. M., and McConnell, H. M. (1979). Fusion of phospholipid vesicles with viable *Acholeplasma laidlawii*. *Proc. Natl. Acad. Sci. USA* **70**, 1238–1240.
Haberer, K., and Frosch, D. (1982). Lateral mobility of membrane-bound antibodies on the surface of *Acholeplasma laidlawii:* Evidence for virus-induced cell fusion in a prokaryote. *J. Bacteriol.* **152**, 471–478.
Hoekstra, D. (1990). Fluorescence assays to monitor membrane fusion: Potential application in biliary lipid secretion and vesicle interactions. *Hepatology* **12**, 61S–66S.
Prakash, G., and Gabridge, M. G. (1981). Influence of the fusogenic agent polyethylene glycol on attachment of *Mycoplasma pneumoniae* to other cells. *Infect. Immun.* **32**, 969–972.
Salman, M., Tarshis, M., and Rottem, S. (1991). Small unilamellar vesicles are able to fuse with *Mycoplasma capricolum* cells. *Biochim. Biophys. Acta* **1063**, 202–216.
Salman, M., and Tarshis, M., and Rottem, S. (1992). Fusion-mediated transfer of plasmids into *Spiroplasma floricola* cells. *J. Bacteriol.* **174**, 4410–4415.
Salman, S., Shirazi, I., Tarshis, M., and Rottem, S. (1993). Fusion of *Spiroplasma floricola* cells

with small unilamellar vesicles is dependent on the age of the culture. *J. Bacteriol.* **175,** 6652–6658.

Tarshis, M., Salman, M., and Rottem, S. (1991). Fusion of mycoplasmas: The formation of cell hybrids. *FEMS Microbiol. Lett.* **82,** 67–72.

Tarshis, M., Salman, M., and Rottem, S. (1993). Cholesterol is required for the fusion of single unilamellar vesicles with *Mycoplasma capricolum*. *Biophys. J.* **64,** 709–715.

C5

MYCOPLASMA MEMBRANE POTENTIALS

Ulrich Schummer and Hans Gerd Schiefer

Introduction

Various kinds of electric potentials are associated with biological membranes. According to location, it is possible to distinguish between transmembrane and membrane surface potentials.

Transmembrane Electric Potential

The transmembrane electric potential is due to an electric field penetrating the membrane. It is defined as the difference in voltage measured between two identical electrodes on opposite sides of the membrane. In order to determine the membrane potentials of cells that are too small to be impaled with microelectrodes, potential-sensitive fluorescent dyes (Freedman and Laris, 1981; Schiefer and Schummer, 1982; Schiefer et al., 1990; Schummer and Schiefer, 1987; Waggoner, 1985) or, even more simply (Schummer and Schiefer, 1980, 1986; Schummer et al., 1980), K^+ are used as probes. The intensity of fluorescence is inversely related to the transmembrane potential. In principle, the membrane potential is compared with a K^+ diffusion potential that is artificially induced by the addition of the K^+-specific ionophore valinomycin. External $[K^+]$ is varied until, after addition of valinomycin, no changes in fluorescence intensity or intracellular $[K^+]$ occur. At this very special $[K^+]$ (the "null point," "critical K^+ concentration") (Freedman and Laris, 1981; Schummer and Schiefer, 1980, 1986, 1987; Schummer et al., 1980; Waggoner, 1985), the membrane potential $\Delta\Psi$ equals the K^+ diffusion potential induced by valinomycin, and can be calculated according to

$$\Delta\Psi = (RT/F)\ln(K_{crit}/K_{int}),$$

where R is the molar gas constant, T is the thermodynamic temperature, F is the Faraday constant, K_{crit} is the external K^+ concentration for which no changes in fluorescence or intracellular K^+ concentration are observed after the addition of valinomycin, and K_{int} is the intracellular K^+ concentration.

Membrane Protonmotive Potential Δp

The membrane potential $\Delta\Psi$ is a parameter of the transmembrane protonmotive potential Δp (Kashket, 1985) which is composed of a transmembrane potential $\Delta\Psi$ and a transmembrane proton-gradient ΔpH according to

$$\Delta p = \Delta\Psi - (2.3RT\Delta pH)/F = \Delta\Psi - (Z\Delta pH),$$

where R is the molar gas constant, T is the thermodynamic temperature, and $Z = 2.3RT/F$ (= 59 mV at $T = 310$ K).

Membrane Surface Potential Φ

Anionic sites on membrane surfaces (Schiefer *et al.*, 1990) give rise to a negative electrostatic potential in the aqueous phase immediately adjacent to the membrane. The membrane surface potential affects the distribution of ions, including protons, at the membrane bilayer–aqueous medium interface. Accordingly, the pH at the surface differs from the bulk pH. The membrane surface potential is calculated from the difference of the bulk and interfacial pH. The pH at the membrane surface is determined by measuring the apparent pK of a fluorescent lipid pH indicator, 4-heptadecyl-7-hydroxycoumarin (HHC), which, because of its hydrophobic side chain, is anchored in the membrane, with the fluorophore being located at the membrane–aqueous medium interface (Fromherz, 1989). Both the excitation spectra and the quantum yield of the ionized and nonionized forms of the dye differ, thus allowing determination of the dissociation degree α and hence the pK of the dye. The membrane surface potential is calculated according to

$$\Phi = -(pK_{ch} - pK_o)2.3RT/F,$$

where pH_{ch} and pK_o denote the apparent dissociation constants of the indicator at a charged and at a neutral (e.g., Triton X-114) interface, respectively; R is the molar gas constant; T is the thermodynamic temperature; and F is the Faraday constant.

In this contribution, studies on *Spiroplasma floricola* membranes are described in detail, serving as an example. Analogous experiments with slightly modified conditions have been done with *Mycoplasma mycoides* subsp. *capri*, *M. gallisepticum*, and *Acholeplasma laidlawii* (Schiefer and Schummer, 1982).

Materials

3,3'-Dipropyl-2,2'-thiodicarbocyanine iodide [DiSC$_3$(5)], commercially available from Molecular Probes (Eugene, OR), and Interchim, F-03100 Montlucon, France. A 1.5 mM solution is prepared in ethanol.
HHC, a generous gift from P. Fromherz (Martinsried, Germany). A 5 mM solution is prepared in ethanol.
Valinomycin (val), from Sigma (St. Louis, MO). A 7.5 mM solution is prepared in ethanol.
5,5'-Dimethyl[2-^{14}C]oxazolidine-2,4-dione (DMO), from Amersham Buchler (Braunschweig, Germany).

Spiroplasma floricola Culture

Spiroplasma floricola is grown at 34°C with the pH kept constant at 7.5 by automatic titration, in HEPES-buffered DSM 4 medium which contains, per 100 ml, 1.5 g PPLO broth base (Difco, Detroit, MI), 8 g sucrose (Merck, Darmstadt, Germany), 2.5 mg phenol red (Merck), 1.32 g HEPES (Merck), 500 mg glucose, and 10 ml of inactivated (30 minutes, 56°C) horse serum [ICN (Flow) Costa Mesa, CA]. Shape and motility of the spiroplasmas are regularly monitored by dark-field microscopy (1250×).

Procedures

Determination of $\Delta\Psi$

Isotonic media are prepared containing 2^n mM KCl + $(2^7 - 2^n)$ mM NaCl + 5 mM HEPES, pH 7.5, with n = 7, 6, 5, ..., 0.

Spiroplasmas are grown as described earlier. After cooling in an ice bath, they are harvested by centrifugation (15 minutes, 4°C, 10,000 g), resuspended in one-tenth of the original volume of fresh DSM 4 medium, and further incubated at 25°C for 1 hour with the pH kept constant at 7.5. For each sample, 1 ml of the spiroplasma suspension is harvested immediately before fluorescence measurements by centrifugation at 8000 g for 1 minute in an Eppendorf 3200 bench centrifuge. They are homogeneously resuspended in 3 ml of the prepared medium samples to an absorbance at 509 nm of about 0.4, corresponding to a final protein concentration of about 30 μg/ml, and filled into quartz cuvettes. One microliter of [DiSC$_3$(5)] dissolved in ethanol is added (final concentration, 0.5 μM). After about 1 minute, when the signal (I_o) is constant, the fluorescence signal is measured with a Hitachi Perkin–Elmer fluorescence spectrophotometer 204 with the excitation wavelength set at 625 nm and the emission at 660 nm.

One microliter of the valinomycin solution is then added (final concentration, 2.5 μM), and the fluorescence intensity I_{val} is recorded.

The intensity of fluorescence on the addition of valinomycin (I_{val}) depends on $[K^+]_{ext}$, i.e., increasing $[K^+]_{ext}$ by successively substituting KCl for NaCl in the suspension medium leads to a smaller decrease or even increase in fluorescence intensity after the addition of valinomycin. When $\Delta I/I_o$ ($\Delta I = I_o - I_{val}$) is plotted vs $\log_2$ of the corresponding $[K^+]_{ext}$ (Fig. 1), the curve obtained has a sigmoid shape [for theory and discussion see Schummer and Schiefer (1986)]. However, within a considerable range, about the center of symmetry, it can be approximated by a straight line which crosses the abscissa at $n = 3.7$, corresponding to $[K^+]_{crit} = 13$ mM. At this point, $[K^+]$ is in Donnan equilibrium.

The intracellular potassium concentration, $[K^+]_{int}$, is determined following a novel technique using measurements of $[K^+]$ and gravimetry (Schummer and Schiefer, 1980; Schummer et al., 1980).

The spiroplasmas are harvested from growth medium (300 ml DSM 4 medium) by centrifugation, resuspended in 20 ml of fresh medium to give a cell

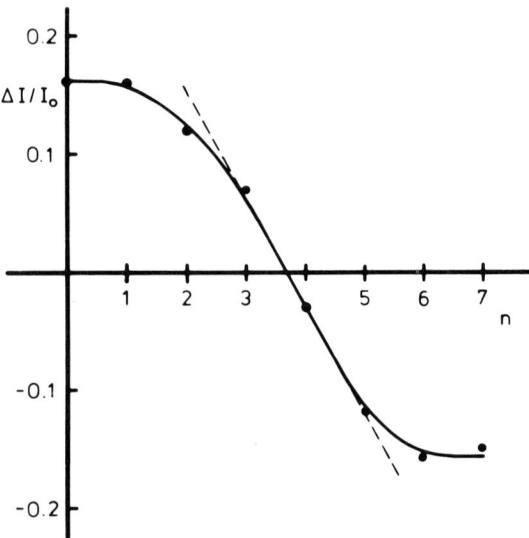

Fig. 1. Changes in fluorescence intensity $\Delta I/I_o$ {$\Delta I = I_o - I_{val}$} of 3,3'-dipropyl-2,2'-thiodicarbocyanine iodide in a suspension of *Spiroplasma floricola* plotted vs $\log_2$ of the corresponding external potassium concentrations. The intersection point of the curve with the abscissa at $n=3.7$ indicates $[K^+]_{crit}=13$ mM. (From Schiefer et al., 1990, and Schummer and Schiefer, 1987, with permission.)

concentration of about 20 mg wet wt/ml, and further incubated at 37°C for 1 hour. The pH is kept constant at 7.5 by the addition of 0.1 M NaOH.

Stock solutions I (150 mM KCl + 5 mM Tris–HCl) and II (150 mM NaCl + 5 mM Tris–HCl) are prepared and adjusted to pH 7.5. Incubation media are prepared from stock solutions I and II as follows: $n/10$ ml of stock solution I are added to $(1 - n/10)$ ml of stock solution II ($n = 0,1,2,\ldots,10$). The composition of the media varies according to $15n$ mM KCl + $15(10 - n)$ mM NaCl + 5 mM Tris–HCl.

For each sample, 1 ml of the cell suspension is centrifuged, and the pellet is resuspended in the respective incubation medium for 1 minute. After centrifugation, the supernatant is removed carefully using a microsyringe and the wet weight of the microorganisms is determined. The cells are then deep-frozen in liquid nitrogen and dried *in vacuo* overnight. After the dry weight is measured, the cells are disintegrated by ultrasonic irradiation in 1 ml of 0.1 M HCl.

The [K$^+$] of the probe is

$$K(K_{out}) = [V_{ext}/(V_L + V_{tot})]K_{out} + [V_{int}/(V_L + V_{tot})]K_{int}, \quad (1)$$

where V_L is the volume of the disintegration solution; V_{ext} and V_{int} are the extracellular and intracellular water volumes, respectively; V_{tot} is the total pellet water volume; K_{int} is the intracellular potassium concentration of native cells; and K_{out} is the potassium concentration of the incubation medium.

Since $V_L \gg V_{tot}$ (≤ 10 μl), Eq. (1) reduces to

$$K(K_{out}) = [V_{ext}/V_L)K_{out} + (V_{int}/V_L)K_{int}. \quad (2)$$

A plot of the measured potassium concentration $K(K_{out})$ vs K_{out} shows a linear relationship between $K(K_{out})$ and K_{out} (Fig. 2). The slope of the resulting straight line is $s = V_{ext}/V_L$. Thus, the external water volume, V_{ext}, can be calculated from the slope of the curve according to $V_{ext} = sV_L$.

The intracellular water volume, V_{int}, is determined as follows: The total mass (m_{ww}, wet weight) and, after drying the pellet in vacuo, the dry mass (m_{dw}, dry weight) are measured. The terms are related by:

$$m_{ww} = V_{ext}\rho_{ext} + V_{int}\rho_{int} + m_{dw}, \quad (3)$$

where ρ_{ext} and ρ_{int} denote the density of the extra- and intracellular fluid, respectively. The densities of the extra- and intracellular fluid can be replaced by the density of water, and Eq. (3) reduces to

$$m_{ww} = (V_{ext} + V_{int})\rho + m_{dw} \quad (4)$$

and we obtain

$$V_{int} = (m_{ww} - m_{dw})/\rho - V_{ext}. \quad (5)$$

Since all terms on the right-hand side of Eq. (5) are known, the intracellular water volume, V_{int}, can be calculated.

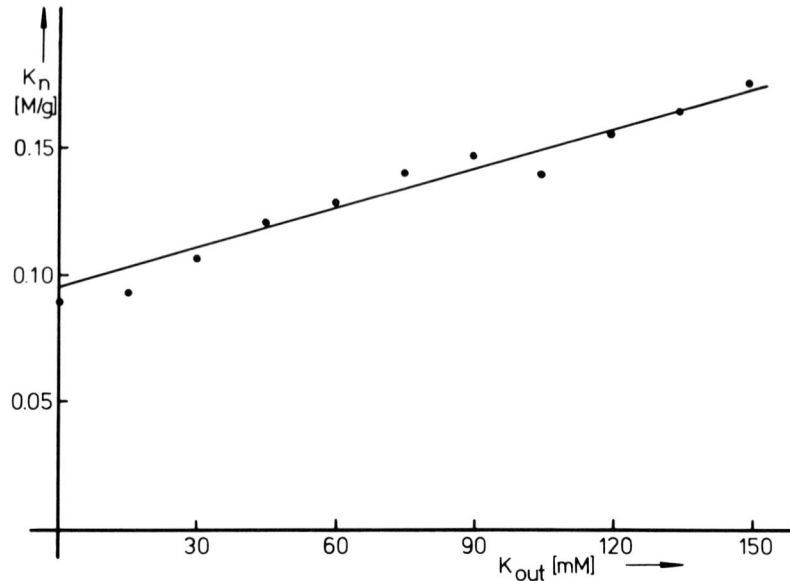

Fig. 2. Plot of the normalized potassium concentration of *S. floricola* vs the potassium concentration of the incubation media. The organisms are incubated for 15 minutes at 20°C. From the intersection point of the curve and the ordinate, $K_N(0)$ is determined. V_{ext} is calculated from the slope of the curve.

Finally, the intracellular potassium concentration, K_{int}, can be calculated from the intersection point of the curve with the ordinate (i.e., $K_{out} = 0$). Equation (2) then gives

$$K(0) = (V_{int}/V_L) K_{int}, \text{ and } K_{int} = (V_L/V_{int})K(0). \tag{6}$$

Determination of Transmembrane Proton Gradient ΔpH

The transmembrane proton gradient ΔpH is measured by the distribution of the weak organic acid, DMO, between the intra- and extracellular water space.

Spiroplasmas are harvested by centrifugation and are suspended in one-tenth of the original volume in isotonic HEPES-buffered sucrose (270 mM) solution. Initially, no nutrient substrates are added. The pH is kept constant at 8.0 by automatic titration. [^{14}C]DMO (final concentration, 0.1 μM) dissolved in saline is added. After taking a first sample, glucose (final concentration, 50 mg/ml) is added and serves as the energy source. The extracellular pH is monitored with a glass electrode and is allowed to drop due to glycolysis of the sugar. Samples are taken at pH 8, 7.5, 7, ..., 5 and are harvested by centrifugation. The extra- and intracellular water volumes of the pellets are determined as described earlier. The radioactivity in the cell pellets and of 100 μl of the supernatants is measured by scintillation spectrophotometry. ΔpH is calculated according to

$$\Delta pH = pH_{int} - pH_{ext} = pK_{DMO} - pH_{ext} + \ln\{[C_{int}/C_{ext}][10^{(pH_{ext}-pK_{DMO})}+1]-1\},$$

where pH_{int} and pH_{ext} are the intra- and extracellular pH; pK_{DMO} is the negative logarithm of the apparent dissociation constant of DMO, approximately 6.2 at 37°C; and C_{int} and C_{ext} are the intra- and extracellular concentrations of DMO.

Starving spiroplasmas produce only minimal amounts of protons and are unable to maintain a transmembrane proton gradient, i.e., ΔpH approximates zero. A few minutes after the addition of glucose the spiroplasmas start to eject protons into the suspension medium. The intracellular pH of metabolizing spiroplasmas remains more alkaline than the extracellular medium (Fig. 3). For example, at an extracellular pH of 7, the intracellular pH is 7.54. When the external pH decreases from 8 to 5 by the metabolic activity of the spiroplasmas, the transmembrane proton gradient ΔpH increases from near 0.5 to about 1.8, whereas the intracellular pH decreases from 8.53 to about 7 (at pH_{ext} of 6) and then remains constant (Fig. 3).

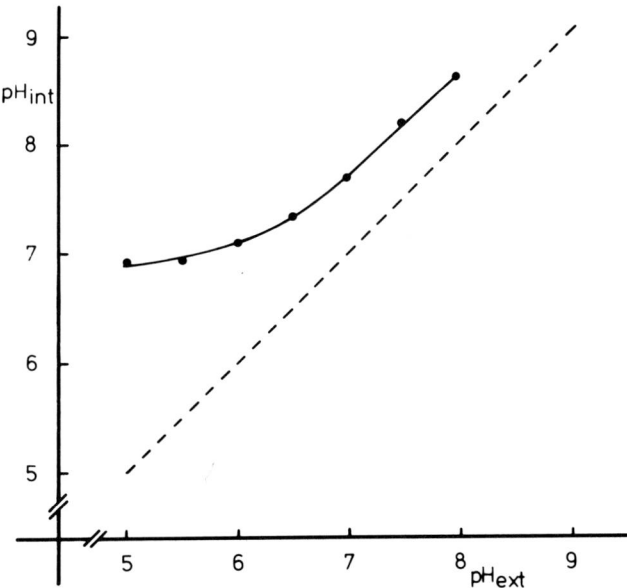

Fig. 3. The intracellular pH of glycolyzing *S. floricola* plotted vs the extracellular pH. The dotted line indicates the "iso-pH curve," i.e., $pH_{int}=pH_{ext}$. (From Schiefer et al., 1990, and Schummer and Schiefer, 1987, with permission.)

Calculation of Transmembrane Protonmotive Potential Δp

Δp is calculated according to $\Delta p = \Delta \Psi - (Z\Delta pH)$. As shown in Fig. 4, $\Delta\Psi$ strongly depends on pH_{ext}, and decreases from -55 mV (at pH_{ext} of 5.5) to -100 mV (at pH_{ext} of 8.5). ΔpH also depends on pH_{ext}, but in the opposite direction, i.e., ΔpH increases from -84 mV (at pH_{ext} of 5.5) to -31 mV (at pH_{ext} of 8.5). Δp is calculated to be $\Delta p = -131$ mV (at pH_{ext} of 8.5), $\Delta p = -118$ mV (at pH_{ext} of 7.5), $\Delta p = -103$ mV (at pH_{ext} of 6.5), and $\Delta p = -139$ mV (at pH_{ext} of 5.5). Δp is kept constant at $\Delta p = -123$ mV ± 16% over a wide range of pH_{ext} values. Its magnitude is distinctly higher in *S. floricola* than in other microbial species (Kashket, 1985; Schiefer and Schummer, 1982; Schiefer *et al.*, 1990; Schummer and Schiefer, 1987).

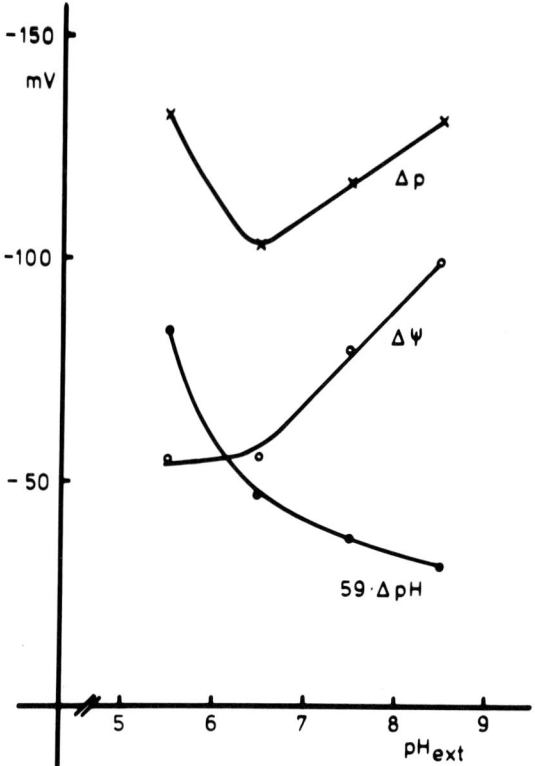

Fig. 4. Effect of pH of the medium on the transmembrane protonmotive potential Δp and its parameters ΔΨ and ΔpH. (From Schiefer *et al.*, 1990, and Schummer and Schiefer, 1987, with permission.)

Determination of Membrane Surface Potential

Spiroplasmas are grown and harvested as described earlier, and the wet weight is determined. For labeling, spiroplasmas are resuspended in cold growth medium without phenol red (spiroplasma concentration: 20 mg wet wt/ml) and are incubated at 34°C for 10 minutes with the pH being kept constant at 7.5. HHC is then added to the suspension (final concentration: 1 µg HHC/mg wet wt of cells), and incubation continues for a further 15 minutes in order to allow for incorporation and distribution of HHC.

The spiroplasmas are then sedimented and resuspended (30 mg wet wt/ml) in ice-cold NaCl/KCl/HEPES medium (115 mM NaCl + 13 mM KCl and 25 mM HEPES, pH 7.5). The spiroplasmas are incubated for another 15 minutes at 34°C with the pH kept constant at 7.5, then cooled and kept at 0°C. NaCl/KCl/HEPES solutions adjusted to pH 7.5, 8, 8.5, ..., 13 are prepared. Measurements at pH 14 are carried out in 0.1 M NaOH. In experiments with 50 mM CaCl$_2$, the NaCl content is reduced to 50 mM; in those with 100 mM CaCl$_2$, no NaCl is present.

Quartz cuvettes are filled with 3 ml of the pH-adjusted solutions, and 100-µl spiroplasma suspension is added immediately before the excitation spectrum is recorded over the 250- to 420-nm range with a Hitachi Perkin-Elmer 204 fluorescence spectrophotometer with the analyzer wavelength set at 450 nm. For peak analysis the excitation wavelength is set at 370 nm.

For calibration, HHC (final concentrations, 7.5 and 0.75 µM) is incorporated into a neutral micellar solution of Triton X-114 (final concentration, 5 mM) in 5 mM Tris buffer. The pH is adjusted to the desired values starting from pH 12.0 through 8.0, with 0.5-unit intervals. Spectra are recorded as described earlier.

The membrane surface potential is calculated according to the formula derived by Fromherz (1989)

$$\Phi = -(pK_{ch} - pK_o)2.3RT/F,$$

where pK_{ch} and pK_o denote the apparent dissociation constants of HHC at a charged and a neutral (Triton X-114) interface, respectively; R is the gas constant; T is the thermodynamic temperature; and F is the Faraday constant. pK_{ch} and pK_o can be obtained by measuring the dependence of HHC's degree of dissociation (α) on the pH of the aqueous bulk medium (pH$_b$), and calculated according to

$$pK = pH_b - \log\{\alpha/(1-\alpha)\}.$$

The degree of dissociation α is calculated from the fluorescence intensity (I) at the excitation wavelength of 370 nm, according to

$$\alpha = I/I_{max},$$

where I_{max} denotes the maximally obtainable fluorescence intensity at 370 nm, when the dye is completely dissociated.

A rapid and complete incorporation of HHC into the spiroplasma membranes is observed. A noxious effect on spiroplasma metabolism and motility is not detected. The experimental results are independent of the HHC concentration used. The membrane-bound HHC rapidly responds to the varied pH in the surrounding media. However, in Ca^{2+}−free media at bulk pH > 11.5, a rapid increase in fluorescence intensity with time is observed due to lysis of the spiroplasmas, as observed under dark-field microscopy. Therefore, for determination of α, the fluorescence intensity at $\lambda_{ex} = 370$ nm is recorded rather than the whole excitation spectrum. In media containing Ca^{2+}, the spiroplasmas are considerably more stable, allowing the whole spectrum to be recorded. A set of excitation spectra recorded from spiroplasmas suspended in media containing 50 mM $CaCl_2$ is shown in Fig. 5.

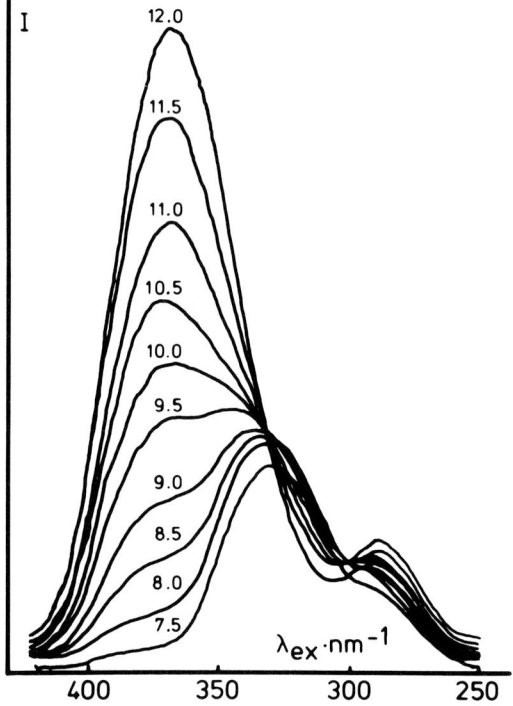

Fig. 5. Fluorescence excitation spectra of the lipoid pH indicator, 4-heptadecyl-7-hydroxy-coumarin, incorporated into *S. floricola* membranes at different pH values (as indicated on curves) of the aqueous suspension medium containing 50 mM $CaCl_2$. Spectra are recorded between 250 and 420 nm with the analyzer wavelength set at 450 nm. (From Schiefer et al., 1990, and Schummer and Schiefer, 1988, with permission.)

When measuring the fluorescence intensity at $\lambda_{em} = 450$ nm, maximal intensity for the undissociated form of HHC is obtained at $\lambda_{ex} = 330$ nm, and for the dissociated form at $\lambda_{ex} = 370$ nm. The excitation spectra overlap in a minimal and negligible way, i.e., when the HHC is in the undissociated form, then $I_{370\ nm} \approx 0$. When the fluorescence intensity of the completely dissociated form at $\lambda_{ex} = 370$ nm is I_{max}, then the degree of dissociation α corresponds to $\alpha = I_{370\ nm}/I_{max}$. For $\alpha = 0.5$, $pK = pH_b$. For calculation of the membrane surface potential, the pH of the suspension medium leading to $\alpha = 0.5$ has to be determined.

Determination of I_{max}

This parameter is obtained when the HHC is completely dissociated, i.e., when the fluorescence intensity does not increase on increasing the pH of the suspension medium. With neutral (Triton X-114) micelles, this point is reached at pH ≥ 11, and for the determination of α, I_{max} is taken as $I_{370\ nm}$ at pH 12. This simple procedure fails with *S. floricola*. When $I_{370\ nm}$ is plotted vs pH of the suspension medium, one can see that I_{max} is not reached even at pH 14 (Fig. 6).

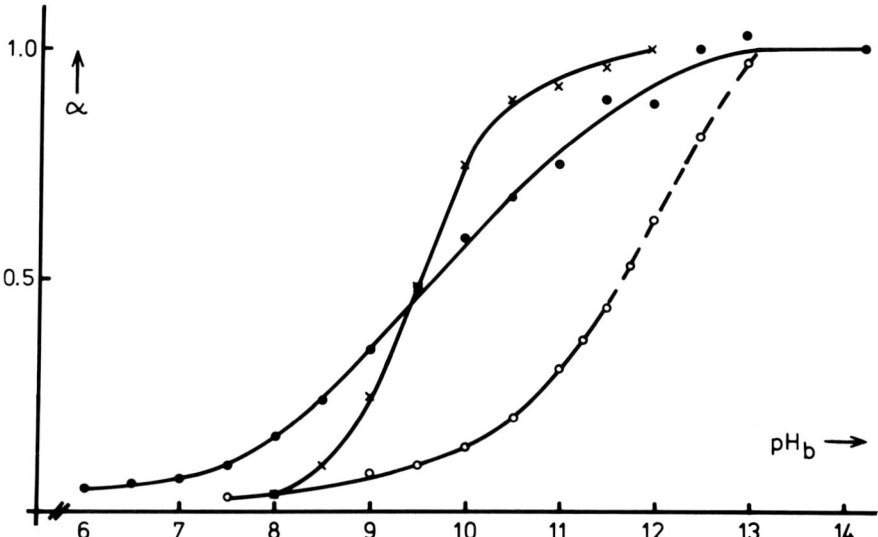

Fig. 6. Dissociation degree (α) of the lipoid pH indicator, 4-heptadecyl-7-hydroxycoumarin, incorporated into *S. floricola* membranes suspended in 25 mM HEPES + 115 mM NaCl + 13 mM KCl (○), or 25 mM HEPES + 100 mM CaCl$_2$ + 13 mM KCl (●), respectively, and neutral (Triton X-114) micelles in 25 mM HEPES + 115 mM NaCl + 13 mM KCl (×) vs pH of the aqueous suspension medium. (From Schiefer et al., 1990, and Schummer and Schiefer, 1988, with permission.)

It proves necessary to try a different approach based on the following considerations and observations. For a given preparation of *S. floricola,* the proportion of the fluorescence observed at λ_{em} of 450 nm of the dissociated and undissociated forms, i.e.,

$$\beta = I_{max}(\text{dissociated}, \lambda_{ex} = 370 \text{ nm})/I_{max}(\text{undissociated}, \lambda_{ex} = 330 \text{ nm}),$$

is constant. If one succeeds, e.g., by the addition of cations, in reducing the negative surface charges, at last partially, complete dissociation of HHC should be achieved at lower pH. Then, after determination of β,

$$I_{max} = I_{max}(\text{dissociated}, \lambda_{ex} = 370 \text{ nm})$$
$$= \beta I_{max}(\text{undissociated}, \lambda_{ex} = 330 \text{ nm}).$$

The negative membrane surface charges can readily be neutralized by substituting NaCl in the suspension media by 100 mM CaCl$_2$. Then I = constant = I_{max} is obtained at pH $\geq$ 12 of the suspension media. For different preparations of spiroplasmas, β is found to vary only between 2.8 and 3.1.

Thus I_{max} can be determined for Ca^{2+}-free suspension media, and the degree of dissociation α is calculated. For *S. floricola* in NaCl/KCl/HEPES, α = 0.5 at pH 11.6 is obtained, i.e., pK_{ch} = 11.6. For neutral (Triton X-114) micelles, pK_o = 9.6. Hence, the membrane surface potential of *S. floricola* is calculated to be Φ = $-(11.6-9.6)2.3RT/F$ = $-2\cdot 59$ mV = -118 mV. The measurements in suspension media containing Ca^{2+} give pK_{ch} = 10.1 (at 50 mM) and 9.6 (at 100 mM), with the corresponding membrane surface potential being Φ = -30 and 0 mV, respectively.

References

Freedman, J. C., and Laris, P. C. (1981). Electrophysiology of cells and organelles: Studies with optical potentiometric indicators. *In* "Membrane Research, Classic Origins and Current Concepts." International Review of Cytology, Suppl. 12, 177–246.

Fromherz, P. (1989). Lipid coumarin dye as a probe of interfacial electrical potential in biomembranes. *In* "Methods in Enzymology" (S. Fleischer and B. Fleischer, eds.), Vol. 171, pp. 376–387. Academic Press, New York.

Kashket, E. R. (1985). The proton motive force in bacteria: A critical assessment of methods. *Annu. Rev. Microbiol.* **39,** 219–242.

Schiefer, H. G., and Schummer, U. (1982). The electrochemical potential across mycoplasmal membranes. *In* "Current Topics in Mycoplasmology" (M. Barile, S. Razin, P. F. Smith, and J. G. Tully, eds.). Rev. Infect. Dis. **4,** S65–S70.

Schiefer, H. G., Schummer, U., and Krauss, H. (1990). Electric properties of spiroplasma membranes. *In* "Recent Advances in Mycoplasmology" (G. Stanek, G. H. Cassell, J. G. Tully, and R. F. Whitcomb, eds.), pp. 131–138. G. Fischer Verlag, Stuttgart/New York.

Schummer, U., and Schiefer, H. G. (1980). A novel method for the determination of electrical potentials across cellular membranes. I. Theoretical considerations and mathematical approach. *Biochim. Biophys. Acta* **600,** 993–997.

Schummer, U., and Schiefer, H. G. (1986). Ion diffusion potentials across mycoplasma membranes determined by a novel method using a carbocyanine dye. *Arch. Biochem. Biophys.* **244,** 553–562.

Schummer, U., and Schiefer, H. G. (1987). Transmembrane protonmotive potential of *Spiroplasma floricola. FEBS Lett.* **224,** 79–82.

Schummer, U., and Schiefer, H. G. (1988). Membrane surface potential of *Spiroplasma floricola. FEBS Lett.* **236,** 337–339.

Schummer, U., Schiefer, H. G., and Gerhardt, U. (1980). A novel method for the determination of electrical potentials across cellular membranes. II. Membrane potential of acholeplasmas, mycoplasmas, streptococci and erythrocytes. *Biochim. Biophys. Acta* **600,** 998–1006.

Waggoner, A. S. (1985). Dye probes of cell, organelle, and vesicle membrane potentials. *In* "The Enzymes of Biological Membranes" (A. N. Martonosi, ed.), 2nd Ed., Vol. 1, pp. 313–331. Plenum, New York/London.

C6

ION FLOW AND CELL VOLUME
Shlomo Rottem

Introduction

Membrane-bound transport processes regulate the ionic environment within cells. These processes control the intracellular concentration of ions by moving ions across the cell membrane and, by doing so, generate electrochemical gradients. These gradients can be used for many essential functions, such as cell volume regulation, regulation of intracellular pH, and uptake of nutrients.

Mollicutes are microorganisms that do not have the peptidoglycan-based rigid cell wall. These organisms are bound by a single membrane, the plasma membrane, and are faced with the problem of regulating their cell volume. Electrolytes, mainly Na^+, Cl^-, and water, diffuse into the cell because of the colloid osmotic effect (Wilson, 1954). This influx must be counteracted by the extrusion of ions (and water) by ion pumps. Mollicutes are the most prevalent parasites of man, animals, plants, and insects. Having various natural habitats, these organisms must have developed primary and secondary ion transport mechanisms in adaptation to different environments.

A primary ion transport mechanism is a process whereby the energy for transport comes directly from the potential energy of cellular metabolites. The transport of ions across a sealed cell membrane, without a compensatory movement of ions, results in the generation of a chemical potential for that ion. For example, an H^+-ATPase would generate an electrochemical potential of H^+, the protonmotive force ($\Delta\mu_{H_+}$), consisting of a chemical gradient of H^+(ΔpH) and an electrical gradient (membrane potential, $\Delta\psi$). Primary transport systems in mollicutes are best exemplified by the membrane-bound ATPase which uses the chemical energy from the hydrolysis of the terminal phosphate bond of ATP to drive a transport process.

The second type of basic process by which cations can be transported across a cell membrane is referred to as secondary transport. In these processes, the electrochemical potential of an ion is coupled to the movement of another molecule across the membrane. The most prevalent forms of secondary transport are those linked to $\Delta\mu_{H+}$, e.g., the Na^+/H^+ antiporter. The Na^+/H^+-antiporter actively drives Na^+ across the cell membrane, against its electrochemical potential. This is accomplished by coupling the movement of Na^+, in the opposite direction to that of H^+ moving down its electrochemical potential, $\Delta\mu_{H_+}$. Through this mechanism, $\Delta\mu_H{}^+$ actively drives the exchange of Na^+ for H^+. Most bacteria studied have this antiporter. In mycoplasmas, a Na^+/H^+-antiporter has been reported in *Mycoplasma mycoides* var. *capri*, in *Acholeplasma laidlawii,* and in *Spiroplasma floricola* (BNR1).

Materials

Cell and Membrane Preparations

Cell suspensions in an isotonic buffer (about 2 mg cell protein/ml) or isolated membrane preparation (1–4 mg membrane protein/ml)

Chemicals

Buffers: 3-[*N*-morpholino]Propanesulfonic acid (MOPS)
Lipids: Crude soybean phospholipids (asolectin); cholesterol
Energy source: D-Glucose; L-arginine
Inhibitors: Acridine orange; nigericin; carbonyl cyanide *m*-chlorophenylhydrazone (CCCP); SF 6847
Silicone oil: Either XF 1792B from Dexter Hysol (Olean, NY) or a 1:1 (v/v) mixture of grades 550 and 556 oil from Dow Corning Corp. (Midland, MI)
Radiochemicals: ^3H_2O, [^{14}C]polyethylene glycol-4000 (PEG); $^{22}NaCl$; ^{42}KCl; and [^{14}C]palmitic acid

Special Equipment

Spectrofluorimeter with excitation at 492 nm and emission at 530 nm
Bath sonicator

Filtration apparatus equipped with 25-mm GF/C fiberglass filters (Whatman, Inc.)
Microfuge centrifuge

Procedure

Measurement of Cell Volume

1. One milliliter of the cell suspension is added to 1.2 ml of an isotonic buffer, e.g., 25 mM Tris–HCl in 225 mM NaCl, adjusted to pH 7.5, containing an energy source (5–10 mM of either glucose or arginine), ^{3}H-labeled water (4–6 μCi/ml), and [^{14}C]PEG (0.1–0.5 μCi/ml; 100 μg/ml).

2. After incubation at 37°C for 15 minutes, 1-ml samples in duplicate are pipetted onto the surface of silicone oil (0.5 ml) in 1.5-ml plastic microfuge tubes and centrifuged at room temperature at 12,800 g for 1 minute. Under these conditions, the cells pass through the silicone oil and form a pellet at the bottom of the tube. The aqueous phase remains above the oil.

3. Samples of the supernatant fluid are taken for radioactivity analysis. Both the aqueous phase and oil are removed by suction, and the tip of the plastic centrifuge tube containing the cell pellet is cut off with a razor blade and placed in a vial containing 10 ml of scintillation fluid. The cell pellet is freed from the tip of the centrifuge tube by vigorous mixing and is taken for radioactivity analysis. ^{3}H-labeled water is used to measure total pellet water. It is assumed that PEG (molecular weight 4000) does not penetrate the cells and could be used to measure extracellular fluid in the cell pellet. The PEG space is approximately 20% of the total pellet water. The water space minus the PEG space is taken as the intracellular water space.

Preparation of Sealed Membrane Vesicles

1. Sealed mycoplasma membrane vesicles are obtained by fusing mycoplasma membranes with liposomes (Cirillo *et al.*, 1987). The liposomes are prepared by intensively mixing, on a vortex mixer, 50 mg asolectin and the desired amount of cholesterol (i.e., 20 mg cholesterol for 40 mol%, dried from chloroform solutions) in 1 ml of 10 mM potassium phosphate buffer, pH 7.4. The milky solution is clarified by sonication in a bath-type sonicator. The resultant lipid vesicle preparation is mixed with a mycoplasma membrane suspension at the desired ratio of asolectin to membrane protein. Fusion is initiated by freezing the mixture in liquid nitrogen. The frozen mixture is thawed at room

temperature and sonicated for 30 seconds. Any additions to be trapped in the fused preparations are added before the freezing step.

Na^+/H^+- or K^+/H^+- Antiporter Activity

Na^+/H^+ or K^+/H^+ exchange activity is tested by imposing a ΔpH (interior acid) across the cell membrane of intact cells or sealed membrane vesicles and by determining the effect of Na^+ and/or K^+ on the ΔpH gradient generated. A ΔpH is generated by the ammonium chloride dilution procedure described previously (Shirvan et al., 1989a).

1. Cells or sealed membrane vesicles are loaded with ammonium ions and diluted into ammonium-free medium. NH_4^+-loaded cells or vesicles are obtained by incubating washed cells in 0.25 M NH_4Cl containing 5 mM Tris–HCl buffer (pH 8.0) and 2.5 mM $MgCl_2$ for 2 hours at 37°C. Sealed membrane vesicles are loaded by including ammonium chloride (0.25 M) in the reaction mixture during the freeze-thaw/sonication cycle.

2. The ΔpH is generated by a 100-fold dilution of the NH_4^+-loaded cells (160 μg of protein per ml) in 2.5 ml of a solution containing 0.25 M choline chloride, 10 mM Tris–HCl (pH 8.0), 2.5 mM $MgCl_2$, and 5 μM acridine orange.

3. The ΔpH is visualized from the quenching of the acridine orange signal. Na^+/H^+- and K^+/H^+-antiporter activity is tested by measuring, at the required temperature, fluorescence levels following the addition of 25 mM NaCl or KCl to the reaction mixture.

$^{22}Na^+$ or $^{42}K^+$ Efflux

1. For efflux measurements, cells are preloaded with $^{22}Na^+$ or $^{42}K^+$ by incubating washed cell suspensions (6 mg of cell protein/ml) for up to 2 hours at the appropriate temperature in 0.45 M sucrose solution in 25 mM Tris–MOPS buffer (pH 7.5) containing 3.0 μCi/ml of either $^{22}Na^+$ or $^{42}K^+$.

2. The loaded cells are diluted 200-fold into a reaction mixture containing either 0.45 M sucrose or 0.225 M NaCl or KCl in 25 mM Tris–MOPS buffer adjusted to the required pH, in the presence or absence of 10 mM glucose or arginine.

3. The reaction mixtures are incubated at the appropriate temperature and, at various time intervals, 10-ml portions are withdrawn and filtered through fiberglass filters (25 nm GF/C, Whatman) under negative pressure (filtration time, 3–10 seconds). The filters are washed twice with a 10-ml volume of cold 0.25 M NaCl, and the radioactivity retained on the filters is determined. The amount of the cells retained on the filters can be determined by measuring the retention of [^{14}C]palmitate-labeled cells.

Primary Na+ Pumping Activity

Primary sodium pumping activity is assayed by monitoring the membrane potential-driven uptake of protons by cells in the presence of an uncoupler. The formation and maintenance of ΔpH due to the uptake of protons are estimated by determining the fluorescence quenching of the dye acridine orange with a spectrofluorimeter with excitation of 492 nm and emission at 530 nm. Cells (final concentration, 30 µg of protein/ml) suspended in 2.5 ml of reaction medium [consisting of 225 mM NaCl, 2.5 mM MgCl$_2$, 10 mM Tris–HCl (pH 8.5), and 5 µM acridine orange] are made permeable to protons by the addition of 0.4 µM of the uncoupler SF 6847. Fluorescence quenching is initiated by the addition of an energy source (10 mM glucose or arginine).

Discussion

The absence of a protective cell wall renders mycoplasmas much more fragile than ordinary bacteria, thus care must be taken not to damage the cells during harvesting and washing. Most mycoplasmas tend to be leaky and lose cofactors upon extensive washings, even in isotonic media. It is suggested, therefore, that the cells be washed only once and resuspended to 2% of the volume of growth medium in cold isotonic buffer containing 10 mM MgCl$_2$.

Table I shows the cell volume of some mollicutes. A study on *M. capricolum* (Romano *et al.*, 1986) revealed that the cell water volumes were rather constant throughout the exponential phase of growth. The cell water volume of stationary phase cells was, however, markedly higher than that of exponential phase cells. This manifestation of aging may be associated with alterations in the composition and physical properties of the cell membrane. Such changes may result in altered permeability or a decrease in transport activities that may lead to cell swelling. In the late stationary phase, cell water volumes are dramatically decreased, apparently due to cell lysis.

TABLE I

CELL VOLUME OF REPRESENTATIVE MOLLICUTES

Organism	Cell volume (µl/mg protein)	Reference
Acholeplasma laidlawii	2.2–2.5	Lelong *et al.* (1989)
Mycoplasma gallisepticum	1.6–1.9	Rottem *et al.* (1981)
M. capricolum	2.7–3.0	Romano *et al.* (1986)
Spiroplasma floricola	3.5–3.8	Shirazi and Rottem (1994)

Changes in membrane lipid composition induced by either varying the cholesterol-to-phospholipid molar ratio or increasing the lipid-to-protein ratio in the membrane may result in significant changes in the cell water volume (Romano et al., 1986). The increase in the cell water volume may be due to changes in the surface area of the cell membrane or can be associated with changes in the physical state of the membrane lipids. The freedom of motion of membrane lipids is shown to affect ion (and water) influx rates, as well as ion pumps (Jinks et al., 1978).

Evidence for a Na^+/H^+-antiporter in *M. mycoides* subsp. *capri* (Benyoucef et al., 1982a,b) and in *Spiroplasma floricola* (Shirazi and Rottem, 1994) came from studies with intact cells and sealed membrane vesicles following the quenching/dequenching of a fluorescent dye and the efflux of $^{22}Na^+$. The addition of glucose to Na^+-loaded cells suspended in a Na^+-free medium results in a transient intracellular acidification along with a rapid Na^+ efflux from the cell. The presence of glucose significantly enhances both the initial rate and the extent of efflux. These results were interpreted to indicate that H^+ is extruded from the cells by a H^+-ATPase, with a concomitant efflux of Na^+ via a Na^+/H^+-antiporter. The initial acidification was proposed to be due to a greater level of H^+ entering the cells from the antiporter then being effluxed by the ATPase. As the Na^+/H^+ exchange decreased or was stopped due to depletion of intracellular Na^+, the intracellular pH became more alkaline.

In *A. laidlawii,* a H^+/cation-antiport activity was reported in intact cells (Lelong et al., 1989). A ΔpH was generated across the cell membrane (inside acid) by the ammonium chloride dilution procedure in which NH_3 diffuses out of NH_4^+-loaded cells, leaving behind H^+ resulting in the presence of high Cl^- and the generation of only a ΔpH (inside acid). The changes in the ΔpH were followed using the fluorescent dye acridine orange. The presence of NaCl or KCl alone does not change the intracellular pH (pH_i), as indicated by lack of a change in the fluorescence in the cell suspension. The presence of both ions together results, however, in intracellular alkalinization. The change in pH_i is independent on the order of addition of the ions, and LiCl could not replace NaCl and RbCl could not replace KCl. In addition, the exchange is not affected by the presence of 100 mM tetraphenylphosphonium (TPP$^+$), a condition which has been reported to collapse $\Delta\psi$, suggesting that the role of K^+ is not to affect the transmembrane potential. In contrast, generation of a ΔpH in sealed membrane vesicles, by the same procedure, does result in intracellular alkalinization by the addition of either NaCl or KCl. The addition of HCl, however, either alone or when added following or preceding the addition of NaCl, has no affect on pH_i. In intact cells, the results are consistent with a H^+/cation antiport activity. The K^+ dependence of exchange activity in intact cells, compared to the activity in sealed membrane vesicles, may suggest that the transporter was modified during the membrane isolation procedure, possibly by proteolysis.

Since the natural habitat of *M. gallisepticum* cells is a high Na^+ environment, they must extrude Na^+ from the cell to survive. However, Na^+ movement across the cell membrane of *M. gallisepticum* was not driven by $\Delta\mu_H{}^+$, nor inhibited by a collapse of $\Delta\mu_H{}^+$ (Shirvan *et al.*, 1989b). Furthermore, *M. gallisepticum* cells grow and remain viable in high salt-containing medium in the presence of an uncoupler (Shirvan *et al.*, 1989a). Therefore, it was suggested that a Na^+/H^+-antiporter is not present in *M. gallisepticum* cells and that Na^+ is extruded by a primary mechanism.

Primary ion pumps were also described in *M. mycoides* subsp. *capri*. Cells that accumulate K^+ from the extracellular medium by a process which consumes ATP were not affected by ionophores (Benyoucef *et al.*, 1982a). In the absence of extracellular Na^+, the level of K^+ accumulation is linearly related to the amplitude of the transmembrane electrical potential across the membrane ($\Delta\Psi$). In contrast, in the presence of extracellular Na^+, K^+ accumulation in the cell exceeds the level that would be expected if $\Delta\Psi$ alone was the driving force. The potassiummotive force ($\Delta\mu_K{}^+$) in fully energized cells suspended in a Na^+-containing medium is approximately 120 mV (inside negative), whereas Δy, under these conditions, is approximately 90 mV. This Na^+ stimulation of K^+ uptake was shown to be dependent on $\Delta\Psi$ and the Na^+ concentration. The apparent K_m for Na^+ is approximately 1 mM and the maximal effect is observed at 10 mM Na^+. The presence of 10 mM Na^+ in the external medium leads to about a threefold increase in intracellular K^+ accumulation.

The kinetics of Na^+ efflux, from $^{22}Na^+$-loaded *M. mycoides* subsp. *capri* cells, significantly differed depending on whether cells were suspended in a Na^+-rich or Na^+-free medium (Benyoucef *et al.*, 1982b). Efflux into a Na^+-rich medium was dependent on the presence of an energy source. In addition, Na^+ efflux against a Na^+ concentration gradient requires the additional presence of K^+. This dependency was specific for K^+. Collapsing the $\Delta\mu_H{}^+$ did not affect the K^+-dependent Na^+ efflux, indicating that the requirement for K^+ is not the result of a possible coupling between a Na^+/H^+-antiporter and a K^+/H^+-antiporter. The results are consistent with a primary Na^+ pump directly associated with K^+. From the similarities in Na^+ and K^+ transport, it was proposed that both active ion fluxes were executed by the same enzyme, a $(Na^+ + K^+)$-dependent ATPase of the type found in higher organisms.

References

Benyoucef, M., Rigaud, J.-L, and Leblanc, G. (1982a). Cation transport mechanisms in *Mycoplasma mycoides* var. *capri* cells. *Biochem. J.* **208,** 529–538.
Benyoucef, M., Rigaud, J.-L., and Leblanc, G. (1982b). Cation transport mechanisms in *My-*

coplasma mycoides var. *capri* cells: The nature of the link between K^+ and Na^+ transport. *Biochem. J.* **208,** 539–547.

Cirillo, V. P., Katzenell, A., and Rottem, S. (1987). Sealed vesicles prepared by fusing *Mycoplasma gallisepticum* membranes and preformed vesicles. *Isr. J. Med. Sci.* **23,** 380–383.

Jinks, D. C., Silvius, J. R., and McElhaney, R. N. (1978). Physiological role and membrane lipid modulation of the membrane-bound ($Mg^{2+} \cdot Na^+$)-adenosine triphosphatase activity in *Acholeplasma laidlawii. J. Bacteriol.* **136,** 1027–1036.

Lelong, I., Shirvan, M. H., and Rottem, S. (1989). A cation/proton antiport activity in *Acholeplasma laidlawii. FEMS Lett.* **59,** 71–76.

Romano, N., Shirvan, M. H., and Rottem, S. (1986). Changes in membrane lipid composition of *Mycoplasma capricolum* affect the cell volume. *J. Bacteriol.* **167,** 1089–1091.

Rottem, S., Linker, C., and Wilson, T. H. (1981). A proton motive force across the membrane of *Mycoplasma gallisepticum* and its possible role in cell volume regulation. *J. Bacteriol.* **145,** 1299–1304.

Shirazi, I., and Rottem S. (1994). Volume regulation in *Spiroplasma floricola:* Evidence for a sodium/proton antiporter. *Microbiology* **140,** 1899–1907.

Shirvan, M. H., Schuldiner, S., and Rottem, S. (1989a). Role of Na^+ cycle in cell volume regulation of *Mycoplasma gallisepticum. J. Bacteriol.* **171,** 4410–4416.

Shirvan, M. H., Schuldiner, S., and Rottem, S. (1989b). Volume regulation in *Mycoplasma gallisepticum:* Evidence that Na^+ is extruded via a primary Na^+ pump. *J. Bacteriol.* **171,** 4417–4424.

Wilson, T. H. (1954). Ionic permeability and osmotic swelling. *Science* **120,** 104–105.

SECTION D
Cell Metabolism

D1

INTRODUCTORY REMARKS
Shmuel Razin

Tests designed to detect specific enzymatic activities in mollicutes, including sugar fermentation, arginine and urea hydrolysis, β-glucosidase, and proteolytic and phosphatase activities, as well as tetrazolium reduction, were described in detail in "Methods in Mycoplasmology" (Razin and Tully, 1983). These tests, which constitute a significant part of the basis for differentiation and classification of mollicutes (Chapter E2, this volume), have not been modified so their previous descriptions are still valid.

Section D in the present volume has a much wider scope, dealing with methodology applied in studies of various aspects of mycoplasma metabolism. The number of enzymatic activities recognized in Mollicutes since the early 1980s has grown considerably, covering the multitude of enzymes of the glycolytic and pentose phosphate pathways, as well as enzymes responsible for the synthesis and interconversions of amino acids, lipids, and nucleic acid precursors (Miles, 1992; Pollack et al., 1995).

Chapter D2 describes in detail the assay of three enzymatic activities, each serving as an example for one of the three major approaches utilized in testing enzymatic activities in mollicutes: the spectrophotometric, radioisotopic, and the chromatographic approach. Microcalorimetric and electroanalytical measurements of mycoplasmal metabolic activities are described in Chapter D3. These are rapid and very sensitive methods, and they require small quantities of organisms, an important advantage considering the generally poor yields of mycoplasma cultures and the fragility of washed organisms. Resembling other bacteria, mycoplasmas have been found to exhibit a stress response to a heat shock. The response is characterized by a significant increase in the expression of a subset of proteins; the heat shock proteins. Methods adapted to the study of the mycoplasmal heat shock response are detailed in Chapter D4.

Enzymes degrading polymeric substrates, such as proteins, nucleic acids, and lipids, play an important role in mycoplasma metabolism. By degrading exogenous macromolecules, they provide a supply of low-molecular-weight metabolites, such as amino acids, nucleic acid precursors, and fatty acids, in a transportable and assimilable form. On the other hand, these mycoplasmal enzymes may cause tissue damage, degrading host nucleic acids, cell proteins, immunoglobulins, and membrane phospholipids. The methods of testing nucleolytic, proteolytic, and lipolytic activities, as adapted to mollicutes, are described in detail and are evaluated in Chapters D5, D6, and D7.

References

Miles, R. J. (1992). Catabolism in mollicutes. *J. Gen. Microbiol.* **138,** 1773–1783.
Pollack, J. D., Williams, M. V., and McElhaney, R. N. (1995). The comparative metabolism of the Mollicutes. Submitted for publication.
Razin, S., and Tully, J. D., eds. (1983). "Methods in Mycoplasmology," Vol. I, pp. 335–396. Academic Press, New York.

D2
METHODS FOR TESTING METABOLIC ACTIVITIES IN MOLLICUTES
J. Dennis Pollack

Introduction

There are about 150–175 cytoplasmic enzymatic activities recognized in Mollicutes (Pollack *et al.*, 1995a,b). These activities are associated with the Embden–Meyerhof–Parnas (glycolysis) and pentose phosphate pathways and routes for the synthesis or interconversion of aromatic amino acids, lipids, purines, pyrimidines, and nucleic acids (Pollack *et al.*, 1995a,b). The mollicutes apparently lack cytochrome pigments, their respiration is characterized as flavin terminated, and they do not have a functional tricarboxylic acid (TCA) cycle (Pollack *et al.*, 1995a).

The metabolism of the mollicutes is unusual in some other respects. In addition to the just-mentioned characteristics and the well-recognized growth requirement for cholesterol in a few genera, some mollicutes require pyrophosphate (PP_i) rather than ATP for phosphorylations. Acholeplasmas and anaeroplasmas require PP_i in the phosphofructokinase reaction; the rate-limiting reaction of glycolysis (Pollack and Williams, 1986), and in acholeplasmas, PP_i is required for the nucleoside kinase reaction in the production of mononucleotides (Tryon and Pollack, 1984; Pollack, 1994). This latter reaction is not known in any nonmollicute cell. Further, all cells in nature are reported to have deoxyuridine triphosphate nucleotidohydrolase activity except for species of *Ureaplasma, Mesoplasma, Mycoplasma, Entomoplasma,* and *Asteroleplasma* (Williams and Pollack, 1984, 1990).

The assessment of individual metabolic activities or enzymes has been generally made by standard biochemical techniques, including spectrophotometric, radioisotopic, and chromatographic procedures. However, there are some exceptions using other approaches, such as ^{31}P nuclear magnetic resonance techniques to study the metabolism of growing *M. gallisepticum* cells (Egan *et al.*, 1986).

The three activities described in this chapter are PP_i-dependent phosphofructokinase (EC 2.7.1.90) (PP_i-PFK), dUTPase (EC 3.6.1.23, dUTP pyrophosphatase), and ATP- or pyrophosphate-dependent deoxyguanosine kinase (EC 2.7.1.113) (ATP/PP_i dGUOk). These particular enzymes were chosen because the analysis of each one emphasizes a different major biochemical technique frequently employed in the study of mollicute metabolism. Also, these particular assays are consequential in a metabolic classification of the mollicutes that parallels a phylogeny based on 16S rRNA sequences (Tully *et al.*, 1993; Pollack *et al.*, 1995a). Table I indicates in part the comparative usefulness of these assays and includes urease; it is adapted from another study (Pollack *et al.*, 1995a).

General Preparation of Cell-Free Extract

The careful attention to the preparation of cell-free extracts of mollicutes for enzymatic assay is of utmost importance, and in almost all cases the procedure is

TABLE I

METABOLIC ACTIVITIES OF *Mollicutes* GENERA

Mollicute	PP_i-PFK	Urease	dUTPase	ATP/PP_i dGUOk
Acholeplasma	+	−	+	−/+
Anaeroplasma	+	−	+	−/+
Ureaplasma	−	+	−	−/−
Spiroplasma	−	−	+	+/+[a]
Mesoplasma[b]	−[c]	−	−	Variable
Mycoplasma[b]	−	−	−[d]	−[e]/−
Entomoplasma[b]	−	−	−	Variable
Asteroleplasma	−	−	−	−/+

[a] *S. mirum* is PP_i-dGUOk negative.
[b] *Mesoplasma* can be distinguished from *Mycoplasma* and *Entomoplasma*, as the mesoplasmas will grow on sterol-free media + Tween 80, whereas the latter two do not (Tully *et al.*, 1993).
[c] *M. florum* GF is PP_i-PFK positive (?).
[d] *M. mycoides* subsp. *mycoides* is dUTPase positive.
[e] *M. fermentans* strains PG-18 and MI are ATP-dGUOk positive.

identical. The exceptions involve differences in media, method of cell breakage, the necessity for anaerobiosis, etc. Many of the associated concerns have been noted before and will only be recounted briefly (Pollack, 1983).

Growth Media

There is no clear preference for media. Economy is a major consideration. We usually find that the log-phase growth from 4 liters of medium is required to yield a pellet of mollicute cells adequate for the three analyses described in this chapter. In order to have replicates, three 4-liter batches of mollicutes are grown per experiment. Horse serum (10%, v/v, maximum) in an Edward-type medium is adequate for the growth of mollicutes, except for *Ureaplasma,* the anaerobes, *M. pneumoniae, M. genitalium,* and a few others (Pollack *et al.,* 1995a). The addition of Bacto-peptone (5%, w/v) to Edward-type media is helpful in reducing the apparent toxicity of some batches of horse serum.

Media must be examined for their potential to suppress the expression of metabolic activities of interest. For example, we believe that the presence of many nucleic acid derivatives in Edward-type media inhibits the expression of pyrimidine synthesis in growing mollicutes (Donelson and Pollack, 1995). Radioisotopic techniques are necessary. Also, some consideration should be made to reduce the amount or kind of putative metabolic repressors. For example, when studying acholeplasmas, the use of tryptose growth medium with 2% (v/v) horse serum may be more effective in the detection of the enzymes of *de novo* pyrimidine synthesis by radioisotopic means than a medium richer in nucleic acid derivatives and precursors, like an Edward-type medium.

Harvesting and Washing Cells

Harvested cells must be carefully washed by centrifugation as previously described (Pollack, 1983). The washing fluid must be buffered (HEPES, 1 mM, pH 7.5), isotonic (0.155 M NaCl), cold (4°C), and contain a reducing reagent, 2-mercaptoethanol (2 mM), like the k-buffer, originally introduced as β-buffer or B2-buffer (Pollack *et al.,* 1965).

A critical point in washing mollicutes for these studies is the strict avoidance of pipetting the cells. Mollicutes will break (shear) at the pipette orifice. We forcefully pipette *at* the cells with cold wash fluid to break up the pellet. Four liters of cells are collected into six 250-ml vessels by centrifuging at 12,000 g for 20 minutes at 4°C. The pellets are disrupted by a stream of cold wash fluid and recollected. After the first wash, the six pellets are resuspended as before and are combined into four vessels, the process is repeated, the next suspension of cells is combined into two vessels, collected, resuspended, and combined into one vessel and centrifuged to form the final washed cell pellet, or crop. During the collection phase, suspensions and pellets are constantly kept on wet ice.

Cell Breakage and Preparation of Cytoplasmic Cell-Free Extract

Washed cells are routinely broken by either osmotic lysis or sonication, taking care to keep the cells cold (wet-ice temperature) (Razin, 1963; Pollack et al., 1965; Pollack, 1983). When employed, sonication in a small test tube is brief: maximum wattage, for no more than four continuous 15 to -45-second periods, and each period is interrupted by rest for an equal period while the sample is kept in wet ice. After breakage by sonication or lysis, the disrupted cell preparation is differentially centrifuged to isolate membranes, cytoplasm, or ribosomes (Pollack, 1983). For the assays described in this section, a cytoplasmic or cell-free fraction was used. The broken cells are centrifuged at 144,000 g for 2 hours at 4°C. The supernatant is filtered through a 0.22-μm pore filter. The filtrate may be frozen for some assays, like nucleoside phosphorylases, at -20 or -75°C for several weeks without apparent loss of activity. However, this is not the case with all enzyme activities. Labile activities may be better preserved during freezing by making the preparation in 20 or 50% glycerol (v/v) prior to storage.

Assays

All reagents are from Sigma Chemical Co. (St. Louis, MO) unless noted. All water is distilled and deionized.

PP_i- (or ATP-) Dependent Phosphofructokinase (Phosphofructotransferase) (PP_i/ATP-PFK) (EC 2.7.1.90) (Pollack and Williams, 1986)

PFK catalyzes the conversion of fructose 6-phosphate (F6P) to fructose 1,6-diphosphate (FdiP). In most cells, including many mollicutes, ATP is the phosphate donor, whereas in acholeplasmas and anaeroplasmas, PP_i is the obligatory donor. The assay we describe is a coupled spectrophotometric assay. In this assay the FdiP formed in the reaction with either PP_i or ATP acting as the phosphate donor is converted by the intervention of aldolase to dihydroxyacetone phosphate (DHAP) and glyceraldehyde 3-phosphate (G3P). The concentration of DHAP and G3P is kept in equilibrium by the triose-phosphate isomerase. In the presence of NADH, α-glycerophosphate dehydrogenase, and G3P, the NADH is oxidized to NAD. The oxidation of NADH is followed at 340 nm. NADH absorbs significantly at 340 nm, whereas NAD does not. The amount of NADH that disappears is stoichiometrically equivalent to the amount of FdiP produced. The rate at which NADH is produced reflects the rate at which FdiP is synthesized and, therefore, the activity of the PFK in the preparation.

MATERIALS

UV recording spectrophotometer (preferably one that can keep the cuvettes at 37°C. However, for our essentially qualitative purposes this is not absolutely necessary). An attached recorder.

Premix: 100 mM imidazole acetate (pH 7.5), 12 mM MgCl$_2$, 4 mM 2-mercaptoethanol. To every 500 µl add 1.6 units aldolase, 2.4 units triose-phosphate dehydrogenase, and 0.2 units α-glycerophosphate dehydrogenase.

4 mM NADH in water

20 mM F6P in water

40 mM tetrasodium pyrophosphate (PP$_i$) in water

40 mM ATP in water

Cell-free Mollicutes extract containing 5–250 µg protein ml^{-1} (Bio-Rad protein reagent assay). It may be necessary to dialyze this preparation to remove cytoplasmic ATP and PP$_i$. The dialysis procedure is described later for the deoxyguanosine kinase assay (see below), in which it is usually required. For this assay, use undialyzed extract initially.

PROCEDURE

1. Final volume is 1.0 ml.
2. Add 500 µl premix to the cuvette, zero spectrophotometer at 340 nm.
3. Add 50 µl of PP$_i$ or ATP solution and approximately 50 µl NADH solution. The $A_{340\ nm}$ should be about 0.8, if it is not, repeat with more or less of the NADH solution.
4. Working rapidly, add a small amount of cell-free extract, perhaps 50 µl in an initial experiment, and water to 950 µl, mix using Parafilm to seal cuvette (steps 5 and 6 may be reversed)
5. Record changes in absorbance for 1–5 minutes to determine the amount and rate of nonspecific NADH oxidation.
6. Add 50 µl F6P solution, mix rapidly, and record changes in absorbance for 1–10 minutes. The PP$_i$-PFK or ATP-PFK reaction rate is calculated as the difference in $A_{340\ nm}$ before and after the addition of F6P. It is very desirable to use as little cell-free extract as possible in order to keep the enzyme saturated and the reactants in excess. Under these conditions the reaction rate will be zero order, indicating that nothing is limiting. This may require some additional testing. As a simple solution, one that is inadequate for critical work, let the slope (downward) of the absorbance difference be no greater than $-20°$, preferably less. Make replicate trials. The decrease in absorbance attributed to PFK activity is converted to moles of NADH and is equivalent to the moles of FdiP produced. Data are presented as the moles of FdiP produced min^{-1} mg^{-1} protein (=specific activity) or perhaps as units. A unit is the amount of enzyme that

converts 1 μmol of F6P to 1 μmol FdiP min^{-1} at 37°C. An additional control substitutes water for the cell-free extract. This control is necessary because the commercial enzymes are frequently contaminated with ATP-PFK.

Deoxyuridine Triphosphate Nucleotidohydrolase (EC 3.6.1.23, dUTP Pyrophosphatase) (dUTPase) (Williams and Cheng, 1979; Williams and Parris, 1987; Williams and Pollack, 1984)

dUTPase is a highly specific enzyme that hydrolyzes only dUTP to dUMP and pyrophosphate. In free-living organisms, dUTPase is located in the cytoplasm. dUTPase is found in all cells in nature except for some mollicutes. The test we describe is radioisotopic and utilizes the observation that DEAE-impregnated filter paper disks will bind dUTP, but not dUMP, using a formic acid wash solution. Care must be used in distinguishing dUTPase activity from enzymes that nonspecifically hydrolyze dUTP, such as alkaline phosphatase and ATPase. ATPase is present in all mollicute membranes tested. Alkaline phosphatase is also membrane associated. Careful disruption of the cells followed by centrifugation can generally separate most of the membrane fragments. We routinely add p-nitrophenol phosphate and ATP to our reaction mixtures. These compounds do not interfere with the dUTPase determinations and act as better substrates than dUTP for alkaline phosphatase and ATPase, respectively.

MATERIALS

DE-81 filter disks, 2.5 cm (Whatman Int, Ltd.)
37°C water bath
2'-[5-^{3}H]deoxyuridine 5'-triphosphate, tetraammonium salt, 5–15 Ci (185–555 GBq)/mmol (Moravek Biochemicals, Inc., 577 Mercury Lane, Brea, CA)
Premix: 2.2 mM p-nitrophenol phosphate, 5.6 mM ATP, 55.6 mM Tris–HCl (pH 8.0), 10.1 mM [5-^{3}H]dUTP (50 μCi/μmol), 2.2 mM 2-mercaptoethanol, 1.1 mM MgCl$_2$, and 0.11 % (w/v) bovine serum albumin. It is very important that the [5-^{3}H]dUTP as purchased be adjusted (diluted to 50 μCi/μmol) with an aqueous solution of nonradioactive dUTP whose concentration is assayed by spectrophotometry rather than by weight. The reason we dilute the [^{3}H]dUTP with dUTP is because at the amount of radioactivity desired for each assay, the actual mass of dUTP in the radioactive material is too low to saturate the enzyme in the following procedure. Therefore, additional non-radioactive dUTP must be added to the [^{3}H]dUTP.
4 M formic acid that is also 1 mM ammonium formate
95% ethanol
Equipment and supplies for assaying radioactivity by liquid scintillation techniques and capability of determining disintegrations per minute (dpm)

Cell-free mollicute extract containing 6–2200 µg ml^{-1} protein (Bio-Rad protein reagent assay)

PROCEDURE

1. Final volume is 100 µl.
2. To 90 µl premix in a small test tube add 10 µl cell-free extract that contains 0.06–22 µg protein or 10 µl of an appropriate control (no enzyme). Incubate at 37°C for various periods of time (5 minutes to overnight). There should be between 75–100 × 10^3 dpm per assay tube.
3. To stop the reaction, spot 50 µl of the reaction mixture on a DE-81 filter disk.
4. Immediately wash the disk in formic acid–ammonium formate solution for 5 minutes with gentle mixing. Repeat the washing step two times.
5. Wash filter disks in 95% ethanol for 3 minutes. Dry disks, count for radioactivity, and determine dpm.
6. The radioactive and nonradioactive dUTP will bind to the filter disk. If there is too much dUTPase activity, no residual [^{3}H]-dUTP will be trapped on the disk. The dUTPase reaction rate is calculated as the decrease in radioactivity remaining on the filter disk after the background or control reaction is subtracted. Data may be reported as specific activity: µmol dUTP hydrolyzed min^{-1} mg^{-1} protein or as units. A unit of dUTPase activity is defined as the amount of enzyme that converts 1 nmol of dUTP to 1 nmol of both dUMP and PP$_i$ min^{-1} at 37°C. The products of the reaction can be determined by thin-layer chromatography (TLC) (Beardsley and Abelson, 1980).

PP$_i$ or ATP-Dependent Deoxyguanosine Kinase (EC 2.7.1.113) (PP$_i$/ATP dGUOk) (Tryon and Pollack, 1984; McElwain and Pollack, 1987; McElwain et al., 1988)

The ability to use PP$_i$ in the nucleoside kinase reactions rather than ATP is known only in some mollicutes. The assay we describe is radioisotopic and chromatographic. The usefulness of the assay is related to the efficiency of the chromatographic (TLC) separation of guanine, deoxyguanosine, and deoxyGMP on polyethyleneimine (PEI)-coated cellulose in one dimension. The ability of the TLC support to separate these compounds is apparently related to the mass of PEI used to impregnate the cellulose. We believe that at least 0.55 mEq of PEI g^{-1} cellulose is required for optimal separation of the reactants and products. Unfortunately, not all batches of PEI–cellulose TLC plates are satisfactory. The separation is somewhat improved by performing the TLC at about 4–10°C.

MATERIALS

PEI–cellulose TLC plates on glass (E. Merck, distributed by Alltech Assoc., Inc., Deerfield, IL). If the plates are yellow, wash them by chromatographing them in either distilled water or methanol:water (1:1 v/v) for almost their entire length. Scrape off and discard the yellow front when the plates are still slightly damp. Box the washed plates and store them in the cold.

TLC glass chromatography tanks

37°C water bath

$2'$-[8-^{3}H]deoxyguanosine, 3–10 Ci(111–370 GBq)/mmol (Moravek Biochemicals, Inc., Brea, CA)

Premix: 200 mM HEPES (pH 7.4), 2 mM MgCl$_2$, and about 250 × 10^3 dpm of [^{3}H]dGUO for every 25 μl of premix

40 mM ATP in water

40 mM tetrasodium pyrophosphate (PP$_i$) in water

Aqueous solution containing 5 mM deoxyguanosine, 5 mM deoxyGMP, and about 5 mM guanine (G-standards). Guanine will not entirely dissolve without heating.

Equipment and supplies for assaying radioactivity by liquid scintillation techniques, having the capability of determining dpm

UV lamp for viewing chromatographed plates

Cell-free mollicute extract containing 1–4 mg ml^{-1} protein (Bio-Rad protein reagent assay). It is generally necessary to dialyze this preparation in changes of wash buffer (κ buffer) to remove cellular ATP and PP$_i$. For example, 10 ml of cell-free extract is dialyzed against 1 liter of κ buffer (containing 100 μM phenylmethylsulfonyl fluoride, a highly toxic proteinase inhibitor) for at least 5 hours at 4–10°C. The dialysis is repeated twice in order to achieve a dilution factor of 10^6.

PROCEDURE

1. Final volume is 100 μl.

2. To 25 μl premix in a small test tube add 10 μl ATP or PP$_i$ solution. Add a total of 65 μl cell-free extract and water if necessary so that the amount of cell-free protein in each assay tube is about 10–40 μg. The water may be added first and the reaction started with enzyme.

3. Incubate three to five replicate samples at 37°C for 2, 6, and 10 minutes (more time periods within 10 minutes may be added, if desired). A reaction blank must be included in the test. A reaction blank contains everything but ATP or PP$_i$. This control is necessary because, even if dialyzed, there may still be some ATP and PP$_i$ in the log-phase cell-free fraction. Therefore, it is obligatory that we determine the magnitude of the background reaction due to this residual

ATP or PP_i. The values obtained in this control assay are subtracted from other experimental values containing ATP or PP_i. An optional control is one containing all ingredients and either ATP or PP_i, but only heat-killed cell-free extract (95°C, 10 minutes).

4. Stop the reaction by heating each tube for 1–2 minutes at 95°C. After heating, the tubes may be stored at -20°C for many weeks without apparent effect if they are very tightly sealed. Loss of fluid from these tubes negates the assay.

5. Spot 3–5 µl of nonradioactive G-standards on prewashed PEI–cellulose TLC plates, with six to seven spots per plate. Dry the spots using a hair dryer.

6. Overspot the applied G-standards with a total of 20.0 µl of one reaction mixture. It is important that the spot be very small. No more than 5 µl, preferably 2.5 µl, should be applied at one time. After the 5 µl has dried (use a hair dryer), apply the second 5 µl, repeat, until exactly 20 µl has been applied. The smaller the spot the better the resolution. Dry. One lane per plate should receive no sample of reaction mixture. This lane is the chromatography blank.

7. Chromatograph the cooled plate at 4–10°C in 0.5 N formic acid that is also 2 M LiCl (about 2–3 hours). Dry in a fume hood.

8. Visualize the spots under UV light and circle with a needle, being careful not to touch or scrape the spot so as to not release radioactive particles. In this system the R_f values for the compounds are guanine, 0.01; deoxyguanosine, 0.52 ± 0.02; and deoxyGMP, 0.62 ± 0.04.

9. Using small funnels, scrape the encircled and radioactive spots directly into scintillation vials. Count for radioactivity and determine dpm. Subtract the values obtained in each of the spots resolved (guanine, deoxyguanosine, or deoxyGMP) in the reagent blank lane (no ATP or PP_i) from the values found in their respective spots in the other lanes (complete reaction assays containing ATP or PP_i).

10. Data may be reported as µmol dGMP synthesized min^{-1} mg^{-1} protein.

Acknowledgments

This study was supported by Public Health Service Grant RO1-A133193 from the NIH. We thank Professor M. V. Williams for help and expertise in developing the dUTPase assay.

References

Beardsley, G. P., and Abelson, H. T. (1980). A thin-layer chromatographic method for separation of thymidine and deoxyuridine nucleotides. *Anal. Biochem.* **105,** 311–318.

Donelson, K., and Pollack, J. D. (1995). Mollicutes can synthesize pyrimidines *de novo*. In preparation.

Egan, W., Barile, M., and Rottem, S. (1986). ^{31}P-NMR studies of *M. gallisepticum* cells using a continuous perfusion technique. *FEBS Lett.* **204,** 373–376.
McElwain, M., and Pollack, J. D. (1987). Synthesis of deoxyribomononucleotides in Mollicutes: Dependence on deoxyribose-1-phosphate and PP_i. *J. Bacteriol.* **169,** 3647–3653.
McElwain, M. C., Chandler, D. K. F., Barile, M. F., Young, T. F., Tyron, V. V., Davis, J. W., Jr., Petzel, J. P., Chang, C.-J., Williams, M. V., and Pollack, J. D. (1988). Purine and pyrimidine metabolism in Mollicutes species. *Int. J. Syst. Bacteriol.* **38,** 417–423.
Pollack, J. D. (1983). Localization of enzymes in mycoplasmas: Preparatory steps. *In* "Methods in Mycoplasmology" (S. Razin and J. G. Tully, eds.), Vol. 1, pp. 327–332. Academic Press, New York.
Pollack, J. D. (1994). Carbohydrate and energy conservation. *In* "Mycoplasmas: Molecular Biology and Pathogenesis" (J. Maniloff, R. N. McElhaney, L. R. Finch, and J. B. Baseman, eds.), pp. 181–200. ASM, Washington, DC.
Pollack, J. D., Razin, S., Pollack, M. E., and Cleverdon, R. C. (1965). Fractionation of mycoplasma cells for enzyme localization. *Life Sci.* **4,** 973–977.
Pollack, J. D., and Williams, M. V. (1986). PP_i-dependent phosphofructotransferase (phosphofructokinase) activity in the Mollicutes (mycoplasma) *Acholeplasma laidlawii. J. Bacteriol.* **165,** 53–60.
Pollack, J. D., Williams, M. V., and McElhaney, R. (1995a). The comparative metabolism of the mollicutes: The utility for taxonomic classification. In preparation.
Pollack, J. D., Williams, M. V., Banzon, J., Donelson, K., Jones, M. A., Harvey, L., and Tully, J. (1995b). Comparative metabolism of *Mesoplasma, Entomoplasma, Acholeplasma,* and *Mycoplasma.* In preparation.
Razin, S. (1963). Osmotic lysis of mycoplasma. *J. Gen. Microbiol.* **33,** 471–475.
Tryon, V. V., and Pollack, J. D. (1984). Purine metabolism in *Acholeplasma laidlawii* B: Novel PP_i-dependent nucleoside kinase activity. *J. Bacteriol.* **159,** 265–270.
Tully, J. G., Bové, J. M., Laigret, F., and Whitcomb, R. F. (1993). Revised taxonomy of the class Mollicutes: Proposed elevation of a monophyletic cluster of arthropod-associated mollicutes to ordinal rank (Entomoplasmatales ord. nov.), with provision for familial rank to separate species with nonhelical morphology (Entomoplasmataceae fam. nov.) from helical species (Spiroplasmataceae), and emended descriptions of the order Mycoplasmatales, family Mycoplaasmataceae. *Int. J. Syst. Bacteriol.* **43,** 378–385.
Williams, M. V., and Cheng, Y.-C. (1979). Human deoxyuridine triphosphate nucleotidohydrolase. *J. Biol. Chem.* **254,** 2897–2901.
Williams, M. V., and Parris, D. S. (1987). Characterization of a herpes simplex virus type 2 deoxyuridine triphosphate nucleotidohydrolase and mapping of a gene conferring type specificity for the enzyme. *Virology* **156,** 282–292.
Williams, M. V., and Pollack, J. D. (1984). Purification and characterization of a dUTPase from *Acholeplasma laidlawii* B-PG9. *J. Bacteriol.* **159,** 278–282.
Williams, M. V., and Pollack, J. D. (1990). The importance of differences in the pyrimidine metabolism of the Mollicutes. *In* "Recent Advances in Mycoplasmology" (G. Stanek, G. H. Cassell, J. G. Tully, and R. F. Whitcomb, eds.), pp. 163–171. Gustav Fischer Verlag, Stuttgart.

D3

RAPID MICROCALORIMETRIC AND ELECTROANALYTICAL MEASUREMENTS OF METABOLIC ACTIVITIES

R. J. Miles

Introduction

The methods described in this chapter enable the monitoring of substrate catabolism by whole cells. In mollicutes, catabolic activities are primarily associated with ATP generation (Miles, 1992). In some cases, energy sources may be identified in growth studies; for example, the ability to produce acid from glucose metabolism or ammonium from arginine hydrolysis, during growth, is used in mollicute identification. However, the routine detection of substrate metabolism accompanying mollicute growth is often not feasible because of low cell yields and/or difficulties in detecting the metabolism of specific substrates in complex media containing a vast array of alternative substrates. In addition, the utilization of di- and polysaccharides of glucose cannot be determined in serum-containing medium as they are rapidly hydrolyzed by serum enzymes.

In the methods described, cells are suspended in inorganic salt solutions and the metabolism of single substrates is monitored: either by enthalpy (heat) changes, using a microcalorimeter; or by oxygen uptake or acid/alkali production measured as changes in the dissolved oxygen tension (DOT) or pH of cell suspensions, respectively. The methods require relatively small quantities of cells and are rapid. They provide kinetic data allowing assessment of the likely significance of substrate metabolism at the concentrations found in natural habitats and have been used to identify energy sources which increase growth yield

(Miles et al., 1988; Taylor et al., 1994). The methods are also applicable to mollicute characterization. Their use has shown major physiological subdivisions among both glucose fermenting (Miles et al., 1991) and nonfermenting *Mycoplasma* species (Miles et al., 1988; Taylor et al., 1994), and patterns of substrate oxidation distinguished specific and subspecific groups within the "*M. mycoides* cluster" (Abu-Groun et al., 1994). Microcalorimetry has also been applied to the characterization of carbohydrate transport-deficient mutants of *M. mycoides* (Lee et al., 1986).

Materials

For Preparation of Cell Suspensions

Liquid (broth) cultures of test strains
Microcentrifuge (MSE Micro Centaur or equivalent)
Eppendorf tubes
Suspension media: 18 g liter^{-1} HEPES and 160 U ml^{-1} catalase (Sigma, C-10) in one-quarter strength Ringer's solution, pH 7.6 (RH solution), or normal saline (0.85%, w/v, NaCl) plus catalase (160 U ml^{-1}), adjusted to pH 7.6.

For Detection of Metabolism

Test substrates, e.g, sugars, amino sugars, organic acids, amino acids, or alcohols (5–500 mM)
Microsyringe(s) for accurate delivery of 5 to 50 μl substrate solution
Thermocirculator
Chart recorder
Water-jacketed, magnetically stirred flat-bottomed reaction vessel
Microcalorimeter (with flow-through facility), oxygen electrode, or pH meter.
  Examples of suitable instruments are:
  Microcalorimeter: 2277 thermal activity monitor (Thermometric AB Sweden; baseline stability ± 0.1 μW), with silicon tubing (1-mm internal diameter) and peristaltic pump (flow inducer).
  Oxygen electrode: Rank Brothers oxygen electrode system (Rank Brothers Ltd., Cambridge, UK); this system incorporates a water-jacketed reaction vessel.
  pH meter: Corning 250 ion analyzer (resolution 0.001 pH units), with a 5-mm-diameter electrode.

  All solutions are sterile; catalase and heat-labile substrates are filter sterilized.

Procedure

1. Preparation of cell suspensions. Broth cultures are harvested toward the end of the exponential growth phase to achieve a maximum yield of metabolically active cells. The procedures described require approximately 10^9 colony-forming units (CFU), and it is convenient to harvest the growth from about 15 ml culture by centrifugation at 14,000 g for 3 minutes in a microcentrifuge at room temperature. Cells are washed twice and resuspended using a Pasteur pipette in 2–3 ml of RH or saline solution and are used immediately. Large batches of cells have also been successfully stored as aliquots in liquid nitrogen prior to use (Miles *et al.*, 1985).

2. Detection of metabolism by microcalorimetry. The fermenter vessel is external to the calorimeter, which is used in a "flow-through" mode. The cell suspension from the fermenter vessel is continuously pumped through silicon tubing to an inlet port of the calorimeter. It is precisely thermostated (typically at 37°C) before passing through a flow cell (0.6 ml in Miles *et al.*, 1985), in which heat evolution is detected via thermopiles as a potential difference and amplified. The suspension emerging from the cell is returned to the fermenter vessel. Output from the calorimeter is fed to a chart recorder and power output (μW) versus time recorded. The instrument is calibrated using internal heaters.

Prior to use, the flow lines and flow cell are disinfected by pumping through ethanol (15 ml hour^{-1}; 15 minutes) followed by sterile distilled water (4 × 15 minutes). Flow lines are then transferred to the fermenter vessel (volume approximately 100 ml) which is magnetically stirred and maintained (typically) at 37°C, preferably using a water jacket with a thermostatic circulator. For aerobic reactions, the stirrer rate should be sufficient to maintain DOT at levels near to air saturation without introducing air bubbles into flow lines. After establishing a baseline in the salts solution (25 ml RH) without cells, 1 ml of mollicute cell suspension (5 × 10^9 CFU in RH) is added. This may result in a transient power output (Fig. 1A), possibly due to metabolism of residual substrates from the growth medium. Test substrates (in approximately 50 μl) are added after the baseline is reestablished (up to 1 hour). Typical data for metabolism of glucose and mannose by *M. mycoides* strain T1 are shown in Fig. 1A. The rate of metabolism is proportional to power and the area under the power–time curve to the enthalpy (heat) change, which can be calculated as kJ (mol substrate)$^{-1}$ and compared to theoretical values for metabolism of the substrate to putative products (determined from standard tables).

3. Detection of metabolism by oxygen uptake. Substrate oxidation is determined from changes in the DOT of cell suspensions measured using a Clark-type oxygen electrode linked to a chart recorder. In the Rank Brothers instrument, the electrodes (working and reference) are mounted beneath the cylindrical reaction vessel (maximum volume 5 ml; diameter 15 mm) and are separated from the

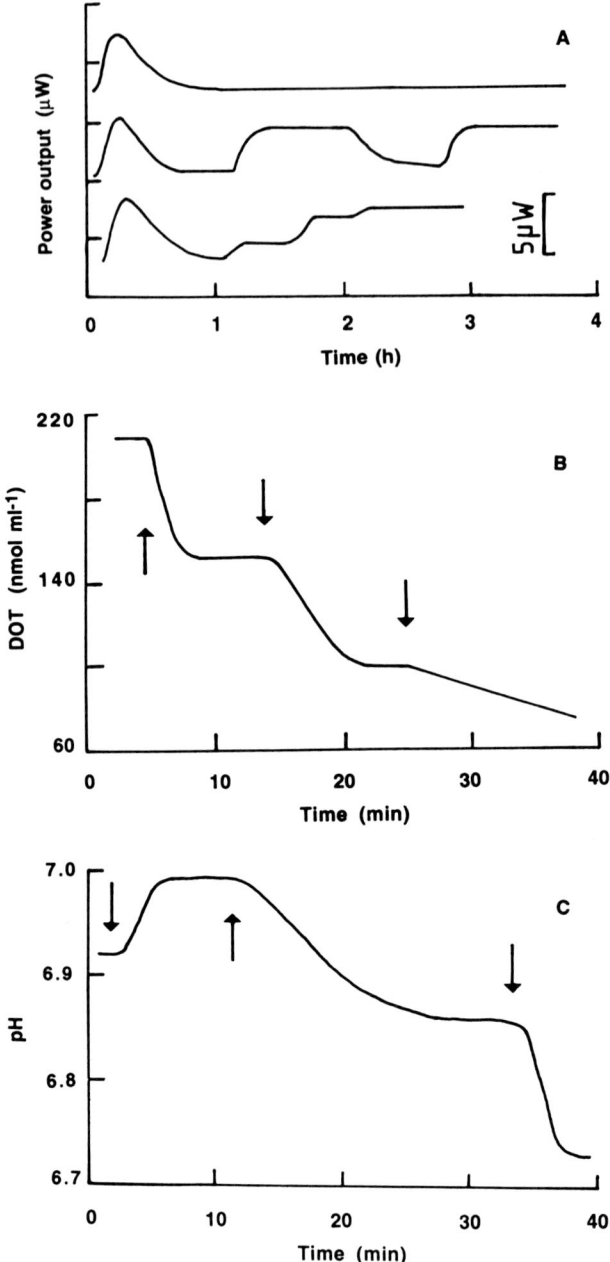

Fig. 1. Substrate metabolism of *Mycoplasma* spp. monitored by microcalorimetry (A) or changes in dissolved oxygen tension (DOT; B) or pH (C). (A) Power–time curves for suspensions of *M. mycoides* subsp. *mycoides* strain T1: top, no substrate addition (control); middle, glucose added, 20 µM at 1 hour and 2 mM at 2.5 hours; bottom, mannose added, 0.4 mM at 1

contents of the vessel by a semipermeable Teflon membrane. Contact between the electrodes is maintained by fine tissue (e.g., lens tissue) saturated with 3 M KCl. The Teflon membrane should be newly installed before use and the reaction vessel thoroughly cleaned and washed with sterile distilled water. Immediately prior to use, the electrode is calibrated; this may be adequately achieved with air-saturated water (210 nmol dissolved O_2 ml^{-1} at 37°C) obtained by simply adding sterile distilled water (that has not been shaken, thus avoiding supersaturation) to the reaction vessel and magnetically stirring. The distilled water is then replaced by 1 to 2 ml of cell suspension (10^9 CFU ml^{-1}), which is air saturated, and maintained at 37°C (or other appropriate temperature) via the water jacket. The suspension (stirred at a constant rate) is then isolated from air using a cylindrical plug with a fine central pore (1 mm diameter). By maintaining the level of the plug such that the meniscus of the cell suspension is within the central pore, oxygen transfer to the suspension is virtually eliminated. Test substrates (2–10 μl of appropriate concentrations) are added to the cell suspension via the central pore of the plug using a microsyringe. The system may be reoxygenated at any time by raising the plug and allowing air into the reaction vessel. The rate of change in DOT is proportional to the rate of metabolism, and reaction stoichiometries can be determined from the total reduction in DOT. Representative data are shown in Fig. 1B.

4. Detection of metabolism by pH change. Cells suspended in saline (approximately 10^9 CFU ml^{-1}) are magnetically stirred and equilibrated at 37°C (or other temperature) in a suitable reaction vessel. Using a 15-mm-diameter vessel, with a 5-mm-diameter pH electrode, 2 ml of cell suspension is adequate. Output from the pH meter is recorded using a chart recorder. Substrates (at appropriate concentrations in a volume of a few microliters) are added by microsyringe after the achievement of a stable baseline (approximately 30 min); typically, pH is approximately 7.0 at this time. Data showing pH changes following the addition of glucose, fructose, and arginine to cell suspensions of *M. fermentans* strain incognitus are shown in Fig. 1C.

5. Substrate concentrations. The concentration necessary to give detectable metabolism varies with the affinity constant (K_m value) of the particular mollicute strain for the selected substrate. Values of K_m for the metabolism of, for

---

hour, 0.8 mM at 1.5 hours, and 2 mM at 2 hours (from Miles et al., 1985, with permission). (B) DOT following additions (arrows) of 25 μM glucose, *N*-acetylglucosamine, and glucosamine at 6, 14, and 24 minutes, respectively, to cells of *M. capricolum* subsp. *capripneumoniae* strain pp goat (from Abu-Groun et al., 1994, with permission). (C) pH change following additions (arrows) of 8 μM arginine, 10 μM glucose, and 10 μM fructose at 3, 12, and 33 minutes, respectively, to cells of *M. fermentans* strain incognitus. Cells were equilibrated in saline plus catalase for approximately 30 minutes before substrate addition (see Taylor et al., 1992).

example, glucose and arginine by *Mycoplasma* strains are typically low ($K_m < 5$ μM, indicating a high affinity). For these substrates, initial substrate concentrations of 10–50 μM will give maximal metabolic rates. However, substantially higher concentrations (up to 1 mM or more) are needed where K_m values are high, for example, mannose utilization by *M. mycoides* subsp. *mycoides* SC strains (K_m approximately 1 mM; Abu-Groun et al., 1994). In general, it is advantageous to use low substrate concentrations since following substrate exhaustion, additions of further different substrates may be made, increasing the information available from each experimental run (see Fig. 1). Stock solutions of test substrates are conveniently prepared at 5, 50, and 500 mM, dispensed in aliquots, and stored at -20°C.

6. Calculation of kinetic parameters of substrate utilization. If it is assumed that the relationship between the rate of substrate metabolism (v) and substrate concentration (s) follows Michaelis kinetics, the saturation constant (K_m value) and maximum rate of substrate metabolism (v_{max}) may be estimated from the plot of $1/s$ against $1/v$, in which the intercepts on the ordinate axis and abscissa are $1/v_{max}$ and $-1/K_m$, respectively. Where K_m values are high, the rate of metabolism (proportional to power output or rate of change in DOT or pH) may be determined at different initial values of s. Alternatively, where K_m values are low, kinetic data may be estimated from the analysis of single curves representing the complete utilization of substrate. It is necessary to assume that a change in DOT, pH, or enthalpy (area under power–time curve) at any time (t) after substrate addition is proportional to the substrate used. Thus, the substrate concentration at time t is (initial substrate concentration)$(a - b)/a$, where a is the total change in DOT, pH, or enthalpy following substrate addition and b is the change at time t.

Discussion

The method suggested for preparing cell suspensions aims to complete centrifugation, washing, and resuspension procedures rapidly, within about 15 minutes, so as to retain maximum metabolic activity. In general, cell suspensions should be used immediately, as their activity declines during storage, even at 0–10°C. Liquid nitrogen-stored cells have been used successfully in microcalorimetric studies of metabolism (Miles et al., 1985); however, cell viability and metabolic activity are critically dependent on the rate of temperature change during freezing/thawing and the presence of osmoprotectants, e.g., glycerol.

Catalase is included in suspension media as, during carbohydrate metabolism, many mollicute species produce substantial quantities of H_2O_2 which may reduce metabolic activity or cell viability; its presence was essential during centrifugation and washing of cells that had been stored in liquid nitrogen with glycerol as an osmoprotectant. The presence of catalase also simplifies the stoichiometric analysis of metabolism measured using either O_2 uptake or microcalorimetric methods, since net oxygen uptake and enthalpy change would otherwise be dependent on the amount of H_2O_2 formed during metabolism, which varies with different species (Miles et al., 1991).

The suspension medium (RH) for microcalorimetric and oxygen uptake studies is buffered to eliminate the effects of pH change (due to metabolism) on metabolic activity. HEPES appears suitable for a wide range of mollicutes, but interferes with the Lowry protein assay. In the pH change method, unbuffered saline is used to maximize pH change due to the production of acid/alkali products.

The *Mycoplasma* and *Acholeplasma* strains that we have used do not appear to possess a significant endogenous metabolism, and detectable O_2 uptake, power output, or acid/alkali production was dependent on the addition of a metabolizable substrate. Metabolism of single substrates is generally in accord with Michaelis kinetics, and K_m and v_{max} values are readily determinable from microcalorimetric and oxygen uptake data (Miles et al., 1985; Taylor et al., 1994; Abu-Groun et al., 1994). Provided that metabolic intermediates do not accumulate during substrate metabolism, the oxygen uptake and enthalpy change will be directly proportional to the substrate metabolized. Where acid/alkali production is determined as pH change this need not be the case, however, since pH is logarithmically related to $[H^+]$ and organic acids produced or metabolized by suspended cells may not be fully dissociated. CO_2 production may also influence pH (as a consequence of carbonic acid formation) although this is related to the amount of substrate metabolized. Nevertheless, the experimental system described can be used to determine the saturating levels of substrate and to compare initial rates of pH change following substrate addition (Miles et al., 1991). Also, presumably because, in the system used, pH changes following the addition of substrate are so small (e.g., for *M. fermentans*, approximately 0.15 and 0.08 units for metabolism of 10 μM glucose and 8 μM arginine, respectively; Fig. 1C), it is possible to estimate kinetic parameters using pH change and rate of pH change as measures of substrate utilized and rate of utilization, respectively. Kinetic data estimated in this way are similar to those obtained by measuring oxygen uptake, where we have compared the two methods using cells derived from a single culture. The ability to demonstrate that cell suspensions are metabolically active and can, for example, produce acid from glucose or alkali from arginine is required to support data showing the inability of suspensions to oxidize substrates.

The metabolic activity of cell suspensions may be related to viable count or cell protein. However, unless growth conditions, inoculum preparation, and cell harvesting procedures are rigorously controlled, the specific activities of cell suspensions prepared on different days will vary. One way of overcoming this is to prepare, from several hundred milliliters of culture, a uniform cell suspension that is dispensed in ampoules and stored frozen in liquid nitrogen (Miles et al., 1985). However, freezing and thawing must be conducted under optimal, defined conditions. A simpler alternative is to initiate all individual experiments by the addition of a low concentration of a rapidly metabolized "control" substrate, e.g., glucose for *M. mycoides* (Abu-Groun et al., 1994), and to express rates of metabolism of substrates added subsequently relative to the rate for this substrate. The control substrate may also be added at the end of experiments to confirm that metabolic activity has not declined. In our experiments, in general, cell activities are fairly constant for about 2–3 hours, allowing metabolism of a number of substrates or of a single substrate at a range of concentrations to be determined. In the oxygen uptake method, rates of oxygen uptake were not affected by DOT in the range of 5–210 nmol ml^{-1}. In the pH change method, rates of metabolism may be affected by the relatively large change in pH following the sequential addition of a number of metabolizable substrates at a low concentration (10 μM) or a single substrate at a high concentration (0.1 mM).

The suitability of each of the methods described for monitoring substrate utilization depends on the particular metabolic pathways present in the test species. Oxygen uptake from carbohydrate metabolism appears, at least in *Mycoplasma* spp., dependent on the oxidation of intracellular pyruvate to acetate plus CO_2. The method is therefore suitable for detecting oxidation of, e.g., organic acids by *M. agalactiae* (Miles et al., 1988) and sugars and carbohydrates by *M. mycoides*. However, it will not enable detection of sugar fermentation by, e.g., *M. fermentans,* which metabolizes glucose to lactate both in the presence and in the absence of oxygen (Miles, 1992). The pH change method can be used to monitor both lactic and acetic acid production from carbohydrates and arginine hydrolysis; however, it is not suitable for monitoring the metabolism of one organic acid to another (e.g., lactate to acetate). Microcalorimetry may be used to monitor all catabolic reactions as these are inevitably exothermic. However, this must be balanced against the very high cost of microcalorimeters. Also, the sensitivity of the technique is greater where oxidations are monitored (as in the data of Fig. 1A) as these are associated with larger enthalpy changes; for example, the theoretical enthalpy change for oxidation of glucose to acetate plus CO_2 is 1050 kJ mol^{-1}, whereas that to lactic acid is only 76 kJ mol^{-1}.

Acknowledgment

The development of the electroanalytical methods described here was supported by a grant from the Wellcome Trust.

References

Abu-Groun, E. A., Taylor, R. R., Varsani, H., Wadher, B. J., Leach, R. H., and Miles, R. J. (1994). Biochemical diversity within the '*Mycoplasma mycoides* cluster'. *Microbiology* **140**, 2033–2042.

Lee, D. H., Miles, R. J., and Beezer, A. E. (1986). Isolation and microcalorimetric characterisation of glucose-negative and pyruvate-negative mutants of *Mycoplasma mycoides* subsp. *mycoides*. *FEMS Microbiol. Lett.* **34**, 283–286.

Miles, R. J. (1992), Catabolism in mollicutes. *J. Gen. Microbiol.* **138**, 1773–1783.

Miles, R. J., Beezer, A. E., and Lee, D. H. (1985). Kinetics of the utilisation of organic substrates by *Mycoplasma mycoides* subsp. *mycoides* a flow microcalorimetric study. *J. Gen. Microbiol.* **131**, 1845–1852.

Miles, R. J., Taylor, R. R., and Varsani, H. (1991). Oxygen uptake and H_2O_2 production by fermentative *Mycoplasma* spp. *J. Med. Microbiol.* **34**, 219–223.

Miles, R. J., Wadher, B. J., Henderson, C. L., and Mohan, K. (1988). Increased yields of *Mycoplasma* spp. in the presence of pyruvate. *Lett. Appl. Microbiol.* **7**, 149–151.

Taylor, R. R., Miles, R. J., Varsani, H., and Abu-Groun, E. A. (1992). Physiology and growth of *Mycoplasma fermentans* strains. *IOM Lett.* **2**, 91.

Taylor, R. R., Varsani, H., and Miles, R. J. (1994). Alternatives to arginine as energy sources for the non-fermentative *Mycoplasma gallinarum*. *FEMS Microbiol. Lett.* **115**, 163–168.

D4

CHARACTERIZATION OF HEAT SHOCK PROTEINS

Christopher C. Dascher and Jack Maniloff

Introduction

Virtually every living organism exhibits some form of heat shock response, characterized by the quantitative increase in expression of a subset of proteins in response to the stress of increased temperatures. Heat shock proteins are thought to carry out a variety of functions within the stressed cell that allow it to adapt to the new environmental conditions. Mycoplasmas possess a heat shock response similar to other eubacteria (reviewed in Dascher and Maniloff, 1992). Mycoplasma heat shock proteins include homologs of two of the most broadly conserved heat shock proteins, DnaK and GroEL. Both of these proteins are involved in protein folding functions and are essential for cell survival. In addition, GroEL has been implicated in pathological immune responses related to bacterial infections (Morrison et al., 1989).

The classic method for investigating heat shock response in bacteria involves isotopically labeling nascent cellular proteins under normal and stressed conditions followed by polyacrylamide gel electrophoresis (PAGE) and autoradiography. The study of mycoplasma heat shock response with these techniques is complicated by both the growth conditions and the characteristics of mycoplasmas. The inability to use defined minimal media necessitates the use of high-specific-activity amino acid mixtures for protein labeling. The relatively slow growth characteristics of mycoplasmas also require extended pulse labeling times to obtain sufficient incorporation of labeled amino acids for protein detection. Therefore, the methods used for the study of mycoplasma heat shock responses are adaptations of methods used previously for other bacteria, taking into account the special characteristics of mycoplasmas (Dascher et al., 1990).

The protocols described in this chapter provide details for identification and analysis of *Acholeplasma laidlawii* heat shock proteins. Experimental modifications for *Mycoplasma* species are discussed later in the chapter.

Materials

For Radioisotopic Labeling of Cells

Logarithmic broth culture of A. *laidlawii* or other mycoplasma
Broth culture medium: for A. *laidlawii*, tryptose broth medium consists of 20 g tryptose (Difco), 5 g Tris, 5 g NaCl, 10 g glucose, and 10 ml of PPLO serum fraction (Difco), per liter, pH 8.0 (Dascher et al., 1990). This is replaced by appropriate media components for other mycoplasmas.
Incubator (37 and 32°C)
Water bath (42°C)
Ice bath
Spectrophotometer and cuvettes
Freeze dry apparatus (Labconco, Kansas City, MO)
10-ml glass vials
Eppendorf centrifuge Model 5412 (Richmond, CA)
1.5-ml microcentrifuge tubes (polypropylene)
^{14}C-labeled L-amino acid mixture (Amersham Corporation, Arlington Heights, IL)

For PAGE and Autoradiography

Glass plates and spacers (17×16×0.1 cm)
Plastic comb (1-cm-wide lanes)
PAGE gel box
Electrophoresis power source
Acrylamide
N,N'-Methylenebisacrylamide
N,N,N',N'-Tetramethylethylenediamine (TEMED)
Ammonium persulfate
2-Propanol

X-Omat AR film (Eastman Kodak, Rochester, NY)
EnHance (Dupont NEN, Boston, MA)
Trichloroacetic acid (TCA)
Acetone
Acetic acid
Liquid scintillation counter (Beckman LS1801, Irvine, CA)
Scintillation vials
Scintillation fluid
Glass fiber scintillation filters
Vacuum apparatus (Hoefer Scientific Instruments, San Francisco, CA)
Slab gel dryer (Bio-Rad Model 483, Richmond, CA)
Filter paper (Whatman No. 1, Clifton, NJ)
Laser densitometer (LKB Ultroscan XL, Bromma, Sweden)
T buffer: 2.1 M Tris, pH 9.18
Low T buffer: 0.54 M Tris, pH 6.1
Cathode buffer: 0.40 M boric acid, 0.41 M Tris, 1% SDS (w/v), pH 8.64
Anode buffer: 20% (v/v) T buffer
Sample buffer: 10 ml cathode buffer, 5 g sucrose, 185 mg EDTA, 5 ml 2-mercaptoethanol, 1.9 g SDS, per 100 ml, pH 8.6
Gel fixer: 50% (v/v) methanol, 10% acetic acid (v/v)
Coomassie blue stain: 50% (v/v) methanol, 10% (v/v) acetic acid, 0.2% [w/v] Coomassie brilliant blue R-250

For Immunoblotting

Nitrocellulose membrane, 0.45 μm (Bio-Rad, Richmond, CA)
Filter paper (Whatman No. 1)
Transfer buffer: 48 mM Tris, 39 mM glycine, 1.3 mM SDS, 20% (v/v) methanol
Phosphate-buffered saline (PBS): 0.13 M NaCl, 2.7 mM KCl, 1.4 mM KH_2PO_4, 0.13 M NaH_2PO_4, pH 7.4
Geletin
Peroxidase-conjugated secondary antibody (specific to species of primary antibody) (Cooper Biomedical, Malvern, PA)
Monospecific antibody to protein of interest (e.g., DnaK)
Semidry electroblotter (Ancos, Olstykke, Denmark)
Heat sealer
Heat-sealable bags
Developer solution: 30 mg 4-chloro-1-naphthol (Sigma, St. Louis, MO) in 10 ml methanol plus 50 ml PBS and 20 μl 30% H_2O_2

Procedure

1. Preparation of radiolabeled amino acids. Prior to pulse-labeling experiments, radioactive label is prepared by adding 50 µCi of a ^{14}C-labeled L-amino acid mixture to a 10-ml glass vial and freeze drying until all liquid is removed. These vials are then capped and stored at 4°C until needed. This procedure allows preparation of the high concentrations of radiolabeled amino acids needed to label mollicute proteins and removes any trace of organic solvents that may inadvertently induce a stress response in the microorganisms being examined.

2. Heat shock and radioisotopic labeling of nascent proteins. Overnight *A. laidlawii* cultures grown in tryptose broth medium at 37°C are used to inoculate 10 ml of fresh medium with a dilution of about 1:1000. Cells are then grown at 32°C to an OD_{610} of 0.1 (usually 18 hours). This absorbance corresponds to about 10^8 colony-forming units (CFU)/ml for *A. laidlawii*. The culture is then split into two equal portions: one portion is placed in a water bath at 42°C to induce heat shock and the other is kept at 32°C as a control. Pulse labeling is done by adding 1 ml of cells growing at either 32 or 42°C to a prewarmed vial containing 50 µCi of radiolabeled amino acids. Pulses are begun at 0, 20, and 40 minutes after the shift to 42°C. The duration of all pulses is 20 minutes. Labeling is stopped by placing the sample on ice. Samples are then centrifuged in an Eppendorf centrifuge for 10 minutes at 4°C. Each cell pellet is resuspended and lysed in 50 µl of sample buffer.

3. Quantitation of acid-precipitable radioactivity. Acid-precipitable radioactivity is assayed by adding a 5-µl lysed cell sample to 5 ml cold 20% TCA (Cardillo *et al.*, 1979), which is then kept on ice for 30 minutes. Precipitates are filtered through glass fiber filters on a vacuum apparatus and are washed twice with 20% TCA and twice with acetone. Filters are then air dried. The dried filters are placed in scintillation vials with 3 ml scintillation fluid. Radioactivity is measured as counts per minute (cpm) in a scintillation counter with parameters set for ^{14}C detection. A total of about 1200 cpm for the 5-µl of cell lysate is a good level of incorporation for subsequent visualization by autoradiography. The cpm per sample may vary between strains of *A. laidlawii*; however, it has been found that the number of viable cells and the rate of label incorporation remain fairly constant for the 1-hour heat shock at 42°C. The entire remaining cell lysate from each time point is therefore used for SDS–PAGE analysis since the protein profile for each time point represents an equal number of cells.

4. PAGE analysis of heat shock proteins. Nystrom and colleagues have found that good one-dimensional resolution of *A. laidlawii* proteins is achieved with an SDS–PAGE system first outlined by Jergil and Ohlsson (Nystrom *et al.*, 1986). The protocol has been adapted to a slab gel system (Johansson, 1985).

The separation gel is 11.1% acrylamide with a 1% cross-link and is prepared as follows: 15 ml acrylamide solution [33% (w/v) acrylamide], 5 ml bisacrylamide solution [1% (w/v) bisacrylamide], 10 ml T buffer, 14 ml water, 1 ml ammonium persulfate solution [5% (w/v) ammonium persulfate], and 23 μl TEMED. A layer of 2-propanol is added after pouring the separation gel solution and is rinsed out prior to pouring the stacking gel. The stacking gel is prepared as follows: 2.5 ml acrylamide solution, 5.0 ml bisacrylamide solution, 2.5 ml low T buffer, 9.6 ml water, 0.4 ml ammonium persulfate solution, and 20 μl TEMED. Total gel dimensions are $17 \times 16 \times 0.1$ cm with 1-cm-wide slots. A separate cathode buffer (diluted 1:10) and anode buffer are used.

Prior to electrophoresis, 5 μl of a 0.5-mg/ml bromophenol blue solution is added to each cell lysate sample and the mixtures are placed in boiling water for 3–5 minutes. Hence, each sample loaded on the gel represents the total cellular protein from about 10^8 cells. Electrophoresis is then carried out at 200 V until the dye front migrates off the gel.

Following electrophoresis, gels are placed in fixer for 2 hours and then stained with Coomassie blue for 1 hour. Gels are destained overnight in acetic acid [10% (v/v)]. Gels containing radioactivity are soaked in autoradiography enhancer (EnHance) for 1 hour and then in water for 1 hour. Gels are then dried onto Whatman filter paper with a vacuum slab drier at 60°C and are exposed to X-Omat AR film for 7–10 days at -70°C. Relative estimates of band intensities on the autoradiograms can be made using a laser densitometer. Increased intensities of protein bands in heat-shocked samples, relative to bands in the control samples, identify heat shock proteins.

5. Immunoblot identification of proteins. Protein samples from heat-shocked cells are prepared as described earlier except that radioactive labeling is not carried out. The proteins are then separated by electrophoresis (as described earlier) and are transferred onto a 0.45-μm nitrocellulose filter using a semidry electroblotter. A sandwich is constructed, in sequence, of nine Whatman No. 1 filter papers, a nitrocellulose filter, the polyacrylamide gel, and another nine Whatman filter papers. All components except the gel are soaked in transfer buffer prior to blotting. The "sandwich" is placed in the blotter with the nitrocellulose filter proximal to the anode and transfer is carried out at 0.8 mA/cm^2 gel for 1 hour at room temperature. After transfer the filter is placed in a heat-sealable bag and soaked for 1 hour with sterile PBS containing 0.5% gelatin. All incubations and washes are performed on a rotary shaker at room temperature. The primary antibody is then diluted 1:1000 in PBS containing 0.25% gelatin, added to the filter, and incubated for 18 hours. The membrane is then washed three times for 10 minutes each in PBS. The peroxidase-conjugated secondary antibody is diluted 1:1500 in PBS containing 0.25% gelatin, added to the filter, and incubated for 1 hour. The filter is then washed as described earlier and developer solution is added until bands are visible. The filter is then rinsed with water and dried in the dark.

Discussion

These techniques have been used successfully to examine the heat shock response of *A. laidlawii* strains JA1 and K2 and *M. capricolum* strain Kid (Dascher *et al.*, 1990). Increased temperature (45°C) was needed to induce heat shock in the case of *M. capricolum* strain Kid due to its higher thermotolerance. The temperatures used for heat shock in our experiments were chosen based on the ability of the strain being tested to maintain a constant cell viability (CFU/ml) during the time course of the experiment (i.e., 1 hour). Loading the entire lysate from 1-ml samples of labeled cells for PAGE and autoradiography (less 5 μl for quantitation of acid precipitable material) will therefore provide a protein pattern reflecting the activity of the same number of viable cells in each sample without having to compensate for changes in cell number during heat shock. A temperature shift of 32 to 42°C for *A. laidlawii* is sufficient to inhibit cell growth and induce a strong heat shock response but not high enough to cause a loss of cell viability and consequent reduction in incorporation of radioactive label below detectable levels. For other mycoplasmas, to maintain a constant number of viable cells during heat shock the appropriate temperatures must be determined from cell viability curves at a range of temperatures (Dascher *et al.*, 1990). Heat shock can also be induced at lower temperatures where the bacteria continue to grow in culture. In these cases it may be necessary to determine the concentration of protein for each time point and adjust to a standard quantity.

The *Mycoplasma* and *Acholeplasma* species tested required the use of 20-minute pulse labeling times. These were necessary to achieve sufficient labeling of proteins for subsequent visualization on autoradiography. Even with these increased pulse times, differential temporal expression of several of the heat shock proteins can be observed (Dascher *et al.*, 1990).

Immunoblot analysis of heat-shocked samples may not effectively demonstrate increased expression of a particular protein due to constitutive expression at temperatures used in control samples. Immunoprecipitation of the heat shock protein from a cell lysate of radiolabeled bacteria (Sambrook *et al.*, 1989) may allow a more accurate estimation of alterations in the synthesis of a particular heat shock protein with high constitutive levels of expression. Another technique that may provide more accurate estimations of changes in protein expression is based on limiting dilution of total cell protein from control and heat-shocked cells followed by transfer to a nitrocellulose membrane by a slot-blot apparatus. The blotted protein samples could then be probed with an excess of antibody to the heat shock protein of interest. This technique would only be applicable if a monospecific or monoclonal antibody to the relevant protein could be obtained.

The identification of cross-reactive proteins in mycoplasmas to eubacterial GroEL and DnaK provides an interesting starting point for the investigation of proteins that may be involved in immunopathology. The GroEL homolog of chlamydia, for example, has been shown to be a potent stimulator of delayed type hypersensitivity in an animal model for trachoma (Morrison *et al.*, 1989). In addition, women with confirmed salpingitis, a serious sequela of pelvic inflammatory disease (PID), possess T cells specific for the chlamydial GroEL (Witkin *et al.*, 1993). Studies such as these might be extended to mycoplasmas given the correlation between PID and genital tract infection with mycoplasmas (Møller *et al.*, 1985).

The techniques described in this chapter provide a relatively simple method for the study of the stress response in mycoplasmas. This research area may be of interest from the standpoint of understanding both mycoplasma cell biology and the potential role of mycoplasma antigens in disease pathology.

References

Cardillo, T. S., Landry, E. F., and Wiberg, J. S. (1979). regA protein of bacteriophage T4D: Identification, schedule of synthesis, and autogenous regulation. *J. Virol.* **32**, 905–916.

Dascher, C. C., and Maniloff, J. (1992). Heat shock response. *In* "Mycoplasmas: Molecular Biology and Pathogenesis" (J. Maniloff, R. N. McElhaney, L. R. Finch, and J. B. Baseman, eds.), pp. 349–354. Am. Soc. Microbiol. Washington, DC.

Dascher, C. C., Poddar, S. K., and Maniloff, J. (1990). Heat shock response in mycoplasmas, genome limited organisms. *J. Bacteriol.* **172**, 1823–1827.

Johansson, K.-E. (1985). "Analysis of Proteins by Electrophoresis in Polyacrylamide and Agarose Gels: A Laboratory Manual." Institute of Biochemistry, Biomedical Center, Uppsala.

Møller, B. R., Taylor-Robinson, D., Furr, P. M., Toft, B., and Allen, J. (1985). Serological evidence that chlamydiae and mycoplasmas are involved in infertility of women. *J. Reprod. Fertil.* **73**, 237–240.

Morrison, R. P., Belland, R. J., Lyng, K., and Caldwell, H. D. (1989). Chlamydial disease pathogenesis: The 57-kD chlamydial hypersensitivity antigen is a stress response protein. *J. Exp. Med.* **170**, 1271–1283.

Nystrom, S., Johansson, K.-E., and Wieslander, A. (1986). Selective acylation of membrane proteins in *Acholeplasma laidlawii*. *Eur. J. Biochem.* **156**, 85–94.

Sambrook, J., Fritsch, E. F. and Maniatis, T. (1989). *In* "Molecular Cloning: A Laboratory Manual," 2nd Ed., Vol. 3, pp. 18.44–18.45. Cold Spring Harbor Laboratory Press, Cold Spring Harbor, NY.

Witkin, S. S., Jeremias, J., Toth, M., and Ledger, W. J. (1993). Cell-mediated immune response to the recombinant 57-kDa heat-shock protein of *Chlamydia trachomatis* in women with salpingitis. *J. Infect. Dis.* **167**, 1379–1383.

D5

NUCLEOLYTIC ACTIVITIES OF MYCOPLASMAS

F. Chris Minion and Karalee J. Jarvill-Taylor

General Introduction

Nuclease activity in mycoplasmas was originally observed in 1957 by Plackett. Since that time, work has progressed sporadically in this area (Minion et al., 1993; Pollack and Hoffmann, 1982; Razin et al., 1964; Roganti and Rosenthal, 1983). DNases and RNases have been identified in all mycoplasma species tested (Minion et al., 1993). These activities do not seem to be due to a single phosphodiesterase, but rather multiple nucleases seem to be present in mycoplasmas. Both soluble (Pollack and Hoffmann, 1982; Razin et al., 1964; Roganti and Rosenthal, 1983) and membrane-associated (Minion and Goguen, 1986; Minion et al., 1993) nucleases have been identified.

Many mycoplasmas are incapable of synthesizing nucleic acid precursors and thus must acquire them from their host or growth medium. Many nucleases are localized to the membrane (Minion et al., 1993) and consequently would provide the enzymatic activities necessary to derive these precursors from exogenous nucleic acids. In addition to providing necessary nutrients, these nucleases might also provide a protective barrier against invading foreign DNA.

Analysis of Nuclease Activity on Agar Surfaces

Introduction

Visualization of nuclease activities on solid agar surfaces can be accomplished using the assays described in this chapter. An advantage of these techniques is the screening for nuclease-deficient mutants. Null mutants might be identified by the lack of a halo around the colony or, conversely, other mutants might be identified with altered halo sizes.

Standard Plate Assay (Razin et al., 1964)

MATERIALS

Standard mycoplasma agar plates containing 0.5–7 mg/ml calf thymus DNA
1 M HCl
Broth culture of mycoplasmas

PROCEDURE

1. Inoculate plates with mycoplasma suspensions to produce isolated colony growth. For best results, space the colonies evenly around the plate by spreading the inoculum with a sterile glass rod. The cell density should be no greater than 200–250 colonies per plate.
2. Incubate the plates at the standard growth temperature until the colonies are visible.
3. Gently flood the plate with 1 M HCl.
4. Clear zones in and around the colonies indicate nuclease activity; cloudy zones contain no nuclease activity.

Methyl Green DNase Assay (Roganti and Rosenthal, 1983)

MATERIALS

Methyl green
Chloroform
Standard mycoplasma plates
Broth culture of mycoplasmas
0.25 M Tris (pH 7.6)
DNA
Agarose

PROCEDURE

Methyl Green Purification

The methyl green solution is prepared by dissolving 0.5 g of the dye in 100 ml water and then extracting the solution with an equal volume of chloroform until the chloroform is colorless. The residual chloroform is evaporated off by heating the solution to 65°C for 10 minutes and the methyl green solution is filter sterilized after cooling.

Preparation of DNA Agarose

1. Dissolve DNA in sterile water at 2 mg/ml.
2. Autoclave 1.0 g of low-melting-temperature agarose (or 0.75 g of Noble agar) in 30 ml of 0.25 M Tris.
3. To cooling agar, add 50 ml of 2 mg/ml DNA and 20 ml of the methyl green solution to give a final concentration of 1 mg/ml of the dye.
4. Melt the agar mixture in a microwave prior to use.

Assay

1. Standard mycoplasma agar plates are inoculated with organisms to produce single colonies (200–250 colonies/plate).
2. Plates are incubated until colonies reach a diameter of 250–500 µm.
3. The plate is overlaid with 10 ml of the DNA–methyl green agarose equilibrated to 40°C. Allow the top agar to harden and place the plate in a 37°C incubator.
4. The plates are incubated until clear zones around colonies are apparent.

DISCUSSION

Screening for nuclease activity in colonies is rapid, but requires a replica plating technique to ensure recovery of viable organisms at the conclusion of the experiment. Two replica techniques may be tried: the filter replicating method of Sundstrom and Wieslander (1990) used with *Acholeplasma laidlawii* and a system where individual colonies are picked, stored in 96-well plates, and replica plated using a 48-pin replicator. The advantage of the first method is that technician time is significantly reduced, but it may not work with many mycoplasmal species. The second method requires considerable technical effort to construct the arrays, but several different types of screens can be performed with the same set of cultures. In both cases, initial results must be confirmed by filter cloning the isolate (Tully, 1983) and repeating the assay.

Spectrophotometric Determination of DNases and RNases (Razin *et al.*, 1964)

DNA, RNA, and their degradation products can be quantitated spectrophotometrically at 260 nm. To analyze the nuclease activity, intact DNA or RNA is precipitated to leave the released nucleotides free in solution for spectrophotometric analysis.

MATERIALS

0.4 M Tris, 0.04 M MgCl$_2$, pH 8.8
0.25 M NaCl
10 mg/ml yeast tRNA or calf thymus DNA solution
Washed organisms or cell extract
0.75% uranyl acetate in 25% perchloric acid (w/v)

PROCEDURE

1. Combine 3 ml of 0.4 M Tris, 0.04 M MgCl$_2$, pH 8.8; 2.4 ml 0.25 M NaCl; 4.2 ml of cell extract or organisms, 1.2 ml of RNA or DNA; and 1.2 ml water. Keep ratios the same if smaller reaction volumes are used.
2. Incubate the suspension at 37°C and sample at various time points.
3. To analyze activity, add 0.5 ml of the uranyl acetate–perchloric acid to 2 ml of reaction mixture and mix well.
4. Incubate the mixture on ice for 10 minutes and centrifuge for 10 minutes at 10,000 g at 5°C.
5. Add 3.5 ml water to 0.5 ml of the supernatant.
6. Read absorbance at 260 nm.
7. Nuclease activity is assessed according to the increase in absorbance compared to controls. Units equal the absorbance increase divided by the milligrams per milliliter of protein.

DISCUSSION

Although this procedure uses several potentially hazardous chemicals, it is relatively simple to perform and allows the use of either RNA or DNA as a substrate. Cell extracts or washed organisms may be used. The assay by itself would not distinguish between cytoplasmic or membrane-bound nuclease activity.

Radioactive Procedure for DNase Measurement (Pollack and Hoffmann, 1982)

Radioactive assays have been the standard assays for DNase activity for many years. Alternatives have been developed, but these alternatives continue to be evaluated against the radioactive methods. The type of radioactive substrate chosen may be altered based on experimental design.

MATERIALS

25 mM Tris–HCl, pH 8.0; 7.5 mM MgCl$_2$; 1.5 mM CaCl$_2$; 5 mM 2-mercaptoethanol
Bovine serum albumin (BSA)
^{14}C-labeled *Escherichia coli* DNA (Du Pont NEN Research Products, Boston, MA)
Cell extract containing 0.2–15.5 μg protein
Calf thymus DNA
5% trichloroacetic acid (aqueous, w/v) (TCA)
1.2 M KOH
Scintillation fluid
DNase control (bovine pancreas DNase I, Bethesda Research Laboratories, Gaithersburg, MD)

PROCEDURE

1. Combine the cell extract, 3 μg ^{14}C-labeled *E. coli* DNA, 100 μg BSA, and the Tris–cation buffer to a final volume of 200 μl.
2. Incubate for a minimum of 30 minutes at 37°C.
3. Stop reaction by adding excess amounts of calf thymus DNA.
4. Add 25 μl of cold 5% aqueous TCA, incubate for 10 minutes on ice, and centrifuge at 1000 g for 10 minutes.
5. Add 50 μl of KOH and 7.5 ml of scintillation fluid to the supernate and determine the counts per minute.
6. Activity is expressed as counts per milligram of cell protein.

Nonradioactive Procedures for DNase Measurements (Minion *et al.*, 1993; Pollack and Hoffmann, 1982)

The substrate chosen for the nonradioactive analysis of nuclease activity may include plasmid DNA (open or closed circular), linear DNA such as λ, or chromosomal DNA. The substrate chosen will determine the end product.

MATERIALS

Phosphate-buffered saline (PBS) or Tris–saline (TS) containing 2 mM MgCl$_2$ and 2 mM CaCl$_2$
DNA substrate, 10 μg/ml in PBS or TS
Agarose
Broth culture of organisms or cell extract

0.089 M Tris–0.089 M boric acid–0.002 M EDTA (TBE buffer)
Tris–borate–EDTA gel-loading buffer (6× TBE buffer–25% Ficoll or 30% glycerol–0.25% bromophenol blue)

PROCEDURE

1. Prepare double dilutions of sample in 25 μl PBS or TS containing cations.
2. Add 25 μl of DNA substrate to each dilution.
3. Incubate at 37°C for a minimum of 30 minutes.
4. Stop reaction by the addition of 7 μl of TBE gel-loading buffer.
5. Analyze reaction products on a 0.7–1.0% agarose gel (Sambrook et al., 1989).

DISCUSSION

Analysis of nucleases by the nonradioactive method is not only convenient but the assay can be modified for many purposes, such as analyzing membrane fractions and cellular fractions. Increased incubation time leads to continued digestion, which may be beneficial when small amounts of protein are assayed. When analyzing membrane-associated nucleases, the use of λ DNA greatly increases the sensitivity of the assay (Minion et al., 1993). The concentration of calcium and magnesium required for maximum activity may vary between species. The membrane-associated nuclease activity of Mycoplasma capricolum was inhibited by calcium and required higher concentrations of magnesium. Other species, such as Mycoplasma pulmonis, require both cations. Zinc has been shown to be inhibitory in some species (Razin et al., 1964). All solutions should be autoclaved prior to use to inactivate residual nuclease activity.

SDS–PAGE Assay of Nucleases (Minion et al., 1993)

Previous research has established that mycoplasmas contain multiple nuclease activities (Minion et al., 1993). Although different nucleolytic proteins are evident by SDS–PAGE, these proteins may be encoded by a single gene displaying size variation (Rosengarten and Wise, 1990) or may represent the expression of several genes. The use of the SDS–PAGE assay may provide useful markers for mycoplasma speciation and as a basis for purification of the individual nucleases.

MATERIALS

Mini-PROTEAN II SDS–PAGE apparatus (Bio-Rad, Richmond, CA)
10% resolving gel containing 10 μg/ml sheared salmon sperm DNA with a 3% stacking layer

Gel-loading buffer (2.5 ml 0.5 M Tris, pH 6.8–2 ml glycerol–2 ml 20% SDS–1 ml 2-mercaptoethanol–0.8 ml 0.05% bromophenol blue–1.7 ml water). Filter sterilize immediately after preparation and store at -20°C to prevent bacterial growth.

5X tray buffer (per liter: 15 g Tris–72 g glycine–5 g SDS)

Protein molecular weight standards

Incubation buffer (final concentrations: 0.04 M Tris, pH 7.5; 0.01% casein; 0.04% 2-mercaptoethanol). The buffer is prepared as a 50 or 100X stock solution without the 2-mercaptoethanol.

0.1 M $CaCl_2$

1 M $MgCl_2$

Ethidium bromide (10 mg/ml in water)

Silver stain kit (Accurate Scientific Chemical Co., Westbury, NY) or a similar procedure (Wray *et al.*, 1981).

50% ethanol/12% acetic acid

40% glycerol

PROCEDURE

Sample Preparation

1. Types of samples: Grow broth culture to mid-log phase, sediment organisms, wash once with PBS, and resuspend in 40% glycerol. Store at -20°C until use. Fresh culture can be used directly by washing organisms and resuspending in sterile water. Cellular extracts can be used if stored at -20°C or used directly after preparation.

2. Samples are mixed 1:1 with gel-loading buffer.

3. Prior to loading gel, samples are boiled for 3 minutes. Alternatively, samples may be treated for 30 minutes at 37°C or for 15 minutes at 65°C.

SDS–PAGE Gel

1. Prepare a 10% resolving gel containing 10 μg/ml sheared DNA. Denature the DNA by boiling prior to adding to the gel mixture.

2. Complete the preparation of the gel by adding a 3% stacking gel containing no DNA.

3. After preparing the samples, load them into the wells.

4. Use 1X tray buffer as the cathode buffer and 0.5X buffer as the anode buffer. The cathode buffer may be supplemented with 0.15% SDS to facilitate separation.

5. Electrophoresis is performed in the cold at 10 mA per gel until the dye front exits the gel. These conditions are essential if nuclease activity is to be maintained. Pay particular attention to the running temperature of the gel; it must not rise above 8°C at any time.

Renaturation and Incubation for DNase Activity

1. Following SDS–PAGE, the gels are washed four times at room temperature for 15 minutes each time in 100 vol of incubation buffer. The gels can then either be incubated for analysis or be maintained overnight in incubation buffer at room temperature.
2. To stimulate DNase activity, 2 mM $CaCl_2$ and 2 mM $MgCl_2$ are added directly to the incubation buffer and the gels are incubated at 37°C for 8 hours or overnight. For more concentrated samples, shorter incubation times may be used.
3. Visualization of activity is performed by removing all but 100 ml of the incubation buffer and adding 100 µl of ethidium bromide. Gels are allowed to stain for 5–10 minutes and are then destained in distilled water for 20–40 minutes.
4. Under UV light, nonfluorescing regions in the gel indicate nuclease activity.

Silver Staining

1. Once the ethidium bromide-stained gels have been photographed, the gels are washed twice for 30 minutes in water to remove residual incubation buffer and are fixed for 1 hour in 50% ethanol in 12% acetic acid. The gel is then washed several times for 2 hours with distilled water to remove acetic acid. Methanol (50%) may be used to facilitate the removal of acetic acid and to reduce background staining. Gels should be equilibrated in water prior to staining.
2. The fixed proteins in the gel are then stained using silver staining methods. We have always used an ammonium hydroxide procedure, but other procedures such as that described by Wray *et al.,* (1981) may be just as effective.
3. Soak gels in water prior to drying.

DISCUSSION

The SDS–PAGE assay has been used to analyze the nuclease activity of cells from over 25 different strains of mycoplasmas, cell extracts, and fractions from ion-exchange columns. Samples for analysis are stable at -20°C for at least a year and at 4°C for a month. Care must be taken to ensure that electrophoresis is performed very slowly and at 4°C. Activity can be lost at electrophoresis temperatures higher than 6°C. Renaturation of the proteins occurs most likely by the removal of the SDS and the incorporation of 2-mercaptoethanol in the buffer. The incubation times for the washed gels may be increased, but should not be decreased. The period of time used to allow DNA digestion depends on the type of sample being used and the amount of protein loaded. Intensity of signal does not necessarily correspond with the amount of protein as some nucleases may be more active than others.

In order to visualize the proteins by the silver stain, the gels must be washed thoroughly to remove the casein from the incubation buffer. The gels will still turn dark due to the staining of the DNA in the gel. Adding larger amounts of the molecular weight standard is often helpful. On the addition of the developer, the solution may turn dark. To avoid this, thoroughly rinse the gel and the dish with water between the silver nitrate and developer steps. When developer is added, agitate the gel rapidly. If the solution does turn dark, remove it promptly, rinse the gel with developer, and then add fresh developer. Continue to stain until molecular weight markers are apparent. The proteins stain differently than in a normal SDS–PAGE gel, and protein profiles may not necessarily correspond to the proteins present in the sample. Many nuclease-containing bands do not appear to stain with either silver or Coomassie blue.

References

Minion, F. C., and Goguen, J. D. (1986). Identification and preliminary characterization of external membrane-bound nuclease activities in *Mycoplasma pulmonis*. *Infect. Immun.* **51**, 352–354.

Minion, F. C., Jarvill-Taylor, K. J., Billings, D. E., and Tigges, E. (1993). Membrane-associated nuclease activities in mycoplasmas. *J. Bacteriol.* **175**, 7842–7847.

Plackett, P. (1957). Depolymerization of ribonucleic acid by extracts of *Asterococcus mycoides*. *Biochim. Biophys. Acta* **26**, 664–665.

Pollack, J. D., and Hoffmann, P. J. (1982). Properties of nucleases of Mollicutes. *J. Bacteriol.* **152**, 538–541.

Razin, S., Knyszynski, A., and Lifshitz, Y. (1964). Nucleases of mycoplasma. *J. Gen. Microbiol.* **36**, 323–331.

Roganti, R. S., and Rosenthal, A. L., (1983). DNase of *Acholeplasma* spp. *J. Bacteriol.* **155**, 802–805.

Rosengarten, R., and Wise, K. S. (1990). Phenotypic switching in mycoplasmas: Phase variation of diverse surface lipoproteins. *Science* **247**, 315–318.

Sambrook, J., Fritsch, E. F., and Maniatis, T. (1989). "Molecular Cloning: A Laboratory Manual." Cold Spring Harbor Laboratory, Cold Spring Harbor, NY.

Sundstrom, T. K., and Wieslander, A. (1990). Plasmid transformation and replica filter plating of *Acholeplasma laidlawii*. *FEMS Microbiol. Lett.* **72**, 147–152.

Tully, J. G. (1983). Cloning and filtration techniques for mycoplasmas. *In* "Methods in Mycoplasmology" (S. Razin and J. G. Tully, eds.), Vol. 1, pp. 173–177. Academic Press, New York.

Wray, W., Boulikas, T., Wray, V. P., and Hancock, R. (1981). Silver staining of proteins in polyacrylamide gels. *Anal. Biochem.* **118,** 197–203.

D6

PROTEOLYTIC ACTIVITIES

Tsuguo Watanabe and Ken-ichiro Shibata

General Introduction

Human mycoplasmas and ureaplasmas digest horse serum proteins (Watanabe et al., 1973), bovine albumin, and casein (Watanabe, 1975). The caseinolytic activity of *Mycoplasma salivarium* was suggested to be due mainly to metalloproteinases and partly to serine proteinases (Watanabe et al., 1984).

Aminopeptidase activity is present in *Ureaplasma urealyticum, Mycoplasma pneumoniae, Mycoplasma hominis,* and *Mycoplasma fermentans,* and carboxypeptidase or endopeptidase activity is also thought to exist in *M. pneumoniae* (Vinther and Black, 1974).

Arginine-specific aminopeptidase (Shibata and Watanabe, 1987) and carboxypeptidase (Shibata and Watanabe, 1988) were purified from *M. salivarium.* The vascular permeability-increasing activity of bradykinin, a chemical mediator of the inflammatory response, is inactivated by the aminopeptidase activity of fermentative and nonfermentative mycoplasmas and by the carboxypeptidase activity of the latter (Shibata and Watanabe, 1989). The carboxypeptidase released the C-terminal arginine residue of tuftsin, a peptide found to stimulate all the known functions of phagocytic cells, abolishing its activity (Shibata and Watanabe, 1988). Therefore, the enzyme is presumed to inactivate complement components such as C3a and C5a as well as carboxypeptidase N (anaphylatoxin inactivator).

Immunoglobulin (Ig) A1 protease activity has been identified in *Ureaplasma urealyticum* (Kilian et al., 1984; Robertson et al., 1984), but not in *Mycoplasma* and *Acholeplasma* species. The just-mentioned peptidase and IgA1 protease activities seem to function *in vivo* to evade the defense system of the host.

Digestion of Serum Proteins

Assay in Agar Plate Cultures

MATERIALS

Hayflick's liquid and agar medium containing 20% (v/v) horse serum, and fresh yeast extract, heated at 60°C for 1 hour
Log-phase broth culture of the test organism
Sterile liquid medium control
Agar medium plates: 6 ml of agar medium plated in petri dishes measuring 5.5 cm in diameter.
$(NH_4)_2SO_4$, saturated
Sterile centrifuge tubes
Refrigerated high-speed centrifuge

PROCEDURE

1. Harvest the cells by centrifugation of the broth cultures (100 ml) at 15,000 g for 15 minutes at 4°C. Treat the sterile liquid medium control similarly.
2. Resuspend the pellet in 0.5 ml of sterile liquid medium.
3. Inoculate agar medium plates with 0.01 or 0.05 ml of the suspension, and incubate at 37°C for 3 to 7 days anaerobically.
4. Overlay the agar plates with saturated $(NH_4)_2SO_4$.

Digestion of serum proteins is indicated by a clear zone around the growth of the organism, visible in contrast to the opaque precipitate of the undigested proteins.

DISCUSSION

All steps must be carried out aseptically. Penicillin G and polymyxin B should be incorporated in the medium used for all procedures. Incubating agar plates under anaerobic conditions suppresses the proteolytic aerobes, such as *Bacillus subtilis,* which may contaminate the mycoplasma cell suspensions.

The density of the mycoplasma cell suspensions, incubation time, and the thickness of the agar affect the results. Occasionally, a clear zone produced around a single colony can be seen.

Normally, the pellet from the sterile liquid medium control is negligible in volume as well as in proteolytic activity.

Assay in PBS–Agar Plates

MATERIALS

Log-phase liquid cultures of the test organism
Sterile liquid medium control
Membrane filter, 0.2-μm pore diameter
Phosphate buffer, 0.01 M, pH 7.2. Prepare solution A: 1.36 g KH_2PO_4 per liter distilled water; solution B: 1.42 g Na_2HPO_4 per liter distilled water. Add solution A to solution B, stirring constantly, until pH 7.2 is obtained.
Sterile phosphate-buffered saline (PBS): Dissolve 8.77 g NaCl in 1 liter of 0.01 M phosphate buffer, pH 7.2. Autoclave at 121°C for 15 minutes.
Bovine serum albumin (BSA), 10% (w/v): Dissolve 10 g BSA in 100 ml PBS. Sterilize by filtration through a membrane filter.
Horse serum: heat at 60°C for 1 hour.
Polymyxin B, 50,000 U/ml: Dissolve 500,000 U polymyxin B in 10 ml of sterile distilled water.
Penicillin G, 100,000 U/ml: Dissolve 1,000,000 U penicillin G in 10 ml of sterile distilled water.
Agar plate: Add 1 g agar to 90 ml PBS (PBS–agar). Autoclave at 121°C for 20 minutes and cool to 50°C. Supplement with 10 ml of horse serum or 10% (w/v) BSA, 1 ml each of penicillin G solution (100,000 U/ml) and polymyxin B solution (50,000 U/ml), mix well, and plate in 17 petri dishes measuring 5.5 cm in diameter.
$(NH_4)_2SO_4$, saturated
Sterile centrifuge tubes
Refrigerated high-speed centrifuge

PROCEDURE

1. Harvest cells by centrifugation of the cultures (100 ml) of the test organism at 15,000 g for 15 minutes at 4°C. Treat the sterile control liquid medium similarly.
2. Wash the pellet three times with PBS and resuspend in 0.5 ml PBS.
3. Place 0.01 or 0.05 ml of the pellet suspension on the agar plates.
4. Incubate the agar plates for 3 to 7 days under anaerobic conditions and overlay with saturated $(NH_4)_2SO_4$.

Hydrolysis of proteins is indicated by a clear zone around a film of the cells, visible in contrast to the opaque precipitate of undigested proteins.

DISCUSSION

All steps should be carried out aseptically. The density of mycoplasma cell suspensions, incubation time, and the thickness of the agar affect the results.

Digestion of Casein

MATERIALS

Washed cell suspensions (1 mg of cell protein/ml) of the test organism
PBS
Casein, 0.5% (w/v): Dissolve 0.5 g casein in 100 ml PBS and boil for 30 minutes.
Trichloroacetic acid (TCA), 10% (w/v)
Lowry's solutions:
   A: Dissolve 0.4 g NaOH in 100 ml distilled water and then add 2 g Na_2CO_3.
   B: Dissolve 1 g sodium citrate in 100 ml distilled water and then 0.5 g $CuSO_4 \cdot 5H_2O$.
   C: Combine 50 ml of solution A and 1 ml of solution B immediately before use.
Folin and Ciocalteu's phenol reagent (Sigma): Dilute to 1 N with distilled water before use.
Spectrophotometer and 1-cm cuvettes
Refrigerated high-speed centrifuge
Water bath, 37°C

PROCEDURE

1. Incubate reaction mixtures consisting of 1.0 ml of 0.5% (w/v) casein and 0.2 ml of a cell suspension at 37°C for appropriate periods (for example, 0, 30, 60, or 120 minutes).
2. Add 1.2 ml of 10% (w/v) TCA to the reaction mixtures to stop the reaction, shake vigorously, and allow to stand at room temperature for 30 minutes.
3. Centrifuge the reaction mixtures at 15,000 g for 30 minutes at room temperature. Separate the supernatants or filter them through Whatman No. 1 filter paper.
4. Add 3 ml of Lowry's solution C to 0.2 ml of the supernatants or filtrates and to distilled water as blank, shake the mixtures vigorously, and keep them at room temperature for 30 minutes.
5. Add 0.5 ml phenol reagent, mix well, and allow to stand at room temperature for 30 minutes. Read in a spectrophotometer at 750 nm.
6. As controls, use reaction mixtures without a cell suspension or without casein.

DISCUSSION

Amino groups liberated by casein digestion over a defined period are measured by means of this procedure.

To avoid serum components affecting the results, PPLO serum fraction (Difco) is recommended as an ingredient of the growth medium for the test organism instead of horse serum. However, PPLO serum fraction may not support the growth of some mycoplasmas, such as *M. pneumoniae*.

Exopeptidase Activity

MATERIALS

For All Methods

Washed cell suspensions (1 mg of cell protein/ml) of the test organism
Refrigerated low-speed centrifuge
Water bath, 37°C
Spectrophotometer and 1-cm cuvettes

Aminopeptidase Activity Assay

p-Nitroanilide derivatives:
　L-Leucine *p*-nitroanilide (LNA) (Protein Research Foundation, Osaka, Japan)
　L-Arginine *p*-nitroanilide (ANA) (Sigma)
TCA, 40% (w/v)
Dimethyl sulfoxide (DMSO)
$MnCl_2$, 0.02 M

Carboxypeptidase Activity Assay

N-Benzoylglycyl-L-arginine (Bz-Gly-Arg) (Protein Research Foundation)
TCA, 20% (w/v)
NaOH, 2.5% (w/v)
Ninhydrin solution: Dissolve 2.0 g ninhydrin and 0.3 g hydrindantin in 75 ml methyl Cellosolve. Combine with 25 ml of 4 M sodium acetate buffer, pH 5.5. (Prepare solution A: Dilute 23 ml of glacial acetic acid in 100 ml distilled water; solution B: 32.9 g anhydrous sodium acetate per 100 ml distilled water. Add solution A to solution B, stirring constantly, until pH 5.5 is obtained.) Prepare immediately before use.
Propanol, 50% (v/v)

Thin-Layer Chromatography

Peptides:
    L-Leu-L-Ala (Aldrich Chemical Co.)
    L-Ala-L-Ala(Ala_2), Ala_3, Ala_4 (Sigma)
    N-Acetyl-Ala_3 and -Ala_4 (Sigma)
    H-Thr-Lys-Pro-Arg-OH (tuftsin) (Protein Research Foundation)
    H-Arg-Pro-Pro-Gly-Phe-Ser-Pro-Phe-Arg-OH (bradykinin) (Protein Research Foundation)

Amino acids (references for thin-layer chromatography):
    L-Arginine
    L-Alanine
    L-Leucine

n-Butanol
Acetone
Acetic acid
Ninhydrin, 0.2% (w/v): Dissolve 0.2 g ninhydrin in 100 ml acetone.
$Cu(NO_3)_2$, saturated: Dissolve 65 g $Cu(NO_3)_2$ in 100 ml distilled water.
HNO_3, 10% (v/v)
Ethanol
Fixing solution: Combine 1 ml of saturated $Cu(NO_3)_2$, 0.2 ml of 10% (v/v) HNO_3, and 100 ml ethanol.
Ion retardation resin (Bio-Rad Laboratories, Richmond CA)
Precoated silica plate (HPTLC Si 50000; E. Merck AG, Darmstadt, Germany)
Developing tanks
Scent-spray type of apparatus

Formulas of Buffers

Borate buffer, 0.1 M, pH 8.0. Prepare solution A: 6.18 g H_3BO_3 per liter distilled water; solution B: 10.60 g Na_2CO_3 per liter distilled water. Add solution B to solution A, stirring constantly, until pH 8.0 is obtained.

Borate-buffered saline (BBS): Dissolve 8.77 g NaCl in 1 liter of 0.1 M borate buffer, pH 8.0.

Phosphate buffer, 0.1 M, pH 6.5. Prepare solution A: 13.60 g KH_2PO_4 per liter distilled water; solution B: 14.2 g Na_2HPO_4 per liter distilled water. Add solution B to solution A until pH 6.5 is obtained.

Sodium acetate buffer, 0.1 M, pH 6.0. Prepare solution A: Dilute 5.75 ml of glacial acetic acid in 1 liter distilled water; solution B: 8.20 g anhydrous sodium acetate per liter distilled water. Add solution A to solution B until pH 6.0 is obtained.

Assay by Spectrophotometry

Aminopeptidase Activity

PROCEDURE

1. Incubate reaction mixtures consisting of 3.0 ml BBS, 0.2 ml of 0.02 M $MnCl_2$, and 0.2 ml of a cell suspension at 37°C for 15 minutes.
2. Add 0.05 ml of 0.1 M LNA or ANA dissolved in DMSO to start the reaction. Shake the mixtures vigorously and then incubate at 37°C for appropriate periods (for example, 0, 30, or 60 minutes).
3. Add 0.1 ml of 40% (w/v) TCA to the reaction mixtures to stop the reaction, shake the mixtures vigorously, and allow them to stand on ice for 30 minutes.
4. Centrifuge the reaction mixtures at 1800 g for 30 minutes and separate the supernatants or pass the reaction mixtures through filter paper.
5. Read the supernatants in a spectrophotometer at 405 nm. One unit of enzyme is defined as 1 μmol of p-nitroaniline liberated per minute at 37°C, taking the molar extinction coefficient at 405 nm as 7620 M^{-1} cm^{-1}.
6. As controls, use reaction mixtures without a cell suspension or without a derivative of p-nitroaniline.

DISCUSSION

When the enzyme is tested in a water-soluble form, p-nitroaniline liberation can be continuously monitored at 405 nm, using a recording spectrophotometer with a thermostated cuvette holder.

Carboxypeptidase Activity

PROCEDURE

1. Incubate reaction mixtures consisting of 0.2 ml of 0.1 M sodium acetate buffer (pH 6.0), 0.2 ml of 5 mM Bz-Gly-Arg dissolved in distilled water, and 0.1 ml of a cell suspension at 37°C for appropriate periods (for example, 0, 30, or 60 minutes).
2. Add 0.5 ml of 20% (w/v) TCA to the reaction mixtures to stop the reaction. Shake the mixtures vigorously and then leave them on ice for 30 minutes.
3. Centrifuge the reaction mixtures at 1800 g for 30 minutes.
4. Add 0.5 ml of the supernatant to 0.5 ml of 2.5% (w/v) NaOH.
5. Add 1.0 ml ninhydrin solution to the mixtures. Boil the mixtures for 15 minutes, then cool under running tap water.

6. Add 5 ml of 50% (v/v) 2-propanol and read at 570 nm in a spectrophotometer.

7. Control reaction mixtures without cell suspension or Bz-Gly-Arg undergo the same procedure.

DISCUSSION

One unit of the activity is defined as the amount of enzyme required to liberate 1 μmol of L-arginine per minute at 37°C. A calibration curve is obtained by treating various amounts of L-arginnine as described earlier.

Assay by Thin-Layer Chromatography

PROCEDURE

1. Incubate reaction mixtures consisting of 0.5 ml each of 0.1 M borate buffer (pH 8.0), a cell suspension, and 0.02 M peptide (5 mM for tuftsin or bradykinin) at 37°C for 2 hours.

2. Add 0.2 ml of 40% (w/v) TCA to the reaction mixtures, shake the mixtures vigorously, then allow to stand at room temperature for 30 minutes, and centrifuge at 15,000 g for 30 minutes.

3. Spot aliquots (2 μl) of the separated supernatants, desalted by ion retardation resin, and mixtures of amino acids as references on precoated silica plates, 1 cm from the bottom edge of the plate.

4. Dry the spots and insert the chromatoplates into the developing tanks.

5. Develop the chromatogram in the solvent system n-butanol–acetone–acetic acid–water (35/35/10/20, v/v/v/v).

6. Remove the plate from the developing tank and dry at room temperature.

7. Spray the plate with 0.2% (w/v) ninhydrin, air dry, and heat at 100°C for 10 minutes. Amino acids and peptides yield blue spots on a white background.

8. Spray the plate with a fixing solution.

9. Control reaction mixtures without cell suspension or with no peptides undergo the same procedure.

DISCUSSION

Arginine cannot be detected in chromatograms of reaction mixtures consisting of nonfermentative mycoplasma cells and Bz-Gly-Arg, tuftsin, or bradykinin. However, citrulline is detectable. This is because arginine is hydrolyzed by the arginine dihydrolase pathway, immediately after liberation from the peptides. Arginine is detected in reaction mixtures consisting of fermentative mycoplasma cells and tuftsin or bradykinin, but not in those of the cells with Bz-Gly-Arg.

References

Kilian, M., Brown, M. B., Brown, T. A., Freundt, E. A., and Cassell, G. H. (1984). Immunoglobulin A1 protease activity in strains of *Ureaplasma urealyticum*. *Acta Pathol. MIcrobiol. Scand. Sect. B* **92,** 61–64.

Robertson, J. A., Stemler, M. E., and Stemke, G. W. (1984). Immunoglobulin A protease activity of *Ureaplasma urealyticum*. *J. Clin. Microbiol.* **19,** 255–258.

Shibata, K., and Watanabe, T. (1987). Purification and characterization of an aminopeptidase from *Mycoplasma salivarium*. *J. Bacteriol.* **169,** 3409–3413.

Shibata, K., and Watanabe, T., (1988). Purification and characterization of an arginine-specific carboxypeptidase from *Mycoplasma salivarium*. *J. Bacteriol.* **170,** 1795–1799.

Shibata, K., and Watanabe, T. (1989). Inactivation of the vascular permeability-increasing activity of bradykinin by mycoplasmas. *FEMS Microbiol. Lett.* **65,** 149–152.

Vinther, O., and Black, F. T. (1974). Aminopeptidase activity of *Ureaplasma urealyticum*. *Acta Pathol. Microbiol. Scand. Sect. B* **82,** 917–918.

Watanabe, T. (1975). Proteolytic activity of *Mycoplasma salivarium* and *Mycoplasma orale 1*. *Med. Microbiol. Immunol.* **161,** 127–132.

Watanabe, T., Mishima, K., and Horikawa, T. (1973). Proteolytic activities of human mycoplasmas. *Jpn. J. MIcrobiol.* **17,** 151–153.

Watanabe, T., Shibata, K., and Totsuka, M. (1984). Aminopeptidase and caseinolytic activities of *Mycoplasma salivarium*. *Med. Microbiol. Immunol.* **172,** 257–264.

D7

PHOSPHOLIPASE ACTIVITY IN MYCOPLASMAS

Shlomo Rottem and Michael Salman

Introduction

Phospholipases are important components of the cellular machinery. The bonds hydrolyzed by phospholipases A_1, A_2, C, and D are shown in Fig. 1. The hydrolysis yields fatty acids, lysophospholipids, diacylglycerol, and phosphatidic acid. These products may serve as precursors for membrane remodeling or as intracellular secondary messengers. In eukaryotic cells, the polyunsaturated fatty acid released by phospholipase A_2 (PLA_2) can be further utilized in the production of specific proinflammatory lipid mediators, e.g., eicosanoids such as prostaglandins, leukotrienes, or thromboxanes (Mayer and Marshall, 1993).

Against the wealth of information available on mammalian phospholipases, little is known about phospholipase activity in mollicutes. A membrane-bound lysophospholipase was described in *Acholeplasma laidlawii* (van Golde *et al.*, 1971) and *Mycoplasma gallisepticum* (Gatt *et al.*, 1982). This enzyme hydrolyzes micellar dispersions of lysophosphatidylcholine and lysophosphatidylglycerol, as well as endogenous lysophospholipids generated by treating membranes with purified preparations of PLA_2. The high levels of free fatty acids in the cell membrane of *Spiroplasma* species were taken to suggest an endogenous phospholipase activity in spiroplasmas (Davis *et al.*, 1985). A phospholipase activity, markedly stimulated by low concentrations of nonionic detergents, was detected in *M. gallisepticum* membranes (Rottem *et al.*, 1986). Phospholipase A and C activities were detected in *Ureaplasma urealyticum* (De Silva and Quinn, 1986)

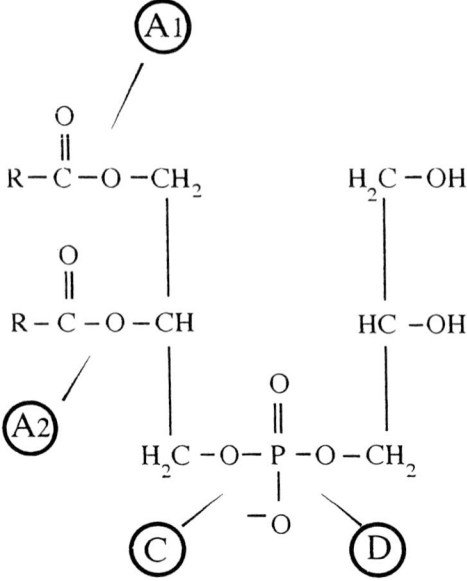

Fig. 1. Chemical structure of phosphatidylglycerol showing the location of the bonds hydrolyzed by PLA_1 (A1), PLA_2 (A2), PLC (C), and PLD (D).

and a phospholipase A_1 activity was demonstrated in the cell membrane of *M. penetrans* (Salman and Rottem, 1995).

Materials

Cell Preparations

Suspensions of intact cells in isotonic buffer (1–5 mg cell protein/ml), cell lysates (1–5 mg cell protein/ml), and isolated membrane preparations (1–2 mg membrane protein/ml) were made.

Chemicals

General chemicals and solvents: $CaCl_2$, EDTA, EGTA, chloroform, methanol, acetic acid, ammonium hydroxide, nitrogen tank, 2-mercaptoethanol, and dimyristoyl phosphatidylcholine (DMPC)

Spray regents: reagents for phosphorus, amino groups, and sugar-containing lipids

Enzymes: pancreatic PLA_2 and snake venom PLA_2

TLC plates: Silica gel G plates, silica gel H plates (Kieselgel 60, Merck, Darmstadt, Germany), and silica gel LK6 plates with concentrating zone (Whatman, Clifton, NJ)

Fluorescent phospholipid substrates: 1-acyl-2-[6-[N-(7-nitrobenz-2-oxa-1,3-diazol-4-yl)]aminocaproyl]phosphatidylcholine (C_6-NBD-PC), 1-acyl-2-[6-[N-(7-nitrobenz-2-oxa-1,3-diazol-4-yl)amino]caproyl]phosphatidylserine (C_6-NBD-PS), and N-(7-nitrobenz-2-oxa-1,3-diazol-4-yl)diacylphosphatidylethanolamine (PE-NBD) (all Avanti Polar Lipids, Alabaster, AL)

Radiochemicals: Radioactive (^{3}H- or ^{14}C-labeled) oleic and palmitic acids, radioactive phosphatidylcholine (PC), and radioactive phosphatidylglycerol (PG)

Special Equipment

Spectrofluorimeter with excitation at 475 nm and emission at 525 nm
Scintillation spectrometer or image analysis system (e.g., FUJIX BAS 1000, Fuji Co., Japan), including phosphorus imaging plates
Probe sonicator (Branson or Heat Systems)
Microfuge centrifuge
TLC tanks

Procedure

Labeling of Mycoplasma Lipids

To label membrane lipids, 0.02 µCi of tritiated oleic or palmitic acid or 0.002 µCi of [^{14}C]oleic or [^{14}C]palmitic acid is added per ml of growth medium. The radioactive fatty acid is added as an ethanolic solution. The final ethanol concentration in the medium should not exceed 0.1%. Cells are then grown to a midexponential phase of growth, harvested, washed, and resuspended in cold 250 mM NaCl solution. The cells are utilized for the preparation of cell lysate, isolation of cell membranes, or lipid extraction.

Lipid Extraction

Lipids are extracted by the method of Bligh and Dyer as previously described (Rottem, 1983). Lipid phosphorus is determined after digestion of the sample in an ethanolic magnesium nitrate solution (10% $MgNO_3$ in ethanol; Ames, 1966).

The total lipid fraction is chromatographed on silica gel G plates using the two-step developing system (Rottem, 1983) to resolve neutral lipids (including free fatty acids and diacylglycerols). For the separation of polar lipids, the lipids are chromatographed on silica gel H plates (Kieselgel 60, Merck) developed first by petroleum ether:acetone (3:1, by volume at room temperature) and then at 4°C by chloroform:methanol:water (65:25:4, by volume). Lipid spots, detected by iodine vapor, are identified by their retention time, compared to lipid standards, and by specific spray reagents (Rottem, 1983).

Generation of Lysophospholipid-Enriched Membranes

Lysophospholipids, mainly lysophosphatidylglycerol or lysodiphosphatidylglycerol, are generated in radiolabeled membranes by treatment with pancreatic PLA_2. Intact or heat-inactivated (15 minutes at 65°C) membranes (200 μg protein) are suspended in 25 mM of a Tris–MOPS (morpholinopropanesulfonic acid) buffer adjusted to pH 5.0 containing 2.5 mM $CaCl_2$ and are incubated for 30 minutes at 37°C with 5 units of pancreatic PLA_2 (final volume 200 μl). The reaction is terminated by the addition of EDTA, pH 7.4 (final concentration 5 mM). The lysophospholipid-enriched membranes, or a lysophospholipid-enriched lipid extract obtained from such membranes, are used to follow lysophospholipase A activity.

Intramembrane Hydrolysis of Lipids

Indication for an endogenous phospholipase activity can be obtained by following the intramembrane hydrolysis of lipids. Cell lysates or membrane preparations (100–500 μg protein/ml) containing labeled lipids are incubated in 25 mM Tris–HCl buffer, pH 7.4, containing 5 mM $CaCl_2$ at 37°C. A reaction mixture containing a dispersion of labeled lipids instead of native membranes is used as a negative control. At desired time intervals, samples are withdrawn, the reaction is stopped by rapid cooling to 0°C, and lipids are extracted. Neutral lipids, e.g., fatty acids and diacylglycerol, are separated on silica gel G plates and phospholipids, including lysophospholipids, on silica gel H plates. Radioactivity in the neutral and polar lipid spots is then determined. A time-dependent increase in radioactivity in either free fatty acids (FFA) or the lysophospholipid spots indicates a phospholipase A activity, whereas an increase in diacylglycerol or phosphatidic acid indicates an activity of phospholipase C or phospholipase D, respectively.

Hydrolysis of Micellar Dispersions of Lipids

Hydrolysis of micellar dispersions of lipids can be measured by utilizing fluorescent or radioactive substrates. Common fluorescent phospholipid substrates are C_6-NBD-PC and PE-NBD. Equal volumes of a fluorescent substrate (1.2 mM in chloroform) and a carrier phospholipid, e.g., DMPC (1.8 mM in chloroform), are mixed. The solvent is then evaporated to dryness under a stream of nitrogen, and the lipids are resuspended in a solution of 150 mM NaCl in 10 mM Tris–HCl (pH 7.4) to a final lipid phosphorus concentration of 300 μM. The lipid suspension is then sonicated for 30 seconds at 4°C in a W-350 Heat Systems sonicator operated at 200 W.

Radioactive phospholipid dispersions are prepared from commercially available radioactive phospholipids (preferably phosphatidylcholine) or from phosphatidylglycerol or diphosphatidylglycerol (DPG) extracted from mycoplasma cells grown in the presence of [^{14}C]palmitate or [^{14}C]oleate. The total cellular mycoplasma lipids are extracted and separated by preparative TLC (Rottem, 1983), and PG and DPG lipid spots, the major phospholipids in most mycoplasmas investigated so far, are scraped off the plates and eluted from the silica gel with chloroform–methanol (1:1, by volume). The solvents are evaporated to dryness under a stream of nitrogen, and the radioactive lipids are resuspended in a solution of 150 mM NaCl in 10 mM Tris–HCl (pH 7.4) to a final lipid phosphorus concentration of 600 μM (containing approximately 2×10^5 dpm/μmol) and are dispersed by sonication as described earlier.

The standard reaction mixture (in a total volume of 100 μl) contained 25–500 μg cell or membrane protein and 20–100 nmol of either the fluorescent phospholipid mixture or the radiolabeled phospholipid dispersion ($2-5 \times 10^4$ dpm) in 150 mM NaCl, 5 mM CaCl$_2$, and 10 mM Tris–HCl, pH 7.4. For a positive control, replace the membrane preparations by a commercial phospholipase, e.g., 5 units of snake venom PLA$_2$. The reaction is carried out at 37°C for up to 1 hour and is terminated by boiling for 5 minutes with or without 5 mM EGTA. For determining phospholipase activity, the entire mixture is applied to Whatman LK6 plates (Whatman, Clifton, NJ), dried thoroughly, and developed in a chloroform:methanol:water (65:35:5, by volume). Under these conditions, C_6-NBD-FA (R_f 0.87), C_6-NBD-PC (R_f 0.45), C_6-NBD diacylglycerol (R_f 0.96), C_6-NBD phosphatidic acid (R_f 0.25), and C_6-NBD-lyso-PC (R_f 0.125) are separated (Martin and Pagano, 1986). The fluorescent bands are then scraped off the plates and are eluted by a mixture of chloroform:methanol:acidic saline (0.9% NaCl in 0.12% HCl solution, 1:1:0.05, by volume), except for C_6-NBD-lyso-PC which is eluted by chloroform:methanol:water (1:3:1, by volume). The fluroscence of the clear extracts is determined in a spectrofluorimeter (excitation at 470 nm and emission at 525 nm).

Radioactive products are separated by TLC on silica gel G and silica gel H plates as described earlier. The bands are scraped off the plates into scintillation vials containing 5 ml of scintillation liquor and radioactivity is then measured in a scintillation spectrometer and expressed as disintegrations per minute (dpm). Alternatively, when ^{14}C-labeled fatty acids are utilized, the TLC plates can be exposed to imaging plates (Fuji Photo Film Co., Japan) for 1–6 hours, the imaging plates are then processed using a radiation image analysis system, e.g., Fujix BAS 1000, Fuji Co., Japan, and radioactivity in each lipid spot is determined. Enzyme activity is expressed as nmol phopholipid hydrolyzed/min/mg membrane protein.

Discussion

As PG and DPG are the major *de novo* synthesized phospholipids in mollicutes, these lipids can be selectively labeled by growing the cells with radioactive fatty acids. In most *Mycoplasma* and *Spiroplasma* species tested, the PG or DPG thus produced has a rather unusual positional distribution of fatty acids, with the unsaturated fatty acids, e.g., oleic acid, being present in position 1 and the saturated fatty acid, e.g., palmitic acid, in position 2 of the molecule (Rottem, 1983). Therefore, to generate radioactive lysophospholipids in membranes by PLA_2 treatment, [^{3}H]oleic acid-labeled membranes should be used. Treating the palmitic acid-labeled membranes with PLA_2 will result in the release of radioactive palmitate and the generation of nonlabeled lysophospholipids. In order to obtain lysophospholipid-rich membranes, it is also important that the activity of the endogenous lysophospholipase is controlled. In *M. gallisepticum*, it was found that the pancreatic phospholipase at pH 5.0 provided close to maximal hydrolysis rates, whereas the activity of the endogenous membranous lysophospholipase was very low (Gatt *et al.*, 1982). The enzyme of *M. gallisepticum* was found to be capable of hydrolyzing membrane lysophospholipids residing in the same membrane (intramembrane hydrolysis) or in adjacent membranes (intermembrane utilization) (Gatt *et al.*, 1982).

Analysis of the end products of reaction mixtures containing mycoplasma membranes and a phospholipid specifically labeled at position 1 or 2, as well as fluorescent PC- labeled at position 2 by NBD-caproic acid, may reveal the nature of the enzyme (phospholipase A_1, A_2, C, or D). Substrate specificity and sensitivity to inhibitors may reveal the properties of the phospholipase and the possible mechanism by which it is controlled. For example, an unusual phospholipase A_1 activity has been demonstrated in *M. penetrans* (Salman and Rottem, 1995). This enzyme is heat labile and, unlike the broad substrate specificity of eukaryotic phospholipases A_1, has a restricted substrate specificity hydrolyzing PC and PG, but not DPG, and has a very low activity toward either the 1-acyl-lyso-PG or the 2-acyl-lyso-PG. The phospholipase A_1 of *M. penetrans* is not affected by SH reagents or by prolonged incubation with dithiothreitol but is substantially inhibited by *p*-bromophenacylbromide. The *M. penetrans* enzyme is sensitive to detergents and does not require free Ca^{2+} ions for maximal activity.

Many phospholipases seem to be regulated by intracellular levels of free calcium ions, cAMP, or pH (Glaser et al., 1993). These regulatory principles, however, do not control the *M. penetrans* activity that is neither stimulated by Ca^{2+} nor inhibited by EGTA. Furthermore, the mycoplasmal activity has a broad pH spectrum and is not affected by cAMP but is inhibited by lyso components, suggesting that it may be regulated by intracellular substrate and/or product levels.

References

Ames, B. N. (1966). Assay of inorganic phosphate, total phospate, and phosphatases. *In* "Methods in Enzymology" (R. A. Cardullo and H. M. Florman, eds.), Vol. 8, pp 15–116. Academic Press, New York.

Davis, P. J., Katzenell, A., Razin, S., and Rottem, S. (1985) *Spiroplasma* membrane lipids. *J. Bacteriol.* **161**, 118–122.

De Silva, N. S., and Quinn, P. A. (1986). Endogenous activity of phospholipases A and C in *Ureaplasma urealyticum. J. Clin. MIcrobiol.* **23**, 354–359.

Gatt, S., Moraq, B., and Rottem, S. (1982). Utilization of membranous lipid substrates by membranous enzymes: Hydrolysis of lysophospholipid by lysophospholipase in membranes of *Mycoplasma gallisepticum. J. Bacteriol.* **151**, 1095–1102.

Glaser, K. B., Mobilio, D., Chang, J. Y., and Senko, N. (1993). Phospholipase A_2 enzymes: Regulation and inhibition. *Trends Biochem.* **14**, 92–98.

Martin, O. C., and Pagano, R. E. (1986). Normal- and reversed-phase HPLC separations of fluorescent (NBD) lipids. *Anal. Bichem.* **159**, 101–108.

Mayer, R. J., and Marshall, L. A. (1993). New insights on mammalian phospholipase A_2(s); comparison of arachidonoyl-selective and nonselective enzymes. *FASEB J.* **7**, 339–348.

Rottem, S. (1983). Characterization of membrane lipids. *In* "Methods in Mycoplasmology" (S. Razin and J. G. Tully, eds.), Vol. I, pp. 269–275. Academic Press, New York.

Rottem, S., Adar, L., Gross, Z., Ne'eman, Z., and Davis, P. J. (1986). Incorporation and modification of exogenous phosphatidylcholines by Mycoplasmas. *J. Bacteriol.* **167**, 299–304.

Salman, M., and Rottem, S. (1995). The cell membrane of *Mycoplasma penetrans:* Lipid composition and phospholipase A_1 activity. *Biochim. Biophys. Acta* **1235**, 369–377.

van Golde, L. M. G., McElhaney, R. N., and van Deenen, L. L. M. (1971). A membrane-bound lysophospholipase from *Mycoplasma laidlawii* strain B. *Biochim. Biophys. Acta* **321**, 245–249.

SECTION E
Taxonomy and Phylogeny

E1

INTRODUCTORY REMARKS
Shmuel Razin

Microbiologists find themselves in the midst of a revolution in bacterial taxonomy driven by the introduction of molecular genetic tools and conserved gene sequences as phylogenetic markers. Since the early 1980s we have evidenced a tremendous influx of information concerning the sequences of bacterial genes, many of which are of the conserved type. Most useful are the 16S rRNA gene sequences, enabling the construction of phylogenetic trees of the microbial world, including Mollicutes (Woese, 1987, 1994; Maniloff, 1992). Data banks now carry 16S rRNA sequences of a large number of mollicutes (see Chapter A2, Vol. II, this series; Larsen *et al.,* 1993). These sequences can be used for taxonomic purposes and for assessment of phylogenetic relatedness among mollicute species. The 16S rRNA sequence can even be used as an indicator for identification of a strain; a striking example is that of the PPAV mycoplasma, isolated from an apple seed and identified as an avian *Mycoplasma iowae* strain according to its 16S rRNA sequence (Grau *et al.,* 1991). The 16S rRNA sequences can also be used as polymerase chain reaction (PCR) targets. Primers selected on the basis of sequences at the variable regions or at the highly conserved regions of the 16S rRNA molecule may serve as probes of high or low specificity, respectively, useful in detection and identification of mycoplasmas (Razin, 1994, Chapters A2, A4, and F4, Vol. II, this series). Recommended procedures for rRNA sequencing and phylogenetic tree building are described and discussed in Chapter E3.

Molecular characterization of the entire mycoplasma genome, including its complete sequencing, would obviously provide a sound basis for taxonomy. Although significant strides have already been made in the development of rapid methods for genome sequencing (Chapter B4, this volume), it will require significant technological breakthroughs before sequencing and computer-aided comparison of entire genomic sequences of mollicutes will become available. Until then we are left with methods that characterize only partially the mollicute genome. Genome size, previously serving as an important taxonomic criterion, separating the mollicutes with the small genome size (*Mycoplasma* and *Ureaplasma* species) from those with the larger genome size (*Acholeplasma, Anaeroplasma,* and *Spiroplasma* species) has lost much of its taxonomic value in light of the finding that there is an overlap in genome sizes of the just-mentioned groups, although *Mycoplasma* and *Ureaplasma* genomes, as well as those of the recently established *Entomoplasma* and *Mesoplasma* species, are clustered in a range of genome sizes definitely smaller than those of the *Acholeplasma* and *Spiroplasma* species (Carle et al., 1995). Methods for genome size determination are described in Chapter B3 (this volume). Genomic base composition (G+C ratio) serves as another taxonomic criterion in mollicute taxonomy (Razin, 1992). Since it is relatively easy to determine (for methods see Carle et al., 1983), it has become an essential requirement in description of new mollicute species (Chapter E2, this volume).

As complete sequencing of mollicute genomes is still in the realm of the future, we have to settle for partial and indirect information on genomic sequences derived from DNA–DNA and DNA–RNA hybridization tests, patterns of genomic cleavage by restriction endonucleases, and Southern blot hybridization of cleaved genomic fragments with conserved genes as probes. According to the recommendations of the International Committee on Systematic Bacteriology (ICSB), bacterial strains belonging to the same species should exhibit at least 70% DNA homology, whereas 60 to 70% homology justifies their separation into subspecies and strains showing 20 to 60% homology can be defined as separate but closely related species (Razin, 1992). DNA hybridization tests are laborious and thus are usually carried out only in laboratories well equipped and experienced in carrying out these tests (for methods see Degorce-Dumas et al., 1983; Bove et al., 1989).

Restriction endonuclease analysis (Chapter E4) provides valuable information on the type and number of specific nucleotide sequences in the genome, considered as genomic fingerprints or restriction fragment length polymorphisms. The relatively easy to perform and cost-effective restriction analysis has become a useful taxonomic tool, facilitating the identification and classification of mollicute isolates, as well as providing means for evaluating genotypic homogeneity or heterogeneity of strains within established species.

Southern blot analysis of restricted genomic DNA hybridized with a cloned conserved gene (most frequently a rRNA gene or genes) has been added to the battery of taxonomic tools. By revealing sequence identity or differences within the conserved gene, it can serve as a sensitive tool in identifying strains within species (Chapter E5). These molecular methods, usually coupled with PCR to increase sensitivity, have been responsible for the dramatic developments in characterization of the unculturable plant and insect mycoplasma-like organisms, enabling for the first time the construction of a tentative classification scheme based on phylogenetic relationships (Chapter E6).

We evidence now a transitional period where the molecular techniques, developing at a rapid pace, gain more and more weight in bacterial taxonomy. Yet, one should not forget that the major role of bacterial systematics has been to expedite the identification of bacteria in diagnostic and research laboratories. For this purpose, phenotypic characteristics are still prevailing, constituting the basic requirements for naming a new mollicute species. Chapter E2 provides a condensed version of an updated document on minimal standards for the description of new species in the class Mollicutes. This document was prepared by the ICSB Subcommittee on the taxonomy of Mollicutes (International Committee, 1995). Chapter E7 describes an updated method for determining cholesterol and Tween 80 growth requirements, properties still important in classifying mollicutes at the order, family, and genus levels.

References

Bové, J. M., Carle, P., Garnier, M., Laigret, F., Renaudin, J., and Saillard, C. (1989). Molecular and cellular biology of spiroplasmas. *In* "Spiroplasmas, Acholeplasmas, and Mycoplasmas of Plants and Arthropods" (R. F. Whitcomb and J. G. Tully, eds.), The Mycoplasmas, Vol. 5, pp. 243–364. Academic Press, San Diego.

Carle, P., Saillard, C., and Bové, J. M. (1983). Determination of guanine plus cytosine content of DNA. *In* "Methods in Mycoplasmology" (S. Razin and J. G. Tully, eds.), Vol. I, pp. 301–308. Academic Press, New York.

Carle, P., Laigret, F., Tully, J. G., and Bové, J. M. (1995). Heterogeneity of genome sizes within the genus *Spiroplasma*. *Int. J. Syst. Bacteriol.* **45**, 178–181.

Degorce-Dumas, J. R., Ricard, B., and Bové, J. M. (1983). Hybridization between mycoplasma DNAs. *In* "Methods in Mycoplasmology" (S. Razin and J. G. Tully, eds.), Vol. I, pp. 319–325. Academic Press, New York.

Grau, O., Laigret, F., Carle, P., Tully, J. G., Rose, D. L., and Bové., J. M. (1991). Identification of a plant-derived mollicute as a strain of an avian pathogen, *Mycoplasma iowae,* and its implication for mollicute taxonomy. *Int. J. Syst. Bacteriol.* **41**, 473–478.

International Committee on Systematic Bacteriology, Subcommittee on the Taxonomy of *Mollicutes* (1995). Proposal for minimum standards for descriptions of new species of the class *Mollicutes*. *Int. J. Syst. Bacteriol.* **45**, 605–612.

Larsen, N., Olsen, G. J., Maidak, B. L., McCoughey, M. J., Overbeek, R., Macke, T. J., Marsh, T. L., and Woese, C. R. (1993). The ribosomal database project. *Nucleic Acids Res.* **21**, 3021–3023.

Maniloff, J. (1992). Phylogeny of mycoplasmas. *In* "Mycoplasmas: Molecular Biology and Pathogenesis" (J. Maniloff, R. N. McElhaney, L. R. Finch, and J. B. Baseman, eds.), pp. 549–559. Am. Soc. Microbiol., Washington, DC.

Razin, S. (1992). Mycoplasma taxonomy and ecology. *In* "Mycoplasmas: Molecular Biology and

Pathogenesis" (J. Maniloff, R. N. McElhaney, L. R. Finch, and J. B. Baseman, eds.), pp. 3–22. Am. Soc. Microbiol., Washington, DC.

Razin, S. (1994). DNA probes and PCR in diagnosis of mycoplasma infections. *Mol. Cell. Probes* **8**, 497–511.

Woese, C. R. (1987). Bacterial evolution. *Microbiol. Rev.* **51**, 221–271.

Woese, C. R. (1994). There must be a prokaryote somewhere: Microbiology's search for itself. *Microbiol. Rev.* **58**, 1–9.

E2

MINIMAL STANDARDS FOR DESCRIPTION OF NEW SPECIES OF THE CLASS MOLLICUTES

Joseph G. Tully and Robert F. Whitcomb

General Introduction

The Subcommittee on the Taxonomy of Mollicutes of the International Commission of Systematic Bacteriology has published (Subcommittee, 1995) a revision of an earlier "Proposed Minimum Standards for Descriptions of New Species of the Order Mycoplasmatales" (Subcommittee, 1979). This new document recommends basic requirements for the naming of new species, designation of a type strain of the species, assignment to order, family and genus of the class, demonstration that the type strain and related strains differ significantly from all previously named species, deposition of the type strain in a recognized culture collection [such as the American Type Culture Collection (ATCC) or the National Collection of Type Cultures (NCTC)], and publication of the description in a journal of wide circulation. If the publication does not appear in the *International Journal of Systematic Bacteriology,* the name should be validated by a literature citation in that journal.

In Table I, the current classification of the Mollicutes into orders, families, and genera is presented (Tully *et al.,* 1993), along with some of the criteria used to distinguish the higher taxa. The following discussion briefly outlines the major criteria necessary to establish a new organism in the class. A more detailed discussion of specific techniques that can be employed in species characterization and appropriate references is given in the revised document (Subcommittee, 1995).

TABLE I
Taxonomy and Characteristics of Members of Class Mollicutes

Classification	Number of recognized species	Guanine + cytosine content (mol %)	Genome size (kbp)[a]	Cholesterol requirement	Habitat	Other distinctive features
Order I: Mycoplasmatales						
Family I: Mycoplasmataceae						
Genus I: *Mycoplasma*	100	23–40	600–1350	Yes	Humans, animals	Optimum growth usually at 37°C
Genus II: *Ureaplasma*	6	27–30	760–1170	Yes	Humans, animals	Urea hydrolysis
Order II: Entomoplasmatales						
Family I: Entomoplasmataceae						
Genus I: *Entomoplasma*	5	27–29	790–1140	Yes	Insects, plants	Optimum growth 30°C
Genus II: *Mesoplasma*	12	27–30	870–1100	No	Insects, plants	Optimum growth 30°C; sustained growth in serum-free medium only with 0.04% Tween 80
Family II: Spiroplasmataceae						
Genus I: *Spiroplasma*	17	25–30	940–2240	Yes	Insects, plants	Helical filaments; optimum growth at 30–37°C
Order III: Acholeplasmatales						
Family I: Acholeplasmataceae						
Genus I: *Acholeplasma*	13	26–36	1500–1650	No	Animals, some plants/insects	Optimum growth at 30–37°C
Order IV: Anaeroplasmatales						
Family I: Anaeroplasmataceae						
Genus I: *Anaeroplasma*	4	29–34	1500–1600	Yes	Bovine/ovine rumen	Oxygen-sensitive anaerobes
Genus II: *Asteroleplasma*	1	40	1500	No	Bovine/ovine rumen	Oxygen-sensitive anaerobes

[a] Genome size, in kilobase pairs, as reported in literature. Reprinted with permission from International Committee on Systematic Bacteriology, 1995.

Strain Cloning

The type strain must be derived from the presumably mixed population of cells occurring in cultures of an uncloned isolate. This is achieved by cloning three times, using gentle filtration of a broth culture through membrane filters with the smallest possible pore diameter (usually 200 to 300 nm). The filtrate and dilutions of it are cultured on solid medium, and isolated colonies are picked from a plate on which few colonies develop. The filtration and cloning sequence is performed three times to optimize the probability that the strain eventually chosen as the type strain derives from a single cell. Triple cloning by limiting dilution in liquid media may be an acceptable alternative method. All tests on which the taxonomic description is based should have been carried out on the cloned strain.

Assignment of Candidate Organism to Class Mollicutes

Absence of Reversion

Tests to exclude the possibility that any new isolate is a bacterial L-phase variant are important. In this procedure, the organism is passaged for at least five consecutive subcultures in broth media devoid of antibiotics or other substances known to induce bacterial L-phase variants. Individual broth passages are plated on conventional blood agar plates and the incubated plates are examined for bacterial-type growth.

Morphology of Limiting Cell Membrane

Placement of the organism in Mollicutes requires examination by appropriate thin-sectioning electron microscopic procedures, showing cells bounded by a single membrane and devoid of a cell wall. This information, in conjunction with failure to revert to a bacterial form and evidence that the organism is not a wall-free archaebacterium, permits initial assignment to the class.

Filterability

Cells of all currently described mollicute species pass through membrane filters of 450-nm pore diameter when broth cultures are used. Most species also will pass through membrane filters with average porosity of 300 and 220 nm, although the titer is usually significantly reduced after passage through membranes with the latter porosity. The minimal pore diameter that permits passage of some organisms should be determined. High-pressure filtration, which can force wall-less cells through pores smaller than their diameter, should be avoided.

16S rDNA Sequence Comparison

Sequence analysis of the 16S rDNA of a candidate organism can provide unambiguous evidence that the organism belongs to the class Mollicutes. The sequences may easily be obtained by sequencing of material amplified by the polymerase chain reaction, using primers specific for eubacterial 16S rDNA (see Chapter E3, this volume). The sequences of many mollicutes and eubacteria are available for comparison in gene data banks.

Classification of Mollicutes into Orders and Families

In proposing the establishment of a new mollicute species, it is essential that the organism be shown to belong to the class Mollicutes and to be significantly different from all previously named species. Determination of the following major characteristics are mandatory in assignment of an organism to an order and family within the class.

Sterol Requirement

Growth requirements for cholesterol are usually determined by direct quantitative comparison of growth responses obtained with an organism in the presence and absence of cholesterol or other sterols (see Chapter E7, this volume).

Gross Cellular Morphology

Determination of the gross morphology of a typical population of the organism in young log-phase cultures is required, usually by either dark-field or phase microscopy. These microscopic techniques should show nonhelical organisms to be pleomorphic, often in the form of small coccoid bodies (0.3–0.8 μm in diameter), bipolar forms, and/or fine branched or unbranched filaments of varying length. Members of the monogeneric family Spiroplasmataceae are helical under most circumstances in liquid media, where they exhibit flexional and, in some cases, translational motility. Since nonhelical as well as helical organisms may exhibit motility, examination of live material is mandatory. Fixatives used in electron microscopic techniques can frequently distort the helical morphology of spiroplasmas, so fixed cells should be examined by dark-field microscopy before embedding and sectioning. It is not possible to identify an organism as a member of the Mollicutes solely on the basis of conventional light microscopy or by scanning electron micrographs of the cells.

Colonial Appearance

It is necessary to define the ability of the strain to grow on solid medium and the morphology of the colonies produced. At least some colonies of most nonhelical mollicutes show a typical "fried egg" morphology characterized by a central zone in which the organisms grow into the medium and a peripheral zone of surface growth. However, motile, helical *Spiroplasma* species do not form such colonies under any currently known cultural condition and frequently exhibit diffuse colonies with peripheral satellite colonies on conventional "soft" agar (0.8–1% purified agar). The use of so-called "hard" agar (2.0–2.5% purified agar) may permit demonstration of "fried egg" spiroplasma colonies. Nonmotile spiroplasma variants will exhibit "fried egg" colonies. Colonial appearance is strongly dependent on the composition of the medium. Most mollicutes can grow beneath the surface of solid medium, and occasionally surface growth may occur without central ingrowth. Therefore, efforts should be made to modify solid media before concluding that a "fried egg" morphology cannot be demonstrated for a (nonhelical) mollicute.

Optimum Growth Temperature Requirement

Temperature requirements for growth are useful in differentiation of some families (see Table I). Tests for assaying optimum temperature and for establishing the range of temperatures (10 to 37°C) at which growth occurs have been described.

Determination of Genome Size

The genome size of mollicutes can be determined with pulsed-field gel electrophoresis (PFGE). PFGE genome size measurements are made directly from whole cell suspensions in agarose, which maintains the essential integrity of the genome (see Chapter B3, this volume). Although genome size cannot be used to differentiate some higher taxa (Table I), genome measurements are useful in some genus assignments.

Requirements for Anaerobiosis

Mollicutes having strict requirements for anaerobic environments, including sensitivity to oxygen, are classified in the family Anaeroplasmataceae, order

Anaeroplasmatales. At present, such organisms are known only from the ovine and bovine rumen. Detailed procedures required for the description of these organisms have been described.

Classification of Organism by Genus

Tests performed to assign mollicute orders and families are also of considerable value in assigning new species to genus. In addition to the aforementioned criteria, the ability to utilize urea should also be examined.

Hydrolysis of Urea

The demonstration of urea hydrolysis, resulting in an increase of pH as carbon dioxide and ammonia are produced, is a mandatory and minimum requirement for assigning a new species to the genus *Ureaplasma*. A standardized technique for the determination of urea hydrolysis in mollicutes has been described.

Specific Characteristics of Genera

GENUS *Mycoplasma*

The genus *Mycoplasma* comprises those nonhelical mollicutes from vertebrates that (i) are not obligately anaerobic (oxygen-sensitive), (ii) require cholesterol or sterols for growth, (iii) have an optimum temperature requirement near 37°C, (iv) are incapable of urea hydrolysis, and (v) have a genome size of around 600 to 1350 kbp. Organisms with similar features, but capable of urea hydrolysis and with a genome size of about 760–1170 kbp, are assigned to the genus *Ureaplasma*.

GENERA OF THE ANAEROPLASMATACEAE

Nonhelical mollicutes from vertebrates that are obligately anaerobic, and that gately anaerobic, (ii) have a sterol growth requirement, (iii) have an optimum growth temperature near 30°C, and (iv) with a genome size in the range of about 790 to 1140 kbp are assigned to the genus *Entomoplasma*. Nonhelical mollicutes from arthropods or plant surfaces that have similar features, but (i) show sustained growth in serum-free medium containing 0.04% Tween 80, and (ii) have genome sizes of about 870–1100 kbp are assigned to the genus *Mesoplasma*. Helical mollicutes from arthropods, plant surfaces, or plant phloem that (i) are not obligately anaerobic, (ii) may or may not have a sterol requirement, and (iii) with a genome size of around 940–2200 kbp are classified in the genus *Spiroplasma*.

GENUS *Acholeplasma*

The genus *Acholeplasma* consists of nonhelical mollicutes from vertebrates, plants, or arthropods that (i) are not obligately anaerobic, (ii) do not require sterols for growth, (iii) exhibit a growth temperature optimum from 30 to 37°C, and (iv) have a genome size in the range of about 1500 to 1650 kbp.

GENERA OF THE ANAEROPLASMATACEAE

Nonhelical mollicutes from vertebrates that are obligately anaerobic, and that have requirements for sterol, are referred to the genus *Anaeroplasma,* whereas obligately anaerobic organisms without sterol needs are referred to the genus *Asteroleplasma.*

Classification of Organism by Species

Definition of Species

The species concept for Mollicutes, as for all prokaryotic taxa, is based on arbitrary criteria. Ideally, mollicute species can be regarded as clusters of morphologically and biologically similar strains whose genomes exhibit a high degree (>70% by DNA/DNA hybridization) of relatedness. It is therefore usually necessary to establish patterns of relationship by determination of alternative phenotypic markers. The following specific tests should be performed to establish the properties of the species, including the specific composition of the medium for the test procedure. If possible, biochemical tests should be performed in less complex media (e.g., in the absence of yeast extract or serum, or with serum replaced by bovine serum fraction) and tests should be carefully controlled by observations of (a) uninoculated medium with substrate, (b) inoculated medium without substrate, and (c) organisms with known requirements.

Serological Relatedness

The candidate organism (putative type strain) should be compared serologically with type or reference strains of all other named species within the presumptive genus. If the strain cannot be classified within one of the already established genera, it should be compared with all previously named species of Mollicutes. The serological examination should include at least two different methods, preferably the standard growth inhibition and the agar plate immunofluorescence tests.

The growth inhibition test is the most useful technique for serologic characterization of new mollicute species. Reciprocal testing of antisera and antigens within the candidate genus is usually not required, providing antigen from the putative new species is tested against specific and potent antisera to all established species in the genus. However, if few species have been described within the genus, reciprocal testing of both antigen and antisera is recommended as additional confirmation of serologic distinctiveness. If the candidate organism cannot easily be grown on agar, the metabolism inhibition test may be used as an alternative method. This procedure has been successfully employed in the serologic analysis of *Ureaplasma* and for *Spiroplasma* species or groups.

The agar plate immunofluorescence test can be performed as either a direct or indirect fluorescent antibody test. As with growth inhibition tests, the minimum comparison should involve the testing of agar colonies of the candidate organism to specific and potent antisera or conjugates of all previously described species within the genus.

The spiroplasma deformation test has special value for screening of candidate helical species against antisera of previously established spiroplasma species or groups. A combined deformation/metabolism inhibition test system has been applied to provide both screening and refined analysis and definition of *Spiroplasma* species/subgroup relationships.

Note: Although other serological procedures have been applied to mollicutes, the techniques recommended here are based on an extensive background of use within the field of mollicute taxonomy and on a unique record of these techniques relative to their sensitivity and specificity in distinguishing species within the class.

Fermentation of Glucose

The ability of the organism to catabolize glucose to acid should be assessed, preferably under both aerobic and anaerobic conditions of cultivation. Medium composition, temperature, and length of incubation should be controlled.

Hydrolysis of Arginine

The ability of the organism to hydrolyze arginine with an increase in pH resulting from the production of ammonia should be determined. A modified test, using arginine concentrations that vary from 2 to 10 g/liter, might also be used as some organisms are inhibited by higher concentrations of arginine. It should also be emphasized that in many cases demonstration of arginine hydrolysis in spiroplasmas can be clearly defined only when glucose or some other energy source is supplied at the same time.

Detection of β-D-Glucosidase

The level of this enzyme in various acholeplasmas can be applied as a useful means for distinguishing species. The test for the enzyme is based on the hydrolysis of esculin or arbutin.

Genetic Characters

As a minimum, it is recommended that determination of genome size and DNA base composition (guanine + cytosine) should be determined on candidate species. DNA base composition should be determined by at least one of the recommended techniques: melting temperature (T_m), buoyant density (B_d), high-performance liquid chromatography (HPLC), or isopycnic centrifugation. Observation of base composition (G + C) values above 40 mol% should raise concerns that the organism might not be a mollicute.

Optional Tests

Useful supporting information at various taxal levels may also be obtained from the following tests: determination of the fermentation of carbohydrates other than glucose and/or related compounds (e.g., mannose, sucrose, and trehalose); hemadsorption of erythrocytes (preferably from guinea pig or sheep cells) to colonies on solid medium; polyacrylamide gel electrophoresis of cell proteins; restriction fragment length polymorphism (see Chapter E4, this volume); random amplified polymeric DNA; and development of polymerase chain reaction techniques that may identify probes specific for higher taxa (see chapter E6, this volume).

References

International Committee on Systematic Bacteriology, Subcommittee on the Taxonomy of Mollicutes. (1979). Proposal for minimum standards for descriptions of new species of the class Mollicutes. *Int. J. Syst. Bacteriol.* **29,** 172–180.

International Committee on Systematic Bacteriology, Subcommittee on the Taxonomy of Mollicutes. (1995). Proposal for minimum standards for descriptions of new species of the class Mollicutes. *Int. J. Syst. Bacteriol.* **45,** 605–612.

Tully, J. G., Bové, J. M., Laigret, F., and Whitcomb, R. F. (1993). Revised taxonomy of the class Mollicutes: Proposed elevation of a monophyletic cluster of arthropod-associated mollicutes to ordinal rank (Entomoplasmatales, ord. nov.), with provision for familial rank to separate species with nonhelical morphology (Entomoplasmataceae, fam. nov.) from helical species (Spiroplasmataceae), and emended descriptions of the order Mycoplasmatales, family Mycoplasmataceae. *Int. J. Syst. Bacteriol.* **43,** 378–385.

E3

RIBOSOMAL RNA SEQUENCING AND CONSTRUCTION OF MYCOPLASMA PHYLOGENIES

William G. Weisburg

Introduction

The use of gene sequences for elucidating the biological relationships among extant organisms and for reconstructing their evolutionary history has experienced an incredible flowering since the mid-1980s. Comparative analysis of a few different genetic loci have contributed to this revolution, but the vast majority of information on the phylogenetic relationships among organisms has come from the sequences of 16S (and 16S-like, such as eukaryotic 18S) ribosomal RNAs. Carl Woese and co-workers (1987) have been the major contributors to this revolution in microbiology, but many researchers around the world are now employing ribosomal RNA sequence determination and analysis in order to better understand relationships among biological species.

In 1980 (Woese *et al.*, 1980), 16S oligonucleotide catalogs were published for 4 mycoplasmas, and in 1985 (Rogers *et al.*, 1985) 5S rRNA sequences were published for 10 mycoplasmas. In 1989 a study reported the complete, or nearly complete, 16S ribosomal RNA sequences from 47 mycoplasmas (Weisburg *et al.*, 1989), including members of the genera *Mycoplasma, Spiroplasma, Acholeplasma, Anaeroplasma, Asteroleplasma,* and *Ureaplasma*. A few of those sequences had been previously reported by various researchers, and since that study, several other sequences have been published or submitted to databases. Three primary results of these studies (Weisburg *et al.*, 1989) are (1) support for the theory of a gram-positive origin of the mycoplasmas, (2) the apparent monophyletic origin of this wall-less cluster of organisms, and (3) clearly defined subgroups among the mycoplasmas.

With the sequence of the rRNA from a mycoplasma of interest in hand (whether 16S, 23S, or 5S) it is possible to (1) determine if it is a new or already classified species, (2) identify its closest relatives, and (3) design specific oligonucleotide probes for diagnostic or epidemiological purposes. This chapter briefly describes the options and methods for the determination of rRNA sequences of mycoplasmas and outlines a general approach to the analysis of these sequences.

There are now several options and decisions which need to be made before evaluating the rRNA sequence from a new mycoplasma or any other bacterium. The first sequences were generated by the laborious process of:

1. Growing a large-scale pure culture
2. Extracting purified DNA
3. Digesting the DNA with restriction enzymes
4. Ligating the cut DNA into a plasmid or phage cloning vector
5. Transforming host *Escherichia coli* cells for cloning
6. Screening recombinants with a rRNA probe
7. Growing desirable clones and purifying plasmid or phage DNA
8. Usually subcloning smaller fragments into M13 vectors
9. Sequencing the rRNA

Current methods allow for the elimination of most of these steps.

The first decision to make is whether to examine the 16S or 23S rRNA sequence of a candidate strain. The 16S is about 1550 bases in length, and the 23S is about 2900 bases. The 5S rRNA, which is 104 to 112 bases long in mycoplasmas (Rogers *et al.*, 1985), is generally too short to yield very useful phylogenetic information. For most purposes, the 16S rRNA is the best choice, unless the investigator is inclined to determine the sequence of both the 16S and 23S. The databases, such as GenBank and EMBL, have a far greater number of 16S entries for mycoplasmas and for other bacteria than 23S entries. The rest of this chapter discusses 16S rRNA sequences.

The next decision is whether to determine the sequence of the 16S rDNA of the mycoplasma, cloned or amplified, or its rRNA, by direct sequencing (Lane *et al.*, 1985) using reverse transcriptase. This author believes that currently the best approach is to amplify the nearly full-length 16S rDNA by the polymerase chain reaction (PCR) and either clone the product or employ one of several direct from PCR sequencing methods. In the author's laboratory, we have had the highest rate of sequence accuracy by cloning the resultant PCR products in a plasmid and sequencing purified plasmid by the primer-extension dideoxynucleotide termination method. To date, the greatest number of mycoplasma rRNA sequences have been determined by sequencing purified RNA from mycoplasma cultures. Detailed descriptions of this method can be found elsewhere (Weisburg *et al.*, 1989; Lane, 1991).

Materials and Procedures

Numerous descriptions of the following specific methods can be found elsewhere, including Section B of this volume; the goal here will be to integrate them. The set of procedures and steps includes:

1. Obtaining and processing a sample to yield usable nucleic acids: This can be done by phenol extraction or, in some cases, samples can be merely boiled in nonionic detergents.
2. Amplifying the 16S rRNA gene by PCR: A set of primers is described later.
3. Assaying for successful PCR by gel electrophoresis.
4. Cloning the appropriately restriction-digested PCR fragment in a plasmid vector and confirming the clone.
5. Purifying plasmid DNA.
6. Sequencing the rDNA by employing primers specific for conserved regions of the 16S rRNA (primers are described later).
7. Analysis of the new sequence by overall similarity, signature analysis, distance treeing, and parsimony analysis.

Steps 4 and 5 can be supplanted by direct from amplicon sequencing methods including asymmetric PCR, using one biotinylated primer and strand separating on streptavidin, exonuclease digesting one strand using one phosphorylated primer, or employing cycle sequencing methods (Reynolds *et al.*, 1993). As stated previously, these methods can be more rapid, but cloning will yield the largest amount of material that can be meticulously sequenced. Equally important is the increased use of PCR to amplify a sequence out of a mixed population. For example, a homopteran midgut may contain a putative mycoplasma, among other bacteria. In such an instance, cloning is necessary.

Broad specificity 16S rDNA PCR amplification primers have been described elsewhere (Weisburg *et al.*, 1991). The primers that have been used effectively for the PCR amplification of mycoplasma 16S rDNA are:

fD1: 5'-ccgaattcgtcgacaacAGAGTTTGATCCTGGCTCAG-3'
rP1: 5'-cccgggatccaagcttACGGTTACCTTGTTACGACTT-3'

These primers have been shown to amplify the 16S rDNA of *Mycoplasma pneumoniae*, *M. gallisepticum*, *M. hominis*, *M. genitalium*, and *U. urealyticum* 16S rRNA genes, as well as a large number of nonmycoplasma bacterial rDNAs

(Weisburg et al., 1991). In *M. gallisepticum,* 1479 bases of 16S are amplified. The entire PCR product, including the linker sequences which are indicated in lowercase lettering, is 1512 base pairs for *M. gallisepticum.*

The fD1 primer contains recognition sequences for *Eco*RI and *Sal*I restriction endonucleases, and the rP1 primer contains *Xma*I, *Bam*HI, and *Hind*III sites. A double digest of *Sal*I and *Bam*HI is generally satisfactory and can be directly cloned into a similarly digested plasmid vector such as pGEM-4 (Promega, Madison, WI). Confirmation of the proper-sized fragment by agarose gel electrophoresis, both post-PCR and postcloning, is recommended.

Sequencing employing primer extension and dideoxynucleotide chain termination is well described. For sequencing mycoplasmal 16S rRNA, whether by reverse transcription off of RNA templates or by cloned material, the same set of primers used on other bacteria should be used. Bacterial 16S sequencing primers have been described elsewhere [for example, Lane (1991)], but this set has been shown also to work with mycoplasmas. The homologous *E. coli* 16S position numbers for the 3' nucleotide are shown in parentheses. All sequences are 5' to 3' using IUPAC designations for mixed bases.

The forward (F) primer set should consist of these oligomers: AGTTGATCMTGGCT (24F), CTCCTACGGGAGGCAGCAG (357F), GTGCCAGCMGCCGCGG (530F), GGATTAGATACCCTRGTAGTC (805F), AAACTCAAATGAATTGACGG (926F), CCGCAACGAGCGCAACCC (1114F), and TGYACACACCGCCCGT (1406F).

The reverse (R) primer set should consist of the oligonucleotide sequences: CTGCTGCSYCCCGTAG (342R), GWATTACCGCGGCKGCTG (519R), TACYAGGGTATCTAATCC (785R), CCGTCAATTCCTTTRAGTTT (907R), GGGTTGCGCTCGTTG (1100R), ACGGGCGGTGTGTRC (1392R), and TACGGYTACCTTGTTACGAC (1494R). If the sequence is to be determined from a cloned fragment, there are usually primers specific for the cloning vector that will aid in sequencing the ends of the clone insert.

Several good commercially available kits for primer extension are available, such as the T7 DNA polymerase kits from United States Biochemical (Cleveland, OH).

Discussion

Several commercially available software packages (for example, Intelligenetics, GCG, and DNAstar) will aid in the compilation and analysis of the sequence data. In addition, the Ribosomal Database Project at the University of Illinois will analyze rRNA sequence data as a grant-supported service to the scientific community.

It is expected that the goal of the scientist in examining the 16S rRNA sequence of a mycoplasma under investigation will generally be to (1) determine if a new isolate is a new species, (2) identify a new species' closest known relatives among the mycoplasmas, or (3) identify unique gene sequences in order to design oligonucleotide probes for diagnostic or epidemiological purposes. These three goals require the analysis of the unknown sequence in the context of a multiple sequence alignment editor. A precise alignment of the new sequence with several available from the database is desirable. The starting point for identifying the likely set of closest relatives which should be used as a template for manual alignment or refinement of a computer-generated alignment would employ signature analysis as described previously (Weisburg *et al.*, 1989).

The first piece of desirable data is the percentage of similarity of a new sequence with its closest relatives. The similarity over the entire 16S should fall between 80 and 100%. A pairwise value over 99% might indicate identity of the species. In this range, it becomes particularly important that the accuracy of the sequence determination is as high as possible. Except for instances of 100% identity, the author does not feel it is prudent to suggest a cutoff value over which sequences demand that their respective strains be called "identical."

Often the examination of a large matrix of pairwise similarity values for the investigated strain's sequence with other mycoplasma sequences will readily suggest closest relatives. Treeing algorithms will aid in the refined analysis and visual interpretation of hierarchical phylogenetic relationships. The methods of optimal tree building can be an intimidating morass. The microbiologist without a strong computational expertise should approach the problem from a pragmatic point of view. If more than one method of tree construction—from the list of distance treeing, parsimony analysis, neighbor joining, maximum likelihood, etc.—gives a similar branching order with respect to the few nodes closest to the investigational sequence, then it is likely to be accurate. Several comprehensive reviews of tree building methods are available [for example, Penny (1991)].

The rRNA sequences of mycoplasmas are readily amenable to use in the design of specific probes. These can be employed in several formats, including PCR, sandwich hybridization, Southern blots, Northern blots, dot blots, and others (see Vol. II, Section A). The key comparative features on which to focus for probe (or primer) design are the consideration of (1) what are the closest mycoplasma relatives which are desirable to exclude and (2) what other species (mycoplasma and other bacteria) are likely to be found in the sample of interest? The author's laboratory has found in most instances that there are several regions of the 16S rRNA sequence with sufficient divergence to design probes which will be specific for a single mycoplasma species, with little or no cross-hybridization to closely related species.

In conclusion, it is likely that in the not so distant future we will see an explosion of newly characterized mycoplasmas, especially within the spiroplasma lineage (and mycoplasma-like organisms). Many of these will be uncultivable. The ability to determine whether a newly identified mycoplasma that may be the etiological agent of a plant or animal disease is the same as (or different from) a previously identified species, without enrichment and cultivation, will be a very powerful tool. It is then desirable to confirm that the sequence one ends up with can be shown to actually be present in the original sample, by hybridization or other methods.

References

Lane, D. L. (1991). 16S/23S rRNA sequencing. *In* "Nucleic Acid Techniques in Bacterial Systematics" (E. Stackebrandt and M. Goodfellow, eds.), pp. 115–175. Wiley, Chichester.

Lane, D. L., Pace, B., Olsen, G. J., Stahl, D. A., Sogin, M. L., and Pace, N. R. (1985). Rapid determination of 16S ribosomal RNA sequences for phylogenetic analysis. *Proc. Natl. Acad. Sci. USA* **82**, 6955–6959.

Penny, D. (1991). From macromolecules to trees. *In* "Nucleic Acid Techniques in Bacterial Systematics" (E. Stackebrandt and M. Goodfellow, eds.), pp. 281–324. Wiley, Chichester.

Reynolds, T. R., Uliana, S. R. B., Floeter-Winter, L. M., and Buck, G. A. (1993). Optimization of coupled PCR amplification and cycle sequencing of cloned and genomic DNA. *BioTechniques* **15**, 462–468.

Rogers, M. J., Simmons, J., Walker, R. T., Weisburg, W. G., Woese, C. R., Tanner, R. S., Robinson, I. M., Stahl, D. A., Olsen, G., Leach, R. H., and Maniloff, J. (1985). Construction of the mycoplasma evolutionary tree from 5S rRNA sequence data. *Proc. Natl. Acad. Sci. USA* **82**, 1160–1164.

Weisburg, W. G., Barns, S. M., Pelletier, D. A., and Lane, D. J. (1991). 16S ribosomal DNA amplification for phylogenetic study. *J. Bacteriol.* **173**, 697–703.

Weisburg, W. G., Tully, J. G., Rose, D. L., Petzel, J. P., Oyaizu, H., Yang, D., Mandelco, L., Sechrest, J., Lawrence, T. G., Van Etten, J., Maniloff, J., and Woese, C. R. (1989). A phylogenetic analysis of the mycoplasmas: Basis for their classification. *J. Bacteriol.* **171**, 6455–6467.

Woese, C. R. (1987). Bacterial evolution. *Microbiol. Rev.* **51**, 221–271.

Woese, C. R., Maniloff, J., and Zablen, L. B. (1980). Phylogenetic analysis of the mycoplasmas. *Proc. Natl. Acad. Sci. USA* **77**, 494–498.

E4

RESTRICTION ENDONUCLEASE ANALYSIS

Shmuel Razin and David Yogev

Introduction

Restriction endonuclease analysis (REA) of the mycoplasma genome provides a convenient and cost-effective means of determining DNA sequence variations among strains of mollicute species. The method involves comparison of the number and size of fragments produced by digestion of the chromosomal DNA with a restriction endonuclease that cuts DNA at a constant position within a specific recognition site, usually composed of 4 to 6 bp. Because of the high specificity of restriction endonucleases, complete digestion of a given genomic DNA with a specific restriction endonuclease provides a reproducible array of fragments. Separation of the fragments by agarose gel electrophoresis and staining with ethidium bromide provide a restriction pattern that can be compared with that of related strains. Variations in the array of fragments generated by a specific restriction endonuclease are called restriction fragment length polymorphisms (RFLPs). RFLPs can result from sequence rearrangements, insertion or deletion of DNA segments, or base substitutions within the restriction endonuclease cleavage sites.

Restriction endonuclease analysis has become a most useful taxonomic tool, facilitating the identification and classification of mycoplasmal isolates as well as providing means for evaluating the degree of genotypic heterogeneity of strains within established species (Razin, 1989, 1991, 1992). In addition, REA provides valuable information on the type and number of specific nucleotide sequences in the genome, serving as a basis for construction of physical genomic maps (Chapter B4, this volume). The great advantage of chromosomal REA is that it is universally applicable and sensitive because the entire genome is evaluated for RFLP and, in addition, the test is relatively easy to perform. Moreover, unlike SDS–PAGE of mycoplasmal cell proteins, REA is not susceptible to contamination by culture medium contaminants and requires very little unlabeled DNA (3 to 5 µg) per test.

Materials

Restriction endonucleases (available from a variety of commercial sources)
Digestion buffers, provided by the supplier of the enzymes
Horizontal slab gel apparatus (30 cm long)
Tris–acetate (TAE) buffer (40 mM Tris acetate; 1 mM EDTA, pH 8.0)
Gel-loading buffer, 6× (0.25% bromophenol blue; 0.25% xylene cyanol; 40% (w/v) sucrose in H_2O)
Ethidium bromide stock solution (10 mg/ml)
Molecular grade agarose

Procedure

1. Prepare a 0.8% agarose gel, using 1× TAE containing 0.5 µg/ml ethidium bromide. In order to improve resolution and obtain a sharp picture of the electrophoresed DNA, the thickness of the gel should be approximately 0.5 cm.
2. Mix sterile deionized water with 3–5 µg of mycoplasmal DNA in a sterile Eppendorf tube to a volume of 17 µl.
3. Add 2 µl of the appropriate 10× digestion buffer and 1 µl of the restriction enzyme (~5 U). Mix by tapping the tube and spin for a few seconds.
4. Incubate at the recommended temperature (usually 37°C) for 2 hours.
5. Stop the reaction by adding 1 µl of 0.5 M EDTA (pH 7.5) and 4 µl of 6× gel-loading buffer. Mix and spin again for a few seconds.
6. Load the digest carefully into a gel slot. In an adjacent slot load a solution of DNA size markers.
7. Electrophorese the DNA at 30–40 constant volts overnight until the front, marked by the blue dye, reaches 20 cm from the origin.
8. Observe the gel under ultraviolet light and photograph it, using polaroid film Type 667 (3000 ASA).

Discussion

Isolation and purification of the tested DNA are critical steps in restriction endonuclease analysis. The DNA must be sufficiently intact and should be free of impurities that may inhibit restriction endonuclease activity and lead to partial or no cutting of the DNA. The lack of a cell wall facilitates mycoplasma cell lysis and DNA release by detergents. Yet, it is not always easy to obtain undegraded DNA of high purity from mycoplasmas. Thus, *Acholeplasma* species are particularly rich in endogeneous DNAses which are activated on cell lysis (see Chapter D5, this volume). By increasing the EDTA concentration in the lysis solution and by processing relatively small pellets of cells, self-digestion of mycoplasmal DNA can be minimized (Razin *et al.*, 1983a). Contaminants (proteins, polysaccharides, phenol, chloroform, ethanol, EDTA, SDS, and improper salt concentration) can inhibit the activity of restriction endonucleases, leading to incomplete digestion of the DNA. The inhibitory effects may be overcome by increasing the number of enzyme units in the reaction, increasing the reaction volume to dilute the contaminants, increasing the time of incubation, or by the successive addition of enzyme after incubation for 1 to 3 hours.

The selection of a restriction endonuclease for use in RFLP analysis should aim at providing restriction fragments in the size range of 1 to 15 kbp, as these are separated well by conventional agarose gel electrophoresis. Yet, the fragments in this size range should not be too numerous to avoid overlapping bands that may obscure differences. By using restriction endonucleases recognizing 6-bp sequences in which at least 4 are either G or C, such as *Bam*HI (G/GATCC), the (G+C)-poor mycoplasma genome will be cleaved into a relatively small number of fragments. Consequently, the electrophoretic patterns exhibit relatively few bands and are easy to compare. However, some of the fragments produced are too large to be resolved by conventional agarose gel electrophoresis (Razin *et al.*, 1983a,b), a deficiency which can now be resolved by applying pulse-field gel electrophoresis (PFGE; Frey *et al.*, 1992). This requires embedding of the mycoplasma cells in agarose blocks and treatment of the blocks with detergent and proteinase K to isolate the DNA, and then treating the blocks with a restriction endonuclease. In this way DNA fragments ranging in size from 250 to 10 kbp can be separated by one of the PFGE techniques (see Chapters B3 and B4, this volume), providing in this way a complete restriction pattern of the genomic DNA tested (Frey *et al.*, 1992). It should also be pointed out that because of the small size of the mycoplasma genome, even restriction endonucleases with (A+T)-rich recognition sites, like *Eco*RI and *Hind*III, which cut the mycoplasma genome in many sites, yield electrophoretic patterns that still exhibit well-defined bands rather than smears (Bove and Saillard, 1979; Razin *et al.* 1983a,b). These multiband patterns, although harder to compare visually, may be taxonomically valuable as they provide more detailed information on genome structure (Razin, 1989).

Resistance to cleavage by restriction endonucleases may also be due to methylation of bases within the recognition sites by prokaryotic DNA restriction modification systems. Differences in DNA methylation patterns that may occur between two otherwise genetically related strains impose some limitation on the selection of the restriction endonuclease suitable for restriction analysis. Restriction endonucleases that are not affected by methylation of nucleotides must be evaluated. As described in Chapter B8 (this volume), REA can be applied to assay methylation of specific DNA sequences, based on the digestion of genomic DNA by pairs of restriction endonucleases (isoschizomers) which recognize the same site, but one of the pair cleaves the DNA when cytosine or adenine is methylated, whereas the other does not.

Comparison of cleavage patterns is usually done visually by direct comparison of the location and intensity of bands on photographs. In our experience, multiband patterns are frequently easier to compare by observing the location of the major dark areas between the bands. In this way one can get an overall picture of the cleavage patterns of the entire gel, facilitating band comparison. Analysis of cleavage patterns can also be done on densitometer tracings of the gel photographs (Saillard and Bove, 1983). However, when bands are too numerous and consequently too closely spaced, the efficiency of the scanning devices may not suffice to provide satisfactory scans, limiting the use of densitometer tracing, particularly when applied to large-scale comparisons of many strains.

Restriction endonuclease analysis serves as an excellent means of testing clonality and identity of strains. Thus, we have used this technique for routine checking of the identity and purity of our *M. genitalium* clones. The replacement by mistake of a *M. genitalium* culture by *M. pneumoniae* was easily detected in this way (Razin, 1985). It should be recalled that it is difficult to distinguish between these two mycoplasmas by growth characteristics and by serology. Restriction endonuclease analysis was also instrumental in identifying the PPAV mycoplasma, isolated from apple seeds, as a strain of the avian pathogen *M. iowae* (Grau *et al.*, 1991). Although the restriction pattern of a strain appears to be a rather stable property, not affected by *in vitro* passage levels (Razin *et al.*, 1983b; Kleven *et al.*, 1988), the cleavage patterns of multiple *M. ovipneumoniae* strains isolated from the same pneumonic lung of a sheep revealed small, but reproducible, differences (Ionas *et al.*, 1991). The occurrence of antigenic variability in pathogenic mycoplasmas based on high frequency mutations (Wise, 1993) leads one to expect rapid, although minor, changes in cleavage patterns. It should be pointed out in this context that REA is more sensitive than SDS–PAGE of cell proteins in revealing strain variations (Ionas *et al.*, 1991).

The frequent presence of extrachromosomal elements (plasmids, viruses) in some mollicutes may affect the interpretation of restriction analysis results. The bands belonging to the extrachromosomal elements introduce variability in the cleavage patterns (Bove et al., 1982). This may constitute a serious problem in the interpretation of results (Razin, 1985).

Restriction analysis of polymerase chain reaction-amplified DNA has been introduced in the detection and differentiation of uncultured mycoplasma-like organisms (MLOs) (Lee et al., 1993). "Universal" primers capable of amplification of 16S rDNA of a broad array of MLOs from infected plants or insects are used. The amplified products are subjected to restriction analysis. The restriction fragments, being very small in size, are separated by electrophoresis in 5–6% polyacrylamide gels. The patterns obtained are used for identification and classification of the uncultured MLOs (see also Chapter E6, this volume). RFLP analyses of amplified MLO 16S rDNA sequences may thus replace direct sequencing, which can become very time-consuming and expensive when numerous strains must be analyzed.

References

Bove, J. M., and Saillard, C. (1979). Cell biology of spiroplasmas. In "The Mycoplasmas" (R. F. Whitcomb and J. G. Tully, eds.), Vol. 3, pp. 83–153. Academic Press, New York.

Bove, J. M., Saillard, C., Junca, P., DeGorce-Dumas, J. R., Ricard, B., Nhami, A., Whitcomb, R. F., Williamson, D., and Tully, J. G. (1982). Guanine-plus-cytosine content, hybridization percentages, and EcoRI restriction enzyme profiles of spiroplasmal DNA. Rev. Infect. Dis. **4**, S129–S136.

Frey, J., Haldimann, A., and Nicolet, J. (1992). Chromosomal heterogeneity of various *Mycoplasma hyopneumoniae* field strains. Int. J. Syst. Bacteriol. **42**, 275–280.

Grau, O., Laigret, F., Carle, P., Tully, J. G., Rose, D. L., and Bove, J. M. (1991). Identification of a plant-derived mollicute as a strain of an avian pathogen, *Mycoplasma iowae*, and its implications for mollicute taxonomy. Int. J. Syst. Bacteriol. **41**, 473–478.

Ionas, G., Clarke, J. K., and Marshall, R. B. (1991). The isolation of multiple strains of *Mycoplasma ovipneumoniae* from individual pneumonic sheep lungs. Vet. Microbiol. **29**, 349–360.

Kleven, S. H., Browning, G. F., Bulach, D. M., Ghiocas, E., Morrow, C. J., and Whithear, K. G. (1988). Examination of *Mycoplasma gallisepticum* strains using restriction endonuclease DNA analysis and DNA–DNA hybridisation. Avian Pathol. **17**, 559–570.

Lee, I.-M., Hammond, R. W., Davis, R. E., and Gundersen, D. E. (1993). Universal amplification and analysis of pathogen 16S rDNA for classification and identification of mycoplasmalike organisms. Phytopathology **83**, 834–842.

Razin, S. (1985). Molecular biology and genetics of mycoplasmas (*Mollicutes*). Microbiol. Rev. **49**, 419–455.

Razin, S. (1989). Molecular approach to mycoplasma phylogeny. In "The Mycoplasmas" (R. F. Whitcomb and J. G. Tully, eds.), Vol. 5, pp. 33–69. Academic Press, San Diego.

Razin, S. (1991). The genera *Mycoplasma, Ureaplasma, Acholeplasma, Anaeroplasma,* and *Asteroleplasma*. In "The Prokaryotes" (A. Balows, H. G. Trüper, M. Dworkin, W. Harder, and K.-H. Schleifer, eds.), 2nd Ed., Vol. 2, pp. 1936–1959. Springer-Verlag, New York.

Razin, S. (1992). Mycoplasma taxonomy and ecology. In "Mycoplasmas: Molecular Biology and

Pathogenesis" (J. Maniloff, R. N. McElhaney, L. R. Finch, and J. B. Baseman, eds.), pp. 3–22. Am. Soc. Microbiol., Washington, D.C.

Razin, S., Harasawa, R., and Barile, M. F. (1983b). Cleavage patterns of the mycoplasma chromosome, obtained by using restriction endonucleases, as indicators of genetic relatedness among strains. *Int. J. Syst. Bacteriol.* **33,** 201–206.

Razin, S., Tully, J. G., Rose, D. L., and Barile, M. F. (1983a). DNA cleavage patterns as indicators of genotypic heterogeneity among strains of *Acholeplasma* and *Mycoplasma* species. *J. Gen. Microbiol.* **129,** 1935–1944.

Saillard, C., and Bove, J. M. (1983). *Eco*RI restriction enzyme analysis of mycoplasma DNA. *In* "Methods in Mycoplasmology" (S. Razin and J. G. Tully, eds.), Vol. 1, pp. 313–318. Academic Press, New York.

E5

SOUTHERN BLOT ANALYSIS AND RIBOTYPING

David Yogev and Shmuel Razin

Introduction

Southern blot analysis of genomic DNA digested by a restriction enzyme and hybridized with a cloned conserved gene or a specific genomic fragment as a probe has become a most useful tool in the identification, classification, and subtyping of mollicutes (Razin, 1992). The plasmid pMC5, carrying the entire 23S and 5S and most of the 16S rRNA genes of *Mycoplasma capricolum* (Amikam *et al.*, 1982), was among the first plasmids to be used as probes in Southern blot analysis of mycoplasmal DNA. The fact that the rRNA operons in the various mollicutes differ in restriction sites within the operon and in the flanking sequences (Amikam *et al.*, 1984; Razin *et al.*, 1984) results in the production of hybridization patterns peculiar to different mollicute species or strains. A restriction enzyme having a 6-bp recognition site will usually cut the one or two mycoplasmal rRNA operons in a few sites so that the hybridization patterns are usually simple and much easier to compare than the multiband restriction patterns obtained by restriction endonuclease analysis (Chapter E4, this volume).

Southern blot hybridization with cloned rRNA genes as probes has been named ribotyping. However, other conserved genes can be employed, such as the *tuf* gene, encoding the elongation factor EF-Tu (Yogev *et al.*, 1988c). The mollicute genome carries only one copy of this gene so that the hybridization patterns obtained with restricted mollicute DNAs are also very simple and easy to compare (Yogev et al., 1988c).

Southern blot analysis can also be used to assess for possible genetic relatedness between different mollicute species. In this case, the entire undigested genomic DNA of one organism is labeled and hybridized with the restricted and electrophoresed total genomic DNA of the other. The appearance of hybridization bands, other than those detected by the rRNA gene probe pMC5, indicates the presence of genomic sequences shared by the two organisms. In this way, the pair of two human pathogens *M. pneumoniae–M. genitalium* and the pair of two

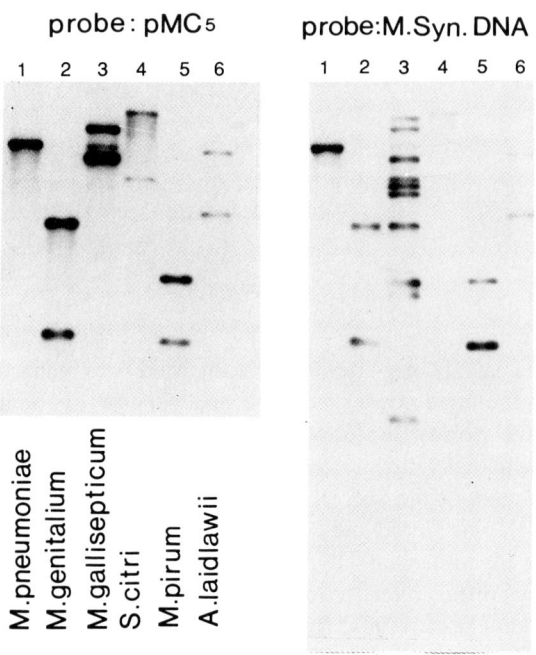

Fig. 1. Demonstration of genomic sequences shared by *Mycoplasma gallisepticum* and *Mycoplasma synoviae*. The Southern blots shown are of *Eco*RI- digested DNAs of various mycoplasmas (numbered 1 to 6) hybridized with the radiolabeled rRNA gene probe pMC5 (left) or with radiolabeled total DNA of *M. synoviae* (right). The bands carrying the conserved rRNA gene sequences reacting with total genomic DNA of *M. synoviae* can be identified by comparison with the bands obtained with the specific rRNA gene probe pMC5. After hybridization with total *M. synoviae* DNA, only *M. gallisepticum* DNA reveals bands additional to those revealed by pMC5, indicating the presence of shared genomic sequences by these two mycoplasmas (from Yogev et al., 1989, with permission).

avian pathogens *M. gallisepticum–M. synoviae* were shown to share genomic regions with common sequences (Yogev and Razin, 1986; Yogev *et al.*, 1989; Fig. 1).

Southern Blotting

Materials

Agarose gel containing electrophoresed restricted mycoplasmal DNA, prepared as described in Chapter E4 (this volume)
Denaturation solution (1.5 M NaCl; 0.5 M NaOH)
Neutralization solution (1.5 M NaCl; 0.5 M Tris–HCl, pH 7.0)
20× SSC (0.3 M trisodium citrate dihydrate; 3 M NaCl, pH 7.0)
Nylon or supported nitrocellulose membrane
Whatman 3MM filter paper sheets

Procedure

1. After photography of the agarose gel with the electrophoresed restriction fragments, cut out unnecessary regions of the gel with a razor blade and measure its dimensions.
2. Denature the DNA by soaking the gel in denaturation solution for 45 minutes with gentle shaking.
3. Pour off the denaturation solution and rinse the gel with distilled water.
4. Neutralize the gel by soaking in neutralization solution for 1 hour with gentle shaking.
5. Cut the nylon or the nitrocellulose membrane to fit the gel size. Soak the membrane in 6× SSC for 10 minutes.
6. Transfer the denatured DNA fragments to the membrane using the upward capillary transfer method (Sambrook *et al.*, 1989) for a minimum of 12 hours (overnight).
7. When transfer is complete, remove the membrane and mark the gel slots and the orientation of the blot with a pen.
8. Rinse the membrane in 6× SSC for 5 minutes with gentle shaking.
9. Place the membrane on a sheet of Whatman 3MM filter paper and allow to dry at room temperature.
10. Immobilize the DNA by baking the nitrocellulose sheet for 2 hours at

80°C or by UV cross-linking of the nylon membrane according to the manufacturer's recommendations.

Hybridization of the Southern Blot with a Labeled DNA Probe

Materials

Prehybridization / hybridization solution (PHS)(6× SSC; 5× Denhardt's solution; 100 μg/ml salmon sperm DNA, denatured by boiling for 10 minutes)
2× SSC; 0.1% sodium dodecyl sulfate (SDS)
0.5× SSC; 0.1% SDS
0.2× SSC; 0.1% SDS
Hybridization bath (56°C)
Sealable bag
DNA to be used as probe (e.g., pMC5; *tuf* gene; mycoplasma genomic DNA)
Random-primed DNA labeling kit
X-ray film

Procedure

1. Label the DNA probe to a specific activity of $>10^8$ dpm/μg. The random oligonucleotide priming technique, available as a kit (Boehringer-Mannheim Biochemicals) is recommended. One of the advantages of this kit is that removal of nonincorporated nucleotides is not necessary prior to the use of the labeled DNA probe in hybridization.

2. Place the membrane carrying the immobilized DNA in a sealable bag containing 250 μl of PHS per 1 cm² of membrane.

3. Incubate the sealed bag in the 56°C hybridization bath with gentle shaking for 6 hours.

4. Remove the PHS and add 25 μl per 1 cm² of the membrane of fresh PHS containing the mycoplasmal DNA probe (make sure to boil the probe for 10 minutes before adding it to the PHS). The concentration of the probe in the hybridization solution should be approximately 10 ng/ml.

5. Incubate as described in step 3 for 18 hours at 56°C.

6. Remove the radioactive PHS, place the membrane in a special container, and perform the following washes: Two washes in 2× SSC, 0.1% SDS; three washes in 0.5× SSC, 0.1% SDS; and two washes in 0.2× SSC, 0.1% SDS. All

washes should be carried out with gentle shaking at 56°C with 200 ml of solution each time.

7. Pour off the wash solution and place the membrane on Whatman paper, allowing it to dry completely at room temperature.

8. Set up an autoradiograph using an X-ray film.

It may be advantageous to add to the PHS containing the mycoplasmal DNA probe a second labeled probe complementary to the DNA size marker used in the agarose gel. This will allow the visualization of the DNA marker bands, facilitating size determination of the mycoplasmal DNA fragments on the film.

Discussion

The conserved genes used as probes may be derived from prokaryotes other than mycoplasmas. Thus, the plasmid pKK 3535 carrying the entire rRNA operon *rrnB* of *Escherichia coli* (Amikam et al., 1984) as well as the cloned *tufA* gene of *E. coli* can be used (Yogev et al., 1988c). A cloned segment of the ATPase (*atp*) operon of *E. coli* was also used effectively in Southern blot analysis of *M. hominis* strains (Christiansen et al., 1987). Identification of many other conserved genes by genetic mapping of mycoplasmas (Chapter B4, this volume) is expected to increase markedly the number and types of conserved genes suitable for Southern blot analysis.

Ribotyping of the uncultured plant mycoplasma-like organisms (MLOs) with cloned rRNA genes, such as pMC5, encounters a problem: the tested MLO DNA purified from the infected plant may contain chloroplast DNA. The rRNA genes of plant chloroplasts show a high degree of homology with prokaryotic rRNA genes and thus hybridize effectively with pMC5, yielding false-positive Southern blot analysis results with DNA of uninfected control plants (Nur et al., 1986). The application of polymerase chain reaction with primers designed on the basis of MLO rRNA genes provides a solution to the problem (Lee et al., 1993). The amplified DNA fragment is subjected to restriction enzyme analysis producing simple restriction fragment length polymorphism (RFLP) patterns useful in the identification and classification of MLOs. Another approach is based on the use of randomly cloned DNA fragments, specific for certain MLOs, as probes. These probes are usually produced by shotgun cloning of restricted genomic DNA followed by selection of the fragments that recognize specific strain clusters of MLOs by dot-blot hybridization. Genetic interrelationships among the MLO strains belonging to the cluster can then be conducted by Southern blot analysis, using these cloned DNA as probes (Lee et al., 1992; Chapter D11, Vol. II). When hybridization reactions are carried out under conditions of moderate stringency, relatedness among the various MLOs is more readily revealed, whereas hybridizations carried out under conditions of high stringency tend to distinguish the MLO strains from one another.

Labeling of the probes has been done by high-energy radioisotopes (e.g., ^{32}P), but the use of nonradioactive reporter molecules, such as biotin or digoxygenin, is gaining popularity as convenient labeling kits have become available and the sensitivity of detection has become closer to that of the isotopically labeled probes (Lee et al., 1992).

Reading of Southern blot analysis results is usually done by visual analysis. If the objective is to compare a limited number of strains to determine whether a strain is different from one or several other strains, simple visual examination of the pattern may suffice. For a more thorough analysis, a precise determination of the relative size of each fragment can be obtained by comparing it with a labeled molecular size standard run on the same gel. Molecular size standards should be run at the ends and the middle of the gel to correct for unequal migration at different gel regions. For ways of analyzing and presenting the RFLP data, the reader is referred to Swaminathan and Matar (1993) and to Lee et al. (1993).

Ribotyping has been successfully applied to detect and identify mycoplasmas infecting cell cultures (Razin et al., 1984), to demonstrate the clustering of the 14 serotypes of U. urealyticum into two genotypically distinct biovars (Harasawa et al., 1991), and to show the genotypic heterogeneity of M. hominis and M. gallisepticum strains, standing in contrast to the relative genotypic homogeneity of M. pneumoniae strains isolated from different epidemics (Yogev et al., 1988a,b). A major disadvantage of ribotyping is that the procedure requires multiple steps and is time-consuming and labor intensive. Use of two or three restriction enzymes simultaneously for ribotyping increases the discriminating power of the method, but also multiplies the work load.

A crucial question is whether the hybridization patterns obtained by Southern blot analysis are species specific. The answer appears to depend on the species tested. Since the concept of a bacterial species is not well defined and uses arbitrary criteria for definition (such as >70% DNA–DNA homology), it is no wonder that established mollicutes species are composed of strain clusters exhibiting different levels of relatedness. This is reflected in sequence divergence among strains of the same species, even in the highly conserved rRNA genes (Yogev et al., 1988a,b). The conclusion is that genomic fingerprints obtained by restriction enzyme analysis and Southern blot analysis may not be good enough to determine species. Yet, the high sensitivity of the RLFP methodology may be useful in distinguishing strains of particular importance, such as the vaccine F strain of M. gallisepticum, indistinguishable by routine serological methods, but distinct from virulent field isolates by ribotyping (Yogev et al., 1988b; Fig. 2). In this respect, restriction enzyme analysis and Southern blot analysis can be considered as effective and sensitive epidemiological tools. The ribotype of a strain appears to be stable after *in vitro* and *in vivo* passage. Therefore this

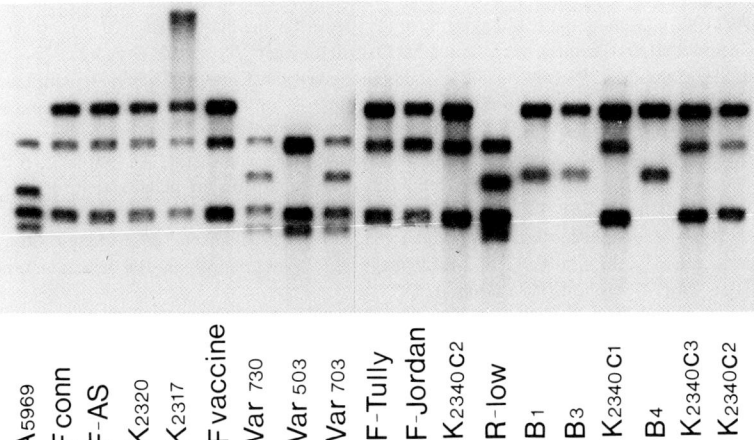

Fig. 2. Hybridization patterns (RFLPs) of BglII-digested DNAs of *M. gallisepticum* strains with the rRNA gene probe pMC5. Four strain clusters can be observed. The largest cluster consists of the vaccine strain F obtained from different laboratories and the K strains isolated from chickens in areas where the F vaccine strain was used for vaccination. The B cluster consists of strains isolated from hosts other than chickens (later classified as a separate species, *M. imitans*). The third cluster, designated "Var," consists of atypical *M. gallisepticum* strains, and the fourth consists of the A5969 and pathogenic R-low strains. The extra band in lane 5 signifies incomplete digestion of the DNA (from Yogev et al., 1988b, with permission).

method may be used for long-term epidemiologic studies. Nevertheless, the occurrence of high-frequency chromosomal rearrangements in mycoplasmas (Dybvig, 1993; Chapter B9, this volume) suggests that such events may affect Southern blot analysis results, even when highly conserved genes are used as probes since changes in RFLP may result from chromosomal sequence changes in regions flanking the conserved genes.

References

Amikam, D., Razin, S., and Glaser, G. (1982). Ribosomal RNA genes in mycoplasma. *Nucleic Acids Res.* **10**, 4215–4222.

Amikam, D., Glaser, G., and Razin, S. (1984). Mycoplasmas (*Mollicutes*) have a low number of rRNA genes. *J. Bacteriol.* **158**, 376–378.

Christiansen, C., Christiansen, G., and Rasmussen, O. F. (1987). Heterogeneity of *Mycoplasma hominis* as detected by a probe for *atp* genes. *Isr. J. Med. Sci.* **23**, 591–594.

Dybvig, K. (1993). DNA rearrangements and phenotypic switching in prokaryotes. *Mol. Microbiol.* **10**, 465–471.

Harasawa, R., Dybvig, K., Watson, H. L., and Cassell, G. H. (1991). Two genomic clusters among 14 serovars of *Ureaplasma urealyticum*. *Syst. Appl. Microbiol.* **14,** 393–396.

Lee, I.-M., Davis, R. E., Chen, T.-A., Chiykowski, L. N., Fletcher, J., Hiruki, C., and Schaff, D. A. (1992). A genotype based system for identification and classification of mycoplasmalike organisms (MLOs) in the aster yellows MLO strain cluster. *Phytopathology* **82,** 977–986.

Lee, I.-M., Hammond, R. W., Davis, R. E., and Gundersen, D. E. (1993). Universal amplification and analysis of pathogen 16S rDNA for classification of mycoplasmalike organisms. *Phytopathology* **83,** 834–842.

Nur, I., Bove, J. M., Saillard, C., Rottem, S., Whitcomb, R. F., and Razin, S. (1986). DNA probes in detection of spiroplasmas and mycoplasma-like organisms in plant and insects. *FEMS Microbiol. Lett.* **35,** 157–162.

Razin, S. (1992). Mycoplasma taxonomy and ecology. *In* "Mycoplasmas: Molecular Biology and Pathogenesis" (J. Maniloff, R. N. McElhaney, L. R. Finch, and J. B. Basman, eds.), pp. 3–22. Am. Soc. Microbiol., Washington, DC.

Razin, S., Gross, M., Wormser, M., Pollack, Y., and Glaser, G. (1984). Detection of mycoplasmas infecting cell cultures by DNA hybridization. *In Vitro* **20,** 404–408.

Sambrook, J., Fritsch, E. F., and Maniatis, T. (1989). "Molecular Cloning, a Laboratory Manual," 2nd Ed. Cold Spring Harbor Laboratory Press, Cold Spring Harbor, NY.

Swaminathan, B., and Matar, G. M. (1993). Molecular typing methods. *In* "Diagnostic Molecular Microbiology" (D. H. Persing, T. F. Smith, F. C. Tenover, and T. J. White, eds.), pp. 26–50. Am. Soc. Microbiol., Washington, DC.

Yogev, D., and Razin, S. (1986). Common deoxyribonucleic acid sequences in *Mycoplasma genitalium* and *Mycoplasma pneumoniae* genomes. *Int. J. Syst. Bacteriol.* **36,** 426–430.

Yogev, D., Halachmi, D., Kenny, G. E., and Razin, S. (1988a). Distinction of species and strains of mycoplasmas (mollicutes) by genomic DNA fingerprints with an rRNA gene probe. *J. Clin. Microbiol.* **26,** 1198–1201.

Yogev, D., Levisohn, S., Kleven, S. H., Halachmi, D., and Razin, S. (1988b). Ribosomal RNA gene probes to detect intraspecies heterogeneity in *Mycoplasma gallisepticum* and *M. synoviae*. *Avian Dis.* **32,** 220–231.

Yogev, D., Sela, S., Bercovier, H., and Razin, S. (1988c). Elongation factor (Ef-Tu) gene probe detects polymorphism in *Mycoplasma* strains. *FEMS Microbiol. Lett.* **50,** 145–149.

Yogev, D., Levisohn, S., and Razin, S. (1989). Genetic and antigenic relatedness between *Mycoplasma gallisepticum* and *Mycoplasma synoviae*. *Vet. Microbiol.* **19,** 75–84.

E6

PHYLOGENETIC CLASSIFICATION OF PLANT PATHOGENIC MYCOPLASMA-LIKE ORGANISMS OR PHYTOPLASMAS

Bernd Schneider, Erich Seemueller, Christine D. Smart, and Bruce C. Kirkpatrick

General Introduction

Sequence analyses of evolutionarily conserved genes such as the 16S ribosomal RNA (rRNA) have provided a detailed picture of the diversity and phylogenetic relationships of prokaryotes (Olsen et al., 1994). Phylogenetic analyses have provided a coherent framework for the classification of diverse taxa, including the Mollicutes (Weisberg et al., 1989). Another very attractive aspect of this type of analysis is the ability to analyze these phylogenetic markers from nonculturable prokaryotes such as plant pathogenic mycoplasma-like organisms (MLOs). Detailed analyses of several MLO 16S rRNA genes (Gunderson et al., 1994; Kuske and Kirkpatrick, 1992; Lim and Sears, 1989; Namba et al., 1993; Seemuller et al., 1994) led to the conclusion that MLOs were a unique, monophyletic group of mollicutes that deserved a more informative and appropriate taxonomic classification. Therefore, the term "phytoplasma" was proposed and it is now being used by many researchers, including the authors of this chapter.

The process of characterizing these phylogenetically conserved genes has been greatly facilitated by the use of polymerase chain reaction (PCR) technologies to amplify and sequence these molecules directly. This chapter presents a phytoplasma taxonomic scheme that is based on the analysis of two evolutionary markers: the 16S rRNA gene and the spacer region that separates the 16S from the 23S rRNA genes. Phytoplasma phylogenetic relationships established by full-length 16S rRNA sequences agree with phytoplasma groups established by spacer region analyses (Kirkpatrick et al., 1994). The information presented in this chapter provides an outline of the procedures that are available for phylogenetically classifying an unknown phytoplasma strain. Many of the specific procedures, such as running and reading a sequencing gel, will not be discussed here and the reader is referred to any of the numerous molecular biology manuals for information concerning such procedures. The procedures presented here will allow those familiar with such technologies to apply them to the classification of an unknown phytoplasma.

Two approaches have been historically used to characterize phytoplasma 16S rRNA genes: restriction fragment length polymorphisms (RFLP) of PCR-amplified 16S rRNA genes and sequence analysis of full-length 16S rRNA genes. Information presented in this chapter also includes phytoplasma phylogenetic relationships based on RFLP analysis and sequence analysis of the 16/23S spacer regions. The procedures used for obtaining and analyzing this type of data are outlined next.

RFLP Analysis of PCR-Amplified Ribosomal DNA

Introduction

The first phylogenetically based classification of the phytoplasmas was based on RFLP analysis of 16S rDNA (Lee et al., 1993; Schneider et al., 1993). This rapid method can provide valuable information on the phylogenetic and taxonomic classification of an unknown phytoplasma. The procedure is based on PCR amplification of phytoplasma rDNA using primers that recognize all phytoplasma strains but which do not amplify products from healthy plants. The primers listed in Table I can be used to amplify most or all of the 16S rRNA gene and the entire 16/23S spacer region. RFLP analysis of the larger PCR fragments is more desirable because they contain more restriction sites, which in turn provides more information than the shorter PCR fragments. The amplification products are digested with frequently cutting restriction endonucleases such as AluI or RsaI. In some cases, further differentiation can be obtained by digesting with additional enzymes such as MseI or HhaI (Lee et al., 1993).

Materials

Template DNA isolated from plants using the "phytoplasma enrichment" method
  (see Chapter B2, this volume)
Thermocycler, such as the Perkin–Elmer (Norwalk, CT) Model 480
A 50-µl PCR mixture containing the following materials:

TABLE I
SEQUENCE OF OLIGONUCLEOTIDE PRIMERS[a]

Primer name	Sequence
Primer 1	5' AAGAGTTTGATCCTGGCTCAGGATT 3'
Primer 3	5' GGATGGATCACCTCCTT 3'
Primer 4	5' GAAGTCTGCAACTCGACTTC 3'
Primer 5	5' CGGCAATGGAGGAAACT 3'
Primer 7	5' CGTCCTTCATCGGCTCTT 3'

[a] These can be used to amplify and sequence the 16S rRNA gene and 16S/23S spacer region.

A mixture of all four dNTPs, final concentration is 150 μM
PCR primers for amplifying phytoplasma 16S rRNA genes, final concentration of each primer is 0.5 μM
1× *Taq* DNA polymerase buffer containing 1.5 mM MgCl$_2$ (supplied by the enzyme manufacturer)
50 ng of phytoplasma DNA (see above)
1 unit of *Taq* DNA polymerase
1 drop of mineral oil to cover reaction mixture

Horizontal gel electrophoresis unit
Molecular biology grade electrophoresis agarose
Electrophoresis buffer, such as Tris/borate/EDTA (TBE)
Molecular size standards such as the 1-kb ladder (GIBCO-BRL, Gaithersburg, MD)
Ethidium bromide solution (1 μg/ml)
UV transilluminator
Restriction endonucleases such as *Alu*I and *Rsa*I
Restriction enzyme buffer (supplied by the enzyme manufacturer)
DNA samples from representative strains in the major phylogenetic phytoplasma clades
Reagents and equipment for polyacrylamide gel electrophoresis (see Chapter E4, this volume)

Procedure

PCR AMPLIFICATION OF PHYTOPLASMA 16S rRNA AND 16/23S
SPACER REGIONS

Primers 1 and 7 are used to amplify a PCR product that contains the entire 16S rRNA gene plus the 16/23S spacer region. However, when the amount of phytoplasma DNA in the template is low, as it often is in woody plant preparations, better amplification may be achieved using primers 5 and 7. The sequences of all the phytoplasma-specific primers are listed in Table I and their relative positions are shown in Fig. 1.

Phytoplasma 16S rRNA genes and/or SRs are amplified from the plant DNA preparation using the following thermocycler conditions: initial strand denaturation for 5 minutes at 95°C, subsequent denaturing steps for 1 minute at 95°C, annealing for 1 minute at 53°C, and extension for 1.5 to 2 minutes (primers 1 or 5 and 7) at 72°C. An adequate amount of PCR product is usually obtained after 30 cycles; however, 35 cycles may be necessary in some very low-titer woody plants.

The quality and the quantity of the PCR product are assessed by electrophoresing approximately 5 µl of the reaction mixture on a horizontal agarose gel. The product is visualized under UV light after staining with the ethidium bromide solution. The amplified phytoplasma product obtained using primer 1 and 7 is approximately 1800 bp, whereas the product from 5 and 7 is approximately 1450 bp.

An appropriate amount (approximately 100 to 200 ng) of the PCR amplification product is digested with the previously described restriction enzymes at 37°C for 4 hours or overnight. The reaction is stopped by adding $\frac{1}{10}$ volume of 10× gel-loading buffer. Approximately one-third of the digest is electrophoresed on a 5 or 8% polyacrylamide gel. Alternatively, the digestion products may be electrophoresed in size-selection gels such as FMC Metaphor or NuSieve. The restriction fragments are visualized under UV light after staining with ethidium bromide, and the restriction profiles of the unknown samples are compared to reference strains. *Alu*I and *Rsa*I profiles from seven major RFLP groups and two subgroups (Schneider *et al.*, 1993) are shown in Fig. 2.

Discussion

RFLP profiles of PCR-amplified, 16S rDNA have been determined for a large number of phytoplasma strains (Lee *et al.*, 1993; Schneider *et al.*, 1993). Based on the similarities and differences of these profiles, several phytoplasma RFLP groups and subgroups were established. Comparisons showed that most of the RFLP groups correlated well with the major groups established by full-length 16S rDNA sequence analysis (Gunderson *et al.*, 1994; Seemuller *et al.*, 1994).

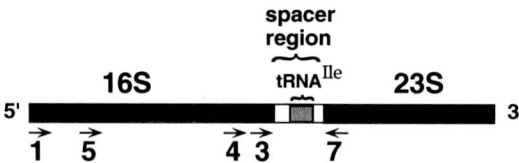

Fig. 1. Map of the 16S/23S rRNA operon found in phytoplasmas. The positions of oligonucleotide primers used in PCR analysis are represented as arrows.

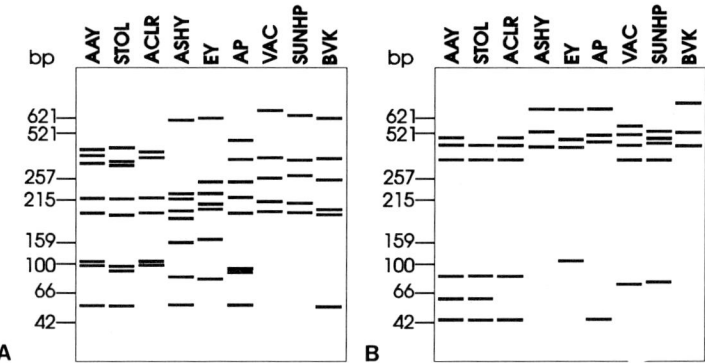

Fig. 2. *Alu*I (A) and *Rsa*I (B) restriction fragment length polymorphism (RFLP) profiles of phytoplasma rDNA amplified with primers 1 and 7. The strains represent the seven major RFLP groups and two subgroups established by Schneider et al. (1993). The phylogenetic positions of the strains are shown in Fig. 3. AAY, American aster yellows—Florida (RFLP group I); STOL, stolbur—Serbia (subgroup of RFLP group I); ACLR, apricot chlorotic leaf roll—Spain (RFLP group II); ASHY, ash yellows—New York (RFLP group III); EY, elm witches' broom—France (RFLP group IV); AP, apple proliferation—Germany (RFLP group V); VAC, vaccinium witches' broom—Germany (RFLP group VI); SUNHP, sunhemp witches' broom—Thailand (subgroup of RFLP group VI); BVK, leafhopper-borne phytoplasma—Germany (RFLP group VII).

However, there are a few cases where classification by RFLP analysis was not as accurate as classification by sequence analysis. For example, strains such as ACLR and AAY clearly differ in their restriction patterns (Fig. 2) but their 16S rDNA sequences are 98.7% homologous. In contrast, there are other instances where two phytoplasma RFLP profiles are less different from each other than those of strains ACLR and AAY, but sequence analysis showed the organisms to be more distantly related. Despite these problems, RFLP analysis can be successfully applied provided that the patterns are compared with strains whose taxonomic status has been determined by 16S rDNA sequence analysis. If an RFLP pattern of an unknown phytoplasma is identical to a standard strain then it can be properly classified. If the pattern differs from the standard stain then sequence analysis will be necessary to determine correctly the taxonomic position of the unknown strain.

Phylogenetic Classification of Phytoplasmas Based on 16S rRNA or 16/23S Spacer Region Sequence Analysis

Sequence analysis of the complete phytoplasma 16S rRNA gene provides the most accurate picture of phylogenetic relationships among phytoplasmas. However, it takes a considerable amount of time and effort to sequence both strands of the full-length 16S rRNA molecule (approximately 1540 nucleotides long). The spacer region (SR) that separates the 16S from the 23S rRNA also contains phylogenetically useful information and, since this region is approximately 280 nucleotide long, it is much easier to sequence. In addition, at the time this chapter was written, full-length 16S rRNA had been determined for approximately 32 phytoplasmas, whereas SR sequences had been determined for more than 60 phytoplasma strains. A manuscript describing the use of SR sequences for classifying phytoplasmas is now being prepared and the SR sequences will soon be available in GenBank.

Once the 16S rRNA or SR sequence is obtained, it is necessary to analyze it using one of the available phylogenetic programs on an appropriate computer. The sequence of the unknown phytoplasma is then compared with the sequences of known phytoplasmas that are available in databases such as GenBank.

Table I presents the sequences of several oligonucleotide primers that can be used to PCR amplify the 16S rRNA gene and/or the 16/23S spacer region. The locations of the primers listed in Table I are shown in Fig. 1.

Materials

See the previous section for materials and equipment required for PCR amplification of phytoplasma rDNA. In addition, the following materials are needed for DNA sequencing operations.

PCR product purification system, such as Wizard Clean Up Columns (Promega, Madison, WI)
Sequencing apparatus and power supply
Cycle sequencing kit, such as dsDNA cycle sequencing kit (GIBCO, BRL, Gaithersburg, MD)

Procedure

PCR AMPLIFICATION OF PHYTOPLASMA 16S rRNA AND 16/23S SPACER REGIONS

The PCR amplification of phytoplasma rDNA has been described earlier. In general, it is most desirable to use primers 1 and 7 to amplify a PCR product that contains the entire 16S rRNA gene plus the 16/23S spacer region, especially if one desires to sequence the 16S rRNA gene. However, if only the phytoplasma SR is to be sequenced, in some instances better amplification is obtained from low-titer hosts using primers 5 and 7 or primers 4 and 7 (Table I, Fig. 1). Phytoplasma 16S rRNA genes and/or SRs are amplified from the plant DNA

TABLE II

GenBank Accession Numbers of 16S rRNA Gene Sequences from MLOs and Culturable Mollicutes

Strain	Accession No.	Reference
Acholeplasma laidlawii	M23933	Weisburg et al. (1989)
Acholeplasma modicum	M23933	Weisburg et al. (1989)
Acholeplasma sp. J233	L33734	Gunderson et al. (1994)
American aster yellows—Florida	X68373	Seemuller et al. (1994)
Apricot chlorotic leaf roll—Spain	X68383	Seemuller et al. (1994)
Apple proliferation—Germany	X68375	Seemuller et al. (1994)
Ash yellows—New York	X68339	Seemuller et al. (1994)
Aster yellows—California	M86340	Kuske and Kirkpatrick (1992)
Aster yellows—Japan	D12569	Namba et al. (1993)
Aster yellows—Maryland	L33767	Gunderson et al. (1994)
Aster yellows—Michigan	M30970	Lim and Sears (1989)
Black alder witches' broom—Germany	X76431	Seemuller et al. (1994)
BVK—Germany	X76429	Seemuller et al. (1994)
Canadian peach X-disease—Canada	L33733	Gunderson et al. (1994)
Clover phyllody—Canada	L33762	Gunderson et al. (1994)
Clover proliferation—Canada	L33761	Gunderson et al. (1994)
Clover yellow edge—Canada	L33766	Gunderson et al. (1994)
Elm witches' broom—France	X68376	Seemuller et al. (1994)
European stone fruit yellows—Germany	X68374	Seemuller et al. (1994)
Flavesence Doree—France	X76560	Seemuller et al. (1994)
Loofah witches' broom—Taiwan	L33764	Gunderson et al. (1994)
Palm lethal yellows—Florida		Unpublished
Peanut witches' broom—Taiwan	L33765	Gunderson et al. (1994)
Pear decline—Germany	X76425	Seemuller et al. (1994)
Pigeon pea witches' broom—Florida		Unpublished
Rice yellow dwarf—Japan	D12581	Namba et al. (1993)
Stolbur—Serbia	X76427	Seemuller et al. (1994)
Sugar cane white leaf—Thailand	X76432	Seemuller et al. (1994)
Sunhemp witches' broom—Thailand	X76433	Seemuller et al. (1994)
Sweet potato witches' broom—Taiwan	L33770	Gunderson et al. (1994)
Tomato big bud—Arkansas	L33733	Gunderson et al. (1994)
Tsuwabuki witches' broom—Japan	D12580	Namba et al. (1993)
Vaccinium witches' broom—Germany	X76430	Seemuller et al. (1994)
Vk grapevine yellows—Germany	X76428	Seemuller et al. (1994)
Western X-disease—California	L04682	Unpublished

preparation using the following thermocycler conditions: initial strand denaturation for 5 minutes at 95°C, subsequent denaturing steps for 1 minute at 95°C, annealing for 1 minute at 53°C, and extension for 1 minute (primers 4 and 7) or 2 minutes (primers 1 and 7 or primer 5 and 7) at 72°C. An adequate amount of

PCR product is almost always obtained after 30 cycles; however, 35 cycles may be necessary in some very low-titer woody plants.

The quality and the quantity of the PCR product are assessed by electrophoresing approximately 5 μl of the reaction mixture on a horizontal agarose gel as previously described. In most cases, there is usually a single band which corresponds to the phytoplasma product. If this is the case then the rest of the PCR reaction mixture is put through the PCR cleanup column to remove the unused primers and buffer. If multiple bands are observed then the reaction mixture will need to be electrophoresed in an agarose gel and the phytoplasma DNA band excised from the gel. The DNA is then removed from the gel matrix using standard procedures such as the GeneClean System (Bio101, LaJolla, CA).

SEQUENCING PCR-AMPLIFIED PHYTOPLASMA 16S rRNA OR 16/23S SPACER REGION

The purified, PCR-amplified phytoplasma 16S rRNA and/or spacer region can either be sequenced directly using a PCR cycle sequencing kit, such as dsDNA cycle sequencing kit (Gibco BRL) or be cloned into a PCR cloning vector, such as the TA cloning kit (Invitrogen, San Diego, CA), and sequenced by standard dideoxy chain-termination reactions. If the full-length 16S rRNA molecule is to be sequenced, then internal sequencing primers will need to be used. Several internal 16S rRNA primers, their sequences, and their locations in the 16S rRNA molecule are presented by Gunderson *et al.* (1994), Namba *et al.* (1993), and Seemuller *et al.* (1994). If the spacer region is to be sequenced then the PCR product can be readily sequenced in both directions using primers 7 and 3 (Table I).

SEQUENCE ANALYSIS OF PHYTOPLASMA 16S rRNA OR 16/23S SR

Spacer region or 16S rRNA sequences are compared and aligned with published sequences using a compilation program such as the Pileup program of the UW-GCG (University of Wisconsin). If the GCG program is used then gap weight should be set to 2.0 and the gap length set to 0.2. The aligned sequences of the known and unknown phytoplasmas can be simply compared and the percentage homology determined or, more informatively, the sequence can be analyzed using a phylogenetic analysis program such as Swofford's Phylogenetic Analysis Using Parsimony (PAUP) version 3.1.1 (1993). This program also contains "bootstrap" analysis which gives confidence values for the branches within the phylogenetic tree. If PAUP is used, bootstrap analysis can be performed using 100 replicates with 10 random additions of taxa per replicate, implementing the tree bisection and reconnection branch-swapping algorithm. Uninformative characters are ignored in this type of analysis.

Table II lists the GenBank accession numbers of the full-length phytoplasma

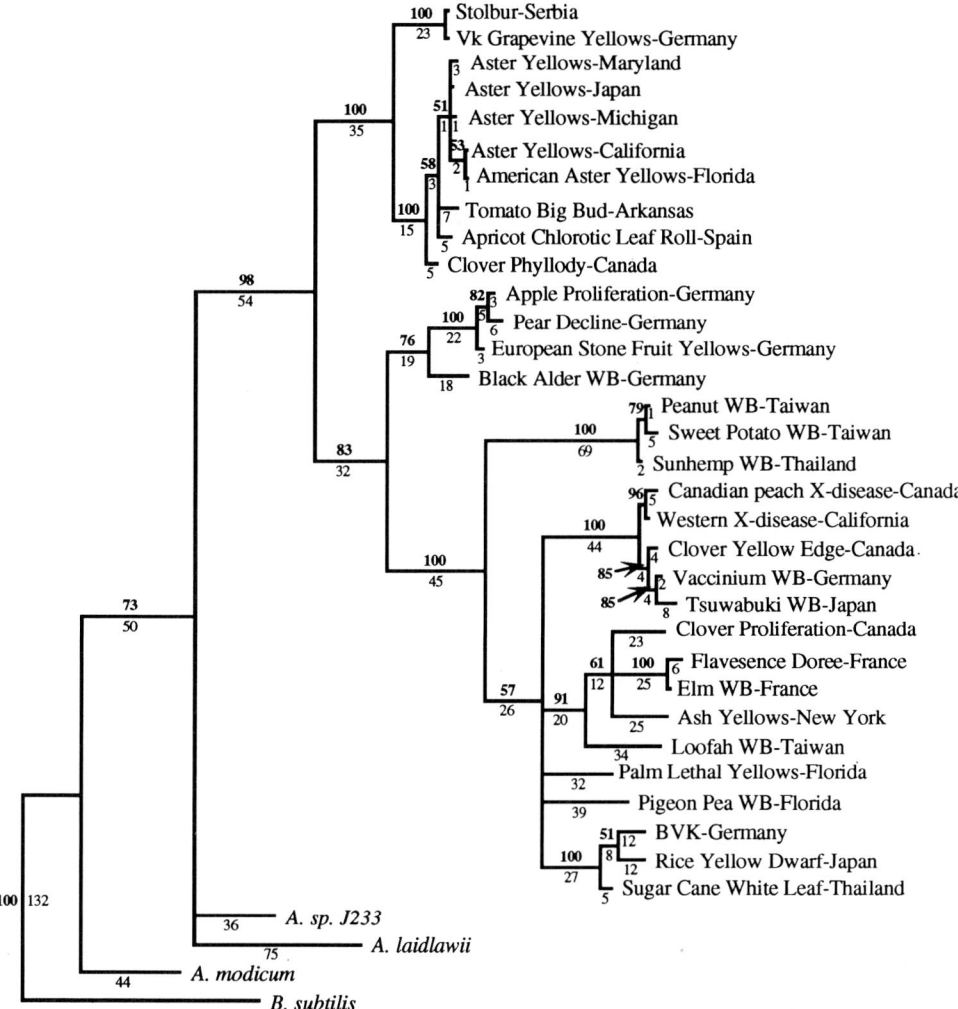

Fig. 3. Phylogram of 16S rRNA sequences generated using a PAUP bootstrap analysis (Swofford, 1993). Numbers in bold above the branches indicate the bootstrap values, whereas numbers below the branches indicate branch length. *B. subtilis* was used as an outgroup.

16S rRNA sequences that were available at the time this chapter was written. A representative phylogenetic tree based on the full-length phytoplasma 16S rRNA sequences that were available at the time this chapter was written is presented in Fig. 3, whereas Fig. 4 shows the relationships of phytoplasmas based on SR sequences.

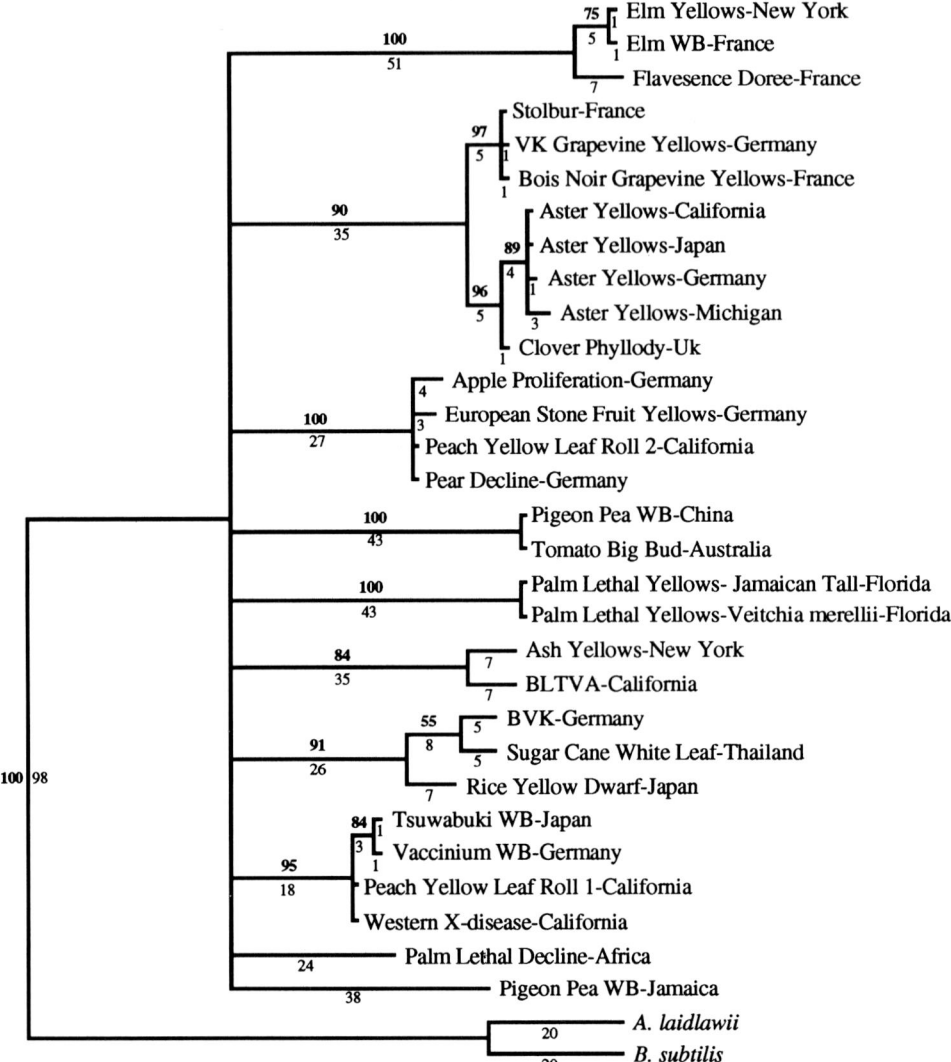

Fig. 4. Phylogram of 16S/23S spacer region sequences generated using a PAUP bootstrap analysis (Swofford, 1993). Numbers in bold above the branches indicate the bootstrap values, whereas numbers below the branches indicate branch length. *A. laidlawii* and *B. subtilis* were used as outgroups.

Discussion

There are advantages and disadvantages associated with both RFLP and sequence analysis of rDNA for classifying phytoplasmas. As previously discussed, a more accurate assessment of the phylogenetic relationship of an unknown phytoplasma strain is obtained from rDNA sequences than from RFLP analysis. However, sequence analysis of either the SR or the 16S rRNA gene requires considerably more time and effort. RFLP analysis of the PCR-amplified 16S rRNA gene is a rapid method to assess the potential affinity of an unknown phytoplasma. However, the total amount of information analyzed by this approach is less than that obtained by sequence analysis. Consequently, RFLP analysis does not provide a detailed picture of the relationships between phytoplasmas within a particular clade.

Similar types of advantages and disadvantages are associated with sequencing either the full-length 16S rRNA gene or the 16/23S spacer region. The SR is comparatively easy to sequence and a large number of phytoplasma SR sequences have been determined (Kirkpatrick *et al.*, 1994; manuscript in preparation). However, the larger size of the 16S rRNA gene allows the most definitive comparisons to be made, and the large database that is now available for both culturable and nonculturable prokaryotes continues to make this marker the most phylogenetically useful gene to analyze.

References

Gunderson, D. E., Lee, I.-M., Rehner, S. A., Davis, R. E., and Kingsbury, D. T. (1994). Phylogeny of mycoplasmalike organisms (phytoplasmas): A basis for their classification. *J. Bacteriol.* **176**, 5244–5254.

Kirkpatrick, B. C., Smart, C. D., Gardner, S. L., Gao, J.-L., Ahrens, U., Maurer, R., Schneider, B., Lorenz, K.-H., Seemuller, E., Harrison, N. A., Namba, S., and Daire, X. (1994). Phylogenetic relationships of plant pathogenic MLOs established by 16/23S rDNA spacer sequences. *IOM Lett.* **3**, 228–229.

Kuske, C. R., and Kirkpatrick, B. C. (1992). Phylogenetic relationships between the western aster yellows mycoplasmalike organisms and other prokaryotes established by 16S rRNA gene sequence. *Int. J. Syst. Bacteriol.* **42**, 226–233.

Lee, I.-M., Hammond, R. W., Davis, R. E., and Gundersen, D. E. (1993). Universal amplification and analysis of pathogen 16S rDNA for classification and identification of mycoplasmalike organisms. *Phytopathology* **83**, 834–842.

Lim, P.-O., and Sears, B. B. (1989). 16S rRNA sequence indicates plant pathogenic mycoplasmalike organisms are evolutionarily distant from animal mycoplasmas. *J. Bacteriol.* **171**, 1233–1235.

Namba, S., Oyaizu, H., Kato, S., Iwanami, S., and Tsuchizaki, T. (1993). Phylogenetic diversity of phytopathogenic mycoplasmalike organisms. *Int. J. Syst. Bacteriol.* **43**, 461–467.

Olsen, G. J., Woese, C. R., and Overbeek, R. (1994). The winds of (evolutionary) change: Breathing new life into microbiology. *J. Bacteriol.* **176**, 1–6.

Schneider, B., Ahrens, U., Kirkpatrick, B. C., and Seemuller, E. (1993). Classification of mycoplasma-like organisms using restriction-site analysis of PCR-amplified 16S rDNA. *J. Gen. Microbiol.* **139,** 519–527.

Seemuller, E., Schneider, B., Maurer, R., Ahrens, U., Daire, X., Kison, H., Lorenz, K.-H., Firrao, G., Avinent, L., Sears, B. B., and Stackebrandt, E. (1994). Phylogenetic classification of phytopathogenic mollicutes by sequence analysis of 16S ribosomal DNA. *Int. J. Syst. Bacteriol.* **44,** 440–446.

Swofford, D. L. (1993). PAUP: Phylogenetic analysis using parsimony, version 3.1.1. Computer program distributed by the Illinois Natural History Survey, Champaign, IL.

Weisberg, W. G., Tully, J. G., Rose, D. L., Petzel, J. P., Oyaizu, H., Mandelco, L., Sechrest, J., Lawerence, T. G., Van Etten, J., Maniloff, J., and Woese, C. R. (1989). A phylogenetic analysis of the mycoplasmas: Basis for their classification. *J. Bacteriol.* **171,** 6455–6467.

E7

DETERMINATION OF CHOLESTEROL AND POLYOXYETHYLENE SORBITAN GROWTH REQUIREMENTS OF MOLLICUTES

Joseph G. Tully

General Introduction

Determination of cholesterol or sterol requirements for growth has been an important element in the classification and taxonomy of newly isolated mollicutes. Although recent reports have established that cholesterol requirements vary within certain phylogenetic groupings of the class Mollicutes (Rose et al., 1993; Tully et al., 1993), tests to measure the growth response to cholesterol and to polyoxyethylene sorbitan (Tween 80) are still part of the recommended techniques in the minimum standards for description of new species within the class (see Chapter E2, this volume; International Committee on Systematic Bacteriology Subcommittee, 1995). Cholesterol requirements are best determined by a quantitative comparison of growth occurring in serum-free media containing various supplements of fatty acids, albumin, and various concentrations (usually 1–20 µg/ml) of solubilized cholesterol. Members of the Mycoplasmataceae and various organisms assigned to the Entomoplasmataceae, Spiroplasmataceae, and Anaeroplasmataceae show minimal or negligible growth in serum-free media, but enhanced growth in the presence of increasing amounts of cholesterol. In addition, a newly described group of mollicutes (genus *Mesoplasma*) has no growth requirement for serum or cholesterol, but shows sustained growth in serum-free media containing low concentrations (0.04%) of Tween 80 (Tully,

1983a; Tully et al., 1993). A supplemental procedure for measuring both serum and polyoxyethylene sorbitan requirements has been reported (Rose et al., 1993).

Growth Response to Cholesterol

Introduction

The direct broth method for quantitating a growth response of mollicutes to cholesterol is performed with eight 100-ml bottles of culture medium (Razin and Tully, 1970; Tully, 1983b). One bottle has a regular serum-containing formulation for growth control; three bottles of serum-free media are for supplement controls; and four bottles contain supplements and various concentrations of cholesterol. Following inoculation with an appropriate test organism, the bottles are incubated until growth is evident. At this point, the cells are sedimented by centrifugation and a standard protein assay is carried out on the cell pellets.

Materials

Young, logarithmic-phase broth culture of test mollicute (10–20 ml): The inoculum should be grown in broth with the lowest possible serum content, preferably in 1% bovine serum fraction broth (Tully, 1984).
Standard broth medium (100 ml): This formulation should be similar to the broth employed for the inoculum or a medium capable of supporting a yield of at least 10^7 colony-forming units (CFU)/ml (or equivalent number of color-changing units per ml) of the test organism. This broth should also contain 0.5% glucose, 0.25% arginine, and 0.002% phenol red indicator.
Serum-free broth medium (800 ml): The recommended formulation is 16.8 g of mycoplasma broth base to 714 ml deionized water. Adjust pH to 7.6–7.8 and sterilize at 121°C for 20 minutes. Add the following sterile supplements: 8 ml of 50% (w/v) glucose solution, 4 ml of 42% arginine solution, 80 ml of fresh yeast extract, and 4 ml of penicillin solution (100,000 U/ml). One hundred milliliters of the complete medium is added to each of seven 125-ml sterile, screw-capped bottles and all bottles are incubated overnight to check sterility.
Stock albumin solution (10%): Add 5 g bovine serum albumin (essentially fatty acid free)(Sigma Chemical A-6003) to 50 ml of heat-sterilized deionized water. Adjust pH to 7.5 and filter sterilize through a 220-nm membrane filter. Store at 4°C.
Palmitic acid stock solution (10 mg/ml): Add 10 ml of 95% ethanol to 100 mg palmitic acid in a sterile screw-capped culture tube. Store at 4°C and warm gently to solubilize the fatty acid just before use.

Polyoxyethylene sorbitan (Tween 80) stock solution (10%): Add 1 ml Tween 80 (Sigma Chemical P-1754) to 9 ml of heat-sterilized deionized water. Filter through a sterile 450-nm membrane filter and store at 4°C.

Cholesterol stock solution (20 mg/ml): Place 200 mg of cholesterol in a heat-resistant glass culture tube and add 10 ml of 95% ethanol. Store at 4°C until ready to use. Warm solution very carefully by heating (with tube cap loose) over a burner until the cholesterol is in solution.

Ethanol, 95% (v/v)

Reagents for protein assay (such as Bio-Rad Laboratories, Richmond, CA)

Wash solution (1 liter), containing 0.25 M NaCl and 0.01 M MgCl$_2$

Procedure

PREPARATION OF MEDIUM SUPPLEMENTS

1. Label bottles containing culture media from A to H, with the serum-containing medium as bottle "A" and the seven bottles of serum-free broth as "B" through "H" (Table I).
2. Add 5 ml of albumin solution and 0.1 ml palmitic acid solution to media bottles "C" through "H" (Table I).
3. Preparation of Reagent C series: In five heat-resistant culture tubes (labeled C1 to C5), add the stated amounts (Table I) of Tween 80 solution, 95% ethanol, and the cholesterol solution. The cholesterol solution should be heated gently to

TABLE I

MEDIUM SUPPLEMENTS FOR TESTING CHOLESTEROL GROWTH REQUIREMENTS[a]

Test bottle	Albumin (ml)	Palmitic acid (ml)	Reagent C series				Final concentration of cholesterol in medium (μg/ml)
			Tube No.	Tween 80 (ml)	Ethanol (ml)	Cholesterol (ml)	
A[b]	0	0		0	0	0	—
B	0	0		0	0	0	0
C	5	0.1		0	0	0	0
D	5	0.1	C1	1.0	1.0	0	0
E	5	0.1	C2	1.0	0.95	0.05	1
F	5	0.1	C3	1.0	0.75	0.25	5
G	5	0.1	C4	1.0	0.50	0.50	10
H	5	0.1	C5	1.0	0	1.0	20

[a] See text for concentrations of ingredients in supplements.
[b] Bottle A is control broth medium containing serum or serum fraction supplements (see text).

solubilize, and a 1-ml glass pipette, which also has been heated gently, should be used to transfer the cholesterol to tubes labeled C2 through C5.

4. Place the C-series tubes in a 56°C water bath to keep the cholesterol in solution. Media bottles "E" through "H" should also be warmed in a 37°C incubator before the addition of C-series components. After media are warm, add 0.2 ml of supplements C1 to C5 to their respective media bottles "D" through "H" (Table I).

INOCULATION, INCUBATION, AND HARVESTING

1. A 1–3% inoculum is generally best, depending on the general growth characteristics of the organism. The critical point is to keep the amount of inoculum as low as possible since it is most often prepared in a serum-containing medium and can, therefore, add significant amounts of cholesterol to the test medium.

2. Incubate bottles at a temperature appropriate for the mollicute and observe growth every 2–3 days. The cultures should be harvested only when the growth in bottles "E" through "H" appears to reach the stationary phase of growth, as determined by pH changes and/or increased turbidity. Fast-growing mollicutes might still require an incubation period of 5–10 days, whereas more fastidious strains may need 14–18 days.

3. Harvest the contents of the individual bottles "A" through "H" by centrifugation of media in 40-ml volume plastic centrifuge tubes at about 20,000 g for 20–30 minutes. Remove supernatant for decontamination and refill respective centrifuge tubes until the total volume of each culture is harvested. Wash each pellet once in the wash solution, repeat sedimentation, and freeze the final cell pellet in the centrifuge tube at $-70°C$

PROTEIN ASSAY

1. Resuspend the cell pellet in 1 ml of 0.1 N NaOH and vortex well to solubilize the total cell material.

2. A variety of protein assays can be used to measure the total amount of protein in each of the respective cell pellets; the Bio-Rad assay is usually the most convenient. A protein standard solution is prepared, usually with three dilutions (1:2, 1:4, and 1:7) that provide a range of protein values from 70, 35, and 20 µg/0.1 ml. The readings, after the addition of a dye component, are taken on a spectrophotometer (750 nm), and values are plotted on graph paper (20 × 20 squares per inch). The readings should yield a straight-line plot, which is then used to calculate the protein content of the cell pellets.

3. A 0.1-ml sample of each of the individual solubilized cell pellets should

first be assayed. If the optical density reading on any one sample is near the upper level of the standard plot, that sample should be diluted 1:10 and retested. Determine the amount of protein (in micrograms per 100-ml medium volume) for each medium formulation by multiplying the protein amount by the dilution factor (20× for undiluted samples and 200× for samples diluted 1:10). See Table II for an example of test results.

Discussion

Sterol or cholesterol growth requirements of mollicutes cannot be assessed simply by a few passages on serum-free agar or broth because of the carryover of cholesterol and sterol in the inoculum. Controls of both sterol-requiring and sterol-nonrequiring strains should always be compared in these tests. Growth yields, especially with spiroplasmas, should not be based on the presence of helical structures since the organisms can show significant yields of nonhelical forms. Sterol-nonrequiring mollicutes, such as the acholeplasmas, will usually show comparable amounts of growth in the various serum-free formulations and in bottles containing additions of cholesterol. Sterol-requiring mollicutes (mycoplasmas, entomoplasmas, and some spiroplasmas) and the mesoplasmas will frequently exhibit the results shown in Table II when Tween 80 levels of 0.01% are employed. However, if the quantity of Tween 80 for medium "D" is raised to 0.04%, the mesoplasmas will exhibit as much growth in this medium as that

TABLE II

Response of Cholesterol-Requiring Mollicute in Broth/Cell Protein Test[a]

Bottle	Sample dilution	Readings at 750 nm	Protein in sample (μg)	Multiplication factor	Final protein in pellet from 100 ml medium (mg)
A	1:10	0.125	25	200	5.00
B	—	0.035	10	20	0.20
C	—	0.125	25	20	0.50
D	—	0.035	10	20	0.20
E	—	0.430	75	20	1.50
F	1:10	0.080	17.5	200	3.50
G	1:10	0.110	22.5	200	4.50
H	1:10	0.125	25	200	5.00

[a] *Mycoplasma citelli*, strain RG-2C, harvested after 5 days of incubation at 37°C. The control medium (A) was mycoplasma medium supplemented with 1% bovine serum fraction. The sample size for the protein test from each solubilized pellet was 0.1 ml. Protein standard readings at 750 nm: 140 μg, 0.82; 70 μg, 0.40; 35μg, 0.19; and 20 μg, 0.10.

obtained in a serum-containing formulation (Rose et al., 1993). A few mollicutes are sensitive to the small amounts of Tween 80 (0.01%) used to keep the cholesterol solubilized; when this occurs the Tween 80 level can be reduced to 0.001%. Some cholesterol may crystallize out in medium containing 20 µg/ml, but this occurrence does not appear to affect the utilization of cholesterol by cholesterol-requiring mollicutes.

Growth Response to Polyoxyethylene Sorbitan

Introduction

Further modification of the standard cholesterol test was prompted by the isolation and description of the mesoplasmas. These organisms appeared to be unique among mollicutes since the group was able to show sustained growth in the absence of added cholesterol or serum, provided rather small quantities (0.04%) of Tween 80 were added to the media (Tully, 1983a; Rose et al., 1993). The technique developed involves the ability of the test organism to maintain sustained growth through 23 serial dilutions in three distinct media formulations: serum-free broth, serum-free broth with 0.04% Tween 80, and serum-containing broth.

Materials

Young, logarithmic-phase broth culture of test mollicute (1–5 ml), preferably in medium containing the lowest possible content of serum supporting at least 10^7 CFU/ml

Fresh yeast extract (25% aqueous solution)(GIBCO, Grand Island, NY)

Stock albumin solution (10%): Add 5 g bovine serum albumin (essentially fatty acid free)(Sigma Chemical A-6003) to 50 ml of heat-sterilized deionized water. Adjust pH to 7.5 and filter sterilize through a 220-nm membrane filter. Store at 4°C.

Polyoxyethylene sorbitan (Tween 80) stock solution (10%): Add 1 ml Tween 80 (Sigma Chemical P-1754) to 9 ml of heat-sterilized deionized water. Filter through a sterile 450-nm membrane filter and store at 4°C.

Fetal bovine serum (Hyclone Laboratories, Logan, UT)

Serum-free broth (100 ml, dispensed in 2-ml volumes in 1-dram (3.6 ml) vials. Prepared as follows: 2 g of mycoplasma broth base, 2 ml of 0.1% aqueous phenol red, 0.5 ml of 1 N NaOH, and 80 ml of deionized water. Sterilize at 121°C for 20 minutes and add the following sterile supplements: 10 ml of 25% fresh yeast extract, 2.5 ml of 10% bovine serum albumin solution, 1 ml of

50% glucose solution, 1 ml of 50% fructose solution, 0.5 ml of 42% arginine solution, and 0.5 ml of 100,000 U/ml penicillin solution.

Serum-free broth with 0.04% Tween 80 (100 ml dispensed as described earlier). The formulation and preparation are similar to regular serum-free broth, except the final medium is also supplemented with 0.4 ml of a sterile 10% aqueous solution of Tween 80.

Serum-containing broth (100 ml dispensed as described earlier). Formulation and preparation are similar to serum-free broth, except the deionized water content is reduced to 65 ml and 15 ml of sterile fetal bovine serum is added to the final medium.

Procedure

1. Label seven vials of each of the three test media, and then add an inoculum of 0.2 ml to the first vial in each of the three test series. Perform serial 10-fold dilutions of each of the three series through the seventh vial (10^{-7} dilution). Incubate at the appropriate temperature for the test organism. Include an uninoculated vial of each test medium for growth comparison (pH changes and/or turbidity).

2. Observe vials at 2- to 3-day intervals. When pH and turbidity changes appear to reach a stationary phase, transfer 0.2 ml from the last 10-fold dilution vial showing growth to a vial of fresh medium of the same formulation. Repeat the serial 10-fold dilutions through another series of seven vials.

3. Continue the 10-fold passage levels until the culture has been maintained through 23 serial 10-fold dilutions. For each of the three test media, record the highest 10-fold dilution that was able to support significant growth, as judged by appropriate changes in the pH or increased turbidity of the test medium.

Discussion

An example of test readings obtained with selected mollicutes is shown in Table III. Sterol-nonrequiring acholeplasmas will usually show sustained growth in both serum-free and serum-containing broth, and a few species may also grow in the 0.04% Tween 80 media. The mesoplasmas are rarely able to show growth beyond 10 serial 10-fold dilutions in serum-free broth, but grow well in broth containing either Tween 80 or serum. Most mycoplasmas, entomoplasmas, and spiroplasmas show sustained growth only in the serum-containing formulation. With both the cholesterol test based on cell protein yields and the 23 serial dilution procedure, some mollicutes with very fastidious growth requirements (i.e., *Mycoplasma genitalium,* etc.) may not grow well enough in the serum- or

TABLE III
Growth Responses of Various Mollicutes to Serial Dilution Tests for Serum/Cholesterol Requirement

	Number of 10-fold serial dilutions maintaining growth in		
Mollicute species/strain	Serum-free medium	Serum-free medium with 0.04% Tween 80	Serum medium
Acholeplasma laidlawii PG9	23	23	23
A. oculi 19L	23	5	23
Mycoplasma mycoides subsp. *mycoides* B3	1	1	23
My. putrefaciens KS-1	2	1	23
Entomoplasma ellychniae ELCN-1	1	1	23
E. lucivorax PIPN-2	2	1	23
Mesoplasma florum L1	7	23	23
Me. seiffertii F7	8	23	23
Me. syrphidae YJS	12	23	23
Spiroplasma citri Maroc R8A2	2	1	23
S. floricola 23-6	23	23	23
S. apis B31	23	5	23
S. taiwanense CT-1	2	0	23

cholesterol-containing control medium described here to clearly establish their sterol requirements. In these instances, it is necessary to test the organism in appropriate base formulations devoid of the serum component and in the same formulations where various quantities of serum are added (Tully *et al.*, 1983).

References

International Committee on Systematic Bacteriology, Subcommittee on the Taxonomy of Mollicutes (1995). Minimum standards for the description of new species in the class Mollicutes. *Int. J. Syst. Bacteriol.* **45**, 605–612.

Razin, S., and Tully, J. G. (1970). Cholesterol requirement of mycoplasmas. *J. Bacteriol.* **102**, 306–310.

Rose, D. L., Tully, J. G., Bové, J. M., and Whitcomb, R. F. (1993). A test for measuring growth responses of mollicutes to serum and polyoxyethylene sorbitan. *Int. J. Syst. Bacteriol.* **43**, 527–532.

Tully, J. G. (1983a). Reflections on the recovery of some fastidious mollicutes with implications of the changing host patterns of these organisms. *Yale J. Biol. Med.* **56**, 799–813.

Tully, J. G. (1983b). Tests for digitonin sensitivity and sterol requirement. *In* "Methods in Mycoplasmology" (S. Razin and J. G. Tully, eds.), Vol. 1, pp. 355–362. Academic Press, New York.

Tully, J. G. (1984). Family Acholeplasmataceae, genus *Acholeplasma*. *In* "Bergey's Manual of Systematic Bacteriology" (N. R. Krieg and J. G. Holt, eds.), Vol. 1, pp. 781–787. Williams and Wilkins, Baltimore, MD.

Tully, J. G., Bové, J. M., Laigret, F., and Whitcomb, R. F. (1993). Revised taxonomy of the class Mollicutes: Proposed elevation of a monophyletic cluster of arthropod-associated mollicutes to ordinal rank (Entomoplasmatales, ord. nov.), with provision for familial rank to separate species with nonhelical morphology (Entomoplasmataceae, fam. nov.) from helical species (Spiroplasmataceae), and emended descriptions of the order Mycoplasmatales, family Mycoplasmataceae. *Int. J. Syst. Bacteriol.* **43,** 378–385.

Tully, J. G., Taylor-Robinson, D., Rose, D. L., Cole, R. M., and Bové, J. M. (1983). *Mycoplasma genitalium*, a new species from the human urogenital tract. *Int. J. Syst. Bacteriol.* **33,** 387–396.

SECTION F
Pathogenicity

F1

INTRODUCTORY REMARKS
Shmuel Razin

Mycoplasma infections are rarely of the fulminant type, but rather follow a chronic course. In humans and animals, mycoplasmas can be considered as surface parasites, colonizing the epithelial linings of the respiratory and urogenital tracts. Adhesion of the mycoplasmas to their target tissue is thus a prerequisite for infection (Razin and Jacobs, 1992). Methods to assess mycoplasma adhesion have been previously described by Kahane and Bredt (1983). Chapter F2 presents new approaches and techniques concerning mycoplasma adhesion, including an adherence–inhibition assay, testing specifically for antibodies inhibiting *Mycoplasma pneumoniae* adhesion in patients' sera, and a capture ELISA test for quantitation of cytadhering mycoplasmas. A new approach for epitope mapping of mycoplasma adhesin molecules is detailed in Chapter F3. This method examines the interaction of oligopeptides, synthesized according to the amino acid sequence of the adhesin molecule, with monoclonal antibodies capable of blocking mycoplasma adhesion. In this way the regions of the adhesin molecule responsible for adhesion to the host can be defined (Opitz and Jacobs, 1992).

The intimate association of the adhering mycoplasmas with the host cells provides an environment in which local concentrations of mildly toxic by-products excreted by the parasite, such as H_2O_2 and superoxide radicals (O_2^-), can build up and cause tissue damage (Razin, 1991). The extensive studies of Almagor *et al.* (1986) have led to the notion that the H_2O_2 and the superoxide radicals penetrate into the host cell. As a result of superoxide accumulation, host cell catalase is inhibited, giving rise to H_2O_2 accumulation. This, in turn, leads to product inhibition of host cell superoxide dismutase. The increased levels of H_2O_2 and O_2^- cause oxidative damage to vital host cell constituents, such as membrane lipids. Chapter F4 describes the methods for testing catalase and superoxide dismutase activities in host cells infected with mycoplasmas.

Host immune reactions appear to play a major role in the pathogenesis of mycoplasma infections, including autoimmune manifestations characteristic of many mycoplasma diseases. The recent molecular definition of many of the immune system components, as well as the availability of reagents and kits for their determination, have resulted in significant advances in our understanding of the interaction of mycoplasmas with immune system components. The larger part of Section F is devoted, therefore, to methods for studying mycoplasma effects on the immune system. Chapter F5 details various procedures for measuring the release of biologically active molecules, such as cytokines, arachidonate metabolites, active oxygen, and nitric oxide, produced in response to macrophage stimulation by mycoplasmas and their products. Chapter F7 describes methods for studying the polyclonal proliferation of lymphocytes induced by mycoplasma mitogens, whereas Chapter F6 deals with the special case of a mycoplasmal superantigen produced by *M. arthritidis,* named MAM. Superantigens act as very potent T-cell activators, including cytotoxic lymphocytes. The activity of the superantigens depends on the expression of certain major histocompatibility complex (MHC) molecules. The current interest in superantigens relates to their ability to modulate the immune system *in vivo* in such a way that may lead to development or enhancement of human autoimmune diseases, such as rheumatoid arthritis. Chapter F6 outlines the criteria for defining proteins as superantigens, focusing on the purification and characterization of the *M. arthritidis* superantigen.

Some mycoplasmas were shown to produce a fraction (named MIaF) that induces major histocompatibility complex class I and class II expression on a variety of animal and human cell types, increasing the antigen presentation function of these cells. Methods for modulation of the expression of MHC molecules by the mycoplasmas are described in Chapter F8. *M. pulmonis* was found to augment natural killer cell activity in mice, playing a possible role in resistance to *M. pulmonis* infections by inducing interferon-γ secretion and activating macrophages that phagocytize and kill the mycoplasmas. Methods for investigating this specific aspect of mycoplasma interaction with immune cells are detailed in Chapter F9.

References

Almagor, M., Kahane, I., Gilon, C., and Yatziv, S. (1986). Protective effects of the glutathione redox cycle and vitamin E on cultured fibroblasts infected by *Mycoplasma pneumoniae. Infect. Immun.* **52,** 240–244.

Kahane, I., and Bredt, W. (1983). Tests for adherence properties of mycoplasmas. *In* "Methods in

Mycoplasmology" (J. G. Tully and S. Razin, eds.), Vol. II, pp. 345–354. Academic Press, New York.

Opitz, O., and Jacobs, E. (1992). Adherence epitopes of *Mycoplasma genitalium* adhesin. *J. Gen. Microbiol.* **138,** 1785–1790.

Razin, S. (1991). The genera *Mycoplasma, Ureaplasma, Acholeplasma, Anaeroplasma,* and *Asteroleplasma. In* "'The Prokaryotes" (A. Balows, H. G. Trüper, M. Dworkin, W. Harder, and K.-H. Schleifer, eds.), 2nd Ed., Vol. II, pp. 1937–1959. Springer-Verlag, New York.

Razin, S., and Jacobs, E. (1992). Mycoplasma adhesion. *J.Gen. Microbiol.* **138,** 407–422.

F2

MYCOPLASMA ADHERENCE TO HOST CELLS: METHODS OF QUANTIFYING ADHERENCE

Itzhak Kahane and Enno Jacobs

Introduction

For many pathogenic mycoplasmas, adherence is a prerequisite for the cytopathic effects. Adhesins and other mycoplasma membrane compounds are involved in the process (for reviews, see Kahane, 1984; Razin and Jacobs, 1992; Kahane and Horowitz, 1993; Baseman, 1993). The characteristics of this process have to be studied for each mycoplasma species, since no general rule can be indicated (Kahane, 1984; Saada et al., 1991). Some of the adhesins have been studied in great detail, e.g., a 170-kDa protein (P1) of *Mycoplasma pneumoniae* and a 135-kDa protein (MgPa) of *M. genitalium*. The adherence process is complex and, at least in some mycoplasmas, auxiliary proteins, such as the high molecular weight proteins of *M. pneumoniae*, are involved, presumably as part of a cytoskeleton (Kahane, 1984) that is apparently responsible for the elongated flask shape of these organisms. The cytoskeleton appears to be involved in the organization of the adhesins and in the terminal tip structure of the other flask-shaped pathogenic species. So far, data indicate that the adhesins and other surface components involved in adherence are not directly responsible for the cytopathic effects, but their proximity to the host cell membrane forms a microenvironment enabling other factors to cause the cytopathic process (see Chapter F4, this volume). Assays to quantitate mycoplasma adherence were developed primarily in two main directions:

1. Adherence of host cells to mycoplasma cells or colonies when the latter are attached to a substrate.
2. Adherence of labeled mycoplasmas to host cells.

Protocols to evaluate the following were designed and detailed in "*Methods in Mycoplasmology*" (Kahane and Bredt, 1983):

1. Adherence of erythrocytes to mycoplasmas attached to a substrate.
2. Adherence of mycoplasmas to cells in suspension.
3. Binding of cell receptors to mycoplasmas.
4. Adherence of mycoplasmas to monolayers of cells in culture.

These procedures are still adequate. In the present chapter, details are provided for modified assays or new ones. These include:

1. Adherence inhibition assay by which adherence blocking antibodies are evaluated. This assay was developed on the basis of the test of adherence of erythrocytes to mycoplasmas attached to substrate (Kahane and Bredt, 1983).
2. Adherence of mycoplasmas to ligands adsorbed on a solid surface matrix.
3. Assessment of adhering mycoplasmas to cells in culture using a cell–enzyme-linked immunosorbent assay (ELISA) system.

Adherence–Inhibition Assay (AIA)

Introduction

The assay for adherence-blocking antibodies is useful, both from the clinical point of view and as a research tool. Since this assay was developed during studies of *M. pneumoniae*, it will be described for this mycoplasma as an example. It can be expanded to other mycoplasmas that resemble *M. pneumoniae*, i.e., *M. genitalium*.

Virulent *M. pneumoniae* strains are able to adhere to different host cells, including erythrocytes. One major adhesin was characterized as a 170-kDa membrane protein concentrated in the tip structure of *M. pneumoniae* (review by Razin and Jacobs, 1992). This adhesin, the P1 protein of *M. pneumoniae*, is not only an important virulence factor of this *Mycoplasma* species, but is also one of the major protein antigens of this human pathogen (Jacobs *et al.*, 1986). Using the Western immunoblot technique, it was found that, especially during the first contact with this pathogen, early and prominent serum antibodies were directed to the P1 protein. From this point of view, the AIA was devised as a routine diagnostic approach to test and quantify human sera for anti-*M. pneumoniae*–adherence antibodies (Jacobs *et al.*, 1985). Moreover, the AIA was useful in screening supernatants of large libraries of monoclonal antibody-producing clones for those monoclonal antibodies with a capacity to inhibit the adherence of *M. pneumoniae* (Gerstenecker and Jacobs, 1990) or *M. genitalium* (Opitz and Jacobs, 1992).

Materials

Flat-bottom sterile microtiter plates with a cover (F-form, Greiner Labortechnics, Nürtingen, Germany)
Roux bottles
Cell scrapers
Hayflick's modification of Edward's medium
Sheep erythrocytes [$3-4 \times 10^6$ cells/ml phosphate-buffered saline (PBS)]
Polypropylene tape for ELISA plate sealing
Alsever's solution (Sigma Chemicals)
Double-distilled water
Phosphate-buffered saline
ELISA photometer reader (414 nm)
Solutions are equilibrated at room temperature

Procedure

PREPARATION OF INOCULUM

Mycoplasmas are grown in Roux bottles containing Hayflick's modified Edward's medium for about 48 hours at 37°C. Only glass-adherent mycoplasmas should be used for an AIA inoculum. It is therefore recommended that the stock be prepared from cultures which show only a slight indicator change from red to orange if phenol red is used as an indicator. The whole supernatant should be removed and the glass-adherent mycoplasmas should be washed with room temperature-equilibrated PBS solution. The mycoplasmas which still adhere to the glass are harvested with a cell scraper in 100 ml of fresh mycoplasma medium and are stored at $-70°C$ in 1-ml aliquots for further use.

PREPARATION OF MICROTITER PLATES

The 1-ml mycoplasma stock suspension is thawed and sonicated by three short pulses to disrupt the microcolonies into single cells. The resulting suspension is further diluted in 50 ml mycoplasma medium and 100 µl is dispensed in each well of the microtiter plate. The plates are covered and incubated overnight at 37°C. The mycoplasma medium of each well is exchanged with fresh (100 µl) medium and the plates are incubated for an additional 24 hours at 37°C. Plates fit for the AIA must show only a slight indicator color change to orange; plates with an indicator change to yellow should be discarded (start assay again with higher stock dilutions).

SHEEP ERYTHROCYTE SUSPENSION

The sheep erythrocyte suspension, stored not longer than 1 week in Alsever's solution, is washed four times in PBS and resuspended in PBS to $3-4 \times 10^8$ cells/ml.

AIA PROCEDURE

Human sera are inactivated at 56°C for 30 minutes. Twofold dilutions of the sera (or monoclonal antibody) in PBS are prepared in a separate microtiter plate. Immediately before starting the test, discard the medium from the microtiter plates containing the mycoplasma. The first row of the ELISA plate is filled only with PBS (without antibodies) to test the maximum adherence of erythrocytes (positive control). The diluted serum (100 µl each) is transferred to the microtiter plate with the pregrown mycoplasmas and incubation takes place for 1 hour at 37°C. The microtiter plates are then washed gently twice with PBS using a microtiter pipette. Do not use washing machines as they can wash the mycoplasmas off. One hundred microliters of the erythrocyte suspension is added per well and plates are incubated for an additional 1 hour at 37°C. After this last incubation, each well is filled to the top with PBS and then sealed with polypropylene tape. The microtiter plate is inverted for 15 minutes to allow nonattached erythrocytes to sink down to the sticky tape, which is removed, and the plates are washed again with 100 µl PBS per well. Adherent erythrocytes are lysed with 100 µl of distilled water per well for 10 minutes. The extinction of lysed sheep erythrocytes is measured at 414 nm. The average absorbance of the first row (positive control with only erythrocytes) of each plate is taken as 100% (maximum adherence). A serum dilution is considered to be positive if the antibodies in it inhibited adherence of erythrocytes by more than one-half (absorbance values less than 50% of the positive control).

Discussion

The advantages of the AIA are that performance of the AIA is fast and, if the laboratory has experience in growing mycoplasmas, the test is easy to perform. Moreover, the test is inexpensive and one is independent of costly commercial tests. Another advantage is the assay is specific. No apparently false-positive reactions occurred in sera from patients with autoantibodies which can lead to false-positive results in the complement fixation test. One major disadvantage of the AIA is that it is less sensitive than the complement fixation test or other serological tests used in serodiagnosis of *M. pneumoniae* diseases (Jacobs, 1993). This limited sensitivity depends on the fact that, although during *M. pneumoniae* diseases prominent specific antibody responses are directed to the P1 protein, many of these antibodies are not directed to the regions of the P1 protein which support adherence, but to other epitopes of the P1 protein with no function in adherence (Jacobs, 1991). Comparison of serum titers measured by the complement fixation test with those titers in the AIA showed that, in most cases, the AIA titers were less than 5% of the complement fixation titers. Therefore, AIA will certainly not replace other serological test to prove *M. pneumoniae* disease. It may be used in cases where serological results in the complement fixation test are questionable, i.e., possible false-positive titers such as those found in sera of patients with known autoaggressive diseases or diseases with cellular necrosis, including pancreatitis and meningitis (Jacobs *et al.*, 1985). In contrast to these limitations in clinical serodiagnosis, the AIA is a useful tool in screening monoclonal antibody-secreting clones for monoclonals with adherence-inhibiting capacity or to follow up the production of adherence-inhibiting antibodies in hybridoma cultures (Gerstenecker and Jacobs, 1990).

Another approach for the assay of adherence-inhibiting antibodies using an ELISA system is described later in this chapter.

Adherence of Mycoplasmas to Ligands Adsorbed to a Solid Surface Matrix

Introduction

The goal of this assay is to identify purified cell components to which mycoplasmas adhere. The advantage of this assay is that the compounds are immobilized on the matrix to allow easy separation between the bound and free mycoplasmas.

Materials

Ligands (glycolipids, glycoproteins isolated from host cells)
Microtiter plates, 96 wells (Falcon 3912, Becton–Dickinson or similar, e.g., Immulon Removeawell plates)
Mycoplasma metabolically labeled by [^{3}H]palmitate (Kahane and Bredt, 1983; Krivan *et al.*, 1989) or [^{35}S]methionine (Saada *et al.*, 1991)
0.01 M sodium phosphate buffer, pH 7.4, containing 150 mM NaCl, 1 mM CaCl$_2$, and 0.01% NaN$_3$–PBS

Methanol containing 0.1 µg/ml cholesterol and phosphatidylcholine (methanol–lipid solution)

Tris–BSA buffer: 50 mM Tris–HCl, pH 7.6, 110 mM NaCl, 5 mM CaCl$_2$, 0.2 mM phenylmethylsulfonyl fluoride, and 1% (w/v) bovine serum albumin (BSA) (TBS)

RPMI 1640 medium containing 1% (w/v) BSA (Sigma, fatty acid-free) and 25 mM HEPES, pH 7.3 (RPMI–BSA)

Scintillation fluid: nontoxic and biodegradable, e.g., Aquasafe (Zinsser Analytic, Frankfurt, Germany)

Procedure

PREPARATION OF IMMOBILIZED LIGAND

Glycolipids

Purified glycolipids (Krivan *et al.*, 1989) are serially diluted in microtiter plate wells in 25 µl of methanol–lipid solution. The solvent is dried by evaporation and the wells are filled with TBS, incubated for 1 hour at 37°C, and rinsed with RPMI–BSA.

Glycoproteins

Glycoproteins, e.g., as detailed by Roberts *et al.* (1989), dissolved in phosphate buffer are aliquoted (100 µl) in microtiter plate wells and are incubated for 16 hours at 4°C. Unbound proteins are removed, and the wells are filled with TBS, incubated for 30 minutes at room temperature, and rinsed with RPMI–BSA.

SOLID-PHASE ADHERENCE ASSAY

1. A labeled mycoplasma suspension (25 µl) is added to every well. For inhibition studies, various ligands in RPMI–BSA are added to the appropriate wells.
2. The plates are incubated for 2 hours at 37°C.
3. The wells are washed five times with PBS.
4. Radioactivity of bound mycoplasmas in each well is quantified by scintillation counting.

Discussion

The major advantage of this procedure is the easy way of separating unbound material. This procedure can also be modified to characterize the adherence of solubilized mycoplasma components. Evaluation of the mycoplasma material can be assessed, not only by radioactivity, but also by specific antibodies which can be detected in an ELISA assay, as detailed in the following section. A variation of this assay was described by Krivan *et al.* (1989) by which glycolipids (or lipids) are separated by thin-layer chromatography which serves as the solid phase. This, in turn, is overlaid with mycoplasmas, rinsed, dried, and autoradiographed.

Assessment of Adhering Mycoplasmas to Cells in Culture Using Cell–ELISA System

Introduction

In this procedure, cytadherent mycoplasmas are detected by specific antibodies and by capture ELISA. The assay system used by Henrich *et al.* (1993) with *M. hominis* will be described as an example.

Materials

96-well flat bottom microplates (Greiner, Frickenhausen, Germany)
HeLa cells grown on these plates as monolayers
Dulbecco's modified Eagle's medium (DMEM) with 10% (v/v) fetal bovine serum
Mycoplasma suspension [about 10^8 colony-forming units (CFU)/ml]

Procedure

HeLa cells are grown in the wells to confluency (about 2–3 days, 10^5 cells per well). The wells are washed twice with DMEM prewarmed to 37°C. A mycoplasma suspension (100 μl, approximately 10^7 CFU) is added and incubated for 2 hours at 37°C. Unbound mycoplasmas are washed away with DMEM. Cytadherent mycoplasmas are detected with mycoplasma-specific murine monoclonal antibodies incubated for 2 hours at 37°C using a capture ELISA test.

Discussion

Instead of using the ELISA assay, the quantity of adhering mycoplasmas can be evaluated employing the protocols for *in vitro* DNA amplification, e.g., polymerase chain reaction, as detailed in Vol. II.

Conclusions

A variety of protocols are currently available for studying the interactions of mycoplasmas with host cells. Adaptations are sometimes required, depending on the mycoplasma to be studied. In addition, although it was earlier assumed that mycoplasmas are external membrane parasites, evidence suggests that at least some mycoplasmas may penetrate into the host cell. This factor should be considered in the design of the experiment.

References

Baseman, J. B. (1993). The cytadhesins of *Mycoplasma pneumoniae* and *M. genitalium*. *In* "Subcellular Biochemistry: Mycoplasma Membranes" (S. Rottem and I. Kahane, eds.), Vol. 20, pp. 243–259. Plenum, London.

Gerstenecker, B., and Jacobs, E. (1990). Topological mapping of the P1 adhesin of *Mycoplasma pneumoniae* with adherence-inhibiting monoclonal antibodies. *J. Gen. Microbiol.* **136,** 471–476.

Henrich, B., Feldmann, R.-C., and Hadding, U. (1993). Cytoadhesins of *Mycoplasma hominis*. *Infect. Immun.* **61,** 2945–2951.

Jacobs, E. (1991). *M. pneumoniae* virulence factors and the immune response. *Rev. Med. Microbiol.* **2,** 83–90.

Jacobs, E. (1993). Serological diagnosis of *Mycoplasma pneumoniae* infections: A critical review of current procedures. *Clin. Infect. Dis.* **17**(Suppl. 1), S79–S82.

Jacobs, E., Bennewitz, A., and Bredt, W. (1986). Reaction pattern of human anti-*Mycoplasma pneumoniae* antibodies in enzyme-linked immunosorbent assay and immunoblotting. *J. Clin. Microbiol.* **23,** 517–522.

Jacobs, E., Schöpperle, K., and Bredt, W. (1985). Adherence inhibition assay: A specific serological test for detection of antibodies to *Mycoplasma pneumoniae*. *Eur. J. Clin. Microbiol.* **4,** 113–118.

Kahane, I. (1984). *In vitro* studies on the mechanism of adherence and pathogenicity of mycoplasmas. *Isr. J. Med. Sci.* **20,** 874–877.

Kahane, I., and Bredt, W. (1983). Tests for adherence properties of mycoplasmas. *In* "Methods in Mycoplasmology" (J. G. Tully and S. Razin, eds.), Vol. II, pp. 345–354. Academic Press, New York.

Kahane, I., and Horowitz, S. (1993). Adherence of mycoplasmas to cell surfaces. *In* "Subcellular Biochemistry: Mycoplasma Membranes" (S. Rottem and I. Kahane, eds.), Vol. 20, pp. 225–241. Plenum, London.

Krivan, H. C., Olson, L. D., Barile, M. F., Ginsburg, V., and Roberts, D. D. (1989). Adhesion of *Mycoplasma pneumoniae* to sulfated glycolipids and inhibition by dextran sulfate. *J. Biol. Chem.* **264,** 9283–9288.

Opitz, O., and Jacobs, E. (1992). Adherence epitopes of *Mycoplasma genitalium* adhesin. *J. Gen. Microbiol.* **138,** 1785–1970.

Razin, S., and Jacobs, E. (1992). Review article: Mycoplasma adhesion. *J. Gen. Microbiol.* **138,** 407–422.

Roberts, D. D., Olson, L. D., Barile, M. F., Ginsburg, V., and Krivan, H. C. (1989). Sialic acid-dependent adhesion of *Mycoplasma pneumoniae* to purified glycoproteins. *J. Biol. Chem.* **264,** 9289–9293.

Saada, A., Terespolsky, Y., Adoni, A., and Kahane, I. (1991). Adherence of *Ureaplasma urealyticum* to human erythrocytes. *Infect. Immun.* **59,** 467–469.

F3

MYCOPLASMA ADHERENCE TO HOST CELLS: EPITOPE MAPPING OF ADHESINS

Enno Jacobs

Introduction

Applications of synthetic peptides can be grouped into two categories: use as immunogens and components of a vaccine and use as antigens to bind antibodies. The latter approach can form the basis for specific serological immunoassays to test for antibodies elicited under natural infections (Jacobs *et al.*, 1990a). In basic research, synthetic peptides were used to characterize B-cell and T-cell epitopes (Jacobs *et al.*, 1990b) or to map the adhesion sites using monoclonal antibodies which inhibit the adhesion of *Mycoplasma pneumoniae* or *M.genitalium* to host cells (Gerstenecker and Jacobs, 1990; Opitz and Jacobs, 1992).

Methods for identifying the regions in a protein responsible for its function used to be time-consuming and limited. The introduction of base-labile fluorenylmethoxycarbonyl (Fmoc)-protecting groups in amino acids for polyamide solid-phase peptide synthesis (Sheppard, 1986), combined with the multipin peptide synthesis strategy (Geysen *et al.*, 1984), has led to a more practical method which allows the systematic mapping of epitopes of a protein. This "Geysen" method consists of the synthesis of synthetic short peptides, i.e., octapeptides, covering the known amino acid sequence of the protein studied. The different peptides are still attached to the support ("pin") which consists of chemically modified polyethylene rods (Geysen *et al.*, 1984). For practical use the pins with the different synthetic antigens can be connected in a tray (Fig. 1) which allows the user to test an antipeptide antibody activity in conventional microtiter plates. One major advantage of this convenient test system is that after

Fig. 1. Epitope mapping with a 96-pin plastic support. Different peptides are synthesized at the tips of the pins and are used as defined antigens in a modified ELISA test.

testing one antibody solution the antibody–antigen binding can be disrupted and the same peptide pins can be reused several times.

To map the regions of the major adhesin of *M.pneumoniae*, the P1 protein, or of the MgPa protein of *M.genitalium*, monoclonal antibodies were produced that are capable of blocking the attachment of mycoplasmas to erythrocytes (Jacobs et al., 1985). Supernatants or ascites fluids from different hybridoma clones were tested for antibody binding to overlapping octapeptides mimicking the known amino acid sequence of the P1 adhesin of *M.pneumoniae* and the MgPa adhesin of *M.genitalium*.

Materials

For Peptide Synthesis

Epitope scanning kit, control pins, control antibodies, and the software program are products of Cambridge Research Biochemicals Ltd., Gadbrook Park, Northwich, Cheshire, England. All other reagents can be purchased from different sources.
Acetic anhydride
Fmoc-L-amino acid active (OPfp- or ODhbt-) esters (stored at −20°C)
Dichloromethane
Diisopropylethylamine
N,N-Dimethylformamide (DMF; highest possible purity)
Ethanedithiol

1-Hydroxybenzotriazole
Methanol
Piperidine
Phenol
Silica gel
Trifluoroacetic acid
Water (HPLC purity)

For Epitope Mapping

Flat-bottom microtiter plates
Blocking buffer (1% bovine serum albumin, 1% ovalbumin, 1% Tween 20 in PBS (0.14 M NaCl, 0.01 M sodium phosphate, pH 7.2)
Washing buffer (PBS, 1% Tween 20)
Secondary antibody (i.e., alkaline phosphatase-conjugated goat anti-mouse immunoglobulins)
Substrate solution (i.e., 1 mg ml^{-1} p-nitrophenyl phosphate in diethanolamine buffer, pH 9.6)
Spectrophotometer ($A_{405\ nm}$ for the substrate recommended earlier)
Disruption buffer (2.5 liter: 1% sodium dodecyl sulfate, 0.1% 2-mercaptoethanol, 0.1 M sodium dihydrogen orthophosphate, adjust to pH 7.2, add 2.5 ml 2-mercaptoethanol)

Procedure

Peptide Synthesis

The main feature of solid-phase peptide synthesis is that the C termini of the first and of the following amino acids are covalently bound to a stepwise elongation of the peptide chain from the C to the N terminus, whereas the newly synthesized peptide is attached at its C terminus to the pin. Before attachment of a succeeding amino acid, the N terminus of the growing peptide chain must be activated by a deprotection step followed by an acylation step between the growing chain and the next amino acid, which by itself is protected at the N terminus to avoid insertion of more than one of the same amino acid into the growing chain. After a washing step to eliminate excess amino acid reactants, the N terminus of the last inserted amino acid of the peptide chain is deprotected. The next amino acid is incubated with the pins to be fixed to the activated N terminus of the peptide chain. These steps are repeated until synthesis of the entire peptide is accomplished. The N terminus of the completed peptide is then acetylated; as a last step the still protected side-chain residues of the different amino acids are deprotected.

1. The software program facilitates the preparation of the different amino acid solutions and the layout of the synthesis of the different peptides. For practical reasons, it is recommended to start with 92 pins, plus two negative controls and two control peptides with amino acid sequences according to the manufacturer's advice (anticontrol peptide monoclonal antibodies are included in the kit). For validation it is recommended that each octapeptide be synthesized twice. With 92 pins 46 different peptides can be synthesized at once. The vials with the different amino acids should be prepared 1 day before synthesis according to the layout of the synthesis.

2. Deprotection and washing. All steps should be carried out in a fume cupboard at room temperature. Wear gloves! To deprotect and activate the linker of a pin, the pin block is incubated for 30 minutes in 20% (v/v) piperidine in DMF, followed by a washing step in DMF and two in methanol. Let pins air dry for 15 minutes and wash once again in DMF for 5 minutes.

3. Dispensing. According to the layout, add dissolved 1-hydroxybenzotriazole to the vials containing the different Fmoc-amino acid active esters immediately prior to dispensing the amino acid solution into the wells of polypropylene microtiter plates according to the layout of the synthesis schedule.

4. Coupling. The deprotected pin block is inserted into the wells and is sealed in a plastic bag to minimize evaporation. After an overnight incubation at room temperature (about 12–18 hours), the pin block is removed and washed once with DMF, four times with methanol, and once again with DMF.

5. Steps 2 to 4 are repeated until all amino acids are linked to the growing peptide chains.

6. Acetylation. Finally, the N terminus of the complete peptide chain is acetylated with DMF:acetic anhydride:triethylamine (5:2:1, v/v/v) for 90 minutes at room temperature and is washed once again with DMF, four times with methanol, and air dried for 10 minutes.

7. Side-chain deprotection. Pins are treated with trifluoroacetic acid:phenol:ethanedithiol (95:2.5:2.5, v/v/v) for 4 hours.

8. Neutralization. The pin block is washed twice in dichloromethane, twice in 5% diisopropylethylamine in dichloromethane, and once in dichloromethane. Pins are air dried, washed in water, then in methanol, and stored until use *in vacuo* over silica gel.

Peptide Mapping with Modified ELISA Technique

In contrast to the conventional enzyme-linked immunoassay, the antigen is not linked directly to the bottom of a microtiter plate but the synthetic peptide antigens are bound to the pins. Since 96 pins are arranged in a plastic tray comparable to the outfit of a microtiter plate (Fig. 1), all different steps, i.e., incubation of the first antibody solution, washing steps, secondary antibody incubation, substrate reaction, and optical evaluation, can be run in microtiter plates.

The protocol to screen hybridoma clones for antipeptide antibodies is as follows: Prior to the use of the newly synthesized pins, the pins have to be precoated in 200 µl blocking buffer for at least 1 hour at 37°C.

1. The supernatant from a well-grown hybridoma clone has to be enriched 1:1 with blocking buffer. Each well of a round-bottom microtiter plate receives 175 µl of this solution. A pin with a distinct peptide antigen is now inserted into each well and incubated overnight at 4°C.

2. The tray with the pins is rinsed and washed three times in washing buffer.

3. In a new microtiter plate, 175 µl of an enzyme-conjugated secondary antibody solution (diluted in blocking buffer as recommended by manufacturers for ELISA testing) is introduced into each well and the pins are incubated for 1 hour at 37°C.

4. The pin tray is washed again three times in washing buffer.

5. After final washing the pins are incubated in 150 µl substrate solution at 37°C. For screening supernatants of hybridoma clones, the incubation time should be extended between 1 and 4 hours. It is recommended, therefore, not to stop the substrate reaction by the addition of NaOH solution but to measure spectrophotometrically the optimal enzyme reaction increase versus background every hour.

6. The epitope scanning kit (IBM) software supports plotting of the absorbance data.

7. Immediately after the substrate reaction is measured, pins should be regenerated for the next ELISA test. The disruption buffer is warmed to 60°C and transferred to a sonication bath. After 30 minutes of sonication the pin block is washed in distilled water at 55°C, changing the hot water bath several times. Immerse the pin block for 2 minutes in a boiling methanol bath (no open flame should be present) followed by an air-drying step. The pins are now ready for another ELISA test or can be stored under vacuum with silica gel.

Discussion

When monoclonal antibodies are produced against native proteins, antibodies can be directed to epitopes which form a linear continuous region or they can be directed to two linear determinants located on different regions of the amino acid chain brought together by protein-folding mechanisms. These regions are regarded as discontinuous immunogenic epitopes. If the binding affinity of the antibody to both parts of a discontinuous epitope is high enough, the monoclonal antibody will bind to two different peptides which might be separated by several amino acids of the mapped protein sequence. Assuming that the hybridoma clone is one unique clone (limiting dilution experiments should be repeated at least five times), it is possible to define more closely the amino acids of both octapeptides on the discontinous epitope. It is necessary, therefore, to synthesize derivatives of both octapeptides again, substituting once in each position a nonsense amino acid (e.g., glycine) within the newly synthesized octapeptides. The monoclonal antibody might bind to a limited number of both constructs. Finally, in the new synthesis the amino acids of both parts can be linked together. The monoclonal antibody will show an elevated affinity to its whole epitope, measured as an increased absorbance affinity compared to the two incomplete binding sites. Alternatively, Cambridge Research Biochemicals (Northwich, Cheshire, Great Britain) offers a "Mimitope Design Kit" which allows the characterization of discontinuous epitopes, if the affinity of the monoclonal antibody to one of the determinants of the discontinuous epitope is too low. Starting with the known amino acid sequence, all different amino acids in a next synthesis are linked to pins with a known sequence. These pins are tested for an increased affinity compared to the known part of the discontinuous sequence. The amino acid sequence with the highest absorbance value is chosen for further extension of the amino acid chain. The elongated peptide can be compared after elongation of the peptide chain with three amino acids having the original sequence to facilitate the next synthesis for completing the discontinuous epitope.

With this strategy of scanning epitopes, the functional sites of the adhesins of *M.pneumoniae* and *M.genitalium* have been characterized and compared (Razin and Jacobs, 1992; Optiz and Jacobs, 1992). Furthermore, this method can be used to examine the application of sequential epitopes of amino acid sequences, e.g., the P1 protein of *M.pneumoniae,* for serology of human diseases caused by this pathogen (Jacobs *et al.*, 1990a).

For the systematic study of T-cell determinants in cell proliferation assays, a modified version of the epitope scanning method has been introduced (Maeji *et al.*, 1990). The major advantage of this "cleavable peptide" method (Cambridge Research Biochemicals, Great Britain) is to synthesize the peptides on pins as described earlier and cleave the products to release peptides in aqueous solutions at physiological pH.

By scanning B-cell and T-cell epitopes of the two mycoplasmal adhesins, it may be possible to construct an effective and defined vaccine against human mycoplasmal diseases.

References

Gerstenecker, B., and Jacobs, E. (1990). Topological mapping of the P1-adhesin of *Mycoplasma pneumoniae* with adherence-inhibiting monoclonal antibodies. *J. Gen. Microbiol.* **136,** 471–476.

Geysen, H. M., Meloen, R. H., and Bartling, S. J. (1984). Use of peptide synthesis to probe viral antigens for epitopes to a resolution of a single amino acid. *Proc. Natl. Acad. Sci. USA* **81,** 3998–4002.

Jacobs, E., Pilatschek, A., Gerstenecker, B., Oberle, K., and Bredt, W. (1990a). Immunodominant epitope of the adhesin of *Mycoplasma pneumoniae. J. Clin. Microbiol.* **28,** 1194–1197.

Jacobs, E., Röck, R., Dalehite, L., and Bredt, W. (1990b). A B-cell T-cell linked epitope of *Mycoplasma pneumoniae. Infect. Immun.* **58,** 2464–2469.

Jacobs, E., Schöpperle, K., and Bredt, W. (1985). Adherence inhibition assay: A specific serological test for detection of antibodies to *Mycoplasma pneumoniae. Eur. J. Clin. Microbiol.* **4,** 113–118.

Maejii, N. J., Bray, A. M., and Geysen, H. M. (1990). Multi-pin peptide synthesis strategy for T-cell determinant analysis. *J. Immunol. Methods* **134,** 23–33.

Opitz, O., and Jacobs, E. (1992). Adherence epitopes of *Mycoplasma genitalium* adhesin. *J. Gen. Microbiol.* **138,** 1785–1790.

Razin, S., and Jacobs, E. (1992). Review article: Mycoplasma adhesion. *J.Gen. Microbiol.* **138,** 407–422.

Sheppard, R. C. (1986). Modern methods of solid-phase peptide synthesis. *Sci. Tools* **33,** 9–16.

F4

OXIDATIVE DAMAGE INDUCED BY MYCOPLASMAS

Itzhak Kahane

Introduction

Mycoplasmas can cause oxidative damage to host cells. By this pathogenic mechanism, the adhering mycoplasmas produce, or induce the cell to produce, reactive oxygen molecules or free radicals. The proximity between the adhering mycoplasmas and the host cells allows the reactive oxygen, which usually has a short half-life, to reach and affect the host cell. The subject has been reviewed by Tryon and Baseman (1992) so only the basics will be summed up here. Many studies suggest that the initial stages of mycoplasma adherence and pathogenesis include the following steps:

1. Adherence of the mycoplasmas to the host cell membrane via specific receptors (sialoglycoconjugates, proteins, or sulfatides) or to hydrophobic sites.
2. Production of superoxide anions by mycoplasmas or induction of enhanced production of superoxide anions by host cells.
3. Partial inhibition of host cell catalase by elevated levels of superoxide anions.
4. Partial inhibition of host cell Cu,Zn-superoxide dismutase (SOD) by H_2O_2 produced by host cells and by the mycoplasmas.
5. Increased oxidative stress in cytoplasmic and membrane components by reactive oxygen.
6. Secondary oxidation and breakdown of fatty acids in membranes and production of the cross-linker malonyl dialdehyde.

7. Cross-linkage of cytoplasmic and membrane components.
8. Membrane becomes leaky.
9. Cytoplasmic components ooze through host cell membrane.
10. Cytadhering mycoplasmas benefit from the nutrients released from host cells.

The events described in steps 2, 3, and 4 can actually proceed in cycles and, therefore, amplify the outcome of all the steps. Altogether, these result in a prolonged, rather than overwhelming, deleterious effect of the mycoplasmas on the host cell, making this model of mycoplasma pathogenesis more attractive since it may explain the chronic nature of mycoplasma diseases.

Methods measuring parameters of oxidative stress will be discussed in this chapter. These include assays for catalase and superoxide dismutase (SOD) activity.

Catalase Activity

The activity of catalase of host cells in culture at various points after infection with mycoplasmas is assayed using an oxygen electrode.

Materials and Equipment

Oxygen electrode with a temperature-regulated chamber (Yellow Springs Instruments Co., Yellow Springs, OH) and a recorder. H_2O_2 solution (30% w/v).

Assay

1. The spontaneous decomposition of a 8 mM H_2O_2 solution is recorded. This value should be determined at least at the beginning and the end of every set of experiments.
2. Cultured cells, or homogenate of cultured cells, suspended in phosphate-buffered saline (PBS) (1 ml) are incubated at 30°C in the oxygen electrode chamber until equilibrium is reached.
3. The reaction is started by the addition of H_2O_2 (8 mM final concentration).
4. The rate of oxygen release is recorded. The sample size of cells or homogenates is adjusted to give about 50- to 100-fold that of the spontaneous decomposition.

Calculations

The specific activity is expressed as nmol of oxygen released/minute/mg cell protein.

Discussion

Care should be taken to try and minimize a possible artifact of carryover of catalase activity from the serum supplement of the culture medium. This can best be conducted if cells and mycoplasmas are grown in serum-free media. Otherwise, sera should be tested for catalase activity, and batches exhibiting minimal catalase activity should be used. The catalytic activity of catalase can also be measured spectrophotometrically at 240 nm.

Superoxide Dismutase Activity

The SOD activity of host cells grown in culture is assayed at various points after infection with mycoplasmas. The SOD activity in cultured cell homogenates is assayed by the inhibition of cytochrome c reduction by superoxide produced by xanthine oxidase in the presence of xanthine, as detailed by McCord and Fridovich (1988).

Materials

Cell suspension in PBS
Cytochrome c
Ethanol-based freezing bath or liquid air bath
Solution A: 5 mM xanthine and 20 mM cytochrome c in 50 mM phosphate buffer, pH 7.8, containing 0.1 mM EDTA. The solution is stable for up to 3 days at 4°C.
Solution B: Fresh solution of 0.2 U/ml of xanthine oxidase in phosphate buffer, pH 7.8, containing 0.1 mM EDTA, and kept at 4°C.
Spectrophotometer plus recorder
Xanthine (Sigma)
Xanthine oxidase (Sigma)

Procedure

The cell suspension in PBS is disintegrated by freeze-thawing (6×) and homogenization at 4°C in a Teflon homogenizer. The SOD in the cell homogenate is assayed spectrophotometrically by the rate of inhibition of cytochrome c reduction as follows:

1. Solution A is equilibrated at 25°C.
2. The assay is conducted in a 3-ml cuvette, to which 2.9 ml of solution A and 50 μl of tested sample are added. The reaction is started by the addition of 50 μl of solution B.
3. The reaction is followed spectrophotometrically at 550 nm.
4. One unit of SOD activity is defined as the amount of protein that inhibits 50% of the rate of cytochrome c reduction.

COMMENT

The method has the limitation characterizing all indirect assays (McCord and Fridovich, 1988), e.g., a decrease in the cytochrome c reduction may appear as SOD activity, thus caution should be taken to conduct it under optimal conditions.

Discussion

Evaluation of oxidation stress is rather difficult since most reactive oxygen molecules and free radicals are short-lived. Thus, the outcome of oxidative stress on cell components is frequently assessed.

The effects of oxidative stress on membrane lipids can be assayed via measurement of malonyl dialdehyde (MDA) production as an indicator of some of the secondary breakdown products and has been used in many studies (Almagor et al., 1986; Kahane, 1984). In essence, and despite its many limitations (Yagi et al., 1992), the level of MDA is determined by the thiobarbituric acid assay of Buege and Aust (1978). MDA reacts with the thiobarbituric acid, and the red-colored product is spectrophotometrically determined at 535 nm. The MDA concentration is calculated using the extinction coefficient of $1.56 \times 10^5\ M^{-1}\mathrm{cm}^{-1}$.

Oxidation of antioxidants in the cell, e.g., glutathione, can also serve as an indicator of the oxidation of cell constituents. This was studied by Almagor et al. (1986) with *Mycoplasma pneumoniae* infecting human fibroblasts. The same may apply to the effect on the addition of antioxidants, e.g., vitamin E, assaying its effect on relief from oxidative stress indicated by the oxidation of membrane components causing membrane leakage of cytoplasmic constituents. Protein oxidation may be studied by the decrease of enzymatic activity, as in the case of catalase activity, or as a general term of oxidation of thiol groups. Another aspect that should be taken into consideration is the possible oxidative effects of nitric oxide (Green and Nacy, 1993).

Finally, in addition to loss of activity, oxidative stress may cause the activation of genes in the host cell. Several regulators are affected, e.g. Nuclear factor κB (NF-κB) which induces the human immunodeficiency virus long terminal repeat (HIV-LTR). This may provide an explanation for HIV-LTR transactivation by several mycoplasmas (Nir-Paz *et al.*, 1992). This is in accord with some observations of Lo *et al.* (1991) and Montaignier *et al.* (1990) supporting the notion that mycoplasmas may also serve as a cofactor in the activation of HIV as well as in inducing immunoregulators of the immune system.

References

Almagor, M., Kahane, I., Gilon, C., and Yatziv, S. (1986). Protective effects of the glutathione redox cycle and vitamin E on cultured fibroblasts infected by *Mycoplasma pneumoniae*. *Infect. Immun.* **52,** 240–244.

Buege, J. A., and Aust, S. D. (1978). Microsomal lipid peroxidation. *In* "Methods in Enzymology" (S. Fleischer and L. Packer, eds.), Vol. **52,** pp. 302–310. Academic Press, New York.

Green, S. J., and Nacy, C. A. (1993). Antimicrobial and immunopathologic effects of cytokine-induced nitric oxide synthesis. *Curr. Opin. Infect. Dis.* **6,** 384–396.

Kahane, I. (1984). *In vitro* studies on the mechanism of adherence and pathogenicity of mycoplasmas. *Isr. J. Med. Sci.* **20,** 874–877.

Lo, S.-C., Tsai, S., Benish, J. R., Shih, J. W.-K., Wear, D. J., and Wong, D. M. (1991). Enhancement of HIV-1 cytocidal effects in CD^{4+} lymphocytes by the AIDS-associated mycoplasma. *Science* **251,** 1074–1076.

McCord, J. M., and Fridovich, I. (1988). Superoxide dismutase: The first twenty years (1968–1988). *Free Radical Biol. Med.* **5,** 363–370.

Montaignier, L., Berneman, D., Guetard, D., Blanchard, A., Chamaret, S., Rame, V., Van Rietschoten, J., Mabrouk, K., and Bahraoui, E. (1990). Inhibition of HIV prototype strains infected by antibodies directed against a peptidic sequence of mycoplasma. *C. R. Acad. Sci. (III)* **311,** 425–430.

Nir-Paz, R., Israel, S., Honigman, A., and Kahane, I. (1992). Activation of HIV-LTR by mycoplasmas. *In* "Proceedings of the 9th International Congress of IOM," Ames, IA.

Tryon, V. V., and Baseman, J. B. (1992). Pathogenic determinants and mechanisms. *In* "Mycoplasmas: Molecular Biology and Pathogenesis" (J. Maniloff, R. N. McElhaney, L. R. Finch, and J. B. Baseman, eds.), pp. 457–471, ASM, Washington, DC.

Yagi, K., Kondo, M., Niki, E., and Yoshikawa, A., eds. (1992). "Oxygen Radicals," Excerpta Medica, Amsterdam/London/New York.

F5

ACTIVATION OF MACROPHAGES AND MONOCYTES BY MYCOPLASMAS

Ruth Gallily, Ann Avron, Gerlinde Jahns-Streubel, and Peter F. Mühlradt

Introduction

Monocytes and macrophages, which belong to the mononuclear phagocyte system, defend the body against invasion by pathogens. Acting as true phagocytes, they engulf and remove old and dead cells as well as cell debris. They may also be considered accessory cells by virtue of their ability to present processed antigen to lymphocytes. Once activated, they also constitute effector cells due to their microbicidal and tumoricidal capacity (Lewis and McGee, 1992).

Macrophages exist in different states: resting, primed, and activated. An important priming substance is immune interferon (IFN γ); the classical activators are microbial cell wall components such as lipopolysaccharides, zymosan, bacterial lipoproteins, and peptidoglycans. Mollicutes, their membranes, and extracts from these membranes can also activate macrophages (Loewenstein et al., 1983; Sher et al., 1990; Mühlradt and Schade, 1991; Gallily et al., 1992; Ruschmeyer et al., 1993; Rosendal et al., 1994). The process of activation is accompanied by morphological, biochemical, and functional changes. Activation involves the release of biologically active molecules, such as cytokines [tumor necrosis factor α (TNFα), interleukin (IL), IL-1, IL-6, IL-8, IL-12], and other mediators or effector molecules (arachidonate metabolites, active oxygen, and nitrogen species). For reviews, see Beutler and Cerami (1988), DiGiovine and Duff (1990), Hirano et al. (1990), and Liew and Cox (1991). This chapter details various procedures used for measuring the release of biologically active molecules produced in response to stimulation by mollicutes or their products.

Materials

Actinomycin D

1. To 5 mg of actinomycin D (Sigma A-1410, Sigma Chemical Co., St. Louis, MO), add 1 ml of absolute ethanol.
2. Add sterile distilled water (9 ml) to give a stock solution (500 μg/ml), which is stored, protected from light, at 4°C (good for at least 3 months).
3. The working solution (4 μg/ml) is prepared fresh by diluting the stock solution 1:125 in Dulbecco's modified Eagle's medium (DMEM)–5% fetal calf serum (FCS).

Crystal Violet Stain

1. To 2 g of crystal violet (also called methyl violet and gentian violet, BDH, Poole, England) add absolute ethanol (20 ml) and leave overnight to dissolve.
2. Add 80 ml tap water (the pH is less acid than that of distilled water) to prepare the 10× stock solution and store at 4°C.
3. For the working stain solution, dilute the stock 1:10 with tap water and filter through Whatman No. 1 filter paper.

Culture Medium

DMEM (high glucose 4.5 g/liter, GIBCO, Grand Island, NY) supplemented with 2 mM L-glutamine, 10 mM HEPES, 100 U/ml penicillin, and 100 μg/ml streptomycin is used as the basal culture medium (DMEM–0% FCS). Fetal calf serum (selected from batches that do not preactivate macrophages (Mϕ) in the tumoricidal assay described later) is inactivated by heating at 56°C for 30 minutes and is added to the basal medium to a final concentration of 5 or 10% (v/v) (DMEM–5% FCS or DMEM–10% FCS, respectively). For growing bone marrow macrophages (BMMϕ; Sher et al., 1990), 15% FCS is added to the basal medium as well as 5% horse serum (inactivated at 56°C for 30 minutes) and 30% L cell-conditioned medium (LCM, see later).

RPMI 1640 medium (GIBCO), supplemented with antibiotics as for the DMEM, 2 mM L-glutamine, and 2.5×10^{-5} M 2-mercaptoethanol and with FCS added to a final concentration of 5 or 10% (RPMI–5% FCS or RPMI–10% FCS, respectively), is used for some of the cell lines described.

MYCOPLASMA GROWTH MEDIUM

For growth of *Mycoplasma capricolum* subsp. *capricolum,* a modified Edward's growth medium is prepared which consists of heart infusion broth (13 g, Difco, Detroit, MI), NaCl (2.5 g), Bacto-peptone (5 g, Difco), and yeast extract (7 g, Difco) in 1 liter. The pH is adjusted to pH 8 with NaOH and autoclaved (1.5 kg/cm^2 for 15 minutes). The growth medium is completed by the addition of penicillin G (4×10^5 units), horse serum (inactivated by heating to 56°C for 30 minutes; 40 ml), and glucose phosphate solution (20 ml). For growth of more fastidious mycoplasma, Hayflick's medium or SP -4 medium should be used.

GLUCOSE PHOSPHATE SOLUTION

The glucose phosphate solution is prepared by separately autoclaving a 37% (w/v) solution of glucose and a 1 M K_2HPO_4 solution and then mixing equal volumes of the two solutions. The glucose phosphate solution is stored at room temperature.

Griess Reagent

Solution A: naphthylethylenediamine dihydrochloride (0.1%, w/v) in distilled water
Solution B: sulfanilamide (1%, w/v) in 5% (w/v) H_3PO_4

To make the working reagent, mix equal parts of solutions A and B.

Luminol

Luminol (5-amino-2,3-dihydro-1,4-phthalazinedione, Sigma A-8511) should be protected from light.

1. A stock solution of 10 mg/ml in dimethyl sulfoxide is prepared and stored in the dark at room temperature for up to 1 week.
2. The working solution (to be prepared fresh) is a 1:25 dilution of the stock solution in phosphate-buffered saline (PBS).

Phosphate-Buffered Saline

PBS is prepared as a $10\times$ stock and is then diluted and autoclaved. The final concentration of the components is Na_2HPO_4, 8.2 mM; KH_2PO_4, 1.8 mM; and NaCl, 0.136 mM; the pH should be 7.2.

Thioglycolate Medium

1. Add fluid thioglycolate medium powder (7.45 g; Difco) to distilled water (250 ml), mix to dissolve, and boil. The color changes from brownish through pink to yellow as the oxygen is removed from the solution.
2. When a clear yellow color is obtained, the thioglycolate medium is quickly dispensed into bottles.
3. Autoclave for 15 minutes at 1.5 kg/cm^2.

The solution, which should have a clear honey color, is stored in the dark at room temperature, for up to 3 weeks before use.

Zymosan

Zymosan (from *Saccharomyces cerevisiae*, Sigma Z-4250) (15 mg) is suspended in 3 ml saline (NaCl, 0.9 g/liter) and is boiled in a water bath for 30 minutes. The zymosan is washed twice with saline (600 g for 10 minutes) and is finally resuspended in PBS (1.8 ml) to give a solution of 8.3 mg/ml. Store at $-20°C$.

Cells and Culture Conditions

CLONE 7 CELLS

A fibroblast-like cell line derived from BALB/c mice, designated clone 7 (ATCC, American Type Culture Collection, Rockville, MD), is grown in 10-ml volumes of DMEM–5% FCS in 80-cm^2 culture flasks (Nunc, Roskilde, Denmark) maintained horizontally in an incubator at 37°C with 5% CO_2. The cells divide about once a day and reach confluence at approximately 1.6×10^7 cells/flask. The number of cells seeded is varied from a minimum of 5×10^5 in order to attain the number required for a certain experiment on a particular day. At confluence, the cells are harvested by replacing the growth medium with 3 ml trypsin/versene solution (0.25% trypsin, 0.05% EDTA). After about 5 minutes the cells can be released from the flask with a Pasteur pipette and harvested. The trypsin is neutralized by transferring the cells to 7 ml DMEM–5% FCS and centrifuging them at 200 g for 5 minutes at 4°C. They are then resuspended in DMEM–5% FCS and are counted in a hemocytometer for either passage or assays.

A9 CELLS

A9 cells (C3H-derived, a clone of the L929 cells described later, ATCC) are grown and harvested as described for clone 7 cells except that DMEM–10% FCS is used as the growth medium.

L929 CELLS AND CONDITIONED MEDIUM

L929 cells (C3H-derived, ATCC) are grown and harvested as described for the A9 cells. For preparation of LCM, 1×10^6 cells are resuspended in DMEM–10% FCS (20 ml) in 80-cm^2 culture flasks. After 4 days of incubation at 37°C, the supernatant (LCM) is removed, filtered through a 0.22-µm filter, and stored in aliquots at $-20°C$.

B9 CELLS, DEPENDENT ON IL-6

B9 plasmacytoma cells (ATCC), which are dependent on IL-6, are maintained in RPMI 10% FCS supplemented with 1 mM sodium pyruvate, nonessential amino acids, and recombinant IL-6 (100 U/ml Genzyme, Cambridge, MA). Alternatively, a standard supernatant, obtained by stimulating thioglycolate-elicited macrophages (TGMφ; Loewenstein et al., 1983) with LPS (1 µg/ml for 24 hours), can be titrated in the cell medium and used as a source of IL-6. The cells are subcultured every 2–3 days by mixing the culture and diluting it 1:10 in fresh medium.

Preparation of Concanavalin A (Con A)-Conditioned Medium as Source of IL-2

1. Remove, aseptically, the spleens from four mice. Cut the spleens into small pieces, suspend these in Hanks' buffered salt solution (HBSS, GIBCO), and passage them through a nylon sieve to give a single cell suspension.

2. Adjust the volume to about 20 ml with HBSS and wash the cell suspension by centrifugation at 300 g for 10 minutes. Repeat the washing step until the supernatant remains clear.

3. Count leukocytes by dilution in Turk's solution (0.01% crystal violet in 3% acetic acid) and adjust cell concentration to 5×10^6 cells/ml in RPMI 1640–5%FCS. The yield should be more than 80 ml of cell suspension.

4. Distribute two 40-ml portions to two 75-cm^2 cell culture flasks, add Con A to a final concentration of 2 µg/ml, and incubate for 20 hours.

5. Harvest conditioned medium by centrifugation at 400 g for 15 minutes and filter through a 0.45-µm sterile filter. For propagation of D10 cells (described later), use the medium at a 1:5 or 1:10 dilution.

Propagation of D10 Cells

PREPARATION OF MITOMYCIN-TREATED FEEDER LAYERS

1. Prepare a single cell suspension from one spleen of a C3H/HeJ mouse (Bomholtgaard, Ry, Denmark, or Jackson LABS, Bar Harbor, ME) as described earlier.

2. Adjust the volume to 6 ml with HBSS and layer two 3-ml portions onto two 4-ml Lymphoprep (Nyegaard & Co., Oslo, Norway) step gradients in 12-ml Falcon tubes. Accelerate slowly to 350 g and centrifuge for 20 minutes without the brake. Collect the interphase on Lymphoprep and combine it with those cells which have moved into the gradient, but discard the pellet.

3. Dilute with HBSS and sediment cells at 350 g for 10 minutes. Wash once with HBSS and adjust to 1×10^7 cells/ml in RPMI–10% FCS.

4. Add mitomycin C to a final concentration of 25 μg/ml. Incubate for 30 minutes and wash twice with RPMI–10% FCS.

5. Adjust the cell concentration to 1×10^6/ml in RPMI–10% FCS including 10% Con A-conditioned medium (described earlier) and 50 mM α-methyl mannoside, and add conalbumin (Sigma) to a final concentration of 200 μg/ml. Transfer 10 ml of the cell suspension to 25-cm² culture flask (feeder cells).

PASSAGE OF D10 CELLS ON FEEDER LAYER

6. Add 1×10^6 D10 cells into the culture and incubate for 5 days.

7. Split 1:1 with propagation medium (RPMI–10% FCS, 10% Con A-conditioned medium). Split again after 2 days or when cells have reached 1.5×10^5/ml.

8. After 2 weeks, transfer D10 cells again to feeder cells as described earlier.

Procedures

Preparation of Mycoplasma and Membrane Fractions

GROWTH OF MYCOPLASMA

1. An inoculum (1%, v/v) of mycoplasma culture, stored frozen at $-70°C$, is added to 500 ml of culture medium in a flat culture bottle and incubated at 37°C without shaking until the OD_{640} reaches 0.35–0.40 (usually 24 hours). The pH of the culture should be pH 6.5–7.0 if there is no contamination by bacteria. Retain aliquots of the original culture stored at $-70°C$ for use as inocula for future cultures. In addition, a sample of the culture is routinely plated on blood agar and incubated at 37°C for 48 hours to detect any contamination of the mycoplasma culture by bacteria.

HEAT-KILLED MYCOPLASMA

Mycoplasma are heat-killed by heating at 60°C for 45 minutes.

PREPARATION OF THE MEMBRANE FRACTION

2. Harvest the mycoplasma culture and centrifuge at 10,000 g for 20 minutes.

3. Discard the supernatant and resuspend the pellet of mycoplasma in 2% of the original volume of wash buffer (25 mM Tris, pH 7.5, 0.25 M NaCl).

4. Centrifuge at 12,000 g for 10 minutes to repellet the mycoplasmas.

5. Resuspend the pellets from 1 liter of mycoplasma culture in 2.5 ml of washing buffer and add 2.5 ml of 2 M glycerol.

6. Incubate at 37°C for 10 minutes.

7. Load the suspension into a syringe, squirt it into 100 ml of distilled water kept at 37°C, and mix thoroughly.

8. Incubate at 37°C for 15 minutes.

9. Centrifuge at 12,000 g for 30 minutes at 4°C to sediment the membranes.

10. Discard the supernatant and wash the membranes by resuspending the pellet well in wash buffer (2.5 ml/liter of original culture) and recentrifuging at 12,000 g for 30 minutes.

11. Combine all the pellets obtained from 1 liter of mycoplasma culture in 6 ml of cold wash buffer.

12. Ensure that the suspension is homogeneous by passage through a 23-gauge needle or by very short sonication.

13. Determine the protein content by the Lowry or other protein assay.

Preparation of Delipidated Detergent Extract of Membrane Fraction (Ruschmeyer et al., 1993)

1. Centrifuge the membrane preparation (10 mg protein) for 30 minutes at 11,000 g and resuspend the pellet in 10 ml acetone or absolute ethanol. Incubate for 1 hour at room temperature and repeat the extraction. Sediment the delipidated membranes by centrifugation at 11,000 g for 30 minutes.

2. Extract the pellet with 20 ml 50 mM octyl glucoside in PBS at 25°C and sediment insoluble material by centrifugation at 11,000 g for 30 minutes.

Preparation of Macrophages

BONE MARROW MACROPHAGES

1. Mice are killed by asphyxiation with CO_2, and the skin and flesh of the hind legs are removed.

2. The femur and tibia are excised, the ends of the bones are cut open, and 5

ml of BMMφ medium is forced through the marrow space with a syringe (25-gauge needle) to wash out the cells.

3. The cells are collected, washed (200 g for 5 minutes), and then resuspended in the BMMφ medium described earlier (2.5×10^6 cells/10 ml) in 9-cm-diameter bacteriological grade culture dishes (10 ml/dish) and incubated in a 37°C incubator with 5% CO_2.

4. An extra 1–2 ml of medium is added every 3–4 days and the cells are ready for use after 10 days in culture (routinely 10–21 days). One day before assay, the medium is removed and replaced with 10 ml of DMEM–10% FCS without LCM.

THIOGLYCOLATE-ELICITED MACROPHAGES

1. For recruitment of Mφ, mice are injected intraperitoneally with thioglycolate solution (1.5 ml). On Day 4, the mice are killed by asphyxiation with CO_2.

2. Expose the peritoneal wall and inject 5 ml of PBS into the peritoneal cavity. Massage the peritoneum and draw the fluid back into the syringe. More cells can be obtained by further rinsing the peritoneum with PBS.

3. Centrifuge the collected exudate (200 g for 5 minutes at 4°C), resuspend the cells in cold DMEM–10% FCS, and count in a hemocytometer after mixing the cells 1:10 with the crystal violet–citric acid stain (10 mg crystal violet, 1.9 g citric acid in 100 ml). This facilitates counting by lysing any contaminating red blood cells.

4. Adjust the cell suspension to 1.3×10^6 cells/ml (taking into account that there will be cells other than macrophages in the exudate) and dispense, either in 96-well flat-bottom tissue culture microtiter plates at 100µl/well or in 9-cm-diameter tissue culture petri dishes at 10 ml/plate.

5. Incubate macrophage preparations for 2 hours in a 37°C incubator, then remove the nonadherent cells by three washes with PBS.

6. Add DMEM–10% FCS (200 µl/well of 96-well plates or 10 ml/petri dish) and incubate the cells at 37°C, at least overnight, before stimulation.

Resident peritoneal exudate cells are prepared as for thioglycolate-elicited macrophages except that naive mice are used.

Blood Monocytes

1. Draw 100 ml venous blood from healthy volunteers, dilute 1:1 with pyrogen-free saline, and layer about 30 ml of this mixture onto 15-ml Lymphoprep (Nyegaard and Co. AS, Oslo).

2. Centrifuge for 20 minutes at 400 g without using the brake.

3. Collect the interphase immediately and wash with saline, once at 800 g for 10 minutes and then twice at 300 g for 10 minutes.

4. Resuspend the cells in 2 ml serum-free medium (M-SFM medium, GIBCO) and keep on ice while the number of monocytes is determined by esterase staining (Tucker et al., 1977) or by FACS analysis.

5. Adjust the concentration of monocytes to 1×10^6/ml in M-SFM medium, distribute in 24-well cell culture plates at 0.5 ml/well, and incubate for 1 hour in a CO_2 incubator.

6. Remove nonadherent cells by three vigorous washings with medium. Add appropriate dilutions of macrophage activator in 0.5-ml volumes, and incubate the cultures for 24 hours. Harvest the culture supernatants for the subsequent testing of cytokines. These samples may alternatively be kept frozen at $-20°C$ or $-70°C$ until use.

Stimuli

Thioglycolate-induced macrophages can be maintained in culture for up to a week with only an increase in the TNFα response to subsequent stimulus and without affecting the chemiluminescence response. However, in cytotoxic assays, freshly harvested cells are used as these are the most efficient in target killing.

Various species of mycoplasmas, alive, heat-killed, or as membrane preparations, can be used to stimulate macrophages. Usually, 10 μg/ml of protein is used, although stimulation is also observed at 1 μg/ml. Bacterial lipopolysaccharide (LPS from *Escherichia coli*) is routinely used at 1 μg/ml as a control stimulant. For measurements of nitric oxide it is necessary to give macrophages two stimulants. Generally this means the addition of interferon-γ (50–75 U/ml) either before or concomitantly with the mycoplasma preparation. The stimulant is added to macrophages or monocytes in the appropriate maintenance medium and the cells are incubated at 37°C. Culture supernatants are collected after 4–24 hours for TNFα and after 24–48 hours for IL-1, IL-6, prostaglandins, and nitric oxide (NO). The supernatants may be kept frozen at $-20°C$ or $-70°C$ until use. For the determination of IL-1 it may be advisable to release cell-bound IL-1 by repeated freeze–thawing of cultures. Cell debris should be removed by 400 g centrifugation.

TNFα Assay (Loewenstein et al., 1983, Flick and Gifford, 1984)

STIMULATION OF Mφ

1. Routinely, TGMφ plated in 96-well plates are used. Alternatively, distribute BMMφ into 96-well microtiter plates (10^6/ml, 100 μl/well) and allow 2 hours at 37°C for adherence.

2. Remove medium and replace with DMEM without FCS (100 μl/well).
3. Add stimulant at the appropriate concentration in 100 μl of DMEM without FCS.
4. Incubate the cells overnight, collect the supernatant, and store at $-20°C$ until use.

PREPARATION OF TARGET CELLS, L929 OR CLONE 7 CELLS

5. Harvest target cells and resuspend in DMEM–5% FCS at 4×10^5 cells/ml.
6. Plate out 100 μl/well into 96-well flat-bottom microtiter plates. It is important to ensure that the cells are distributed as homogeneously as possible by frequent mixing of the cell suspension, avoiding the introduction of air bubbles.
7. Incubate the plates in a 37°C incubator with 5% CO_2 overnight.

DILUTION OF TEST SAMPLES

8. Add 50 μl of test supernatant [from (4)] in duplicate or triplicate to each well of the first column (A1–H1).
9. Dilute the test sample across the plate by repeated mixing of the solution in the well followed by transfer of 50 μl to the next well. In the penultimate well (e.g., A11), the dilution is mixed, and then 50 μl is removed and discarded to leave the last column (A12–H12) untouched as control for total growth, i.e., cells without test solution. Care should be taken not to disturb the cell monolayers during the dilution step.
10. Immediately add actinomycin D solution (2 μg/ml, 100 μl/well). This inhibits RNA synthesis and consequently the production of new proteins, induced by TNFα, which antagonize the cytotoxic effect.
11. Incubate the plates for 18–20 hours in a 37°C incubator.

STAINING PLATES

12. Tip off the medium, blot the plates on absorbent paper, and add 100 μl crystal violet stain to each well.
13. Incubate for 10 minutes at room temperature.
14. Discard the stain and wash the plates in running tap water. Blot the plates vigorously on absorbent paper and leave to dry. Spills of color can be cleaned with ethanol.
15. Assess the color intensity by reading the plates at 550 nm in an ELISA reader (microELISA reader, Artek, Dynatech, Farmingdale, NY). The intensity of the color reflects the number of cells remaining adherent to the plastic and therefore not killed by TNFα in the test dilution.

If the cell layer is not homogeneous (due to technical difficulties in either plating or dilution), the color can be extracted from the cells by the addition of 0.5% sodium dodecyl sulfate (100 μl/well). The plates are incubated at room temperature overnight and the color intensity is read at 630 nm.

TREATMENT OF RESULTS, CALCULATION OF S50, AND RECIPROCAL OF DILUTION OF TEST SUPERNATANT AT WHICH HALF THE TARGET CELLS ARE KILLED (SHER ET AL., 1990)

16. The S50 is obtained from the best-fit regression line of the graph of OD_{550} versus log(dilution), where S50 is the reciprocal of the dilution at which half the target cells are killed, i.e., when

$$OD_{550} = OD_{550}(\text{total cells, i.e., A12–H12})/2$$

This can be converted into picograms of TNFα by comparison with a standard of recombinant TNF run under the same experimental conditions. The calculation is as follows: Since by definition both standard and sample contain the same concentration of TNFα at their respective S50, the test sample contains $S50_{(sample)} \times$ pg TNFα at $S50_{(standard)}$.

Additional Information

1. A number of alternative target cells may be used including WEHI 164 fibrosarcoma (ATCC).
2. An alternative method for assessing the number of viable cells adhering to the plates after exposure to the TNF samples uses MTT [3-(4,5-dimethylthiazol-2-yl)-2,5-diphenyltetrazolium bromide, Sigma M-2128] reduction as described next for the IL-1 assay.
3. There are also a number of commercial ELISA plates on the market which can be used to assay TNFα according to the manufacturer's instructions.

IL-1 Assay (Günther et al., 1989)

Although commercial ELISAs for IL-1 are available, we prefer a biological assay that is based on IL-1-dependent induction of the T cell growth factor IL-2 in the conalbumin-specific, I-A^k-restricted helper T cell line D10.G4.1 (Kaye et al., 1983). IL-2 causes autocrine growth of the D10 cells. Growth is monitored by MTT reduction. The IL-2-dependent cell line D10 requires stimulation by antigen every 2 weeks on syngeneic feeder cells and is passaged in an IL-2-containing propagation medium.

IL-1-DEPENDENT D10 GROWTH ASSAY

1. Prepare twofold serial dilutions of the IL-1 samples in 50 μl medium (RPMI 1640–10% FCS, 0.1 μg/ml Con A) in a 96-well flat-bottom tissue culture plate. This should include a positive control with murine recombinant IL-1 (rIL-1) starting with 125 pg/ml in the first well. Add triplicates of negative controls (medium only). Equilibrate in a CO_2 incubator.
2. Adjust D10 cells which have not been split or transferred for 3 days to 4×10^5 cells/ml, and add 50-μl aliquots to equilibrated samples and controls. Incubate for 3 days.
3. Add 10 μl of a 5-mg/ml MTT (Sigma) solution in PBS to each well and incubate for 4 hours.
4. Add a solution of 5% formic acid in 2-propanol, 120 μl/well, mix well, sonicate in a sonic bath, and read absorption at 577 nm in a microplate reader. Calculate IL-1 from the calibration curve with rIL-1, using half-maximal MTT reduction as read-out points. Positive samples should yield at least twice as much OD_{577} as negative controls. The signal to noise ratio depends on the state of the D10 cells.

IL-6 Assay

Human and murine IL-6 can both be assayed by a growth assay using an IL-6-dependent cell line, and units per milliliter IL-6 are defined as the dilution of sample yielding half-maximal growth (van Snick *et al.*, 1986). Easier to perform, but less sensitive, are capture ELISAs using commercially available pairs of monoclonal antibodies (MAb) specific for either murine or human IL-6 which can be used according to the manufacturer's instruction. As an example the procedure for testing murine IL-6 (Rosendal *et al.*, 1994) will be detailed.

BIOASSAY FOR IL-6

1. Centrifuge B9 cells in log growth phase, at 200 g for 5 minutes, and wash once with RPMI.
2. Resuspend in the original volume of complete medium but without the IL-6 added and incubate overnight to deplete intracellular stores of IL-6.
3. Harvest the cells and wash twice with RPMI as described earlier. Resuspend at 5×10^4 cells/ml in complete medium but without added IL-6. Store at room temperature until the dilutions are completed.
4. Put 100 μl of medium into each well of a 96-well plate.
5. Dilute test supernatants 1000 times and add 100 μl to row A. This gener-

ally gives a dilution within the range of the assay for TGMϕ stimulated with mycoplasma but can be adjusted for various types of cells and stimuli.

6. Double dilute 100 μl down the plate to row G, leaving row H for control without added test supernatant.

7. Add 100 μl of the B9 cell suspension from (step 3) to all wells.

8. Incubate for about 60 hours and then add 0.5μCi of [^{3}H]thymidine (specific activity 5 Ci/mM, 1 mCi/ml) and incubate for a further 18 hours.

9. Harvest the cells onto filters with a cell harvester and count in a β-counter according to the manufacturer's instructions.

One unit of IL-6 activity is the amount that causes half-maximal proliferation of B9 cells. Otherwise the results can be related to nanograms of IL-6 by comparison with the results obtained with a standard of recombinant IL-6.

IMMUNOASSAY FOR IL-6

1. Immunoplates (Maxisorp F96, Nunc) are coated overnight with 1 μg MAb from clone MP5-20F3 (Endogen) in 100 μl PBS–1% BSA/well by incubation at 4°C in a moist plastic case.

2. Wash four times with TBS-T (0.01 M Tris–HCl, pH 7.5, 0.05% Tween 20 in physiological saline).

3. Add 2% blocking reagent (Boehringer Mannheim, FRG) in PBS and incubate for 4 hours at 4°C. Wash four times as described in step 2.

4. Add a 100-μl twofold serial dilution of an IL-6 standard of known activity, followed by three medium controls and three 100-μl replicates of test samples. If required, these are prediluted to the appropriate range. After a 2-hour incubation at room temperature, wash the plate as described earlier.

5. Add 0.2 μg biotinylated detection MAb (from clone 6B4, Jackson ImmunoResearch, West Grove, PA) in TBS-T (100 μl/well) and incubate for 1 hour at room temperature.

6. Wash the plate as before.

7. Add 100 μl streptavidin-conjugated alkaline phosphatase (Jackson ImmunoResearch) at a dilution of 1:2000 in PBS–1% BSA, and incubate for 30 minutes at room temperature.

8. Wash the plate four times as before.

9. Add 0.1 mg 4-nitrophenyl phosphate in 100 μl diethanolamine buffer (97 ml diethanolamine, 100 mg $MgCl_2$ per liter).

10. Stop the reaction after about 20 minutes by adding 3 N NaOH (50 μl/well).

11. Read the absorbance at 405 nm and calculate the amount of IL-6 from the linear section of the standard curve.

Assay for Nitric Oxide in Cell Cultures (Ruschmeyer et al., 1993)

Nitric oxide has a short half-life, decaying into nitrite (NO_2^-) and nitrate (NO_3^-) in about equimolar amounts in less than a minute. For most purposes it is sufficient to determine nitrite, especially if the cultivation time is extended to 48 hours. However, to increase sensitivity, e.g., to economize on the number of NO-producing cells, or to shorten incubation times, e.g., in order to determine NO and cytokines in the same culture supernatant collected after 24 hours of incubation, the sum of both products can be determined after the reduction of nitrate to nitrite. Griess reagent is then used to detect the resulting total nitrite in the samples. The assay is described for direct use in 100-μl volume tissue cultures in 96-well microtiter plates. It can equally be used for assaying 100-μl volumes of supernatants from cultures with larger volumes for simultaneous measurement of cytokines or cytotoxicity. Note that NO is only formed after double stimulation with IFNγ plus mycoplasma-derived components and that the addition of IFNγ will approximately triple the yield of cytokines. Human monocytes do not produce NO under these conditions.

1. Use a medium that is nitrate-free, such as DMEM–5% FCS, including 2.5×10^{-5} M 2-mercaptoethanol. Prepare a 96-well flat-bottom tissue culture plate as follows: Leave 6 wells in row A free to receive 100-μl duplicates of blanks (medium) and controls (cells with or without IFNγ in a total of 100 μl, see later). Prepare serial twofold dilutions of macrophage activator in 50-μl volumes of nitrate-free medium in rows B–F. It is advisable to include 1 row with a positive sample of known activity as an internal positive control.

2. While the plate is equilibrated in a CO_2 incubator, prepare peritoneal exudate cells (PEC) from C3H/HeJ endotoxin low-responder mice (Bomholtgaard, Ry, Denmark, see earlier), and adjust cells to 2×10^6 cells/ml. Remove a small aliquot as a control for unstimulated cells and add 100 to 150 U/ml recombinant murine IFNγ to the remaining cells.

3. Add 50 μl cells to the serial dilutions and incubate for at least 24 hours, optimally 48 hours. Alternatively, the dilutions in step 1 may be done on plates of either BMMφ or TGMφ prepared as for the TNFα assay and the plates incubated for 24–48 hours.

4. Prepare stock solutions of 0.2 mM $NaNO_2$ and 0.2 mM KNO_3 in the same medium as the cell supernatants to be tested. Add two serial dilutions of nitrite and nitrate to rows G and H to calibrate the test. Nitrate is reduced by the addition of 10 μl nitrate reductase prepared freshly in 1 mM NADPH to give a final concentration of 20 mU enzyme/well. Incubate for 10 minutes at 25°C.

5. Stop the reaction and determine total nitrite by the addition of 100 μl Griess reagent to each well. Incubate for 10 minutes and read absorbance at 550 nm in a microplate reader. A red color indicates the presence of nitrite; the

phenol red in the medium does not interfere with the assay as it turns yellow in the acidic reagent.

6. The amount of nitrite (proportional to the amount of nitric oxide) can be calculated by comparison of the curve of absorbance against the concentration of nitrite in the standard samples. The serial dilution of nitrate monitors the efficiency of the nitrate reductase. A good test should give marginal NO production with IFNγ alone and should reach OD_{550} values of 1.2 with a good stimulating agent. Omitting the reduction yields about half the OD_{550}.

Nitrate and nitrite can also be assayed on supernatants from cells stimulated as described for the other cytokine activities with or without the reduction step (step 4.) or by the addition of 100 μl of supernatant to 100 μl of Griess reagent in step 5.

Radioimmunoassay for Prostaglandin E_2

The assay described here is based on the dextran charcoal radioimmunoassay using rabbit antiprostaglandin E–BSA serum (BioMakor, Nes Ziona, Israel). The radioimmunoassay is a competition assay between radiolabeled prostaglandin E_2 (PGE_2) and cold PGE_2 present in the samples, for antiprostaglandin antibody which is then precipitated on activated charcoal and counted in a liquid scintillation counter.

1. Prepare a solution of PGE_2 (1 μg/ml, Sigma) in absolute ethanol and dilute an aliquot in assay buffer [0.01 M sodium phosphate buffer, pH 7.4, 0.15 M NaCl, 0.1% NaN_3, 0.1% bovine serum albumin (BSA)] to a concentration of 10 ng/ml. Prepare serial double dilutions of this to give a total of seven standard solutions.

2. Pipette 100 μl of standards or supernatants from stimulated cells into plastic assay tubes. It is necessary to include a zero control, a blank, and a tube for the total amount of radiolabel in the assay, each containing 100 μl of assay buffer.

3. Add 0.5 ml of the working dilution (in assay buffer) of rabbit antiprostaglandin E_2–BSA serum (BioMakor) according to the manufacturer's instructions.

4. Incubate all the tubes at 4°C for 30 minutes.

5. Add 100 μl of [^{3}H]prostaglandin E_2 (DuPont, 169.5 Ci/mmol, 10^6 dpm/ml) to all tubes and incubate at 4°C for 60 minutes.

6. Keeping the tubes on ice, add 200 μl dextran-coated charcoal solution [1% (w/v) Norit A-activated charcoal (Sigma), 0.1% (w/v) dextran T −70 (Sigma) in assay buffer] to all tubes except for the tube for the total amount of radiolabel in the assay, to which 200 μl assay buffer is added instead.

7. Mix all tubes, incubate for 10 minutes at 4°C, and centrifuge at 1000 g for 15 minutes.

8. Remove 250 μl of supernatant, mix with an appropriate volume of scintillation liquid, and count in a liquid scintillation counter.

9. Calculation of results is

$$\% \text{fraction bound} = (S - B/Z - B) \times 100,$$

where S is the cpm in the sample, B is the cpm in the blank, and Z is the cpm in the zero control.

The amount of PGE_2 in each sample is compared to the standard curve prepared by plotting the percentage of label bound against the log dose of standard.

Induction of Tumoricidal Activity

1. One day before the assay, A9 cells growing in log phase in 250-ml flasks are pulsed by the addition of 0.5 μCi/ml[^{3}H]thymidine (specific activity 5 Ci/mM, 1 mCi/ml) for 24 hours.

2. Harvest pulse-labeled target cells and resuspend at 10^5cells/ml.

3. To 100 μl of cells (six samples) add 100 μl of 1% SDS and leave for at least 1 hour at 37°C. This is the control for total radiolabeled cells used in the experiment.

4. To the wells of a 96-well microtiter plate containing either TGMφ (see earlier) in 100 μl DMEM–10% FCS (freshly changed before the assay) or BMMφ plated at 10^5 cells/100 μl/well in the same medium add: (i) 100 μl DMEM–10% FCS including test stimulant, at the appropriate concentration, or control, and (ii) 100-μl radiolabeled target cells.

5. For each test it is necessary to add a control for nonspecific killing in the absence of Mφ. Therefore, add to a parallel well without Mφ: (i) 100 μl medium, (ii) 100 μl medium with stimulant or control, and (iii) 100 μl radiolabeled target cells.

6. Incubate the cells for 3 days. During this time, Mφ activated to cytotoxicity will lyse the target cells which will release the radiolabel into the medium. Since neither TGMφ nor BMMφ divide under these conditions, the macrophages do not incorporate the released radiolabel.

7. Transfer the supernatants (300 μl) to scintillation vials, add 2.5 ml Quicksafe (Zinsser Analytic, Frankfurt, Germany), and vortex to mix.

8. Count samples in a β scintillation counter.

9. Calculate the percentage of specific cytolysis by applying the equation:

$$\% \text{ specific cytotoxicity} = (E-S)/(T-S) \times 100 \%,$$

where E is cpm or dpm of supernatant from cocultures of target cells and Mϕ (average of triplicates); S is spontaneous release, cpm or dpm of control culture, without Mϕ; and T is cpm or dpm of total radiolabel added (and released with SDS) (average of six samples).

Chemiluminescence

1. Chemiluminescence plastic tubes (Berthold, Wildbad, Germany) are sterilized by washing once with ethanol (1 hour), then twice with sterile PBS, and once with sterile DMEM–10% FCS.
2. Harvest cells: TGMϕ are harvested from 9-cm tissue culture petri dishes with a cell scraper, centrifuged (200 g for 5 minutes), and resuspended at 5×10^5/ml in DMEM–10% FCS and dispensed (0.5 ml/tube). BMMϕ are harvested from bacteriological dishes as described for the TNF assay, resuspended at 5×10^5/ml in DMEM–10% FCS, and dispensed (0.5 ml/tube). The tubes are covered with sterile aluminum foil and are incubated overnight at 37°C. Mycoplasma preparations (routinely, at 10 μg protein/ml) are added at various times (5–18 hours) before assay.
3. Switch on the luminometer (Biolumat Berthold LB 9500 T, Wildbad, Germany, or any equivalent instrument) and allow to warm up to 37°C.
4. Tip the medium out of one of the tubes, wash twice with 0.5 ml HBSS without phenol red (the color interferes with the assay) and add luminol (10 μl). It is important to adjust the pH of the HBSS to pH 6.5 with CO_2. Insert the tube into the machine and initiate counting. For a positive control for the capacity of the test macrophages to produce a measurable chemiluminescence response, zymosan (10 μl) and luminol are added to macrophages without mycoplasma. The zymosan gives rise to a much stronger chemiluminescence response than that elicited by the mycoplasma preparations.
5. Emission of chemiluminescence by each sample is followed for at least 30 minutes until a plateau of activity is reached. The results are assessed by plotting activity against time.

References

Beutler, B., and Cerami, A. (1988). The common mediator of shock, cachexia and tumor necrosis. *Adv. Immunol.* **42,** 213–231.
DioGiovine, S. F., and Duff, G. W. (1990). Interleukin 1: The first interleukin. *Immunol. Today* **11,** 13–20.
Flick, D. A., and Gifford, G. E. (1984). Comparison of *in vitro* cytotoxic assays for tumor necrosis factor. *J. Immunol. Methods* **68,** 167–175.
Gallily, R., Salman, M., Tarshis, M., and Rottem, S. (1992). *Mycoplasma fermentans* (incognitus

strain) induces TNF α and IL-1 production by human monocytes and murine macrophages. *Immunol. Lett.* **34,** 27–30.

Günther, C., Röllinghoff, M., and Beuscher, H. U. (1989). Proteolysis of the native murine IL 1β precursor is required to generate IL 1β bioactivity. *Immunobiology* **178,** 436–448.

Hirano, T., Akira, S., Taga, T., and Kishimoto, T. (1990). Biological and clinical aspects of interleukin 6. *Immunol. Today* **11,** 443–449.

Kaye, J., Porcelli, S., Tite, J., Jones, B., and Janeway, C. A. (1983). Both a monoclonal antibody and antisera specific for determinants unique to individual cloned helper T cell lines can substitute for antigen and antigen-presenting cells in the activation of T cells. *J. Exp. Med.* **158,** 836–856.

Lewis, C. E., and McGee, J. O'D. (1992). "The Macrophage." IRL Press, Oxford Univ. Press, Oxford.

Liew, F. Y., and Cox, F. E. G. (1991). Nonspecific defense mechanism: The role of nitric oxide. *Immunol. Today* **12,** A17–A21.

Loewenstein, J., Rottem, S., and Gallily, R. (1983). Induction of macrophage-mediated cytolysis of neoplastic cells by mycoplasmas. *Cell. Immunol.* **77,** 290–297.

Mühlradt, P. F., and Schade, U. (1991). MDHM, a macrophage-stimulatory product of *Mycoplasma fermentans,* leads to *in vitro* interleukin-1 (IL-1), IL-6, tumor necrosis factor, and prostaglandin production and is pyrogenic in rabbits. *Infect. Immun.* **59,** 3969–3974.

Rosendal, S., Levisohn, S., and Gallily, R. (1995). Cytokines induced *in vitro* by *Mycoplasma mycoides* subsp. *mycoides,* large colony type. *Vet. Immunol. Immunopathol.* **44,** 269–278.

Ruschmeyer, D., Thude, H. J., and Mühlradt, P. F. (1993). MDHM, a macrophage-activating product from *Mycoplasma fermentans,* stimulates murine macrophages to synthesize nitric oxide and become tumoricidal. *FEMS Immunol. Med. Microbiol.* **7,** 223–230.

Sher, T., Rottem, S., and Gallily, R. (1990). *Mycoplasma capricolum* membranes induce tumor necrosis factor α by a mechanism different from that of lipopolysaccharide. *Cancer Immunol. Immunother.* **3,** 86–92.

Tucker, S. B., Pierre, R. V., and Jordan, R. E. (1977). Rapid identification of monocytes in a mixed mononuclear cell preparation. *J. Immunol. Methods* **14,** 267–269.

VanSnick, J., Cayphas, S., Vink, A., Uyttenhove, C., Coulie, P. G., and Ruber, M. R. (1986). Purification and NH_2-terminal amino acid sequence of a T-cell-derived lymphokine with growth factor activity for B-cell hybridomas. *Proc. Natl. Acad. Sci. USA* **83,** 9679–9683.

F6

IDENTIFICATION, CHARACTERIZATION, AND PURIFICATION OF MYCOPLASMAL SUPERANTIGENS

Barry C. Cole and Curtis L. Atkin

Introduction

Superantigens (SAg) are a group of microbial proteins that interact with the immune system in a unique manner. They can be secreted products such as the staphylococcal and streptococcal exotoxins (Marrack and Kappler, 1990) or can be a part of the cell wall such as the streptococcal M proteins (Tomai et al., 1991), *Yersinia enterocolitica* (Stuart and Woodward, 1992), and others. As reviewed by Acha-Orbea and Palmer (1991), endogenous and exogenous murine retroviruses also have SAg activity.

SAgs are T-cell mitogens that are dependent on major histocompatibility complex (MHC) accessory molecules. However, unlike classic antigens, they do not require processing by accessory cells but bind directly to MHC molecules outside of the antigen-binding groove. The MHC/SAg complex interacts with the β chain segments of the variable region (V_β) of the α/β T-cell receptor for antigen (TCR), largely irrespective of other restricting TCR elements. The term superantigen was coined since all T cells bearing a specific superantigen-reactive V_β chain segment become activated (White et al., 1989). This contrasts to the very small percentage of T cells that interact with traditional antigens in which specificity is determined by multiple TCR elements.

The *Mycoplasma arthritidis* superantigen, MAM, was first recognized by its ability to induce cytotoxic lymphocytes and lymphocyte proliferation depending on the expression of certain MHC molecules (Cole et al., 1981). As reviewed

(Cole, 1991), MAM has been shown to possess all of the properties of SAgs, including the "superantigen bridge" by which CD4$^+$ T$_H$ cells can trigger T-cell-dependent polyclonal B-cell activation (Tumang et al., 1990). MAM is presented to T cells primarily by the α chains of the murine H-2E and human HLA.DR molecules. However, both H-2A and HLA.DQ molecules can also present MAM (unpublished observations). Murine T cells bearing V$_\beta$5.1, V$_\beta$6, V$_\beta$8.1, 8.2, and 8.3 or their human equivalents are the most common TCR elements which recognize MAM. In mouse strains lacking these elements because of clonal or genomic deletions, other V$_\beta$ TCR segments can be functional (Cole et al., 1993b).

The current interest in superantigens relates to their ability to modulate the immune system *in vivo* in a V$_\beta$-specific manner resulting in expansion or deletion of specific V$_\beta$-bearing T cells as well as a V$_\beta$-specific polyclonal B-cell activation with enhanced Ig production. It has been hypothesized that these pathways may lead to the development or enhancement of human autoimmune diseases such as rheumatoid arthritis, lupus erythematosus, and multiple sclerosis (MS) (Heber-Katz and Acha-Orbea, 1989; Marrack and Kappler, 1990; Tumang et al., 1990; Cole, 1991, 1993a; Posnett, 1993). The triggering and exacerbation of experimental autoimmune arthritis (Cole and Griffiths, 1993) and experimental allergic encephalomyelitis, an MS model (Schiffenbauer et al., 1993), have already been demonstrated.

This chapter describes the criteria by which proteins are classified as superantigens and briefly describes or references established methods to achieve this goal. The chapter will then focus on the quantitation and purification of MAM or MAM-like superantigens.

Characterization of Superantigens

Although there are a few exceptions, characteristic properties of the superantigens include: induction of T-cell proliferation; requirement for accessory cells bearing MHC molecules which act as receptors; lack of processing by accessory cells; T-cell recognition is not MHC restricted and is mediated primarily via the β chains of the variable region V$_\beta$ of TCR. The following methods briefly describe how tests for these properties may be performed. Details of individual assays can be found in Coligan et al. (1991).

T-Cell Activation Requiring MHC-Bearing Accessory Cells (AC)

A typical proliferation assay using uptake of ^{3}H-labeled thymidine to measure DNA synthesis is detailed under Quantitation. To determine whether T cells are activated, they can be removed from murine splenocyte suspensions by treatment with anti-Thy1 antibody and C', and the effect on ^{3}H uptake is determined. Human T cells can be removed from human peripheral blood cultures by their ability to bind to sheep red cells forming "rosettes" which can then be collected by centrifugation.

To determine the requirement for accessory cells, the latter can be removed from lymphocyte suspensions by (a) "panning" or absorption to petri dishes coated with antibodies to immunoglobulins or to Class II MHC molecules or (b) depleting MHC-bearing cells from suspensions by treatment with antibodies to Class II, MHC molecules in the presence of C'; lysed cells and membrane fragments are then removed by centrifugation. A combination of (a) and (b) may be necessary.

Binding to MHC Molecules

In the original studies, an absorption assay was used to demonstrate the binding of MAM to MHC-bearing cells (Cole et al., 1982). Splenocyte suspensions of 5×10^7 cells per ml from different strains of mice were incubated for 1 hour at 37°C with a dilute solution of MAM. These and solutions treated with medium alone were centrifuged to remove the cells, the absorption was again repeated, and supernatants were then assayed for mitogenic activity. It was found that lymphocytes which expressed H-2E and underwent proliferation in response to MAM totally absorbed out the mitogenic activity of MAM solutions. In contrast, the lymphocytes from congenic or recombinant mice that were identical, except that they lacked expression of H-2E, failed to remove mitogenic activity from MAM solutions. Since lymphocytes treated with paraformaldehyde could also remove MAM from solution, it was apparent that there was no requirement for internalization and processing of MAM. Now that homogenous MAM is becoming available, future studies to measure binding will be developed using ^{125}I or biotinylated MAM preparations.

Processing of Superantigens by Accessory Cells Not Required

These studies were initially conducted using MAM-reactive, HEL (hen egg lysozyme) antigen-specific T hybridoma cells (2Hd-11.2) derived from a BALB/c (H-2^d) mouse (Cole et al., 1986). Accessory cells consisted of MHC-bearing syngeneic B-cell lymphoma cells (2PK-3). Activation of T cells results in the secretion of IL-2, the amount of which is therefore used as a measure of the degree of activation. Since concanavalin A (Con A) is not dependent on MHC-bearing cells for T-cell activation, it will induce T hybridoma cells to secrete interleukin-1 (IL-2) in the absence of accessory cells or MHC molecules. MAM and other superantigens will not activate T hybridoma cells in the absence of these additives or other costimulatory molecules.

To test the processing requirement for MAM, 2PK-3 AC are tested live, fixed by paraformaldehyde, disrupted, or in purified membrane form for their ability to induce IL-2 by T hybridoma cells. Whereas the T cells can only respond to HEL antigen in the presence of live AC, all preparations of 2PK-3 cells effectively present MAM to the T cells. Furthermore, whereas HEL antigen requires a prolonged period of incubation with AC before presentation (i.e., a processing step), MAM activates T cells immediately after addition to the AC. Since continuous cell lines are used in these assays (and are thus subject to mycoplasma contamination), important controls consist of T hybridoma cells alone and T cells with AC in the absence of any inducers. Con A is also used as a positive control since T hybridomas can lose their ability to become activated.

T-Cell Recognition of Superantigens Not MHC Restricted

Antigen-specific T cells recognize antigen only when associated with MHC molecules on syngeneic accessory cells. Superantigens can be presented by any MHC molecule as long as that molecule has the necessary SAg-binding site. Thus, HEL antigen is presented by H-2^d (BALB/c) AC to BALB/c-derived T cells but is not presented by H-2^k, i.e., C3H or CBA-derived AC. In contrast, MAM, for example, can be presented to BALB/c-derived T cells by both H-2^d and H-2^k-bearing AC since both of these express the H-2E MHC molecule that is a receptor for MAM. Using methodology described in the previous section, antigen-specific, SAg-reactive T-cell hybridomas and T-cell lines are incubated with syngeneic AC or allogeneic AC in the presence of antigen or superantigen and are tested for the secretion of IL-2. In these experiments, AC can consist of splenic lymphocytes that have been depleted of T cells and gamma-irradiated to inhibit DNA synthesis, but not protein metabolism. IL-2 secretion, as before, is used as a measure of activation.

T-Cell Activation Is V_β Specific

Different superantigens interact with a characteristic set of V_β TCR-bearing T cells. Thus, by using a panel of T-cell hybridomas, each expressing a different TCR V_β chain, most superantigens can readily be distinguished from each other as illustrated in Table I. Again, IL-2 is used as a measure of T-cell activation. Con A, as a positive control, will activate all T hybridoma cells irrespective of V_β TCR expression.

An alternative method to detect V_β usage involves activating splenic or peripheral blood lymphocytes by exposure to the unknown superantigen or to Con A. When microscopic observation reveals the presence of activated cells, the

TABLE I

Characteristics of Superantigens Measured against Murine T Hybridomas

Superantigen[a]	IL-2 produced in T hybridomas					
	$V_\beta 1$	$V_\beta 3$	$V_\beta 7$	$V_\beta 6$	$V_\beta 8.2$	$V_\beta 11$
MAM	−	−	−	+	+	−
SEA	+	+	−	−	−	+
SEB	−	−	+	−	+	−
Con A	+	+	+	+	+	+

[a] SEA, Staphylococcal enterotoxin A; SEB, staphylococcal enterotoxin B.

cultures are then washed and further expanded in the presence of IL-2. Under these conditions, unactivated cells will die, but activated cells will continue to proliferate. When sufficient cells have been obtained, they can be separated from dead cells by Ficoll gradient centrifugation and analyzed for V_β expression by flow cytometry using antibodies to specific V_β chains of the TCR, by polymerase chain reaction using V_β-specific primers or by the RNA protection assay (Singer et al., 1989). All V_β TCRs tested should be expressed in the Con A-expanded lymphocytes, whereas a more restricted number of V_β TCR-bearing cells or transcripts will be present in SAg-expanded cultures.

B-Cell Activation by "Superantigen Bridge"

Some superantigens, such as MAM and TSST-1 (toxic shock syndrome toxin), can activate B cells to proliferate and secrete immunoglobulins (Mourad et al., 1989; Tumang et al., 1990). However, it is necessary to differentiate this pathway from that mediated by B-cell mitogens that are present in mycoplasmal membranes and bacterial cell walls. To achieve this it is necessary to show that SAg-mediated activation of B cells is T-cell dependent.

Resting B cells are purified from murine splenocytes or from human peripheral blood or tonsils. T cells may be purified from lymph nodes or blood or may consist of T-cell lines, generated by exposure to MAM or to other antigens (Ag), and cloned. Separate CD4+ and CD8+ T-cell lines can be generated in this manner. To prevent overgrowth, T cells are gamma-irradiated, a process that does not affect T helper functions. Cultures are established consisting of B cells alone, B cells and T cells, and B cells plus T cells plus SAg. Additional controls are B cells plus T cells alone and B cells or T cells plus SAg. The cultures are incubated for 7–10 days at 37°C in 5% (v/v) CO_2 in air after which culture supernatants are collected and assayed by ELISA for IgG or IgM secretion.

MAM or other SAgs should induce Ig secretion only in the presence of both B cells and T cells. Furthermore, the T cells must express a T-cell V_β chain receptor that is used by the SAg tested. Also, only $CD4^+$ T helper cells, not $CD8^+$ cells, can result in B-cell activation (Tumang et al., 1990).

Purification of *M. arthritidis* Superantigen MAM

Quantitation

MAM, like most of the bacterial superantigens, is a secreted or released molecule with toxic properties when systemically administered *in vivo*. In order to develop an effective purification scheme, it was first necessary to devise a method to quantitate the amount of MAM present in a given preparation. This method has been detailed elsewhere (Atkin et al., 1986) and is summarized here.

Lymphocyte proliferation assay: RPMI 1640 medium (GIBCO, Chagrin Falls, OH) supplemented with 2% (w/v) L-glutamine, 5% (v/v) heat-inactivated human serum, and antibiotics (50 U penicillin G and 50 μg streptomycin per ml) is used throughout. Splenocyte suspensions from CBA/J mice (Jackson Laboratories, Bar Harbor, ME) 6–12 weeks of age were treated with 0.83% NH_4Cl to remove erythrocytes are washed and adjusted to contain 2.5×10^6 cells/ml, and are then distributed into the microliter wells of 96-well Falcon 3072 microtiter plates (Becton–Dickinson, Oxnard, CA). Inducers are added in 20-μl volumes to triplicate wells and medium alone is also added to triplicate wells to ascertain spontaneous activity. Three wells containing 5 μg Con A are used as a positive control. After 2 days at 37°C in 5% CO_2 in air, each well is pulsed with 1 μCi[^{3}H]thymidine (2 Ci per mmol). After an additional 24 hours of incubation, DNA is harvested using a Skatron Basic 96 harvester and the resulting filter mats are counted on a Skatron beta counter.

For the initial purification studies, a MAM standard called $\Sigma 1$ consisted of a mycoplasma-free supernatant of *M. arthritidis* strain 14124 P10 (MAS) that had been extensively dialyzed overnight against phosphate-buffered saline to remove inhibitory substances. Numerous aliquots of 0.2 ml were stored at $-70°C$. Future standards will consist of titrated homogenous MAM preparations. Figure 1 shows titration curves for a twofold dilution series of $\Sigma 1$ standard and of an unknown sample from Sephadex gel filtration chromatography. MAM, at 1 unit/ml, is arbitrarily defined as giving 50% of maximal lymphocyte proliferation in the just-mentioned standardized assay. Thus, the reciprocal of the dilution of $\Sigma 1$ standard that gives 50% of the maximal response, using the linear portion of the log-dose/response curve, represents the number of units in the preparation. The dilution of the unknown sample that gives the same proliferative response is then calculated and the number of units is deduced.

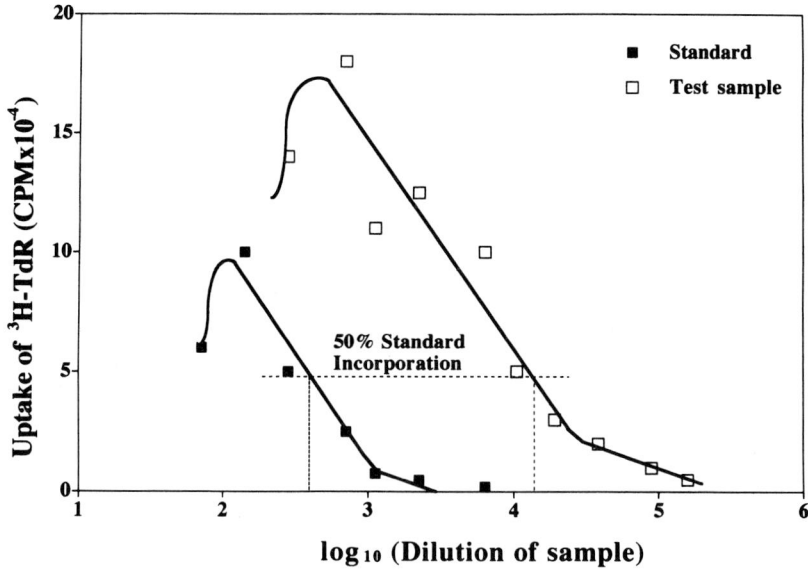

Fig. 1. Titration of MAM for calculation of units of mitogenic activity. Dilutions of Σ1 standard (■) and test sample (□) were examined for uptake of [^{3}H]thymidine by CBA murine splenocytes. The titration curves (solid lines) include straight portions from least-square fittings. Broken lines represent interpolated values for half-maximal incorporation.

Materials for MAM Purification

Buffer 1: 1 M $(NH_4)_2SO_4$, 10 mM Tris–HCl, adjusted to pH 8.3 with aqueous ammonia
Buffer 2: 10 mM KH_2PO_4, adjusted to pH 7.2 with KOH
Buffer 3: 0.5 M KH_2PO_4 adjusted to pH 7.2 with KOH
Buffer 4: 2 M $(NH_4)_2SO_4$, 50 mM KH_2PO_4, adjusted to pH 7.2 with KOH
Buffer 5: 50 mM KH_2PO_4, adjusted to pH 7.2 with KOH
$(NH_4)_2SO_4$ (Schwarz–Mann ultrapure)
UV monitor and recorder with flow cuvette for continuous monitoring at A_{280} and A_{214}
Sephadex G-50 medium grade (Pharmacia-LKB, Piscataway, NJ) in a 2-liter chromatography column (diameter:height = 1:5–10)
Fast flow Sepharose-S cation-exchange resin (Pharmacia-LKB) in a 30-ml column (diameter:length = 1:5–10)
Mono S cation-exchange resin (Pharmacia-LKB) in a 1-ml HR 5/10 FPLC column
Alkyl-Superose resin (Pharmacia-LKB) in a 1-ml HR 5/10 FPLC column
Centricon-10 tubes (Amicon, Danvers, MA) for centrifugal desalting and concentration

Procedure of Purification

1. Adapt *M. arthritidis* to growth in autoclaved modified Edward–Hayflick (AMEH) medium by serial passage and freeze multiple aliquots at −70°C. The AMEH medium is detailed elsewhere (Atkin *et al.*, 1994). Briefly, after autoclaving the modified Edward–Hayflick medium (Atkin *et al.*, 1986) the coagulated proteins are removed by centrifugation and filtration and the medium is supplemented with 0.5% (w/v) arginine hydrochloride and 1% (w/v) HEPES (Sigma Chemical Co., St. Louis, MO). After readjusting to pH 7.0, the medium is reautoclaved, cooled, and supplemented with 1000 U/ml penicillin G (Apotheon, Pinceta, NJ).

2. Add two 5-ml frozen cultures each to two 2-liter volumes of AMEH medium. Gently rotate on a shaker at 37°C and use for MAM purification protocol when culture is in senescence, i.e., when OD at A_{600} declines over maximal recorded value; this will occur after approximately 50 to 54 hours of incubation.

3. Conduct $(NH_4)_2SO_4$ fractionation on whole *M. arthritidis* culture. Dissolve the 50 to 80% precipitate in 60 ml distilled H_2O and eliminate solid residue by centrifugation.

4. Pass resulting material over a Sephadex G-50 column that has previously been equilibrated with Buffer 1. Elute with Buffer 2 and collect the 200-ml fraction that corresponds to a molecular weight of 30,000. Adjust electrolytes to Buffer 2 by dialysis.

5. Pass the Sephadex G-50 fraction over a Sepharose-S column that has previously been equilibrated with Buffer 2. Wash with Buffer 2 until the OD returns to baseline and frontal elute with Buffer 3. Collect the single 280-nm-absorbing peak and adjust electrolytes back to Buffer 2.

6. Pass adjusted material over a Mono S cation-exchange column previously equilibrated with Buffer 2. Wash with Buffer 2, as before, then elute with a 20-ml linear gradient to 40% (v/v) of Buffer 3. Collect fractions which possess most lymphoproliferative activity and freeze at −70°C or use immediately in the next step.

7. Adjust Mono S fractions to 2 M $(NH_4)_2SO_4$ and pass over an alkyl-Superose hydrophobic interaction column equilibrated with Buffer 4. Wash and elute with a 20-ml inverse linear gradient of Buffer 4 to Buffer 5. MAM elutes as the last peak at approximately 1.4 M $(NH_4)_2SO_4$.

Stabilization of MAM

MAM is stable at 4°C in high ionic strength buffers such as Buffer 4; however, freezing in this buffer destroys activity. Alkyl-Superose fractions can be desalted in Centricon-10 tubes in the presence of 5–10 mg bovine serum albumin or 2.5%

heat-inactivated serum and frozen at $-70°C$. Since repeated thawing and freezing results in loss of activity, numerous aliquots should be frozen and preferably used once and the surplus discarded.

Discussion

Much of the earlier work on the detection of lymphocyte-activating substances from mycoplasmas was conducted using whole organisms, membranes, or cell extracts. In view of the fact that MAM is a secreted or released factor, culture supernatants of other mycoplasma species should now be screened for similar soluble substances. It should be noted, however, that the procedure developed here for the purification of MAM may not be appropriate for other mycoplasmal mitogens.

A major problem in the purification of MAM is its highly basic nature that results in binding to culture medium components and nucleic acids, as well as to glass and plastic surfaces. It is also heat (56°) and acid (<pH 7.0) labile and is destroyed by repeated thawing and freezing (Atkin *et al.*, 1986, 1994). Low concentrations of MAM in the absence of carrier proteins are rapidly destroyed or lose biologic activity. A number of approaches were used to overcome these problems. First, quantitation of MAM activity in terms of units per milligram of protein was important in the location of active fractions and in determination of the efficacy of each of the purification steps. Next, the organisms were grown in AMEH medium which lacked most of the large medium proteins. In view of the high p*I* of MAM, gel filtration was conducted under high-salt conditions which also allowed the selective adsorption of MAM to cation-exchange resins. A main key to MAM purification was to perform most of the steps sequentially and without interruption once the location of the active fractions had been established. When necessary, the 50–80% $(NH_4)_2SO_4$ precipitate could be stored for a few days at 4°C as could the active fractions from the Mono S column. The latter fractions were also stable on freezing at $-70°C$. The addition of $(NH_4)_2SO_4$ directly to the mycoplasma culture avoided a time-consuming centrifugation step.

Since MAM is now known to be half-maximally active at $\geq 10^{-14} M$ (Atkin *et al.*, 1994), caution should be taken in order to avoid trace contamination of membranes and other cell preparations from different mycoplasma species with MAM-like biologically active superantigens.

Acknowledgments

The authors' work described in this chapter was supported by grants from the National Institute of Allergy and Infectious Disease (AI12103), the National Institute of Arthritis and Metabolic Diseases (AR02255), and by a grant from the Nora Eccles Treadwell Foundation.

References

Acha-Orbea, H., and Palmer, E. (1991). Mls—A retrovirus exploits the immune system. *Immunol. Today* **12,** 356–361.

Atkin, C. L., Cole, B. C., Sullivan, G. J., Washburn, L. R., and Wiley, B. B. (1986). Stimulation of mouse lymphocytes by a mitogen derived from *Mycoplasma arthritidis*. V. A small basic protein from culture supernatants is a potent T cell mitogen. *J. Immunol.* **137,** 1581–1589.

Atkin, C. L., Wei, S., and Cole, B. C. (1994). The *Mycoplasma arthritidis* superantigen MAM: Purification and identification of an active peptide. *Infect. Immun.* **62,** 5367–5375.

Cole, B. C. (1991). The immunobiology of *Mycoplasma arthritidis* and its superantigen MAM. *Curr. Top. Microbiol. Immunol.* **174,** 107–119.

Cole, B. C., Ahmed, E. A., Araneo, B. A., Shelby, J., Kamerath, C., Wei, S., McCall, S., and Atkin, C. L. (1993a). Immunomodulation in vivo by the *Mycoplasma arthritidis* superantigen MAM. *Clin. Infect. Dis.* **17**(Suppl.), S163–S169.

Cole, B. C., Araneo, B., and Sullivan, G. J. (1986). Stimulation of mouse lymphocytes by a mitogen derived from *Mycoplasma arthritidis*. IV. Murine hybridoma cells exhibit differential accessory cell requirements for activation by either *Mycoplasma arthritidis* T cell mitogen, concanavalin A, or hen eggwhite lysozome. *J. Immunol.* **136,** 3572–3578.

Cole, B. C., Balderas, R. A., Ahmed, E. A., Kono, D., and Theofilopoulos, A. N. (1993b). Genomic composition and allelic polymorphisms influence V_β usage by the *Mycoplasma arthritidis* superantigen. *J. Immunol.* **150,** 3291–3299.

Cole, B. C., Daynes, R. A., and Ward, J. R. (1981). Stimulation of mouse lymphocytes by a mitogen derived from *Mycoplasma arthritidis*. I. Transformation is associated with an H-2-linked gene that maps to the I-E/I-C subregion. *J. Immunol.* **127,** 1931–1936.

Cole, B. C., Daynes, R. A., and Ward, J. R. (1982). Stimulation of mouse lymphocytes by a mitogen derived from *Mycoplasma arthritidis*. III. Ir gene control of lymphocyte transformation correlates with binding of the mitogen to specific Ia bearing cells. *J. Immunol.* **129,** 1352–1359.

Cole, B. C., and Griffiths, M. M. (1993). The *Mycoplasma arthritidis* superantigen, MAM, triggers and exacerbates autoimmune arthritis. *Arthritis Rheum.* **36,** 994–1002.

Coligan, E. C., Kruisbeek, A. M., Margulies, D. H., Sheuadi, E. M., and Strober, W. (1991). "Current Protocols in Immunology." Green Publishing Associates and Wiley-Interscience, New York.

Heber-Katz, E., and Acha-Orbea, H. (1989). The V-region disease hypothesis: Evidence from autoimmune encephalomyelitis. *Immunol. Today* **10,** 164–169.

Marrack, P., and Kappler, J. (1990). The staphylococcal enterotoxins and their relatives. *Science* **248,** 705–711.

Mourad, W., Scholl, P., Diaz, A., Geha, R., and Chatila, T. (1989). The staphylococcal toxic shock syndrome toxin 1 triggers B cell proliferation and differentiation via major histocompatibility complex-unrestricted cognate T/B cell interaction. *J. Exp. Med.* **170,** 2011–2022.

Posnett, D. N. (1993). Do superantigens play a role in autoimmunity? *Semin. Immunol* **5,** 65–72.

Schiffenbauer, J., Johnson, H. M., Butfiloski, E. J., Wegrzyn, L., and Soos, J. M. (1993). Staphylococcal superantigens can reactivate experimental allergic encephalomyelitis. *Proc. Natl. Acad. Sci. USA* **90,** 8543–8546.

Singer, P. A., Balderas, R. S., McEvilly, R. J., Bobardt, M., and Theofilopoulos, A. N. (1989). Tolerance-related V_β clonal deletions in normal CD4$^-$8$^-$, TCR-α/Vβ^+ and abnormal lpr and gld cell populations. *J. Exp. Med.* **170,** 1869–1877.

Stuart, P. M., and Woodward, J. G. (1992). *Yersinia enterocolitica* produces superantigenic activity. *J. Immunol.* **148,** 225–233.

Tomai, M. A., Alion, J. A., Dockter, M. E., Majumdar, G., Spinella, D. G., and Kotb, M. (1991).

T cell receptor V gene usage by human T cells stimulated with the superantigen streptococcal M protein. *J. Exp. Med.* **174,** 285–288.
Tumang, J. R., Posnett, D. N., Cole, B. C., Crow, M. K., and Friedman, S. M. (1990). Helper T cell-dependent human B cell differentiation mediated by a mycoplasmal superantigen bridge. *J. Exp. Med.* **171,** 2153–2158.
White, J., Herman, A., Pullen, A. M., Kubo, R., Kappler, J. W., and Marrack, P. (1989). The Vβ-specific superantigen staphylococcal enterotoxin B: Stimulation of mature T cells and clonal deletion of neonatal mice. *Cell* **56,** 27–35.

F7

MYCOPLASMAL B-CELL MITOGENS
Yehudith Naot

Introduction

A number of plant lectins and bacterial, viral, and mycoplasmal cell components are capable of stimulating lymphocytes in a nonspecific, mitogenic manner. Whereas antigens stimulate only those lymphoid cells bearing antigen-specific receptors, mitogens activate a large proportion of T cells, B cells, or both B and T lymphocytes, resulting in the polyclonal expansion of cells which differentiate into effector cells or give rise to memory cells. Following lymphocyte stimulation by B-cell mitogens, distinctive differentiation features are observed at the ultrastructure level. Quiescent B lymphocytes are activated into growing and proliferating lymphoblasts. Ultimately, many B-cell blasts mature into terminally differentiated plasma cells secreting immunoglobulins of specificities not related to the stimulating agent. A variety of *Mycoplasma*, *Acholeplasma*, and *Spiroplasma* species have been shown to induce mitogenic stimulation of human, mouse, rat, guinea pig, and hamster lymphocytes. Among these mitogenic mycoplasmas, *M. pneumoniae*, *M. fermentans*, *M. orale*, *M. pulmonis*, *M. neurolyticum*, *M. arthritidis*, *A. laidlawii*, *M. hyorhinis*, and *M. arginini* act as polyclonal B-cell activators (Cole *et al.*, 1985; Ruuth and Praz, 1989). This chapter describes methods used to study mycoplasmal B-cell mitogens. The activity of mycoplasmal B-cell mitogens can be determined by the extent of lymphocyte proliferation and/or by measuring the production and secretion of immunoglobulins not related to mycoplasmal antigens.

Materials

Note: All solutions and equipment for handling cultures must be sterile.

Germ-free rats (200–250 g) or mice (18–20 g) reared under specific-pathogen-free (SPF) conditions
Dulbecco's phosphate-buffered saline (D-PBS)
Dulbecco's modified Eagle's medium supplemented with 2 mM L-glutamine, 1 mM sodium pyruvate, 100 units/ml penicillin, 100 μg/ml streptomycin, and 50 μM 2-mercaptoethanol (complete DMEM)
Mycoplasma, Acholeplasma, or *Spiroplasma* lysed cells, suspended at 1 mg/ml in a 1:20 dilution of β buffer
β buffer: 0.15 M NaCl, 0.05 M Tris(hydroxymethyl)aminomethane, 0.01 M 2-mercaptoethanol, pH 7.4
Mycoplasma-free serum: fetal calf serum, heat inactivated at 56°C for 60 minutes
Concanavalin A, 1 mg/ml in D-PBS
Lipopolysaccharide (LPS) from *Escherichia coli* 055:B5, 1 mg/ml in D-PBS
Pokeweed mitogen, 1 mg/ml in D-PBS
Ethanol, 70% (v/v)
Ethanol, 95% (v/v)
Tris(hydroxymethyl)aminomethane hydrochloride, 0.1 M, pH 7.2, containing 0.8% NH_4Cl
Heparin, preservative free
Ficoll–Hypaque solution, density 1.077 g/ml
Trypan blue, 0.2% in D-PBS
[*methyl*-^{3}H]Thymidine, sterile aqueous solution 5 Ci/mmol, 1 mCi/ml
Scintillation fluid
Sheep red blood cells (SRBC) suspension: 10% (v/v) in complete DMEM supplemented with 10% heat-inactivated fetal calf serum
Trinitrophenylated sheep red blood cell (TNP-SRBC) suspension: 20% (v/v) in complete DMEM supplemented with 10% heat-inactivated fetal calf serum
2× Basal Eagle's medium
Agarose, 1% (w/v) in water
Guinea pig serum: 50% (v/v) in complete DMEM supplemented with 10% heat-inactivated fetal calf serum
Refrigerated centrifuge, 1000 g with adapter for microtiter plates
Forceps and scissors kept in sterile beaker with 70% ethanol
Syringes
Tissue culture petri plates, 60×15 mm
Conical plastic centrifuge tubes, 50 ml
Pasteur pipettes and a rubber bulb

Pipettes, different volumes
Dispensing adjustable pipettes with tips
Stainless-steel wire mesh (40×40) (Sigma-Aldrich Co. Inc., Milwaukee, WI) or nylon mesh screen, 200-μm pore size (Tetco Inc., Elmsford, NY)
Hemacytometer and coverslips
Filter paper No. 1
Tissue culture plates: 96 wells, U-shaped microplates with lids
Hood for handling tissue cultures
Automatic cell harvester (Dynatech AG, Zug, Switzerland)
Glass fiber filter strips for cell harvester
Incubator for tissue cultures set at 37°C with 5% CO_2 in air
β-Scintillation counter
Scintillation vials
Water bath prewarmed to 42°C
Microscope
Dissecting microscope 10× magnification
Tissue culture test tubes: plastic disposable, various volumes
Test tubes: 5-ml glass test tubes prewarmed at 42°C
Microscope slides: Glass, frosted at one end
Glass petri dishes, 15 cm
Vortex mixer

Procedures

Preparation of Lymphoid Cell Suspensions

Lymphoid cells may be obtained from lymphoid organs such as spleen, lymph nodes, and thymus, and from lymph and peripheral blood. This section describes preparation of cell suspensions from spleen and lymph nodes of rats and mice and from human peripheral blood.

Note: Sterile conditions must be maintained for solutions, equipment, and procedures. When working with human blood or with radioisotopes, biosafety level 2 or 3 procedures should be followed.

Rat and Mouse Lymphoid Cell Suspensions

Rats and mice are sacrificed by cervical dislocation or by euthanasia with ether. The skin of the animal is swabbed with 70% ethanol. To remove lymphoid organs, a small incision through the skin of the left flank is made, and the abdominal skin is cut through and pulled apart. A cut of the peritoneal wall exposes the spleen which is removed and placed in a tissue culture plate containing 3 ml of D-PBS. Axillary, bronchial, inguinal, and mesenteric lymph nodes are exposed in a similar manner, gently removed, and placed in another tissue culture plate containing D-PBS. Fatty and connective tissues attached are excised and the lymphoid organs are transferred to another set of plates containing 5 ml D-PBS. Organs are cut into small fragments with scissors. The suspensions are gently pressed through a stainless-steel wire screen or nylon mesh with the aid of a 5-ml plastic syringe plunger. Small volumes (1–2 ml) of D-PBS are often poured through the mesh to ensure wetness and easy dripping. The suspensions (separate for spleen and lymphoid organs) are transferred to 50-ml conical plastic test tubes and are centrifuged at 300 g for 10 minutes at 4°C. Supernatant fluids are discarded.

To lyse red blood cells, spleen cells are suspended in 5–8 ml (for one spleen) of 0.1 M Tris–HCl, pH 7.2, containing 0.8% NH_4Cl and are incubated for 10 minutes at room temperature with occasional agitation. Both spleen and lymph node cell suspensions are centrifuged at 300 g for 10 minutes at 4°C and are washed once with 40 ml D-PBS and then with Dulbecco's modified Eagle's medium supplemented with 2 mM L-glutamine, 1 mM sodium pyruvate, 100 units/ml penicillin, 100 μg/ml streptomycin, and 50 μM 2-mercaptoethanol (complete DMEM). Finally, cells are suspended in 10 ml complete DMEM, counted using a hemacytometer, and cell viability is determined by Trypan blue exclusion (Ginsburg and Nicolet, 1973; Naot and Ginsburg, 1978).

Human Peripheral Blood Mononuclear Cell Suspensions

Human mononuclear cell suspensions are prepared from blood of healthy donors having no indication of past infection with mycoplasmas and serologically negative for antibodies to the tested mycoplasma species as well as to other common human mycoplasmas (Biberfeld and Nilsson, 1978). Collect about 40 ml venous blood with 50 units/ml heparin. Incubate for 2 hours at room temperature to allow separation of plasma from red blood cells. In a conical centrifuge tube, place Ficoll–Hypaque (density 1.077 g/ml). Six milliliters of Ficoll is needed for each 10 ml of plasma. Carefully place plasma over the Ficoll–Hypaque layer. Centrifuge at 800 g for 30 minutes at room temperature. With a Pasteur pipette, remove the upper layer which contains platelets and plasma components. Using a second pipette, collect the mononuclear cell layer into a centrifuge tube. Mononuclear cells are washed by centrifugation at 300 g for 10 minutes at 4°C, twice in D-PBS, and then with complete DMEM. Finally, cells are suspended in 10 ml complete DMEM, counted using a hemacytometer, and cell viability is determined with Trypan blue.

Determination of Cell Viability

The viability of cells is assessed with Trypan blue dye which is excluded by viable lymphoid cells, but stains dead cells. Trypan blue solution should be freshly prepared. Dissolve 0.1 g Trypan blue in 50 ml D-PBS by heating. Cool and filter through a No. 1 filter paper. To 0.2 ml fresh Trypan blue solution, add 0.2 ml cell suspension ($2-4 \times 10^6$ /ml) and mix. Transfer the suspension to a hemacytometer. It is crucial to complete counting within 3 minutes to avoid the uptake of dye by viable cells. Count 150–200 cells.

Determination of Mycoplasma-Induced Lymphocyte Proliferation

1. Following the counting of lymphocytes and determination of cell viability, centrifuge cells (300 g, 10 minutes at 4°C) and suspend them at $2-4 \times 10^6$ cells/ml in complete DMEM supplemented with 10% heat inactivated (1 hour, 56°C) fetal calf serum. Cultures are set up in U-shaped microtiter plates by dispensing 0.1-ml cell suspension into each of 96 wells.

During dispensing, agitate the suspension frequently to ensure homogeneity. To sets of triplicate wells containing lymphoid cells, add 0.1 ml/well of stimulating agents at various dilutions, control buffers, and control media. Concanavalin A, a known mitogen of T lymphocytes, serves as a positive control for the competence of rat, mouse, and human lymphoid suspensions to respond to mitogenic stimulation. Prepare concanavalin A dilutions of 5, 10, and 20 µg/ml in complete DMEM and add 0.1 ml/well of these solutions, each to triplicate wells.

In the same manner, prepare dilutions of the tested mycoplasmal mitogens at concentrations varying from 2 to 200 µg/ml in complete DMEM and dispense. Dilutions of 1:100 and 1:10,000 of β buffer in complete DMEM, as well as complete culture medium, serve as negative controls without mitogens. To determine the proliferative responses of lymphocytes at different time periods of cultures, prepare three identical culture plates to be further assayed at different times. Incubate cultures at 37°C in a humidified atmosphere containing 5% CO_2 in air.

2. At 24, 48, and 72 hours of culture add, to all wells in one plate, 1 µCi [^{3}H]thymidine by dispensing a 10-µl solution of tritiated thymidine diluted to 100 µCi/ml in complete DMEM. Incubate the pulsed plate for an additional 24 hours under the same conditions. Harvest cells onto glass fiber strips using an automated multichannel cell harvester. Wash the wells of the plate thoroughly with water by filling and aspirating about 10 times. Wash the filter strips with 95% ethanol and air dry. Transfer filter discs to scintillation vials and add 3 ml scintillation fluid to vials.

Count samples in a β-scintillation counter. Calculate the mean counts per minute (cpm) of each set of triplicate wells. Variations of more than 15% between individual samples of a triplicate indicate technical errors (Naot and Ginsburg, 1978).

Determination of Mycoplasma-Stimulated Antibody-Producing Cells

1. Lymphoid cell suspensions, 5×10^6 cells/ml in complete DMEM supplemented with 10% heat-inactivated fetal calf serum, are dispensed (0.1 ml/well) into triplicate wells of U-shaped, 96-well tissue culture plates. To each set of three wells, add 0.1 ml of mycoplasmal mitogens diluted to concentrations of 2–200 μg/ml in serum-supplemented complete DMEM. When mouse lymphoid cells are studied, *E. coli* LPS is employed as a positive control for the mouse B-cell polyclonal activator.

Pokeweed mitogen is used as a positive control in experiments employing rat and human B lymphocytes. From the appropriate control B-cell mitogen prepare 10-, 20-, and 40-μg/ml solutions in serum-supplemented DMEM and add to triplicate wells (0.1 ml/well). As a negative control use the T-cell mitogen–concanavalin A. Prepare concanavalin A dilutions of 5, 10, and 20 μg/ml in serum-supplemented complete DMEM and add to wells.

In the same manner, prepare dilutions of 1:100 and 1:10,000 of β buffer in serum-supplemented complete DMEM, as well as serum-supplemented complete medium, and dispense into wells to serve as negative controls without stimulating agents. Incubate cultures for 5, 6, and 7 days at 37°C in a humidified atmosphere of 5% CO_2 in air.

2. Wipe microscope slides with 70% ethanol and air dry. Boil 1% agarose and dilute 1:10 in water (0.1% final concentration). Apply a thin film of 0.1% agarose to cleaned slides by quickly dipping the slides and air drying. At 5, 6, and 7 days of culture, centrifuge plates at 300 *g* for 10 minutes at 4°C. Carefully remove and save 0.15 ml supernatant fluid from each well. These culture fluids can be further tested in enzyme-linked immunosorbent assays (ELISA) or other serologic assays for antibodies to various antigens not related to mycoplasma antigens.

3. Discard the remaining fluids by inverting the plate on a sterile tissue placed in a culture hood. Add 0.2 ml of serum-supplemented medium to each well. Centrifuge again and discard supernatants. Add 0.2 ml of serum-supplemented medium and resuspend cells in wells by agitation.

4. Mix equal volumes of 1% agarose with 2× basal Eagle's medium in a 50-ml centrifuge tube. Dispense 0.4 ml of 0.5% agarose into glass tubes prewarmed to 42°C in a water bath. Add 1 drop (about 50 μl) of a 10% suspension of SRBC or 1 drop of a 20% suspension of TNP-SRBC in serum supplemented DMEM to

each tube (Rittenberg and Pratt, 1969). Add 0.1 ml of lymphoid cultures from the culture plates to each tube. Vortex tubes and immediately pour and spread suspensions containing agarose, lymphoid cells, and red blood cells onto agarose-precoated slides. Allow to harden. Place slides in glass petri dishes containing a damp tissue at their bottom. Four slides are placed in a 15-cm petri dish. Incubate for 90 minutes at 37°C in a CO_2 incubator.

5. Dilute 50% guinea pig serum 1:10 with cold D-PBS. Remove the damp tissue from the petri dishes and add 8 ml (2 ml/slide) of freshly diluted 5% guinea pig serum. Incubate for 90 minutes at 37°C in a CO_2 incubator. Remove slides and count plaques under 10× magnification of a dissecting microscope (Rittenberg and Pratt, 1969; Biberfeld and Gronowicz, 1976).

Discussion

A number of experimental methods can be employed to study the polyclonal proliferation of lymphocytes induced by mycoplasmal mitogens. Enumerating cells before and after exposure to mitogen, counting the number of lymphoblasts developed in cultures, and determinating DNA, RNA, and protein synthesis by measuring the uptake of radiolabeled precursors are all adequate means to assess proliferative responses (Ginsburg and Nicolet, 1973; Biberfeld and Gronowicz, 1976; Naot and Ginsburg, 1978).

This section describes the most commonly used method of measuring [^{3}H]thymidine uptake into DNA in cultures of lymphocytes, in the presence or absence of mycoplasmal mitogens. It should be noted, however, that despite being effective and useful, the method described here provides a rapid estimate of the overall DNA synthesis in an entire lymphoid cell population containing B and T lymphocytes, large granular lymphocytes, and monocytes/macrophages.

To examine the potential of mycoplasmal mitogens to activate purified populations of B cells, whole lymphoid cell suspensions can be further fractionated using methods for the enrichment of B lymphocytes. Positive selection by panning or negative selection by cytotoxic antibodies and complement makes use of commercially available specific antibodies to different cell populations. In addition, nude athymic mice or thymectomized, bone marrow-reconstituted animals are also used to study the responses of cultured B lymphocytes to mitogenic mycoplasmas (Biberfeld and Gronowicz, 1976; Biberfeld and Nilsson, 1978; Naot and Ginsburg, 1978; Naot et al., 1979).

We have chosen to present here yet another approach used to investigate the polyclonal activation of B cells by mycoplasmas. This experimental approach measures the differentiation of stimulated B lymphocytes into antibody-producing plasma cells. Examination of this aspect is in fact essential since certain mycoplasmas may activate polyclonal antibody production (as measured by increased number of plaque-forming cells) without an apparent increase in DNA synthesis (Biberfeld and Nilsson, 1978).

A key factor in studies of mycoplasmal-induced mitogenic stimulation of lymphocytes is the ability to differentiate between specific anamnestic immune response and polyclonal activation. This issue must, therefore, be addressed when investigating the mitogenic potential of mycoplasma species commonly isolated from the lymphocyte donor. The medical history of healthy human donors should exclude any past infection with mycoplasmas and blood samples must yield negative antimycoplasmal serologic results. Moreover, since seronegativity cannot be used as an absolute criterion that excludes possible past infections in studies of human lymphocytes, cells from 6 to 10 different human individuals must be used before concluding that a certain mycoplasma is indeed mitogenic (Biberfeld and Gronowicz, 1976; Biberfeld and Nilsson, 1978).

To obtain conclusive differentiation between antigenic versus mitogenic stimulation of rat and mouse lymphocytes, studies with unprimed germ-free animals are advocated. To ascertain that an animal colony is actually germ free, animals are tested by cultural isolation, serologic assays, and histological staining for mycoplasmas (Naot and Ginsburg, 1978).

A major pitfall of *in vitro* studies with eukaryotic cells and cell lines is their possible contamination with mycoplasmas, which leads to erroneous results (see Section F, Vol. II). Sera used in tissue cultures, the most common source for contamination, should be pretested to be mycoplasma free and should be able to support lymphocyte viability and proliferation. Fetal calf serum is often used to study the proliferation of lymphocytes from mice, rats, and human donors. However, horse serum and pooled human AB-positive serum are as satisfactory and less expensive for usage with rat and human cell cultures, respectively.

In preliminary studies of mycoplasmal mitogenicity, nonviable, lysed cells are the appropriate source for crude mycoplasmal mitogens. Live replicating organisms may deplete the culture medium of essential nutrients and may induce cytopathic and/or cytotoxic effects on eukaryotic cells (Ginsburg and Nicolet, 1973; Naot and Ginsburg, 1978; Cole *et al.*, 1985). Following initial demonstration of a mitogenic capacity, further studies with purified mycoplasmal cell components can be performed. Heat treatment or sonication of mycoplasmal preparations should be avoided in view of the possible involvement of heat-labile proteins in the mitogenic activity of the tested mycoplasma species (Cole *et al.*, 1985; Ruuth and Praz, 1989). When applying the methods described in this section, a multitude of technical variables may affect the results obtained. To maintain the viability of lymphoid cell suspensions, pelleted cells are best kept on crushed ice. The yield of mononuclear cells from human peripheral blood is $1-2 \times 10^6$ cells/ml blood; 60–70% are lymphocytes with ≥95% viability. From spleens of rat and mouse, 300 and 100×10^6 cells are obtained, respectively, with viability of ≥95%. Since the ratio of B lymphocytes to T lymphocytes is higher in the spleen than in blood or lymph nodes, this organ should be preferred, when possible, as a source of lymphocytes in studies of B-cell mitogens.

The division of B lymphocytes and their differentiation into plasma cells are both affected by the concentration of lymphoid cells in cultures ($1-5 \times 10^5$ cells/well), by the dose of the mitogen (0.1–10 μg/well), and by the length of culture incubation periods. Two, 3, and 4 days of incubation are usually applied in [^{3}H]thymidine incorporation assays, whereas 4- to 8-day cultures should be included in the preliminary experiments of hemolytic plaque-forming cells. The geometry of the well in tissue culture plate affects the interactions of lymphoid cells and mitogens. Temperature, humidity, and CO_2 levels are also variables that influence the viability of cells and the magnitude of response and should therefore be monitored.

Under optimal conditions, the mitogenic stimulation of B cells results in a 3- to 10-fold increase in cell numbers during 3–5 days of culture.

Thymidine incorporation assays often show a background incorporation of 0.2–2 and $5-100 \times 10^3$ cpm in stimulated cultures. The extent of the mitogenic response of lymphocytes is usually presented either directly as experimental minus background cpm or expressed as a stimulation index (SI), which is obtained by dividing experimental cpm by background cpm.

The expected range of hemolytic plaque-forming cells detected using SRBC or TNP-SRBC as indicators is 10–200 PFC/culture in unstimulated cultures, whereas mitogen-stimulated cultures are expected to yield 200–5000 PFC/culture following 4–5 days of incubation of lymphoid cell cultures.

It should be noted that because of the ability of immunoglobulin M (IgM) antibodies to fix complement, the direct plaque-forming hemolytic assay measures mainly the production of IgM antibodies by plasma cells. To determine the secretion of IgG and IgA antibodies, rabbit antibodies specific to these immunoglobulins can be added to the hemolytic assay prior to the addition of complement.

Acknowledgment

The development of certain procedures was supported by Grant No. 864-305 from the Colleck Fund for Research.

References

Biberfeld, G., and Gronowicz, E. (1976). *Mycoplasma pneumoniae* is a polyclonal B-cell activator. *Nature (London)* **261,** 238–239.

Biberfeld, G., and Nilsson, E. (1978). Mitogenicity of *Mycoplasma fermentans* for human lymphocytes. *Infect. Immun.* **21,** 48–54.

Cole, B. C., Naot, Y., Stanbridge, E. J., and Wise, K. S. (1985). Interactions of mycoplasmas and their products with lymphoid cells *in vitro*. *In* "The Mycoplasmas: Mycoplasma Pathogenicity" (S. Razin and M. F. Barile, eds.), Vol. 4, pp. 203–257. Academic Press, New York.

Ginsburg, H., and Nicolet, J. (1973). Extensive transformation of lymphocytes by a mycoplasma organism. *Nature (London) New Biol.* **246,** 143–146.

Naot, Y., and Ginsburg, H. (1978). Activation of B lymphocytes by mycoplasma mitogen(s). *Immunology* **34,** 715–720.

Naot, Y., Merchav, S., Ben-David, E., and Ginsburg, H. (1979). Mitogenic activity of *Mycoplasma pulmonis*. I. Stimulation of rat B and T lymphocytes. *Immunology* **36,** 399–406.

Rittenberg, M. B., and Pratt, K. L. (1969). Antitrinitrophenyl (TNP) plaque assay: Primary response of Balb/c mice to soluble and particulate immunogen. *Proc. Soc. Exp. Biol. Med.* **132,** 575–581.

Ruuth, E., and Praz, F. (1989). Interactions between mycoplasmas and the immune system. *Immunol. Rev.* **112,** 133–160.

F8

MODULATION OF EXPRESSION OF MAJOR HISTOCOMPATIBILITY COMPLEX MOLECULES BY MYCOPLASMAS

P. Michael Stuart and Jerold G. Woodward

Introduction

Mycoplasma organisms have been shown to be pathogenic for a wide variety of plant and animal species, including humans (Maniloff et al., 1992). These organisms are typically found in close association with eukaryotic cells because of their ability to adhere and colonize the membrane of these cells (Razin and Jacobs, 1992). In addition, the mycoplasmas have diverse and multiple effects on immunocompetent cells, including T and B lymphocytes, natural killer cells, and macrophages (Cole et al., 1985; Stuart et al., 1989, 1990a). Our laboratory has shown that at least six different mycoplasma species have the ability to produce a factor, which we have termed MlaF, that induces major histocompatibility complex (MHC) class I and class II expression on a variety of cell types. These cells include primary murine bone marrow macrophage cultures (Stuart et al., 1989), cultures of several different murine and human macrophage cell lines (Stuart et al., 1989, 1993), and at least one murine glioma cell line, G26-20 (Stuart et al., 1993). Following induction with MlaF, we have also observed a concomitant increase in the ability of these cells to present antigen to MHC class II-restricted T cells (Stuart et al., 1990b), indicating that increases in MHC expression also increases the antigen presentation function of these cells. The mechanism whereby MlaF induces MHC class II expression is not known. However, we have reported that it does not appear to act indirectly by first inducing the production of interferon-γ (IFN-γ), interleukin-4 (IL-4), or granulocyte monocyte colony-stimulating factor (GM-CSF) (Stuart, 1993).

Materials

Mycoplasma Cells

We have demonstrated MlaF activity in *Mycoplasma arginini, M. arthritidis, M. pulmonis, M. fermentans* incognitus strain, *M. pneumoniae,* and *M. hominis.* As a consequence, any of these species can be used; however, it is easier to use one of the first three species because they either are easier to grow than the *M. fermentans* incognitus strain or do not produce factors that have deleterious side effects on the indicator cell line (personal observations). In order to grow *M. arginini, M. arthritidis,* and *M. pulmonis,* a modified Hayflick's medium is used. This medium consists of *Mycoplasma* broth (BBL 11458), 1% phenol red (Sigma P5530), 0.5% dextrose (Baker 1916-01), and 5–17% heat-inactivated fetal bovine serum. The pH of the medium is adjusted to 7.5. For growth of *M. arginini* and *M. arthritidis,* the medium is supplemented with 2 g/liter of arginine monohydrochloride (Baker B577-05). In addition to media, culturing bottles (Schott bottles, 500 or 1000 ml), centrifuge tubes, or bottles that are designed for use in a superspeed centrifuge (up to 20,000 rpm) are required. When performing ammonium sulfate precipitation of culture supernatants (CSN), material is mixed in Erlenmeyer flasks whose size is dependent on the volume of CSN to be precipitated. Dialysis tubing (6000 to 8000-Da pore size) and Erlenmeyer flasks for dialyzing the ammonium sulfate precipitate are also required. The amount of tubing required is determined by the volume of mycoplasma to be grown. For lysis of the mycoplasma cells, we use either a sonicator or a French press. In order to determine the protein concentration of samples, some sort of protein determination system has to be adopted; we use the BCA system (23225) of Pierce Chemical Co. (Rockford, IL).

Mammalian Cells

It has been demonstrated that murine cultures of WEHI-3 and G26-20, as well as human cultures of U-937 and HL-60, are responsive to MlaF (Stuart *et al.,* 1989, 1993). One could also use primary murine bone marrow-derived macrophages; however, the isolation of these cells is somewhat time consuming. Most of these cell lines can be obtained from the American Type Culture Collection (Rockville, MD). The cell lines are grown in RPMI 1640 medium with sodium bicarbonate, pH 7.2 (we purchase this from Atlanta Biologicals D 30210), which is supplemented with 10% fetal calf serum (FCS), 10 mM L-glutamine, 10 mM HEPES, and 5×10^{-5} M 2-mercaptoethanol (antibiotics are optional). T-25 or T-75 tissue culture flasks are used, depending on the volume of cells to be cultured. For harvesting cells we use sterile conical centrifuge tubes (either 15-

or 50-ml volume) that fit in a swinging bucket centrifuge (0–5000 rpm). For washing of cells we use phosphate-buffered saline, pH 7.2–7.4 (PBS).

Antibodies and Flow Cytometry

The monoclonal antibodies (MAbs) used to detect increases in murine MHC class II antigen expression are MKD6 (anti-I-A^d) and 14-4-4 (anti-I-E^d) for WEHI-3 cells and 34-5-3 (anti-I-A^b) for G26-20. For detection of human MHC class II antigen expression, use L234 (anti-HLA-DR) or 9.3 F10 (anti-HLA-DR,DQ). MHC class I expression is determined by using 34-1-2 (murine) or W6/32 (human) MAbs. (The hybridomas that produce these antibodies can be obtained from American Type Culture Collection.) As negative controls, use an irrelevant MAb that does not react with murine or human determinants. The positive control consists of cells induced with recombinant interferon γ (rIFN-γ) (Life Technologies, 3284). For flow cytometry, these MAbs are counter stained with FITC-labeled rabbit anti-mouse IgG (Cappel, 55497). For washing of cell samples, use PBS, supplemented with 1% FCS and 0.1% sodium azide (PBS–FCS–azide). For fixing the samples, prepare a 2% solution of paraformaldehyde (Sigma, P6148).

RNA Extraction

Guanidine thiocyanate solution (100 ml) consists of 60 g guanidine thiocyanate (Sigma, G6639), 0.5 g N-lauroylsarcosine (Sigma, L5125), and 5 ml of 1 M LiCl (Aldrich, 21320-9). The CsCl solution consists of 96 g CsCl (Fisher, BP210-500) in 100 ml, 0.1 M EDTA, pH 7.0. Other reagents include 5 M NaCl, 2.5 M sodium acetate, and 95% ethanol.

Northern Blot Hybridization and Probes

To make 500 ml of 10× gel buffer, add 23.13 g MOPS (Sigma, M9381) + 10 ml 2.5 M sodium acetate + 10 ml 0.5 M EDTA and adjust to pH 7.0. The gel consists of 1% agarose (FMC BioProducts, 50101) dissolved in 1× gel buffer containing 2.2 M formaldehyde. The denaturation buffer consists of 60 μl 10× gel buffer + 100 μl 37% formaldehyde (Baxter, 2106-01) + 300 μl formamide (Baxter, 4028-01). To make 1 liter of 20× SSC, mix 175.3 g NaCl + 88.23 g sodium citrate in H$_2$O. The probes we use for Northern blot analysis have been previously described (Stuart et al., 1989) and consist of murine MHC class I gene, pH2IIa (which can be acquired from Dr. Lee Hood, University of Washington, Seattle, WA); MHC class II genes, pEAC11 (E α), pAβ 1, and pAAC6 (Aα) (which can be acquired from Dr. Christophe Benoist, Institut de Chimie Biologique, Strasbourg, France); and human MHC class I gene, MHC class II genes (which can be acquired from Dr. Sherman Weisman, Yale University, New Haven, CT).

Equipment

Two 37°C incubators are required; one to grow the mycoplasmas and one with a mixed CO_2 atmosphere for growing mammalian cells. A tissue culture hood is needed for handling mycoplasma and mammalian cell cultures. (Note: If the same hood is used for the manipulation of both mycoplasmal and mammalian cultures, be careful to clean the hood completely after using it for mycoplasma cultivation.) For flow cytometry, any type of flow cytometric analyzer is adequate; we use a FACScan (Becton–Dickenson, Palo Alto, CA). For RNA analysis, a submarine gel apparatus and a power supply are required as well as film cassettes and some form of X-ray film developing system.

Procedure

Growth of Mycoplasmas

Mycoplasmas are grown in a modified Hayflick's medium until the color of the medium turns orange (*M. pulmonis*) or until it becomes turbid (*M. arginini, M. arthritidis*). The cultures are then harvested by centrifugation at 8000–10,000 g for 20 minutes at 4°C. Save the CSN and the cells because the MlaF activity is associated with both fractions. The CSN can be concentrated by doing an ammonium sulfate precipitation at 80% saturation. Mycoplasma membranes are prepared by washing the cell pellet once in PBS and then resuspending it in distilled H_2O. This suspension is then either sonicated five times at 80% of maximal setting on ice or put through a French press three times. The resulting preparation is then centrifuged at 7000 g for 20 minutes at 4°C to remove intact cells. The supernatant is centrifuged again at 25,000 g to collect the membrane fraction. The CSN and membrane preparations are both filter sterilized (0.1-μm filters) and irradiated (10,000 rad) to ensure inactivation of the mycoplasmas. Protein determinations are performed on all samples. Samples are aliquoted and stored at $-80°C$ until tested for MlaF activity in the assay described next by titration analysis.

Induction Conditions and Fluorescence Assays

The initial mammalian cell concentration for these induction assays depends on several factors. In general, cultures should be initiated so that there are at least

4×10^6 cells for flow cytometric analysis or $10-20 \times 10^6$ cells for RNA extraction at the time of harvesting. Consequently, factors such as time of culturing, growth characteristics of the cells to be induced, and effects of the inducing agent on the growth characteristics of cells must be taken into consideration. Once these parameters are determined, grow the cells in the presence of MlaF (amount determined by titration analysis), IFN-γ (100 U/ml) (positive control), or uninoculated mycoplasma medium (negative control). Typically, if flow cytometric analysis is performed, incubate these cultures for 3–4 days. If RNA analysis is performed, incubate these cultures for 2–3 days. Previous studies have shown that these incubation times result in the optimal induction of MHC expression (Stuart *et al.*, 1989, 1993). Following incubation, the cells are harvested by centrifugation, washed once in PBS, and counted. For flow cytometric analysis, 1×10^6 cells are aliquoted to individual tubes and washed once with PBS–FCS–azide. Samples are then stained with MAbs by adding excess anti-MHC class I, anti-MHC class II MAb, or an irrelevant MAb (negative control) to separate individual tubes. The cells are then incubated for 45 minutes on ice and are washed twice to ensure removal of the primary antibody. If the primary antibody is already conjugated with a particular fluorochrome, the pellet should be resuspended in 1 ml of PBS–FCS–azide and analyzed in a flow cytometer. If the primary MAb is not conjugated with a fluorochrome, then a secondary fluorochrome-labeled anti-mouse IgG antibody is required. Incubation conditions are the same as described for the primary MAb. If the samples cannot be analyzed on the same day, then fix the cells with 1% paraformaldehyde by adding equal volumes of PBS–FCS–azide (first to resuspend the cell pellet) and 2% paraformaldehyde (to fix the cells). Once fixed, the samples are stable in the dark at 4°C for at least 2 weeks.

RNA Extraction and Northern Blot Analysis

Total cellular RNA is extracted from cells by the guanidine thiocyanate procedure (Chirgwin *et al.*, 1979). Briefly, cell pellets are resuspended in disruption buffer (guanidine thiocyante, LiCl) by forcing the solution through a syringe fitted with an 18-gauge needle. The lysate is layered on CsCl and is centrifuged at 36,000 rpm for 16 hours using an SW55Ti rotor in order to pellet the RNA. The RNA is resuspended in RNase-free water, precipitated by ethanol, and quantitated by reading the optical density at 260 and 280 nm. For Northern blot analysis, 20 μg of RNA is electrophoresed on a 1% agarose gel containing 2.2 M formaldehyde, and is then transferred to nitrocellulose or nylon filters with $20\times$ SSC by blotting overnight. The filters are prehybridized at 42°C overnight and are then hybridized to 2×10^6 cpm/ml of ^{32}P-labeled probe in 10 ml of hybridization buffer for 24–48 hours at 42°C. Blots are then washed four times in $2\times$ SSC with 0.1% SDS at 25°C, followed by three washes with $0.1\times$ SSC and

0.1% SDS at 55°C. The blots are dried, subjected to autoradiography, and quantified by densitometry. Each blot can be sequentially hybridized with probes for MHC class I, MHC class II, and actin genes. Between each hybridization, the preceding probe is removed by boiling the blot in $0.1 \times$ SSC and 0.1% SDS for 5 minutes at least twice.

Discussion

The ability of mycoplasmas to produce MlaF does not seem to be absolutely dependent on the type of media that they are grown in. We have grown mycoplasmas in both SP4 and Hayflick's media and both support the production of this factor. We have also reduced the serum concentration in the culture medium to as low as 5% and MlaF was detectable (Stuart et al., 1993). It is therefore our belief that as long as the mycoplasmas can grow, they will produce MlaF. However, it should be pointed out that while most preparations of MlaF that we have produced have some measurable MlaF activity, the amount of MlaF detected varies. Therefore, a titration of each preparation is required in order to determine the precise amount of MlaF activity that a particular preparation contains.

The storage conditions of samples to be tested for MlaF activity are highly critical. These samples must be stored at $-70°C$ in order for activity to be maintained. We have stored samples at 4 and $-20°C$ and the samples do not remain active beyond a week under these conditions.

Two very important considerations need to be kept in mind concerning the indicator cell lines used. The first regards the inherent instability of the cell lines in terms of their ability to respond to either MlaF or IFN-γ. We have found that some lines of WEHI do not respond to these factors. Additionally, we have observed that a previously responsive line of WEHI becomes unresponsive with continuous culturing. Another important consideration is to make sure that the indicator line does not become contaminated with mycoplasmas. Should that happen, the cell line must be either discarded and new cells obtained or treated with antibiotics to eliminate the mycoplasma contaminants. Consequently, aliquots of those cells known to respond to MlaF or IFN-γ should be frozen so that there are backup cells available in case the line loses responsiveness or becomes contaminated.

References

Cole, B. C., Naot, Y., Stanbridge, E. J., and Wise, K. S. (1985). Interactions of mycoplasmas and their products with lymphoid cells *in vitro*. *In* "The Mycoplasmas" (S. Razin and M. F. Barile, eds.), Vol. IV, pp. 203–257. Academic Press, New York.

Chirgwin, J. M., Pazbyla, A. E., MacDonald, R. J., and Rutter, W. J. (1979). Isolation of biologically active ribonucleic acid from sources enriched in ribonuclease. *Biochemistry* **18,** 5294–5299.

Maniloff, J., McElhaney, R. N., Finch, L. R., and Baseman, J. B., eds. (1992). "Mycoplasmas, Molecular Biology and Pathogenesis." Am. Soc. Microbiol., Washington, DC.

Razin, S., and Jacobs, E. (1992). Mycoplasma adhesion. *J. Gen. Microbiol.* **138,** 407–422.

Stuart, P. M. (1993). Mycoplasma induction of cytokine production and major histocompatibility complex expression. *Clin. Infect. Dis.* **17,** S187–191.

Stuart, P. M., Cassell, G. H., and Woodward, J. G. (1989). Induction of class II MHC antigen expression in macrophages by Mycoplasma species. *J. Immunol.* **142,** 3392–3399.

Stuart, P. M., Cassell, G. H., and Woodward, J. G. (1990a). Differential induction of primary bone marrow derived macrophages by mycoplasma involves the production of GM-CSF. *Infect. Immun.* **58,** 3558–3563.

Stuart, P. M., Cassell, G. H., and Woodward, J. G. (1990b). Mycoplasma induction of class II MHC antigens, possible role in autoimmunity. *In* "Recent Advances in Mycoplasmology" (G. Stanek, G. H. Cassell, J. G. Tully, and R. F. Whitcomb, eds.), pp. 570–577. Gustav Fisher Verlag, Stuttgart.

Stuart, P. M., Egan, R. M., and Woodward, J. G. (1993). Characterization of MHC induction by *Mycoplasma fermentans* (incognitus strain). *Cell. Immunol.* **152,** 261–270.

F9
INTERACTION OF MYCOPLASMAS WITH NATURAL KILLER CELLS
Wayne C. Lai and Michael Bennett

Introduction

Natural killer (NK) cells have been identified in virtually every mammalian species and also in invertebrates. NK cells are a subpopulation of lymphoid cells which spontaneously lyse a limited number of tumor cells and virus-infected cells *in vitro,* without prior sensitization of the host. NK cells, unlike T and B cells, are independent of thymus, bursa, and gut-associated lymphoid tissues, and do not have rearrangement of antigen receptor genes.

Several viruses, bacteria, a few fungi, and protozoa are known to augment the NK cell system in mice. *Mycoplasma pulmonis* causes one of the most common diseases of laboratory rats and mice, and also stimulates NK cell function (Lai *et al.,* 1987). Interferon (IFN) and interleukin-2 (IL-2) are also potent inducers of the system in mice. Therefore, *M. pulmonis* may augment NK cells in mice by inducing the secretion of interferon and/or interleukin-2.

If NK cells pay a role in *M. pulmonis* infections, several predictions should or may be realized: (1) *M. pulmonis* infection should stimulate NK cell function; (2) the stimulated NK cells should have anti-*M. pulmonis* activity; (3) animals depleted of T and B cells (such as SCID mice) should establish some, albeit reduced, resistance to infection; (4) depletion of NK cells or their products (e.g.,

IFNγ) should reduce the resistance of mice, especially SCID mice, to *M. pulmonis* infection; and (5) NK cells may directly lyse or inhibit the growth of *M. pulmonis* and/or may act indirectly by activating macrophages with IFNγ.

Materials

Specific-pathogen-free, mycoplasma-free C57BL/6 and C.B17 scid/scid (SCID) mice
M. pulmonis UAB or CT (T2) strains
Poly(I:C) = polyinosinic:polycytidylic acid (Sigma Chemical Co., St. Louis, MO)
YAC-1 lymphoma cells
Chalquest's agar medium (Olson *et al.*, 1963)
Chromium-51 ($Na_2{}^{51}CrO_4$) (New England Nuclear, Boston, MA)
Nylon wool columns
Rabbit antiasialo-G_{M1} serum (Wako Chemical Co., Dallas, TX)
Mouse anti-Thy-1.2 monoclonal antibodies (MAb) (New England Nuclear)
Rabbit complement (C) (Cedarlane Laboratories Limited, Hicksville, NY)
Fetal calf serum (FCS)
RPMI 1640 medium (GIBCO Laboratories, Grand Island, NY)
Lysing agent ZAP-OGLOBIN II (Coulter Diagnosis, Coulter Electronics, Inc., Hialeah, FL)
Microtiter U-bottom plates (Costar, Cambridge, MA)
Benchtop centrifuge
Gamma counter (TM Analytic, Inc., Elk Grove Village, IL)
5-Fluorodeoxyuridine (FUdR) (Sigma Chemical Co.)
[^{125}I]iododeoxyuridine ([^{125}I]UdR) (ICN Radio Chemicals, Irvine, CA)
Cyclophosphamide
PK136 anti-NK 1.1 MAb (Dr. Bennett, Dallas, TX)
HB-170 anti-Thy-1 MAb (New England Nuclear)
3A4 IgM anti-NK MAb (Dr. Bennett, Dallas, TX)
Hemacytometer
Recombinant human interleukin-2 (rhIL-2) (Amgen Biologicals, Thousand Oaks, CA)
WEHI-279 tumor cells
FACS cell sorter (Becton–Dickinson, Mountain View, CA)
B16F10 melanoma cells

Augmentation of NK Cell Activity in Mice by *M. pulmonis* Infection

Procedure

ANIMALS

Four- to 6-week-old specific-pathogen-free C57BL/6J mice are used. These mice were selected because NK cells are moderately good at lysing YAC-1 cells and because of the extensive use of these mice in immunologic research. Mice should be free of serum antibodies to murine viruses, free of *M. pulmonis* by culture of oropharyngeal samples, and free of *M. pulmonis* and *M. arthritidis* antibodies as determined by the enzyme-linked immunosorbent assay (ELISA). All test mice should be delivered in filtered shipping flow hoods and held throughout the experiments in laminar flow hoods in rooms completely separate from all other animal facilities. They should be housed in sterilized polycarbonate cages on sterile hardware bedding and should be provided with a sterile commercial diet and water *ad libitum*. The animal room is maintained at 20–22°C with a 12-hour light–dark cycle. Mice should be held for 1 week prior to initiating experiments. At the end of these experiments, control mice should remain free of the just named pathogens, applying the same methods of detection. *M. pulmonis* should be identified only in experimentally infected animals. Mice are anesthetized with ether and exsanguinated prior to necropsy.

ANIMAL INOCULATION

M. pulmonis strain UAB-6510 is used as the inoculum. This isolate, recovered originally from rats, causes lesions in the trachea, lungs, tympanic bullae, and nasopharynx of mice. Seventy-two mice are divided into three groups of 24 mice per group. Group 1 mice are injected intraperitoneally (IP) with a 4×10^8 colony-forming unit (CFU) *M. pulmonis* in 0.2 ml phosphate-buffered saline (PBS); group 2 mice are injected IP with 100 μg polyinosinic–polycytidilic acid [poly(I:C)] 1 day prior to testing; and group 3 mice are injected IP with 0.2 ml PBS 1 day prior to testing. Three mice from each group are tested for splenic NK cell activity on days 1,3,5,7,10,14,21, and 28 postinjection of *M. pulmonis*.

CONFIRMATION OF INFECTION

Peritoneal swabs and tracheolung and nasal washings are collected from each mouse at the time of necropsy to determine if *M. pulmonis* is present in inoculated animals and absent in uninoculated animals. Specimens are enriched in PPLO broth, inoculated onto modified Chalquest's agar medium, and incubated at 37°C and 95% humidity. Plates are examined daily for colony formation for a maximum of 14 days before being considered negative. Suspected *M. pulmonis* colonies are identified by the growth inhibition test.

TARGET TUMOR CELLS

YAC-1 lymphoma cells are used as target cells in the NK assay. The tissue cultures are checked to be free of mycoplasma contamination (see Section F in Vol. II). The tumor cells are cultured in complete medium composed of RPMI 1640 medium supplemented with 25 mM *N*-2-hydroxyethylpiperazine-*N*-2-ethanesulfonic acid buffer, 10% heat-inactivated FCS, 100 U/ml penicillin, and 100 μg/ml streptomycin. Cells are transferred into fresh, complete medium 24 hours before use in chromium-51 release assays.

NK CELL ASSAYS

The methodology for the NK cell cytotoxicity assay is an adaptation of a procedure described previously (Herberman *et al.*, 1975). YAC-1 target cells are labeled by incubating 10×10^6 cells/ml RPMI 1640 with 200 μCi of $Na_2{}^{51}CrO_4$ at 37°C for 90 minutes in an atmosphere containing 5% CO_2. Washed target cells are counted by trypan blue exclusion and are adjusted to 2×10^5 cells/ml with RPMI 1640 plus 5% FCS. Spleen effector cells are added to labeled target cells in microtiter plates to achieve effector:target (E:T) ratios of 50:1, 25:1, and 12.5:1. After 4 hours of incubation, plates are spun gently and the supernatants harvested from each well are counted in a gamma counter. The percentage specific lysis is calculated using the formula

$$^{51}Cr\ (cpm) = [(test - spontaneous\ release)/(maximum\ release - spontaneous\ release)] \times 100.$$

Maximum release of ^{51}Cr is determined by lysing tumor cells with ZAP-OGLOBIN II.

PREPARATION OF EFFECTOR SPLEEN CELLS

Spleen cell suspensions are prepared by gently squeezing spleens between the frosted ends of two sterile glass slides. The resulting cells and debris are suspended in 20 ml of sterile Hanks' balanced salt solution (HBSS), and the debris is allowed to settle for 5 minutes at 4°C. The cell suspension is centrifuged at 1000 g for 10 minutes and washed in HBSS. Cell counts are adjusted to 1×10^7 cells/ml with RPMI 1640 containing 5% FCS.

MACROPHAGE, B-CELL, AND T-CELL DEPLETION FROM SPLEEN CELLS

Purified NK cell preparations are prepared to test the specificity of the spleen cell-mediated YAC-1 cytolysis. Erythrocytes are removed by treating spleen cell suspensions with NH_4Cl (Nabavi *et al.*, 1985). Adherent macrophages and B cells are selectively depleted from spleen cell suspensions by nylon wool column fractionation. Cells are washed in complete medium, added to nylon wool columns, and incubated at 37°C for 1 hour. Nonadherent cells are eluted with warm (37°C) complete medium, washed twice, resuspended in complete medium, and used as effector cells in NK cell assays.

Nylon wool-treated spleen cells are further selectively depleted of T cells using a mouse monoclonal IgM anti-Thy-1.2 antibody and baby rabbit serum complement (C) diluted 1:8 with PBS. Treated cells are washed in HBSS, resuspended in complete medium, adjusted to 1×10^7/ml, and used as effector cells in the NK cell assays.

SELECTED DEPLETION OF NK CELLS FROM SPLEEN CELLS

Rabbit antiasialo-G_{M1} serum and C are used to deplete NK cells from nylon wool-treated spleen cell suspensions. Briefly, 2×10^7 treated spleen cells in 1 ml HBSS are incubated with 10 μl of rabbit antiasialo-G_{M1} in small centrifuge tubes at 4°C for 45 minutes. The tubes are centrifuged at 800 g for 5 minutes and the supernatant is removed. The spleen cells are suspended in 1 ml of complete medium and 1 ml of 1:8 baby rabbit complement and incubated for 45 minutes at 37°C. Cells are washed twice with complete medium, adjusted to 1×10^7/ml, and assayed for NK cell function.

NK CELL SORTING

Effector spleen cell suspensions prepared as described in previous sections are passaged over nylon wool columns and the nonadherent cells are fractionated by Percoll density gradient centrifugation. The low density cells ($\delta = 1.072$) are incubated at 4°C with the anti-NK-1.1 (1/20 dilution) followed by a fluorescein isothiocyanate-conjugated $F(ab)_{12}$ goat anti-mouse reagent (1/20). The stained cells are then analyzed and sorted into positive and negative populations using a FACS II flow cytometer (Becton–Dickinson, Mountain View, CA). By postsort analysis, the NK-1.1+ cells should be >95% pure and are used in a 4-hour ^{51}Cr-release assay.

Discussion

M. pulmonis infection augments NK cell activity, peaking at Day 3 postinoculation (PI), and gradually returning to normal levels by Day 14 PI based on a 4-hour ^{51}Cr-release assay using YAC-1 tumor target cells (Lai et al., 1987). Selective cell depletion experiments showed clearly that the cells mediating cytolysis of YAC-1 tumor cells had characteristics of murine NK cells. Activated macrophages and monocytes do not appear to be involved in the cytolysis of YAC-1 tumor cells because removal of these cells from spleen cell suspensions by nylon wool columns did not reduce the cytolytic effect. The involvement of cytotoxic T lymphocytes was ruled out because treatment with anti-Thy-1.2 plus complement did not alter YAC-1 cell cytolysis. The exposure of spleen cell suspensions to antiasialo-G_{M1} ganglioside plus complement significantly abrogated spleen cell-mediated cytolysis. Natural killer cells are characterized by the presence of asialo-G_{M1} ganglioside determinants on their surfaces. Although it has been demonstrated previously that at least some NK cells express Thy-1, we were unable to detect a significant reduction in cytotoxic activity following anti-Thy-1.2 plus complement treatment. It is conceivable that the readjustment of viable cell numbers after treatment with anti-Thy-1.2 plus complement masks a reduction in cytotoxic activity. Selective cell depletion was further demonstrated in 4-hour ^{51}Cr-release assays using NK cells highly purified by FACS cell sorting.

Inhibition of *M. pulmonis* and Tumor Growth and Antitumor Metastatic Activity by Augmented NK Cells

PULMONARY CLEARANCE ASSAY

A modification of methodology described previously (Riccardi *et al.*, 1978) is used. ^{51}Cr-labeled YAC-1 tumor cells (2×10^6) are injected via the tail vein into panels of five infected and control mice. The animals are killed humanely 2.5 hours later, the lungs are removed, and the radioactivity is measured. NK activity is inversely correlated with lung radioactivity. The percentage of retention of labeled cells is calculated by: (radioactivity in lungs at 2.5 hours/radioactivity in inoculum) $\times$ 100.

PULMONARY TUMOR COLONY ASSAY

An additional method can be used to detect NK cell activity in the lungs of infected and control mice. The lung colony assay is an accurate method for evaluating resistance to metastases of B16F10 melanomas, a property of NK cells. The B16F10 melanoma cells are maintained as monolayer cultures free of mycoplasma contamination. Mice are inoculated IP with 5×10^4 viable B16F10 melanoma cells. The number of tumor foci in unsectioned whole lungs is determined using a magnifying glass 21 days after melanoma cell challenge.

IN VITRO GROWTH INHIBITION ASSAY

Spleen and lung cell suspensions are prepared as described earlier; adjusted to 8×10^7, 8×10^6, or 8×10^5 cells/ml in RPMI 1640 medium with 10% FCS; and used as effector cells. The concentration of *M. pulmonis* is adjusted to 4×10^5 CFU/ml in PPLO broth used as target cells. An aliquot (0.1 ml) of the mycoplasma suspension is added to a sterile tube followed by 0.5 ml of the effector cell suspension, at E:T ratios of 10:1, 100:1, and 1000:1. The controls contain 0.1 ml of the *M. pulmonis* suspension and 0.5 ml of culture medium. The tubes are shaken gently in a shaker, incubated at 37°C for 4 hours, titrated on Chalquest's agar medium, and incubated at 37°C for 7 days (triplicate samples). The *M. pulmonis* colonies are counted, and the percentage of growth inhibition is calculated as: $(A - B/A) \times 100$, in which A is the mean number of CFUs grown from the tube with medium and B is the mean number of CFUs grown from the tube containing target and effector cells.

IN VIVO GROWTH INHIBITION ASSAY

M. pulmonis suspensions are centrifuged at 17,000 g for 30 minutes, washed twice with RPMI, and resuspended in RPMI 1640 medium; 0.5 ml of this suspension is injected via the tail vein into each mouse. The mice are killed 0, 6, 12, or 24 hours later; the spleen and lungs are collected and homogenized in 2 ml of PPLO broth. The cell suspensions are serially diluted (1:10), and 1.0 µl of each dilution is plated in triplicate on Chalquest's agar medium. The agar plates are incubated, and colonies are counted as previously described (triplicate samples) (Lai *et al.*, 1986).

PULMONARY CLEARANCE OF *M. pulmonis*

Radiolabeled *M. pulmonis* are used as target cells. Organisms are cultured, harvested as described earlier, washed, and resuspended in 1 ml of RPMI 1640 medium. FUdR (26 mg/100 ml of medium; Sigma) is added (0.1 ml), and the suspension is incubated at 37°C. FUdR can inactivate thymidylate synthase of bacteria. One-half hour later, 30–50 µCi of 5-[^{125}I]iododeoxyuridine ([^{125}I]UdR; ICN Radiochemicals) is added, and the suspension is continuously shaken for 4 hours at 37°C. The labeled organisms are washed twice, resuspended in medium, and infused IV into panels of five infected and five control mice. The radioactivity recovered from the lungs of five control mice at Time 0 is considered as 100% retention. The radioactivity of the lungs removed from experimental and control animals should be counted 3 hours after injection. The mean counts per minute (cpm) are calculated for each group of five mice. The percentage of lung clearance is calculated by: radioactivity in lungs at 3 hours/radioactivity in lungs at Time 0 × 100.

TREATMENT OF MICE AND CELL SUSPENSIONS

Mice are injected with 300 mg/kg cyclophosphamide IP for 3 days, with 50 µl of rabbit antiasialo-G_{M1} serum, or with 0.2 ml of mouse PK136 anti-NK1.1 MAb IV 2 days before the assay to temporarily suppress NK cell function. Polyinosinic:polycytidylic acid (100µg) is injected IP 1 day before the assay to stimulate NK cells. Mice are infected with *M. pulmonis* 1–14 days before assay as we have observed enhanced NK cell function in infected mice during the first few days. Mice are injected with 0.5 ml of affinity-purified HB-170 MAb to IFNγ 2 days before infection. Spleen cells or lung cells are treated with baby rabbit serum at 1:8 dilutions as a source of complement, with C + 1:50 dilution of rabbit antiasialo-G_{M1} serum or with MAb 3A4 reagent to destroy NK cells, or with a monoclonal rat anti-Thy-1.2 reagent that lyses T cells and a portion of NK cells. SCID mice are treated with HB-170 MAb against IFNγ or its vehicle (medium) and are infected with *M. pulmonis*, strain T2 intratracheally, and the survival times are recorded and compared.

MECHANISM OF *M. pulmonis* INHIBITION BY NK CELLS

Spleen cell suspensions from SCID mice (devoid of T and B cells) are grown for 7 days in a RPMI 1640 containing recombinant human IL-2 to generate a large population of pure NK cells (Tutt *et al.*, 1987). Mixtures containing NK cells (1×10^7 cells/well), macrophages (1×10^5 cells/well) from normal C57BL/6 spleens, *M. pulmonis* (4×10^4 CFU/well), and/or anti-IFNγ were incubated at 37°C for 24 hours. An aliquot of the supernatant is titrated for viable *M. pulmonis* on Chalquest's agar medium. To assess the amount of IFNγ produced, the remaining culture supernatants are exposed to UV light to inactivate viable *M. pulmonis* and are then added to IFNγ-sensitive WEHI-279 tumor cells. The amount of IFNγ produced is assessed by the inhibition of DNA synthesis in WEHI-279 tumor cells (Table I).

Statistical analyses of variance tests are used to determine the statistical significance among infected and control plates and can be performed on a DEC-10 computer.

Discussion

The mice infected with *M. pulmonis* stimulate NK cells capable of lysing YAC-1 lymphoma cells *in vitro* and *in vivo* and inhibiting B16F10 melanoma cells from developing lung metastases (Lai *et al.*, 1990). In addition to the well-characterized antitumor functions of NK cells, we observed that NK cells of infected mice could directly inhibit *M. pulmonis* colony formation *in vitro* and could eliminate viable *M. pulmonis* from the lungs following IV infusion (Lai *et al.*, 1990). These results indicate that the NK cells stimulated by infection with the mycoplasma can indeed have an antimycoplasmal activity.

Mycoplasma can elicit IFNα and -β and IL-2 production (Rinaldo *et al.*, 1974; Sokley *et al.*, 1977; Levin *et al.*, 1985). IL-2 and IFNα and -β are well recognized as potent inducers of NK function in mice. These could have augmented NK cell function in acutely infected mice.

The mechanism of *M. pulmonis* killing by NK cells is unknown; at least two

TABLE I

Purified NK Cells from SCID Mice Inhibiting Growth of *M. pulmonis* and Stimulating Secretion of γ Interferon[a]

Group	Contents of plates				Growth of *M. pulmonis* (CFU/μl)	WEHI-279 uptake [H₃]thymidine (cpm × 10⁻³)
	M. pulmonis	Mφ	NK	Anti-IFNγ		
1	+	−	−	−	1000 ± 25	815 ± 39
2	+	−	+	−	15.4 ± 6.0	12.1 ± 44
3	+	+	−	−	1000 ± 25	750 ± 35
4	+	+	+	−	1.7 ± 1.7	7.9 ± 0.6
5	+	+	+	+	100 ± 9	145 ± 11
6	+	−	+	+	150 ± 4.0	145 ± 11
7	−	−	−	−	ND	690 ± 61
8	−	−	+	−	ND	504 ± 19

[a] Group 1, *M. pulmonis* alone; group 2, *M. pulmonis* + NK (to determine if NK cells or cytokines released from NK directly kill the mycoplasma); group 3, *M. pulmonis* + Mφ (to determine the killing ability of normal macrophages); group 4, *M. pulmonis* + Mφ + NK (to determine if Mφ activated by cytokines released from NK can kill the mycoplasma; group 5, *M. pulmonis* + Mφ + NK + anti-IFNγ (to determine if IFNγ released from NK is involved in the killing of the mycoplasma); group 6, *M. pulmonis* + NK + anti-IFNγ (to determine if IFNγ released from NK is involved in the killing of the mycoplasma); group 7, media control; and group 8, NK cell control. NK cells + *M. pulmonis* (group 2), but not macrophages + *M. pulmonis* (group 3), directly inhibited the growth of *M. pulmonis*. The most striking inhibitory effect was provided by NK cells in association with macrophages (group 4), and the inhibitory effect was partially blocked by anti-IFNγ serum (group 5). However, there was a moderate killing effect in the group consisting of NK cells plus *M. pulmonis* without macrophages (group 6). These experiments suggest that IFNγ is involved in NK cell killing of *M. pulmonis* but is not the only factor. However, it is not yet clear whether the IFNγ produced by NK cells directly kills the mycoplsms or whether the IFNγ stimulates macrophages to phagocytose the mycoplasma.

different processes exist. Proteinase inhibitors interfere with NK cytotoxicity, and other phospholipase A₂ inhibitors depress the production of lysolecithin, which is associated with lower NK activity. Whether proteinase, lysolecithin, or both are directly responsible for the killing of *M. pulmonis* is still not known. As has been shown (Lai *et al.*, 1986; Mardh *et al.*, 1973), lysolecithin is a powerful killing factor for mycoplasmas. This might explain how NK cells may play a role in resistance to mycoplasmal infection.

The most convincing evidence that NK cells are activated by acute infection with *M. pulmonis* is based on the response of C.B-17 SCID mice. These mice have normal numbers of NK cells, macrophages, and granulocytes but no T or B lymphocytes. The augmented NK cells of SCID mice inhibited *M. pulmonis* growth in spleen and lung within 6–24 hours after infusion. The ability of the cells to inhibit mycoplasma growth is abrogated by treatment with either antiasialo-G_{M1} serum or anti-IFNγ antibodies. Antiasialo-G_{M1} serum eliminates NK cells in SCID mice, and the anti-IFNγ MAbs presumably block the action of IFNγ secreted by NK cells. There is strong evidence that IFNγ can be secreted by NK cells and consequently has the capacity to activate mouse macrophages *in vitro* and *in vivo;* these activated macrophages can inhibit or kill organisms. When IFNγ was administered *in vivo,* significant protection against *Toxoplasma gondii* was observed in mice; a similar phenomenon would be expected in the mycoplasma-infected mice. We have observed that purified NK cells from SCID mice will secrete IFNγ on exposure to *M. pulmonis* (Table I). The IFNγ secreted from NK cells activates macrophages to phagocytose and kill the mycoplasma (group 4, Table I) and the inhibitory effect was partially blocked by anti-IFNγ serum (group 5, Table I). A moderate killing effect was also seen in the group consisting of NK cells plus MP without macrophages (group 6, Table I). Thus, IFNγ may be an important primary factor in host resistance against *M. pulmonis* infection. The increased susceptibility of SCID mice treated with anti-IFNγ to the lethal effects of *M. pulmonis* supports this conclusion. Moreover, clearance of *M. pulmonis* in lungs and spleens was impaired in mycoplasma-infected SCID mice treated with anti-IFNγ antibodies. IFNα and -β augment NK cell function and are stimulated by *M. pulmonis* infection. It will therefore be important to determine any effects of anti-IFNα and -β antibodies on host defense against *M. pulmonis* mediated by NK cells.

Radiolabeled *M. pulmonis* infused into infected mice that had augmented NK cells were eliminated two times faster than in control mice; this effect was ablated completely with the administration of antiasialo-G_{M1} serum. These experiments support the contention that the NK cells and at least one product, IFNγ, play a role in host resistance to *M. pulmonis* infection.

We have reported (Lai *et al.,* 1989) that humoral and cellular immunity is inhibited in *M. pulmonis*-infected mice but that NK cell function is enhanced within 24 hours after infection. Further, it has been reported that the mouse macrophages derived from the peritoneal cavity of mice injected with sodium alginate were able to kill *M. pulmonis* in the absence of antibodies. As observed previously, the resistance to *M. pulmonis* infection in the mouse model suggests that the primary effect of IFNγ *in vivo* appears to be its ability to activate macrophages since activated macrophages can inhibit or kill *M. pulmonis* without collaboration of any antibody. Our results confirm that IFNγ contributes to *M. pulmonis* infection. However, since NK cells inhibit the mycoplasmas directly, it is conceivable that IFNγ secreted by NK cells directly inhibits the growth and/or survival of *M. pulmonis* in the absence of macrophages. Indeed, preliminary experiments indicate that recombinant murine IFNγ inhibits the growth of this mycoplasma.

References

Herberman, R. B., Nunn, M. E., and Lavin, D. H. (1975). Natural cytotoxic reactivity of mouse lymphoid cells against syngeneic and allogeneic tumors. I. Distribution of reactivity and specificity. *Int. J. Cancer* **16,** 216–29.

Lai, W. C., Bennett, M., Pakes, S. P., Kumar, V., Steutermann, D., Owusu, I., and Mikhael, A. (1990). Resistance to *Mycoplasma pulmonis* mediated by activated natural killer cells. *J. Infect. Dis.* **161,** 1269–1275.

Lai, W. C., Pakes, S. P., Lu, Y. S., and Brayton, C. F. (1987). *Mycoplasma pulmonis* infection augments natural killer cell activity in mice. *Lab. Anim. Sci.* **37,** 299–303.

Lai, W. C., Pakes, S. P., Owusu, I., and Wang, S. (1989). *Mycoplasma pulmonis* depress humoral and cell-mediated response in mice. *Lab. Anim. Sci.* **39,** 11–15.

Lai, W. C., Pakes, S. P., Stefanu, C., and Lu, Y. S. (1986). Comparison of Chalquest and Hayflick media, with and without ammonium reineckate, for isolating *Mycoplasma pulmonis* from rats. *J. Chin. Microb.* **23,** 817–821.

Levin, D., Gershon, H., and Naot, Y. (1985). Production of interleukin-2 by rat lymph node cells stimulated by *Mycoplasma pulmonis* membrane. *J. Infect. Dis.* **151,** 541–544.

Mardh, P. A., and Taylor-Robinson, D. (1973). The differentiated effect of lysolecithin on mycoplasmas and acholeplasmas. *Med. Mikrobiol. Immunol.* **58,** 259–266.

Nabavi, N., and Murphy, J. W. (1985). *In vitro* binding of natural killer cells to *Cryptococcus neoformans* targets. *Infect. Immun.* **50,** 50–57.

Olson, N. O., Kerr, K. M., and Campbell, A. (1963). Control of infectious synovitis: Preparation of an agglutination test antigen. *Avian Dis.* **7,** 310–317.

Riccardi, C., Puccetti, P., Santoni, A., and Herberman, R. B. (1978). Rapid *in vivo* assay of mouse natural killer activity. *J. Natl. Cancer Inst.* **63,** 1041–1045.

Rinaldo, C. R., Jr., Cole, B. C., Overall, J. C., Jr., and Glasgo, L. A. (1974). Induction of interferon in mice by mycoplasma. *Infect. Immun.* **10,** 1296–1301.

Sokley, J., Solovieva, A. J., and Vasilieva, V. I. (1977). Induction of interferon in mice cell culture by *Mycoplasma pneumoniae*. *Acta Viral (Praha)* **21,** 485–489.

Tutt, M. M., Schuler, W., Kuziel, W. A., Tucker, P. W., Bennett, M., Bosma, M. J., and Kumar, V. (1987). T cell receptor genes do not rearrange or express functional transcripts in natural killer cells of scid mice. *J. Immunol.* **138,** 2338–2344.

Index

A

Activation, macrophages and monocytes, by mollicutes, 421–438
Adherence
  epitope mapping of mollicute adhesins in, 407–413
  quantifying host cell attachment, 397–405
Adhesins, localization on mollicute surfaces, 89–98
Adhesion, mollicute, 16
Antigenic variation, membranes, 12

B

B-cell mitogens, of mollicutes, 451–460

C

Cell metabolism, in mollicutes, 275–331
Cell volume, in mollicutes, 265–272
Cholesterol, in growth requirements of mollicutes, 381–389
Cultivation, mollicutes, 29–64
Culture media
  insect-based, 55–64
  for mollicutes, 33–64
  for spiroplasmas, 41–64

D

DNA, isolation from mollicutes of plants/insects, 105–117

E

Electric potential, mollicute membranes, 251–263
Electroanalytical methods, for testing metabolic activities, 287–295
Enzymatic analysis, in metabolic activities of mollicutes, 277–286
Epitope mapping, mollicute adhesins, 407–413
Extrachromosomal elements of mollicutes, characterization, 159–166

G

Gene transfer, in mollicutes, 10
Genetic mapping, in mollicutes, 6, 133–157
Genomes
  mollicute
    characterization, 101–211
    size determination, 4, 119–131
    viral characterization, 159–166
Growth measurements, for mollicutes, 65–71

H

Habitat, mollicutes, 2
Heat-shock proteins, characterization in mollicutes, 297–303
Hybridization, *in situ*, 81–87

I

Immune system activation, by mollicutes, 18
In vitro culture, mollicutes, 3
Insect-based culture media, for mollicutes, 55–64
Intracellular location, mollicutes, 73–87
Ion flow, through mollicute membranes, 265–272
Isolation, mollicutes, culture media used for, 33–39

L

Luminometry, for measuring mollicute growth, 65–71

M

Macrophages, activation by mollicutes, 421–438
Major histocompatibility complex, effects on expression by mycoplasmas, 461–467
Membranes
  mollicute, 11
    antigenic variation, 12
    cell volume and ion flow, 265–272

481

482 Index

electric potential, 251–263
  variable proteins in, 227–241
fusion, 243–249
proteins, posttranslational modification, 217–226
Metabolic activity
  electroanalytic methods in metabolic testing, 277–286
  enzymatic analysis in testing, 277–286
  microcalorimetric methods in metabolic testing, 277–286
Metabolism, mollicute, 13
Microcalorimetric methods, for testing metabolic activities, 287–295
Molecular properties, mollicutes, 1–25
Mollicutes
  activation of macrophages/monocytes by, 421–438
  adhesins, 16
  application of Southern blot analysis, 361–368
  B-cell mitogens, 451–460
  cell metabolism, 275–331
  cell volume and ion flow in membranes, 265–272
  cultivation, 29–64
  detection of intracellular growth, 73–87
  DNA isolation in plants/insect derived, 105–117
  epitope mapping in host cell adherence, 407–413
  genetics, 101–211
  gene transfer in, 10
  genome, characterization of viruses in, 159–166
  genome mapping, 6
  genome size, 4
    characterization in unculturable mollicutes, 119–131
    determination, 119–131
  growth measurements, 65–71
  growth requirements for cholesterol, 381–389
  habitat, 2
  heat-shock proteins, 297–303
  host immune activation by, 18
  interaction with natural killer cells, 469–479
  in vitro culture, 3
  location in eukaryotic tissues, 73–87
  measuring host cell adherence, 397–405
  mechanisms of pathogenicity, 16
  membrane electric potential among, 251–263
  membrane fusion among, 243–249
  membranes, 11
    characterization, 215–272
    variable proteins in, 227–241
  metabolism, 13
  modulation of major histocompatibility complex by, 461–467
  molecular properties, 1–25
  morphology, 2, 73–98
  nucleolytic activities, 305–314
  oxidative damage by, 415–419
  pathogenic characteristics, 391–479
  phospholipase activities in, 325–331
  physical and genetic mapping, 133–157
  plant-pathogenic, phylogenic classification, 369–380
  protein synthesis by, 8
  proteolytic activities in, 315–323
  restriction endonuclease analysis in, 355–360
  restriction mapping, 133–157
  ribotyping with cloned genes of, 361–368
  RNA sequencing in establishing phylogenies, 349–354
  standards for description of new species, 339–347
  superantigens, identification and characterization, 439–449
  taxonomy and phylogeny, 14, 335–389
  testing metabolic activity, 277–286
  tip structure and antigens, 89–98
Monocytes, activation by mollicutes, 421–438
Morphology, mollicutes, 2, 73–98
Mycoplasmal B-cell mitogens, 451–460
Mycoplasmal superantigens, identification and characterization, 439–449

N

Natural killer cells, interaction of mollicutes with, 469–479
Nucleolytic activities, in mollicutes, 305–314

O

Oxidative damage, mollicute induction of, 415–419

Index

P

Pathogenicity, mollicutes, 391–479
  mechanisms, 16
Phospholipase activities, in mollicutes, 325–331
Phylogeny
  mollicutes, sequence analysis in assessing, 349–354
  plant-pathogenic phytoplasmas, 369–380
Phytoplasmas, phylogenic classification, 369–380
Plant-pathogenic mollicutes, phylogenic classification, 369–380
Plasmids, of mollicutes, characterization, 159–166
Polyoxyethylene sorbitan, in growth requirements of mollicutes, 381–389
Posttranslational modification, membrane proteins, 217–226
Protein synthesis, in mollicutes, 8
Proteins
  membrane, posttranslational modification, 217–226
  in mollicute membranes, 215–272
  variations, 227–241
Proteolytic activities, in mollicutes, 315–323
Pulse field gel electrophoresis, genome size measurements by, 119–131

R

Restriction endonuclease analysis, in mollicutes, 355–360
Restriction mapping, mollicutes, 133–157
Ribosomal RNA sequencing, in assessing phylogenetic relationships, 349–354
Ribotyping, in mollicutes, 361–368

S

Southern blot analysis, application in mollicutes, 361–368
SP-4 culture medium, 33–39
Spiroplasmas, culture media, 41–64
Superantigens, mycoplasmal, identification and characterization, 439–449

T

Taxonomic standards, for description of new mollicute species, 339–347
Taxonomy and phylogeny, mollicutes, 14, 335–389
Tissues, location of mollicutes in, 73–87

V

Viruses, of mollicutes, genome characterization, 159–166